White Dwarfs

NATO ASI Series

Advanced Science Institutes Series

A Series presenting the results of activities sponsored by the NATO Science Committee, which aims at the dissemination of advanced scientific and technological knowledge, with a view to strengthening links between scientific communities.

The Series is published by an international board of publishers in conjunction with the NATO Scientific Affairs Division

A Life Sciences	Plenum Publishing Corporation
B Physics	London and New York
C Mathematical	Kluwer Academic Publishers
and Physical Sciences	Dordrecht, Boston and London
D Behavioural and Social Sciences	
E Applied Sciences	
F Computer and Systems Sciences	Springer-Verlag
G Ecological Sciences	Berlin, Heidelberg, New York, London,
H Cell Biology	Paris and Tokyo
I Global Environmental Change	

NATO-PCO-DATA BASE

The electronic index to the NATO ASI Series provides full bibliographical references (with keywords and/or abstracts) to more than 30000 contributions from international scientists published in all sections of the NATO ASI Series.
Access to the NATO-PCO-DATA BASE is possible in two ways:

– via online FILE 128 (NATO-PCO-DATA BASE) hosted by ESRIN, Via Galileo Galilei, I-00044 Frascati, Italy.

– via CD-ROM "NATO-PCO-DATA BASE" with user-friendly retrieval software in English, French and German (© WTV GmbH and DATAWARE Technologies Inc. 1989).

The CD-ROM can be ordered through any member of the Board of Publishers or through NATO-PCO, Overijse, Belgium.

Series C: Mathematical and Physical Sciences - Vol. 336

White Dwarfs

edited by

Gérard Vauclair
Observatoire Midi-Pyrénées,
Toulouse, France

and

Edward Sion
Department of Astronomy,
Villanova University,
Villanova, Pennsylvania, U.S.A.

Springer-Science+Business Media, B.V.

Proceedings of the NATO Advanced Research Workshop
Seventh European Workshop on White Dwarfs
Toulouse, France
September 3–5, 1990

Library of Congress Cataloging-in-Publication Data

```
European Workshop on White Dwarfs (7th : 1990 : Toulouse, France)
    White dwarfs : proceedings of the NATO advanced research workshop
 "Seventh European Workshop on White Dwarfs" Toulouse, France,
Septembre 3-5, 1990 / edited by Gérard Vauclair and Edward Sion.
     p.    cm. -- (NATO ASI series. Series C, Mathematical and
physical sciences ; vol. 336)
    "Dedicated to Evry Schatzman"--Pref.
    Includes index.
    ISBN 978-94-010-5423-2    ISBN 978-94-011-3230-5 (eBook)
    DOI 10.1007/978-94-011-3230-5
    1. White dwarfs--Congresses.   I. Vauclair, Gérard, 1943-
II. Sion, Edward M.  III. Schatzman, Evry L.  IV. Title.  V. Series:
NATO ASI series.  Series C, Mathematical and physical sciences ; no.
109.
QB843.W5E95   1990
523.8'87--dc20
                                                           91-650
```

ISBN 978-94-010-5423-2

Printed on acid-free paper

Dedication

These Proceedings are dedicated to Evry Schatzman on the event of his seventieth birthday, in recognition of his contributions to research on white dwarf stars.

TABLE OF CONTENTS

Preface   xiii

Introduction   xv

List of participants   xvii

1-PRE-WHITE DWARF EVOLUTION AND WHITE DWARF COOLING

New pre-white dwarf evolutionary tracks
Blöcker, T., Schönberner, D.   1

Influence of the phase diagram in the cooling of white dwarfs
Isern, J., Garcia-Berro, E., Hernanz, M., Mochkovitch, R.   5

PG1159 stars and the PNN-white dwarf connection
Barstow, M.A., Tweedy, R.W., Werner, K.   17

Analysis of the soft X-ray data from the central star of NGC 7293
Tweedy, R.W., Barstow, M.A.   29

Planetary Nebulae Nuclei with white dwarf spectra
Napiwotzki, R., Schönberner, D.   39

Atmospheric parameters of subluminous B stars
Saffer, R.A., Bergeron, P., Liebert, J., Koester, D.   53

White dwarf space densities and birth rates reconsidered
Weidemann, V.   67

A spectroscopic determination of the mass distribution of DA white dwarfs
Bergeron, P., Saffer, R.A., Liebert, J.   75

The age and formation of the Galaxy : clues from the white dwarf
luminosity function
Wood, M.A.   89

Early results from the ROSAT Wide Field Camera
Barstow, M.A, Abbey, A.F., Cole, R.E., Denby, M., Page, C.G.,
Pankiewicz, G.S., Pounds, K.A., Pye, J.P., Sansom, A.E., Sims, M.R.,
Spragg, J.E., Watson, D.J., Wells, A.A., Willingale, R., Courtier, G.M.,
Gourlay, J.A., Harris, A.W., Kent, B.J., Reading, D.H., Richards, A.G.,
Swinyard, B.M., Wright, J.S., Goodall, C.V., Bentley, R.D., Breeveld, E.R.,
Guttridge, P.R., Huckle, H.E., McCalden, A.J., Bewick, A., Rochester, G.K.,
Sumner, T.J.   99

The stellar component of the Hamburg Schmidt Survey
Heber, U., Jordan, S., Weidemann, V.   109

White dwarfs in the Hamburg Schmidt Survey
Jordan, S., Heber, U., Weidemann, V.   121

2-ASTEROSEISMOLOGY OF WHITE DWARFS

Asteroseismology of white dwarf stars with the Whole Earth Telescope
Winget, D.E.   129

A measurement of the evolutionary timescale of the cool white dwarf
G117-B15A with WET
Kepler, S.O., Kanaan, A., Winget, D.E., Nather, R.E., Bradley, P.A.,
Clemens, J.C., Kleinman, S.J., Claver, C.F., Provencal, J.L., Grauer, A.D.,
Fontaine, G., Bergeron, P., Wesemael, F., Wood, M.A., Vauclair, G.,
Marar, T.M.K., Seetha, S., Ashoka, B.N., Mazeh, T., Leibowitz, E.,
Dolez, N., Chevreton, M., Barstow, M.A., Sansom, A.E., Tweedy, R.W.,
Hine, B.P., Solheim, J.-E., Emanuelsen, P.-I.   143

On the interpretation of the dP/dt measurement in G117-B15A
Fontaine, G., Brassard, P., Wesemael, F., Kepler, S.O., Wood, M.A.   153

The boundaries of the ZZ Ceti instability strip
Wesemael, F., Bergeron, P., Fontaine, G., Lamontagne, R.   159

Long term variations in ZZ Cetis : G191-16 and HL Tau 76
Auvergne, M., Chevreton, M., Belmonte, J.A., Vauclair, G., Dolez, N.,
Goupil, M. J.   167

Predicting the white dwarf light curves
Serre, T., Buchler, J.R., Goupil, M. J.   175

A wavelet analysis of the ZZ Ceti star G191 16
Goupil, M.J., Auvergne, M., Baglin, A.   185

An adiabatic survey for ZZ Ceti stars based on a finite element code
Brassard, P., Fontaine, G., Pelletier, C., Wesemael, F.   193

A study of period change rates in post-AGB stars I. PG 1159-035
Stanghellini, L., Cox, A.N.   205

Nonadiabatic nonradial pulsations for DAV white dwarf stars
Cox, A.N., Hollowell, D.E.   211

3-ATMOSPHERES AND ENVELOPES

NLTE Analysis of four PG1159 stars
Werner, K., Heber, U., Hunger, K.   219

A search for trace amounts of hydrogen in DB stars
Shipman, H., Thejll, P., Bhatia, S., Liebert, J.   229

Abundances of trace heavy elements in hot DA white dwarfs
Vennes, S., Thejll, P., Shipman, H.L.   235

New results on radiative forces on iron in hot white dwarfs
Chayer, P., Fontaine, G., Wesemael, F.   249

The effective temperature of the DBV's, and the sensitivity of DB model
atmospheres to input physics
Thejll, P., Vennes, S., Shipman, H.L.   257

The modified hydrostatic equilibrium equations for stratified high gravity
stellar atmospheres
Unglaub, K., Bues, I.   267

The DBAQ G35-26
Thejll, P., MacDonald, J., Shipman, H.L.   275

LP 790-29 : preliminary model atmospheres for this strongly polarized
carbon white dwarf
Bues, I.   285

Some effect of the UV radiation from white dwarfs on the accretion
of interstellar hydrogen
Verdon, C.P., McCrory, R.L., Epstein, R., Van Horn, H.M., Savedoff, M.P.   295

Convection in white dwarfs : application of CM theory to helium
envelope WDs
Mazzitelli, I., D'Antona, F.   305

Abundances in cool DZA and DAZ white dwarfs : new results using
laboratory damping constants
Hammond, G.L., Sion, E.M., Kenyon, S.J., Aannestad, P.A.   317

Evidence for fractionated accretion of metals on cool white dwarfs
Dupuis, J., Fontaine, G., Wesemael, F.   333

A new look at old friends : 40 Eri B and GD 323
Koester, D.   343

The Lyman Alpha line wing in hydrogen-rich white dwarf atmospheres
Allard, N., Kielkopf, J.   353

Atmospheric parameters for DA white dwarfs in the vicinity of the
ZZ Ceti instability strip
Dolez, N., Vauclair, G., Koester, D.   361

Space Telescope observations of white dwarf stars
Shipman, H.L.   369

4-WHITE DWARFS IN BINARIES

A deep spectroscopic survey of white dwarfs in common proper
motion binaries
Oswalt, T.D., Sion, E.M., Hintzen, P.M., Liebert, J.W.   379

Double degenerate common proper motion binaries
Sion, E.M., Oswalt, T.D., Liebert, J., Hintzen, P.   395

Close binary white dwarfs
Liebert, J., Bergeron, P., Saffer, R.A.   409

New results on cataclysmic variable white dwarfs
Sion, E.M.   417

Whole Earth Telescope observations of the interacting
white dwarf binary system AM CVn : first results
Solheim, J.-E., Emanuelsen, P.-I., Vauclair, G., Dolez, N., Chevreton, M.,
Barstow, M., Sansom, A.E., Tweedy, R.W., Kepler, S.O., Kanaan, A.,
Fontaine, G., Bergeron, P., Grauer, A.D., Provencal, J.L., Winget, D.E.,
Nather, R.E., Bradley, P.A., Claver, C.F., Clemens, J.C., Kleinman, S.J.,
Hine, B.P., Marar, T.M.K., Seetha, S., Ashoka, B.N., Leibowitz, E.M.,
Mazeh, T.   431

IUE observations of V803 Cen in high and low states
Ulla, A.M., Solheim, J.-E.   441

Whole Earth Telescope observations of PG1346+082
Provencal, J.L., Winget, D.E., Nather, R.E., Clemens, J.C., Hine, B.P.,
Henry, G., Kepler, S.O., Vauclair, G., Chevreton, M., O'Donoghue, D.,
Warner, B., Grauer, D.A., Ferrario, L.   449

On the origin of LMXRBS : the ONEMG case
Isern, J., Canal, R., Labay, J.   457

Index   465

PREFACE

The European Workshop on White Dwarfs was initiated by Prof. V. Weidemann, with the first meeting organized in Kiel (FRG) in 1974. Almost every two years, an increasing number of astronomers met to share their results and projects in the subsequent workshops : Frascati (1976), Tel Aviv (1978), Paris (1981), Kiel (1984), Frascati (1986). In the mean time, two major IAU colloquia (No. 53 in Rochester, NY, 1979; and No. 114 in Hanover, NH, 1988) emphasized the importance of these stars for our understanding of stellar evolution.

The informal organization of the white dwarf community has been the starting point for large cooperative projects of which the Hubble Space Telescope "White Dwarf Consortium" and the "Whole Earth Telescope" are the most spectacular examples. But many other successful collaborations have also been born during the very exciting discussions conducted in the last 16 years on the occasion of our regular meetings.

The 7th European Workshop on White Dwarfs took place the year of the seventieth birthday of Prof. Evry Schatzman, whose pioneering work has been the inspiration for many of the new ideas in the white dwarf community. The Scientific Organizing Committee has agreed to dedicate the workshop to him on this occasion. We are pleased to publish as an introduction to the workshop, the text of the talk delivered by Prof. Schatzman at the workshop's banquet.

We are grateful for the help we received from our colleagues of the Scientific Organizing Committee : Francesca D'Antona, Jim Liebert and V. Weidemann. The practical support of Josiane Jobard is gratefully acknowledged.

On behalf of all the participants, we thank the Director of the NATO Scientific Affairs Division, the "Ministère de l'Education Nationale", the University Paul Sabatier of Toulouse and the Director of the Observatoire Midi-Pyrénées for granting financial support for a fruitful and stimulating Advanced Research Workshop.

Gérard Vauclair and Edward Sion

February, 1991

INTRODUCTION

Should I first express the pleasure of being in Toulouse? My grand-mother has lived in Toulouse, her parents having a shop in the Place du Capitole where they sold clothes. From here came a very nice lullaby, used for rocking children, young children, to sleep:

Some some
Beni Beni somme
La some some es arribada
A cavale su une cabra
Rebendra du mamati
A cavale su une pouli

This meeting, for me, is very moving. It is a remarkable meeting, with young people, and not so young ones, bringing to our attention an incredible amount of new results, new data, of a very high quality and high signal to noise ratio. Each talk brings to my mind the idea of a variety of physical problems, each one being more fascinating than the other and perhaps more difficult. Lots of people will have to tackle these problems during the coming years and to solve them.

But this meeting takes me back 47 years, when I was hiding in the Observatory of Haute Provence, near the small village of St Michel l'Observatoire. Life was difficult, but nevertheless living was an absolute necessity, and after all, as I was studying in the observatory on the advice of one of my professors. I was there to continue to work as a physicist, which meant continuing to study and to learn, and, if possible, to produce. Two facts decided what happened. I read, probably in January 44, in a monthly magazine of popular science, Science et Vie, a paper explaining the theory of Hans Bethe of the thermonuclear origin of stellar light. It was fascinating, as it revealed to me the junction between astrophysics and physics. I ordered by mail a microfilm of the 1939 paper of Hans Bethe from the CNRS, and soon discovered in the Zeitschrift für Astrophysik a review paper of George Gamow which explained the whole thing, the crossing of the potential barrier and the effect of temperature.

Another fact came from an idea of Fehrenbach, who was then the vice-director of the Observatory. Having organized a seminar for the five scientists who were there, he asked me to give a review of part of the proceedings of the 1938 meeting in Paris on white dwarfs, Novae and Supernovae. Henry Norris Russell, in his introductory speech, tells that, trying to classify 40 Eri B, he discovered that it did not fall into the dwarf branch. He was desperate, having the feeling that his classification was falling apart. He discussed the question with Pickering, "his benefactor", who told him: "this is the kind of contradiction which is at the origin of great discoveries". This was in 1915. I spoke about white dwarfs. It turned out that I knew just enough statistical physics to understand the problem of the equation of state of a fully degenerate electron gas. I guess that I found in the supplement volume of the Handbuch der Astrophysik some information on the internal structure of white dwarfs. The book of Chandrasekhar (An Introduction of the Study of Stellar Structure) was not then available.

Very soon I realized that the chemical composition of white dwarfs could not be the same as solar composition. The rate of thermonuclear reaction was incompatible with the luminosity of white dwarfs. The gravitational separation of hydrogen, floating at the surface of white dwarfs, came immediately to my mind and was the subject of my first publication on white dwarfs, in the Compte Rendus de l'Académie des Sciences, fall 1944.

I expressed my thanks to the astronomers who had hidden me, by accepting to act as a night assistant. But I swore also that, being a physicist, I would never become an astronomer. As James Bond said, "never say never again". However, it was not until the fall of 1946, when Daniel Barlier gave me the possibility of staying a few months with Bengt Strömgren in Copenhagen, that I decided to remain with the astronomers, but was still possessed by a love of physics.

When in 1955 or thereabouts, Hantz van de Hulbt, on behalf of North Holland, asked me to write a book on white dwarfs, I started immediately to work on the subject. At the time, Michael Seaton, the well known atomic physicist, was spending one year at the Institut d'Astrophysique in Paris, and I took that opportunity to ask him to correct my poor English, which he kindly did. But when I sent the manuscript to North Holland, they gave it for correction to an American scientist who was staying in The Netherlands and he again corrected the manuscript, including the corrections of Michael Seaton: I have not kept track of the linguistic fight that ensued, and this is too bad. I still kept in touch with the white dwarf problem without getting myself into the subject. Beginning in 1962, A. Baglin gave a complete description of non-zero temperature WD, including the general relativity effects, around the maximum mass of WD, opening the way to the description of gravitational collapse and SN production; twelve years later I began this work with Ramon Canal on the problem of non-explosive collapse, and at the time of the 1978 meeting on dense matter, brought Robert Mochkovitch in on this same subject. I remained attracted by the new physical problems which were arising.

A question which I clearly met for the first time some 40 years ago, was the well known question: how useful is what I am doing? Considering the energy production of white dwarfs, even if it drove me to the problem of the effect of pressure on the potential barrier in 1947, did not seem to me to have any application of any kind. It took me many many years to realize that this was a wrong question. The level of culture and the development of scientific research, always leads to unpredictable discoveries and, nonetheless, unpredictable applications. To carry on scientific work, having in mind the aim of finding useful applications, is not doing science. The sentence itself is self-contradictory and we have to make understood to the layman as well as to the politicians, that such a statement is meaningless.

We are enjoying this evening, with a slight change in the weather. The political weather has been changing during the last weeks. It is impossible for me, when this meeting, so well organized and bringing back memories of events of the last 50 years, not to think of the present threat to our world. Let us hope that international law and human rights will prevail in this crisis and be respected with devotion by all nations of the world.

Evry Schatzman

LIST OF PARTICIPANTS
Numbers in parentheses refer to the group photography

Dr. N. ALLARD (4)   LAM, Observatoire de Meudon
92195 Meudon Principal Cedex, France

Dr. M. AUVERGNE (36)   DASGAL, Observatoire de Paris-Meudon
92195 Meudon Principal Cedex, France

Dr. M.A. BARSTOW (12)   X-Ray Astronomy Group, Physics Dpt.
University of Leicester, University Road
Leicester, LE1 7RH, U.K.

Dr. J.-A. BELMONTE (6)   Instituto de Astrofisica de Canarias
38200 La Laguna, Tenerife, Spain

Dr. P. BERGERON (35)   Steward Observatory, The University of Arizona
Tucson, Arizona 85721, U.S.A.

Dr. P. BRASSARD (8)   Département de Physique, Université de Montréal
P.O. Box 6128, succ. "A" Montréal, PQ H3C 3J7
Canada

Dr. I. BUES (23)   Dr. Remeis-Sternwarte Bamberg
Astron. Institut der Universitat
Erlanger-Nürnberg Sternwartstrasse 7
D 8600 Bamberg – F.R.G.

Mrs. A. BRUVOLD (38)   University of Tromsö, Institute of Mathematical
and Physical Sciences, N-9000 Tromsö, Norway

Dr. G. CHABRIER (37)   Laboratoire de Physique, ENS Lyon
46, Allée d'Italie 69364, Lyon Cedex 07, France

Dr. P. CHAYER (9)   Département de Physique, Université de Montréal
P.O. Box 6128, succ. "A" Montréal, PQ H3C 3J7
Canada

Dr. F. D'ANTONA (27)   Osservatorio Astronomico, 00040 Monte Porzio
Rome, Italy

Dr. N. DOLEZ (13)   Observatoire Midi-Pyrénées
14, avenue Edouard Belin
31400 Toulouse, France

Dr. J. DUPUIS (11)   Department of Physics & Astronomy
Wilder Lab., Dartmouth College
Hanover, N.H. 03755, U.S.A.

Dr. G. FONTAINE (10)   Département de Physique, Université de Montréal
P.O. Box 6128, succ. "A" Montréal PQ H3C 3J7
Canada

Dr. M.-J. GOUPIL (24)   DASGAL, Observatoire de Paris Meudon
92195 Meudon Principal Cedex, France

Dr. G. HAMMOND (19)   Dept. of Mathematics (Astro. Prog.)
                       University of South Florida, Tampa,
                       FL 33620, U.S.A.

Dr. U. HEBER (14)   Institut für Theoretische Physik
                       und Sternwarte der Universität, Olshausenstrasse 40
                       D-2300 Kiel, 1 F.R.G.

Dr. J.B. HOLBERG (20) Lunar and Planetary Laboratory-West
                       9th Floor, Gould-Simpson Building
                       University of Arizona, Tucson, AZ 85721, U.S.A.

Dr. J. ISERN (5)   Centre d'Estudis Avançats de Blanes
                       Cami de Santa Barbara, 17300 Blanes (Girona)
                       Spain

Dr. S. JORDAN (17)   Institut für Theoretische Physik
                       und Sternwarte der Universität, Olshausenstrasse
                       2300 Kiel, F.R.G.

Dr. S.O. KEPLER (16)   Département de Physique, Université de Montreal
                       P.O. Box 6128, succ. "A" Montréal PQ H3C 3J7
                       Canada

Dr. D. KOESTER (40)   Dept of Physics and Astronomy
                       266 Nicholson Hall, Louisiana State Univ.
                       Baton-Rouge, LA 70803-4001, U.S.A.

Dr. J. LIEBERT (28)   Steward Observatory, The University of Arizona
                       Tucson, Arizona 85721, U.S.A.

Dr. T. D. OSWALT (42) Dept. Physics & Space Sciences
                       Florida Institute of Technology
                       Melbourne, FL 32901, U.S.A.

Dr. C. A. OXBORROW (25) Nordita, Blegdamsvej 17
                       DK -2100, Köbenhavnö, Denmark

Dr. J. PROVENCAL (22) Department of Astronomy, University of Texas
                       Austin, TX 78712, U.S.A.

Dr. E. POULIN (45)   Departement de Physique, Université de Montréal
                       P.O. Box 6128, Succ A Montréal P.Q. H3C 3J7
                       Canada

Dr. R. SAFFER (34)   Steward Observatory, University of Arizona
                       Tucson, AZ 85721, U.S.A.

Dr. E. SCHATZMAN (3)   DASGAL, Observatoire de Meudon
                       92195 Meudon Principal Cedex, France

Dr. D. SCHÖNBERNER (18) Institut für Theoretishe Physik
                       und Sternwarte der Universität, Olshausenstrasse
                       D -2300 Kiel, F.R.G.

Dr. T. SERRE (22)   DASGAL, Observatoire de Paris Meudon
                       92195 Meudon Principal Cedex, France

Dr. H. SHIPMAN (31)   Physics Department, University of Delaware
                          Sharp Lab., Newark, DE 19716, U.S.A.

Dr. E. SION (7)   Dept. of Astronomy and Astrophysics
                          Villanova University, Villanova,
                          Pennsylvania 19085, U.S.A.

Dr. J.E. SOLHEIM (41)   Institute of Mathematical and Physical Sciences
                          University of Tromsö, P.O. Box 953,
                          N-9001 Tromsö, Norway

Dr. L. STANGHELLINI (29) Osservatorio Astronomico di Bologna
                          Via Zamboni 33, 40146 BOLOGNA, Italy

Dr. P. THEJLL (46)   Nordita, Blegdamsvej 17
                          DK-2100, Köbenhavnö, Denmark

Dr. R. TWEEDY (1)   Department of Physics, University of Leicester
                          University Road, Leicester LE1 7RH, U.K.

Dr. K. UNGLAUB (43)   Dr. Remeis-Sternwarte Bamberg
                          Astronomisches Inst. der Universität
                          Erlangen-Nürnberg, D-8600 Bamberg, F.R.G.

Dr. A. M. ULLA (30)   Instituto de Astrofisica de Canarias
                          IAC, 38200 La Laguna, Tenerife, Spain

Dr. H. Van HORN (15)   Dept. of Physics and Astronomy
                          University of Rochester, Rochester, NY 14627
                          U.S.A.

Dr. G. VAUCLAIR (2)   Observatoire Midi-Pyrénées
                          14, avenue E. Belin, 31400 Toulouse, France

Dr. S. VENNES (44)   Physics Department, University of Delaware
                          Sharp Lab. Newark, DE 19716, U.S.A.

Dr. V. WEIDEMANN (26)   Institut für Theoretische Physik
                          und Sternwarte der Universität Olshausenstrasse
                          2300 Kiel, F.R.G.

Dr. F. WESEMAEL (39)   Département de Physique, Université de Montréal
                          P.O. Box 6128, Succ. A Montréal P.Q. H3C 3J7
                          CANADA

Dr. D. E. WINGET (32)   Department of Astronomy, Univeristy of Texas
                          Austin, TX 78712, U.S.A.

Dr. M. WOOD (33)   Département de Physique, Université de Montréal
                          P.O. Box 6128, succ. "A" Montréal, PQ H3C 3J7
                          Canada

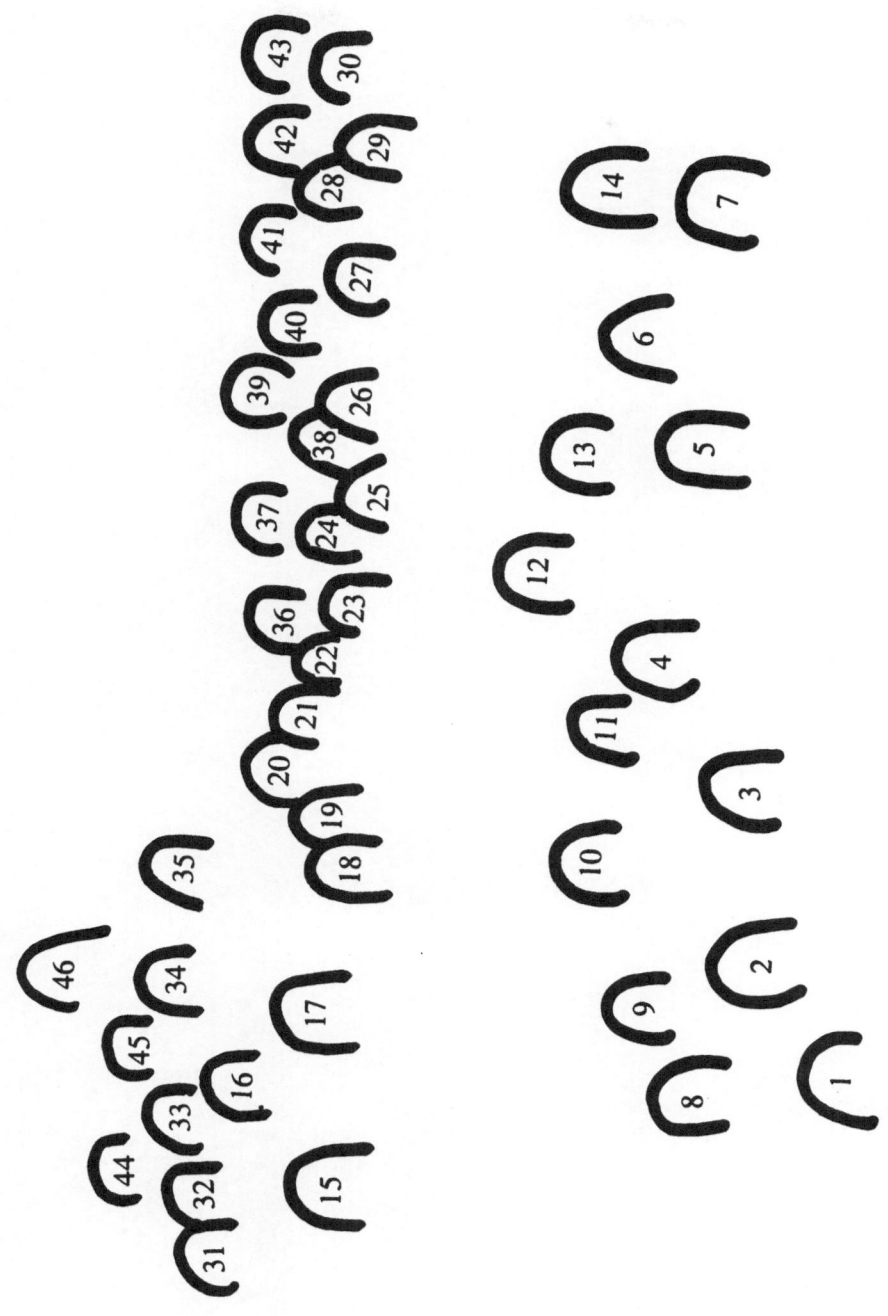

NEW PRE-WHITE DWARF EVOLUTIONARY TRACKS

T. BLÖCKER AND D. SCHÖNBERNER
Institut für Theoretische Physik und
Sternwarte der Universität Kiel
Olshausenstr. 40, D-2300 Kiel 1
Federal Republic of Germany

ABSTRACT. We present new post-AGB evolutionary tracks that have been derived by the evolution of stellar models of 3 and 5 $M_\odot$ from the main sequence through the AGB with consideration of thermal pulses and of mass loss which increases rapidly with increasing luminosity. From $M_{ZAMS} = 3M_\odot$ a remnant of 0.605 $M_\odot$ emerged after 17 thermal pulses. The corresponding remnant from $M_{ZAMS} = 5M_\odot$ leaves the AGB after only 9 thermal pulses. The evolutionary tracks of both remnants are given in Fig.1 together with some more information in the figure's caption. Whereas the 0.605 $M_\odot$ model behaves as expected, i.e. it reaches $\approx 10^2 L_\odot$ within 40000 yrs, the 0.836 $M_\odot$ remant fades much slower and needs about twice as much time to reach the same luminosity.

This behaviour is different to what has been assumed so far, namely that the larger the core's mass the faster is the post-AGB evolution *both* on the "plateau" *and* on the cooling track which means that the least luminous of the observed central stars of planetary nebulae should also be the most massive ones with, say, $M > 0.8M_\odot$ (cf. Renzini, A.: 1977, in *Stars and Star Systems*, B.E. Westerlund, Ed., Reidel, p.155; Renzini, A.: 1981, in *Physical Processes in Red Giants*, I. Iben Jr. and A. Renzini, Eds., Reidel, p.431). But recent observational investigations cannot confirm these expectations (cf. Kwitter, K.B., Jacoby, G.H.: 1989, *Astron.J.* 98, 2159). For instance, Jacoby and Kaler's (Jacoby, H.G., Kaler, J.: 1989, *Astron.J.*, 98, 1662) investigation of nuclei in optically thick planetary nebulae, which are thought to be the most massive ones, does not give evidence for many objects with $M > 0.8M_\odot$.

The reason for this discrepancy is the fact that the large and hot cores of more massive AGB stars posess a relatively high release of gravitational-internal energy, L_{Grav}, exceeding $10^3 L_\odot$. For comparison, a core of $\approx 0.6M_\odot$ releases only $250L_\odot$ by contraction. When hydrogen burning has ceased, it is just this gravitational-internal energy release that determines the final fading time scale of a post-AGB star.

On the other hand, the amount of L_{Grav} is ruled by the core mass $M_H(1)$ at the first thermal pulse, by the total number of thermal pulses and by the thermal-pulse cycle (which is here ≈ 0.5 in both cases). $M_H(1)$ depends only slightly on the initial mass up to $M_{ZAMS} \approx 3M_\odot$ for a Pop I composition and no convective overshooting (cf. Lattanzio, J.C.: 1986, *Astrophys.J.* 311, 708) whereas it increases between $M_{ZAMS} = 3$ and $5M_\odot$ from about 0.53 to $0.8M_\odot$ with a considerable increase of L_{Grav} leading to a strong dependence of a remnant's evolution on M_{ZAMS}. Furthermore, L_{Grav} steadily increases with increasing number of thermal pulses for $M_{ZAMS} = 3M_\odot$ whereas it decreases for $M_{ZAMS} = 5M_\odot$. Additional detailed calculations suggest that, in any case, the lighter model probably has to suffer more than 90 thermal pulses and the massive one more than 50 before the internal-gravitational energy releases of both models will become equal.

Because of these circumstances we want to emphasize that it is indispensable for obtaining realistic post-AGB models to take both the prior evolution into account and to choose a consistent combination of M_{ZAMS} and $M_H(1)$. We conclude that the final evolution of central stars does not only depend on their core masses (and on their thermal pulse-cycle as well), but also on their history, i.e. on their initial mass M_{ZAMS} and the mass-loss law on the AGB.

1

G. Vauclair and E. Sion (eds.), White Dwarfs, 1–3.

Finally, on the reasonable assumption that more massive remnants belong also to more massive parent stars (see Weidemann, V., Koester, D.: 1983, *Astron. Astrophys.* **121**, 77) one deduces immediately that the final fading time scale increases with increasing remnant mass. Thus we predict that the least luminous central stars have masses between ≈ 0.65 and $\approx 0.6 M_\odot$. The recent observational findings mentioned above seem to be consistent with our conclusions. A more detail discussion of this subject is given somewhere else (Astronomy and Astrophysics, in press).

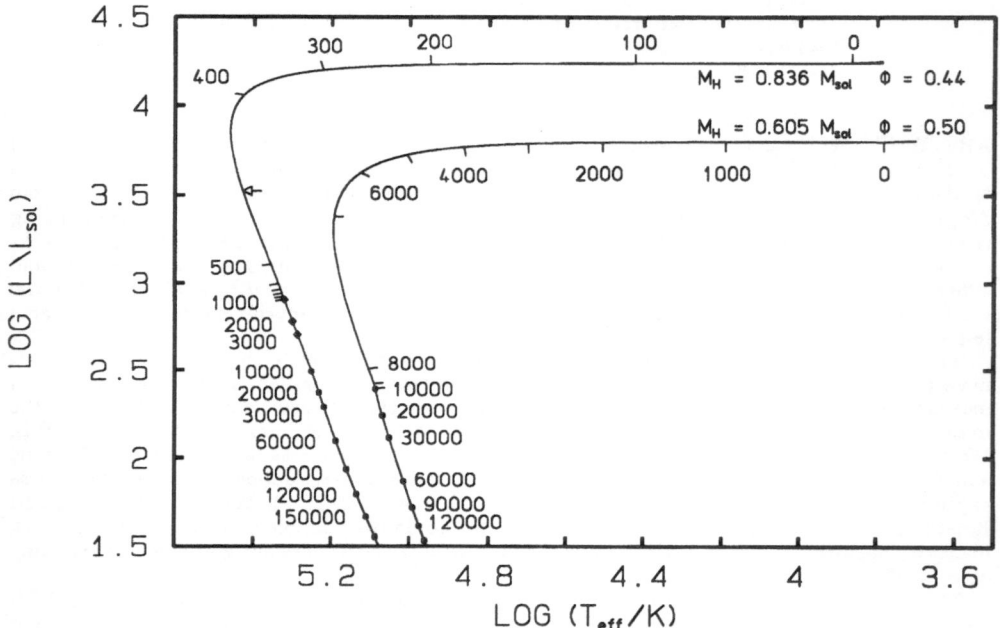

Figure 1: Evolutionary tracks of two hydrogen-burning post-AGB models of $M_H = 0.605$ and $M_H = 0.836 M_\odot$. The numbers refer to the ages given in years. Age zero corresponds to the radial pulsational period (fundamental mode) $P_0 = 50$ d, calculated from the stellar parameters according to Ostlie and Cox (Ostlie, D.A., Cox, A.N.: 1986, *Astrophys. J.* **311**, 864). Mass loss is calculated by the Reimers' formula until the value of the radiation-driven wind theory of Pauldrach et al. (Pauldrach, A., Puls, J., Kudritzki, R.P., Mendez, R., Heap, S.R.: 1988, *Astron. Astrophys.* **207**, 123) is reached. Beyond this point (at $\log(T_{eff}/K) = 4.43$ and $t = 1694$ yrs for $M_H = 0.605 M_\odot$ and $\log(T_{eff}/K) = 4.28$ and $t = 87$ yrs for $M_H = 0.836 M_\odot$, resp.) our adapted formula yields $\dot{M} \sim L^{1.9}$, which provides maximal values of $1.5 \cdot 10^{-8}$ and $1.0 \cdot 10^{-7} M_\odot/\text{yr}$, respectively, along the horizontal part of the evolutionary tracks. During the computations the time steps amount several years on the horizontal part and in the region around the "knee", in the lower region of the cooling track they do not exceed 200 yrs. The arrows indicate the loci where the luminosity of the hydrogen burning shell, L_H, drops beneath the gravitational luminosity, L_{Grav}.

DISCUSSION

SHIPMAN :

Are there any other factors - beyond chemical composition, which you mentioned - which might influence the mass loss rate plus broaden the Mi -Mg relation?

SCHÖNBERNER :

Since we don't yet have a proper understanding of mass-loss processes on the AGB, your question is difficult to answer. One factor I can think of is certainly rotation.

Van HORN :

Does H-burning ever turn off in your latest evolutionary sequences after the gravitational luminosity source begins to dominate?

SCHÖNBERNER :

Hydrogen does continue to burn for a long time, but at a very low level. For instance, at $\log(L_{surf}/L_\odot) \cong -1$, the hydrogen luminosity amounts only to $\approx 10^{-5} L_{surf}$ for the 0.836 $M_\odot$ model, but a few $10^{-2} L_{surf}$ for the less massive one of 0.605 $M_\odot$. For a larger surface luminosity of, say, 10 $L_\odot$, the corresponding figures are 3.10^{-3} L_{surf} and 0.1 L_{surf}, respectively. This different behaviour of the hydrogen burning shell is a direct consequence of the residual envelope mass : At the turn-around point we have 7.10^{-5} $M_\odot$ in the case of the 0.836 $M_\odot$ model and $1.7 \ 10^{-4}$ $M_\odot$ in the case of the 0.605 $M_\odot$ model.

INFLUENCE OF THE PHASE DIAGRAM IN THE COOLING OF WHITE DWARFS

J. ISERN[1,2], E. GARCIA-BERRO[3,2], M. HERNANZ[1,2], R. MOCHKOVITCH[4]

1) Centre d'Estudis Avançats Blanes (CSIC)

2) Grup d'Astrofísica (IEC)

3) Departament de Física Aplicada (UPC)

4) Institut d'Astrophysique de Paris

ABSTRACT. White dwarf interiors are made of a mixture of carbon, oxygen and other minor species, ^{22}Ne being the most abundant one. In this paper we show that the gravitational settling induced by the freezing can delay the cooling process during ~ 1 Gyr depending on the adopted phase diagram. Furthermore, we show that ^{22}Ne can also settle down at the center as an outcome of solidification. In this case, the gravitational energy released keeps the white dwarf warm for 2 Gyrs. A similar effect can be expected from any other chemical component with higher atomic number present in the white dwarf interior, although its contribution would be probably minor because of its low abundance.

1. Introduction

For the last decades, plasma physics developments have led to a better understanding of the physical conditions in white dwarfs interiors. Following the pioneering work of Mestel (1952), the problem of white dwarf cooling has been a subject of continuous interest until the present time. In white dwarf interiors, elements are completely ionized and form what is called a Coulomb plasma which can be completely characterized by the parameter Γ, the plasma coupling constant, that is a measurement of the importance of Coulomb interactions as compared to thermal energy.

5

G. Vauclair and E. Sion (eds.), White Dwarfs, 5–16.
© 1991 Kluwer Academic Publishers.

$$\Gamma = \frac{Z^2 e^2/a}{kT} \cong \text{Coulomb energy per particle/Thermal energy}$$

where a is the distance between particles. The importance of Coulomb interactions in the dense plasma which forms the white dwarf interior was recognized, in the early sixties, by Kirzhnits (1960), Abrikosov (1960), and Salpeter (1961). A first-order transition from liquid to solid phase was predicted and the resultant release of latent heat was shown to somewhat affect the cooling rate (Mestel and Ruderman; 1967, Van Horn, 1968; Koester, 1972; Lamb and Van Horn, 1975; Shaviv and Kovetz, 1976; Sweeney, 1976). Detailed descriptions of such a plasma can be found in Hansen, Torrie and Vieillefosse (1977) and Ichimaru, Iyetomi and Ogata(1988).

White dwarf interiors, however, are not only made of one chemical element but of a mixture of elements and, as the solidification starts, there is always some degree of chemical separation, even in the case where the components are miscible in all proportions. This effect is not negligible. For instance, if carbon and oxygen were not miscible in the solid phase, pure carbon flakes would form an rise towards the surface, melting there. Meanwhile, pure oxygen flakes would sink and settle at the center. If this process would continue all the star would become completely differentiated and $\approx 10^{47}$ erg of gravitational energy would be released. As this would happen at $L \approx 10^{-4} L_\odot$, the cooling process would be delayed ≈ 8 Gyr.

Since white dwarfs are the final remnants of low and intermediate mass stars and since they can be very old, it is possible, at least in principle, to extract information both on the white dwarf themselves and on the structure and evolution of the Galaxy (García-Berro et al 1989b, Iben and Laughlin 1989, Isern et al 1989, Hernanz et al 1990, Mochkovitch et al 1990, Noh and Scalo 1990, Tamanaha et al 1990). The necessary tool to extract such information is the luminosity function of white dwarfs. The first reliable luminosity function observationally obtained is due to Liebert, Dahn and Monet (1989), and it is characterized by a monotonic increase in the number of white dwarfs with magnitude, followed by an abrupt shortfall at a visual magnitude

$M_V \cong 16$. Due to the lack of reliable model atmospheres, the bolometric correction is uncertain, and the luminosity can only be placed somewhere in the range $-4.2 \leq \log(L/L_\odot) \leq -4.6$. The necessary condition for doing that is to have a good model of the cooling process of white dwarfs. The purpose of the present paper is to study the influence of the physical properties of dense plasma mixtures at freezing in the global cooling process.

2. Influence of the phase diagram in the cooling process

2.1 INFLUENCE OF THE C/O MIXTURE

Obviously, the cooling process not only depends on the thermal contents of white dwarfs, but also on the structure of the envelope (mass of the helium layer, metal contents and so on). As here we are only interested on the behaviour of the interior, we can mimic the properties of the envelope by adopting a relationship between the temperature of the core (which is assumed to be isothermal) and the luminosity:

$$L/L_\odot = \mathcal{L}(T_c) \cdot M_{WD}/M_\odot$$

$$\log[\mathcal{L}(T_c)] = 53.810 - 41.080\tau + 10.215\tau^2 - 1.107\tau^3 + 0.046\tau^4$$

where $\tau = \log T$ and $\mathcal{L}(T_c)$ is a numerical fit obtained from the models of Lamb and Van Horn (1975), which assume a pure carbon atmosphere. Of course, this is not realistic because white dwarf envelopes are made of helium and, sometimes, hydrogen. Because of the gravitational field, both elements are strongly stratified and the envelopes have very low metal contents. Furthermore, these structures can change during the cooling process due to accretion from the interstellar medium and to convective mixing. However, in order to compare the different effects induced by crystallization this model of envelope is precise enough. In any case, the nature of the adopted envelope is an open problem and any change in $\mathcal{L}(T_c)$ can introduce dramatic changes in the cooling time

(D'Antonna and Mazzitelli 1990).

In order to make comparisons, we have adopted as standard model (Case A in Table 1) a white dwarf of 0.6 $M_\odot$ composed of an homogeneous mixture of carbon and oxygen with mass fractions $x_c = x_o = 0.5$. The adopted phase diagram for the liquid-solid transition was that of Barrat, Hansen and Mochkovitch (1988). This diagram, which is of the spindle form, predicts a slight oxygen enrichement of the central regions of the star. If the ratio of the equilibrium abundances of oxygen in the solid and liquid phases is $x_o^s/x_o^l = 1 + \varepsilon$, the final distribution of oxygen is given by:

$$x_o^s = x_o^l(0)(1 + \varepsilon)\left(1 - \frac{M_s}{M_{WD}}\right)^\varepsilon$$

where it has been assumed that the liquid phase is always homogeneous due to the convective motions induced by the chemical gradients. Notice that in the previous expression it must be taken into account that ε is a function of the chemical composition. As it can be seen from Table 1, case A, the time needed to reach a luminosity of $10^{-4.5}$ $L_\odot$ is 10.9 Gyr. Here it is necessary to be careful because any small change in the luminosity translates into a big change in the age of the white dwarfs. The uncertainties introduced by the bolometric correction amount to 4 Gyr, for instance.

The values quoted in Table 1 depend on the detailed chemical composition of the core which, in turn, depends on the density and temperature profile of the star during its evolution (D'Antona and Mazzitelli 1990). The relative abundances of carbon and oxygen are governed by the relative ratios of the 3α and $C\alpha$ reactions near the exhaustion of helium. Low temperatures tend to favor the $C\alpha$ reaction and in this case the final oxygen abundance is higher than that of carbon. The inner part of the degenerate core was built during central helium burning phase, while the intermediate layers were produced during the thick helium shell burning, and the outer core during the flashing epoch. As a result of the different temperatures at which these phases happen, the core is already stratified at the birth of white dwarf, being the oxygen abundance much higher in the central regions than in the outer ones. The total amount of oxygen and carbon and the shapes of their distribution depend on the treatment given to the overshooting, semiconvection or helium spikes in the mass of the star (D'Antona and Mazzitelli 1990) and, of course, on the adopted values for the reaction rates. For instance, if the reaction rates for the $C\alpha$ reaction of Harris et al (1983) are adopted, the total abundance of ^{16}O, in the case of a 3 $M_\odot$ star, is ~ 75%. If those of Fowler et al (1975) are used, the final abundance is ~ 50%. As a consequence of the higher average molecular weight per particle, the higher freezing temperature and the smaller release of gravitational energy, the white dwarf cooling is more rapid in this last case. From Table 1, case B, it can be seen that the time necessary to arrive at L = $10^{-4.5}$ $L_\odot$ is 9.8 Gyr.

It is useful to compare these results with those obtained in the case in which the sedimentation induced by freezing is neglected. Model C, in Table 1, assumes that the white dwarf core is composed by half carbon and half oxygen in mass and that the phase diagram is an infinitely narrow spindle. In this case, the time necessary to reach a luminosity L = $10^{-4.5}$ $L_\odot$ is 10.1 Gyr, 0.8 Gyr less than in case A. This is not an enormous difference, if compared with the uncertainties introduced by the poorly known properties of the envelope and the values of the bolometric correction, but it is a positive correction that must be always taken into account in order to use white dwarfs as indicators of the age of the galactic disk. This is especially important in view of the present discrepancy with the age of the globular clusters.

Recently, Ichimaru et al (1988) have discovered that the internal energy of a mixture due to Coulomb effects is a linear combination of the Coulomb energies of each species weighted by the abundance by number of each species. Furthermore, they have found that the entropy is smaller in the liquid phase that in the solid phase. As the liquid phase is favored, the phase diagram adopts an azeotropic form (the solidification temperature versus the chemical composition goes through a minimum). The azeotropic point for the carbon oxygen mixture is

$T/T_c = 0.94$ and $X_A = 0.16$, where T_c is the freezing temperature for pure carbon and X_A is the oxygen abundance by number. When the abundance of carbon and oxygen is placed at the right of the azeotropic point (positive slope), the final distribution of oxygen after freezing is similar to that obtained with the Barrat, Hansen and Mochkovitch (1988) diagram. Due to the smaller freezing temperatures involved, the delay introduced in the cooling process is larger. From model D, Table 1, it can be seen that the time necessary to reach $L = 10^{-4.5} L_{\odot}$ is 11.5 Gyr, a quantity that would not be so far from the age of globular clusters provided that the present knowledge of the properties of the envelope were reliable.

TABLE 1. Cooling times

$-\log(L/L_{\odot})$	t (Gyr)			
	Case A	Case B	Case C	Case D
3.00	0.84	0.90	0.84	0.84
3.20	1.14	1.10	1.13	1.14
3.30	1.31	1.27	1.31	1.32
3.41	1.54	1.48	1.53	1.55
3.53	1.82	1.87	1.82	1.84
3.62	2.10	2.30	2.28	2.09
3.72	2.83	2.87	2.90	2.40
3.82	3.77	3.60	3.70	3.41
3.95	4.94	4.54	4.70	4.80
4.02	5.62	5.11	5.31	5.61
4.17	7.24	6.49	6.76	7.47
4.26	8.19	7.34	7.62	8.55
4.36	9.27	8.31	8.61	9.74
4.47	10.47	9.44	9.74	11.06
4.59	11.88	10.76	11.05	12.53
4.74	13.49	12.32	12.59	14.14
4.93	15.37	14.22	14.42	16.08
5.16	17.66	16.58	16.66	18.29
5.51	20.52	19.63	19.53	20.97

The temperature at which the phase transition happens is also very important because of the strong dependence of the luminosity on the core temperature. As the luminosity scales like $L \sim T_c^b$, where b takes values in the range 2.5 - 3.5 depending on the adopted model of envelope, any small change in the temperature at which the energy is released translates into a large delay in the cooling process. Table 2 shows the effect of changing the value of Γ defining the liquid solid transition from 155 (Hansen et al 1977) to 180 (Ogata and Ichimaru 1987).

TABLE 2: Effect of changing the value of Γ

$-\log(L/L_\odot)$	t (Gyr)		Δt (Gyr)
	$\Gamma = 155$	$\Gamma = 180$	
-4.26	7.99	8.19	0.20
-4.47	10.13	10.49	0.36
-4.74	12.94	13.49	0.55
-4.93	14.73	15.37	0.64

2.2 THE ROLE OF THE MINOR SPECIES

The existence of an azeotrope in the phase diagram opens, however, new and very interesting possibilities for obtaining additional sources of energy (Garcia-Berro et al 1988b). Since any binary mixture with an atomic number ratio larger than that of the carbon and oxygen mixture will also display an azeotropic behavior at freezing, the phase diagram (freezing temperature against abundance of the heaviest element) will have a negative slope in the region of the smaller abundances of the heavier element. In this case, the liquid remnant will be denser than the solid and the last will migrate outwards, thus a chemical stratification of the star, with the heaviest elements being concentrated in the innermost regions, will result.

White dwarfs are made of carbon, oxygen, and a mixture of minor components reflecting the initial metallicity of the parent star. Among them, ^{22}Ne is the most important one. It comes from helium burning of the ^{14}N left by the CNO cycle. Its total amount as well as its distribution throughout the white dwarf interior depend on the initial metallicity and mass of the progenitor. Despite of its low abundance, the importance of ^{22}Ne as a source of gravitational energy comes from the fact that white dwarf structures are very sensitive to the shape of the electron mole number distribution Ye (Mochkovitch 1983). A white dwarf with all the neon concentrated in the innermost regions is far more compact than a white dwarf with the same amount of this element uniformly distributed throughout the entire star.

The phase diagrams for C-Ne and O-Ne mixtures as computed by Isern et al (1990) are displayed in Figures 1 and 2, respectively. Both show the existence of an azeotropic point at X_{Ne}=0.053, T/T_c=0.982 and X_{Ne}=0.04, T/T_0=0.995 respectively (or x_{Ne}^a=0.09 and x_{Ne}^a=0.05, if mass fractions are preferred), where T_c and T_0 are the solidification temperatures for pure carbon and pure oxygen. As it was mentioned in Isern et al (1990), these results are strongly dependent upon the values taken by the entropy of the solid and the fact that in white dwarfs the mixtures are not binary but at least ternary.

The ^{22}Ne distribution is modified by the process of solidification in the following way. If the neon abundance is smaller than the azeotropic value, a neon poor solid forms which, being lighter, rises towards the surface and melts somewhere. As a consequence, the liquid mixture becomes progressively more neon rich. When the liquid reaches the azeotropic abundance, a solid with this last composition forms. This solid, being denser than the mixture, sinks and a neon rich core progressively forms. The final structure is a core containing all the ^{22}Ne of the star surrounded by a mantle free of neon. Obviously, the size of such a core is:

$$M_{core} = \frac{x_{Ne}^0}{x_{Ne}^a} M_{wd},$$

where x_{Ne}^0 and x_{Ne}^a are the initial and the azeotropic mass fractions.

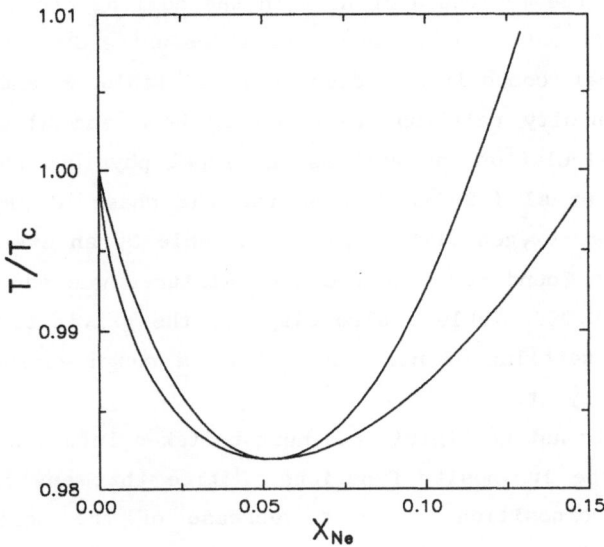

Figure 1. The fluid-solid phase diagram for a binary C-Ne mixture. The freezing temperature compared to that of pure carbon are represented as a function of the neon abundance by number.

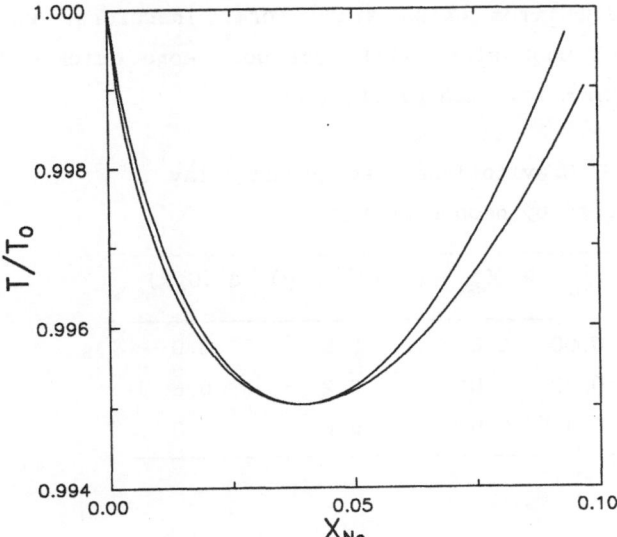

Figure 2. The fluid-solid phase diagram for a binary O-Ne mixture. The freezing temperatures compared to that of pure oxygen are represented as a function of the neon abundance by number.

To evaluate the influence of neon in the cooling process of white dwarfs, we have computed the structure of three white dwarfs of 0.6 $M_{\odot}$, with the chemical compositions displayed in Table 3 and the core temperature-luminosity relationship given by Wood and Winget (1989). The method of calculation, as well as the input physics, can be found in Garcia-Berro et al (1988a), except for the phase diagram. In the case of the carbon-oxygen mixture given in Table 3, an average of the azeotropic values found for C-Ne and O-Ne mixtures was adopted, i.e.: $X_{Ne}=0.045$, $T/T_{CO}=0.989$. Table 3 also displays the gravitational energy released by the settling of neon as well as a rough estimate of the delay introduced by it.

This delay is not negligible and must be taken into account in the calculation of the luminosity functions. Notice that any increase of the azeotropic composition and/or a decrease of the solidification temperature of the azeotropic mixture would result in a dramatic enhancement of the cooling time. It is difficult, however, to predict how this effect translates into the white dwarf luminosity function, because the abundance of ^{22}Ne depends on the metallicity of the interstellar medium from which stars form. Therefore, white dwarfs coming from metal poor stars will cool down more quickly than white dwarfs coming from metal rich populations.

TABLE 3: Gravitational energy and delay introduced by neon settling.

X_C	X_O	X_{Ne}	$\Delta B/10^{46}$(erg)	Δt(Gyr)
0.99	0.00	0.01	5.5	3.0
0.00	0.99	0.01	3.2	0.6
0.455	0.455	0.01	5.8	2.0

3. Conclusions

We have shown that the gravitational energy released by the settling of heavier species during the cooling of white dwarfs can delay the process during 1 to 3 Gyr, depending on the adopted phase diagram and on the abundance of the chemical species considered. We have also shown that the gravitational settling of minor species, in particular that of neon, is a potentially very important source of energy for white dwarfs and deserves further studies. Among other points, it would be necessary to clarify:

a) The precise form of the phase diagram as well as the position of the azeotrope in it. The results strongly depend on the entropy of the solid as compared to that of the liquid. b) The way the progressive increase of the interstellar metallicity influences the luminosity function of white dwarfs. c) The role played by the other minor chemical species.

Acknowledgements: This work has ben supported by the CICYT grant PB87-0304

4. References

Abrikosov A.A. (1960) Soviet Phys-JETP **12**, 1254.

Barrat J.L., Hansen J.P., Mochkovitch R., (1988), Astron. Astrophys. **199**, L15.

D'Antonna F., Mazzitelli I. (1989) Astrophys. J. **347**, 934.

D'Antonna F., Mazzitelli I. (1990) Ann. Rev. Astron. Astrophys. in press.

Fowler W., Caughlan G.R., Zimmerman B.A. (1975) Ann. Rev. Astron. Astrophys. **13**, 69.

Garcia-Berro E., Hernanz M., Mochkovitch R., and Isern J., (1988a), Astron. Astrophys, **193**, 141.141.

Garcia-Berro E., Hernanz M., Isern J., and Mochkovitch R. (1988b), Nature **333**, 644.644.

Hansen J.P., Torrie G.M., Vieillefosse J.P. (1977) Phys. Rev. **A16**, 2153.

Harris M.J., Fowler W.A. Caughlan G.R., Zimmerman B.A. (1983) Ann. rev. Astr. Astrophys. **21**, 165.

Hernanz M., Garcia-Berro E., Isern J., Mochkovitch R. (1990) in "Astrophysical Ages and Dating Methods", Ed. E. Vangionne-Flam (Ed Frontières), p.171.

Iben I. and Laughlin G. (1989) Astrophys. J. **341**, 312.

Ichimaru S., Iyetomi H., Ogata S. (1988), Astrophys. J. **334**, L17.

Isern J., Garcia-Berro E., Hernanz M., Mochkovitch R. (1989) in "White Dwarfs", Ed G.Wegner, Lecture Notes in Physics 328 (Springer-Verlag) p.27.

Isern J., Mochkovitch R., García-Berro E., Hernanz M. (1990) Astron. Astrophys. In press.

Kirzhnits D.A. (1960) Soviet Phys-JETP **11**, 365.

Koester D. (1972) Astron. Astrophys. **16**, 459.

Lamb D.Q., Van Horn H.M. (1975) Astrophys. J. **200**, 306.

Liebert J., Dahn C.C. and Monet D.G. (1989), in "White Dwarfs". Ed. G.Wegner, Lecture Notes in Physics 328, (Springer-Verlag) p.15.

Mestel L. (1952) MNRAS **112**, 583.

Mestel L., Ruderman M.A. (1967) MNRAS **136**, 27.

Mochkovitch R., (1983), Astron. and Astrophys **122**, 212.

Mochkovitch R., Garcia-Berro E., Hernanz M., Isern J., Panis J.F. (1990), Astron. Astrophys. **233**, 456.456.

Noh H.R., Scalo J. (1990), Astrophys. J. **352**, 605.

Ogata S., Ichimaru H. (1987), Phys. Rev. A36, 5451.

Salpeter E.E. (1961) Astrophys. J. **134**, 669.

Shaviv G., Kovetz A. (1976) Astron. Astrophys. **51**, 383.

Sweeney M.A. (1976) Astron. Astrophys. **49**, 375.

Tamanaha C., Silk J., Wood M.A., Winget D.E. (1990), Astrophys. J. **358**, 164.

Van Horn H.M. (1968) Astrophys. J. **151**, 227.

Winget D.E., Hansen C.J., Liebert J., Van Horn H.M., Fontaine G., Nather R.E., Kepler S.O. and Lamb D.Q. (1987), Astrophys. J. (Letters), **315**, L77.

Wood M.A. and Winget D.M. (1989), in "White Dwarfs". Ed. G.Wegner, Lecture Notes in Physics 328, (Springer-Verlag) p.282.282.

Yuan J.W. (1989), Astron. Astrophys. **224**, 108.

PG1159 STARS AND THE PNN–WHITE DWARF CONNECTION

M.A.BARSTOW AND R.W.TWEEDY
Department of Physics and Astronomy
University of Leicester
Leicester LE1 7RH, UK.

K.WERNER
Institut für Theoretische Physik und Sternwarte
der Universität Kiel
2300 Kiel, Germany

ABSTRACT. The results of atmospheric composition measurements in PG1159 stars, by LTE analysis of soft X-ray observations and NLTE analysis of optical data, are in complete contradiction. Study of the model atmosphere codes used shows that the discrepancies arise from the invalidity of the assumption of LTE for these stars and also from differences between the model atoms utilised by the calculations. While doing these tests we were able to demonstrate agreement between two completely independent NLTE codes. Finally, the soft X-ray data are reanalysed using a new grid of NLTE models.

1. Introduction

The spectroscopic subgroup of very hot He-rich white dwarfs, the PG1159 stars (after the prototype PG1159-035) may be of great importance in solving some of the puzzles of white dwarf evolution. A clear circumstantial link exists between these objects and the planetary nebulae nuclei (PNN). For example, K1-16 which is the central star of a high-excitation planetary nebula, exhibits spectroscopic similarities to the PG1159 objects (Grauer and Bond, 1984). An evolutionary sequence has been proposed (Sion, Liebert and Starrfield, 1985), in which PG1159 stars are transitional objects between the 'OVI' PNN and the hottest DO white dwarfs. In the primordial theory of white dwarf evolution (see eg. Shipman, 1989), the PG1159 stars would merely be the hottest members of the non-DA white dwarf branch. However, in the mixing model they would be an important channel for **both** DA and non-DA stars. Consequently, gaining a better understanding of the temperatures and compositions of the PG1159 stars and related objects is an important piece in establishing the more general picture.

Recently several lines of work have begun to establish the temperatures and abundances of the PG1159 stars. Barstow and Holberg (1990) used LTE model atmospheres including H,He,C,N and O to analyse soft X-ray data from the EXOSAT and Einstein satellites. Latterly, Werner et al (1991) have fitted the He, C and O line features of high resolution optical spectra with synthetic profiles generated with a non-LTE code. While both sets of authors obtain similar temperature estimates their abundance values differ

17

G. Vauclair and E. Sion (eds.), White Dwarfs, 17–27.

widely. Barstow and Holberg find abundances not very different from the solar values whereas the results of Werner et al suggest that the abundances of C and O are 1 or 2 orders of magnitude greater. LTE models with the abundances found by Werner et al predict no measureable soft X-ray flux. Yet X-ray emission is observed in these stars. However, the non-LTE code should yield the most reliable representation of the stellar atmospheres. The aim of this paper is to resolve this discrepancy. In doing so, we will briefly review the soft X-ray and optical results, compare and contrast the results of several model atmosphere calculations and finally reanalyse the soft X-ray data with some new NLTE models.

2. Soft X-ray observations of PG1159 stars

Several PG1159 type objects have been observed at soft X-ray wavelengths by satellites such as the Einstein and EXOSAT observatories. The emergent photospheric X-ray fluxes of white dwarfs are extremely sensitive to their atmospheric composition. For example, broad band flux measurements can be used to determine the abundance of trace He in DA white dwarf atmospheres (eg. Paerels and Heise, 1989), if the white dwarf atmospheres are presumed to be homogeneous. An alternative treatment by Koester (1989) shows that the X-ray data can equally well be explained by stratified model atmospheres; in this case the compositional parameter is the mass of a thin H layer lying on a pure He envelope. Although the question of which explanation is correct remain open, it is clear that X-ray observations are important.

In DA white dwarfs X-ray emission is apparent in stars as cool as 25,000K. This is primarily a consequence of the low soft X-ray opacity of hydrogen. However, in the He or metal rich objects, where the materials have much higher opacities the presence of detectable X-ray emission is indicative of very high temperatures. A total of 5 PG1159 objects (including K1-16) and 9 PNN have been detected at soft X-ray wavelengths. In this work we concentrate on the PG1159 star observations and two PNN that seem to have a particular relationship to this group of stars.

THE PG1159 STARS OBSERVED

H1504+65 appears to be the hottest known PG1159 object (Nousek et al, 1986; Barstow and Tweedy, 1990). It shows many optical spectroscopic features found in PG1159 stars. In addition to the normally observed absence of H lines no He features can be seen either. This may be a result of a real absence of He in the atmosphere but the high effective temperature may have excited and ionised the He to such an extent that none of the energy levels which give rise to observable transitions are populated. PG1159-035 is the prototype of the group and is also a non-radial pulsator. Indeed, the pulsations have also been observed in the soft X-ray flux (Barstow et al, 1986). The star PG1144+005 appears to be spectroscopically identical to PG1159-035 and should have a similar temperature. However, this object shows no evidence for optical pulsations at amplitudes well below those seen in PG1159-035. K1-16 is the central star of a low surface brightness, high excitation planetary nebula. It is optically very similar to the PG1159 stars. Pulsations are also seen in this object but the mode structure is more complex than in PG1159-035 and may not necessarily be driven by the same mechanism. An important object that is closely related to the PG1159 group is the DOZ white dwarf KPD0005+5106. It has some spectroscopic features that are found in the PG1159 stars but clearly a more evolved object. Unlike the PG1159 stars it is cool enough for the far-UV spectral slope to provide a temperature measurement $(80,000 \pm 20,000K)$ rather than only a lower limit. Therefore, this object may represent the cool end of the PNN/PG1159/DO track that feeds into the WD cooling sequence proper. The details of the soft X-ray and UV observations for all these objects are included in the LTE analysis of Barstow and Holberg (1990). Further general information on these stars together with the primary references can also be found in that paper.

We also consider here two important PNN which were not included in the analysis of Barstow and Holberg. NGC246 is an example of the He-rich OVI group of PNN that are thought to be the immediate progenitors of the PG1159 stars. Spectroscopically, NGC246 and K1-16 are almost identical objects but NGC246 does not appear to pulsate. Its temperature has been determined from an analysis of the optical spectrum using NLTE H+He atmospheres (Kudritzki and Mendez, 1989) yielding a value of $130,000 \pm 15,000$K. Soft X-ray count rates for NGC246 were taken from the literature (Tarafdar and Apparao, 1989; Apparao and Tarafdar, 1988). NGC7293 is a member of the H-rich family of PNN and as such is a possible progenitor for the DA WD sequence. However, its temperature $(90,000 \pm 10,000$K; Kudritzki and Mendez, 1989), is somewhat higher than the hottest known DA stars ($\approx 70,000$K) and so the connection remains uncertain. Observations of this object are discussed in detail in a second paper (Tweedy, Barstow and Werner, 1991).

LTE ANALYSIS OF THE SOFT X-RAY DATA

Determination of photospheric effective temperature and composition for any star requires the comparison of fluxes predicted by a model atmosphere with observational results. The technique that we apply here has been outlined in detail several times before (eg. Barstow and Tweedy, 1990; Barstow and Holberg, 1990). However, it is necessary to include a brief summary here in order to understand how the results that we will discuss were obtained.

For a given temperature it is possible to calculate a predicted soft X-ray count rate for any star by folding a model spectrum, normalised to the V magnitude of the star, through the instrument response of each band in which observations were made. It is also necessary to include the absorbing effect of the interstellar medium in this calculation at soft X-ray energies. Count rates are predicted for each model temperature and a finely space grid of N_H. The resulting tables are then interpolated to determine the values of T and N_H that correspond to the observed count rates. For any model (defined by the composition), a family of constant count rate curves can be displayed in the T/N_H plane as depicted in figure 1 for K1-16. The three curves shown for each EXOSAT filter with which this star was observed correspond to the measured count rates and their associated observational errors. A model can be said to be consistent with the data if there is at least one region on the plot where all the sets of contours overlap. At this stage N_H is a free parameter but in some cases its value has been determined by IUE observations and this information provides a further constraint on the acceptability of any model. The far-UV spectra further help the analysis since ranking the objects by their observed far-UV slope also orders them by temperature.

The results of this LTE analysis are summarised in table 1. An important result is the limitation placed on the abundance of CNO in the stellar photospheres. In most cases CNO:He must be no greater than the solar value although H1504+65 can have CNO:He=$2\odot$. Only for PG1159-035 are there any constraints on the H abundance. There is a weak trend of decreasing He and metal abundance as the objects cool (get older). Such a trend could be consistent with the idea that DA white dwarfs could utimately form from these objects by gravitational settling.

Figure 1. Curves of constant count rate, calculated for a single LTE model, corresponding to the observed values for K1-16 and $\pm 1\sigma$ errors.

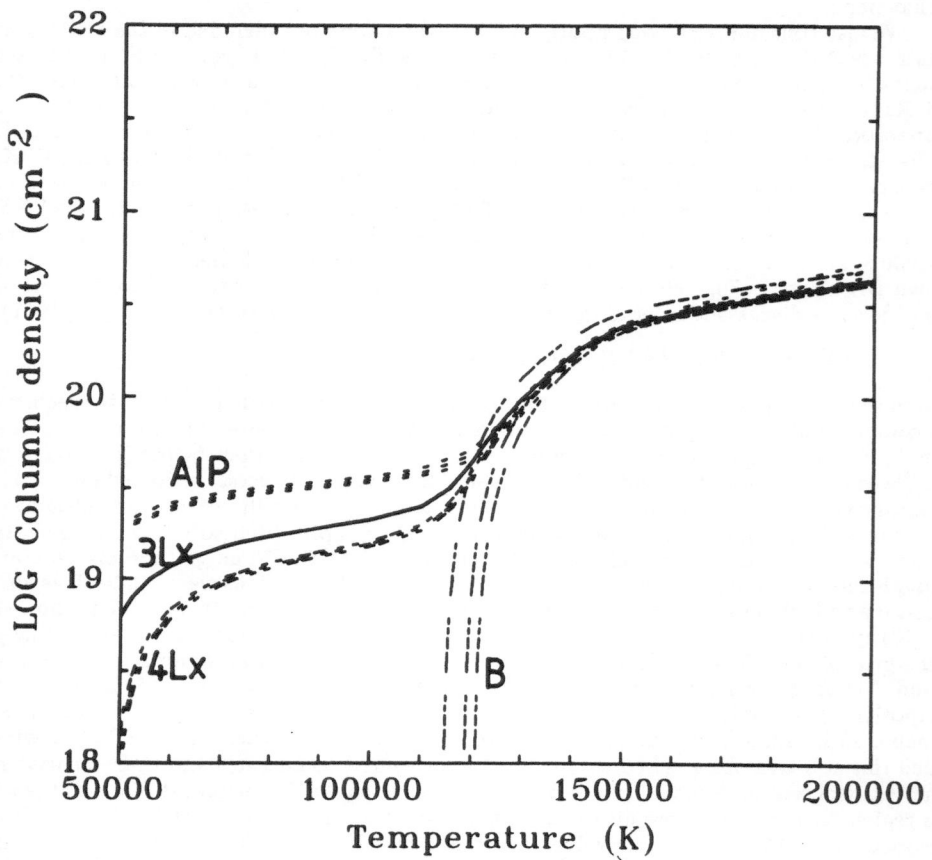

Table 1. Effective temperature and abundance for the PG1159 stars determined with LTE model atmospheres. Most likely values are in bold type.

Object	Temperature (K)	H:He	CNO:He
H1504+65	150000-**180000**-185000	$\odot$-2$\odot$	
PG1159-035	123000-**124000**	1$\odot$	1$\odot$
PG1144+005	93000-124000	1-10$\odot$	0.02$\odot$-$\odot$
K1-16	103000-**110000**	0.5$\odot$-1$\odot$	0.05$\odot$-0.1$\odot$
KPD0005+5106	83000-**97500**-100000	1$\odot$-10$\odot$	**0.1$\odot$**-$\odot$

3. Optical observations of PG1159 stars

A detailed analysis of high resolution optical spectra of four PG1159 stars has been carried out by Werner et al (1991). They observed four stars - PG1159-035, PG1520+525, PG1424+534 and PG1707+427 - but only PG1159-035 was also in the soft X-ray sample. These authors compared synthetic line profiles of HeII, CIV and OVI calculated in NLTE using the computer code PRO2. This NLTE analysis derives a temperature of $\approx 140,000$K for PG1159-035, a little higher than that obtained from the LTE analysis of the soft X-ray data ($\approx 125,000$K), but within the $\approx 10\%$ errors in each measurement. However, the measured abundances - H:He< 1, C:He$\approx 100\odot$ and O:He$\approx 12\odot$ - were several orders of magnitude larger than the LTE result.

Two important questions are raised by these contrasting results. First, what is the cause of the vastly different abundance determinations? Since the NLTE models have a better physical basis than the LTE computations it is likely that departures from the LTE approximation at the temperatures involved are significant. However, this needs to be tested. Secondly, are the soft X-ray fluxes predicted by the NLTE models consistent with the observations? The following sections address these queries and provide the answers.

4. A comparison of LTE and NLTE model atmosphere calculations

To study the effect of departures from the LTE approximation at the temperatures and gravities found in PG1159 stars we have compared the synthetic spectra computed by three completely independent model atmosphere codes. The ATMOS code assumes LTE and was used to generate the model spectra for analysis of the soft X-ray data (Barstow, 1990). PRO2 is the powerful NLTE program used by Werner et al (1991) to interpret their optical spectra. In addition to these two programmes, there is also the code TLUSTY which was written by Hubeny (1988) and made freely available to the astronomical community. TLUSTY computes NLTE model atmospheres but in its original published form was only able to treat H and He. Provision for extending the calculations to other elements is built into the programme by inserting additional data into the input file that drives it. However, to perform NLTE calculations treating the high ionisation states of C, N and O found in PG1159 stars required some additional routines. A different formula for computation of the partition function, based on the scheme used in ATMOS, and some new subroutines for formulation of the C and O opacities were added.

The results of model atmosphere calculations using each of the three programmes were compared in several ways. LTE and NLTE models generated with PRO2 were compared with the LTE results of ATMOS. It is clear that departures from LTE have a significant effect at temperatures in excess of 100,000K when log g=7. This is illustrated in figure 2, where LTE and NLTE spectra calculated by PRO2 for a solar composition, T=140,000K and log g=7 are displayed. A further difference is found between the PRO2 and ATMOS LTE models calculated for the same parameters (figure 3). It appears from this result that departures from LTE could explain some of the difference between the compositions determined from the soft X-ray and optical observations. However, there is some other difference between the PRO2 and ATMOS calculations that also contributes to the discrepancy.

In parallel, initial calculations made using TLUSTY were tested against the NLTE H+He and H+He+CNO models published by Husfeld et al (1984). TLUSTY was found to be able to reproduce their single model with solar abundance He, C, N and O (T=100,000K and log g=5) exactly. However, the spectrum of the equivalent PRO2 model was completely different showing a much larger OVI edge at 3×10^{16}Hz (figure 4).

Having three different model atmosphere codes each giving different results but with

22

Figure 2. A comparison of LTE (- - - -) and NLTE (———) spectra generated by PRO2 for T=140,000K and log g=7. The abundances correspond to the NB model of table 3.

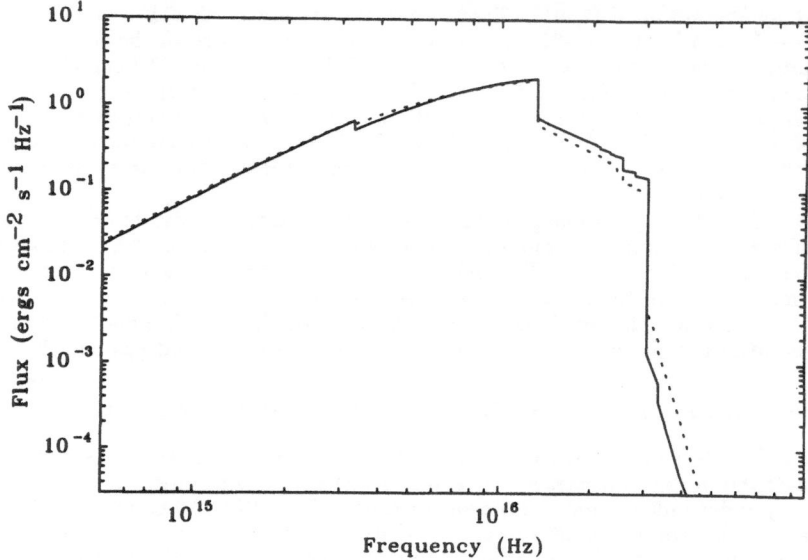

Figure 3. A comparison of LTE models generate by ATMOS (- - - -) and PRO2 (———) for T=140,000K, log g=7 and solar abundances.

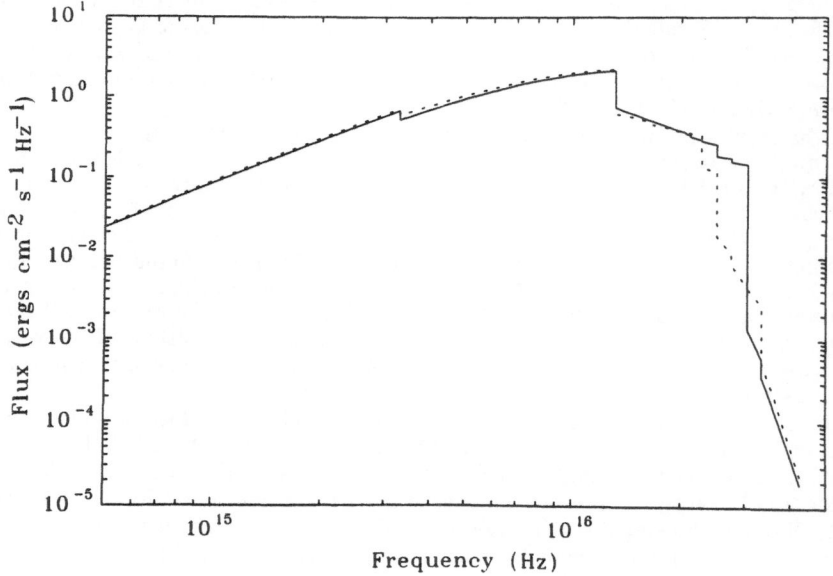

one (TLUSTY) agreeing with other published data, it became necessary to look at them in more detail to discover which was correct. Since the major observed difference arises in one of the OVI absorption features either the opacity or the populations of the levels involved in the transition must be larger in the PRO2 model. Investigation of the O model atoms included in the programmes reveals some important differences. Both ATMOS and TLUSTY incorporate an O model atom constructed from the same sources as Husfeld et al (1984). In fact the model, treating a total of 9 NLTE levels is identical to that used by Husfeld et al to generate their H+He+CNO spectrum. In contrast the PRO2 programme considers 44 NLTE O levels and in particular 36 OVI states. Table 2 summarises the details of the model atoms used by Husfeld et al (ATMOS and TLUSTY) and PRO2. It would seem likely that the observed difference between the programmes are related to the increased detail in the treatment of the OV and OVI ions. To test this a PRO2-like O model atom was included in TLUSTY, replacing the original version. Although it would be possible use TLUSTY with the full PRO2 model atom, TLUSTY does not have the computational speed of PRO2. In order to ensure that the TLUSTY NLTE calculation remained tractable within sensible CPU times the number of OVI levels included was reduced. All the levels up to and including principal quantum number of 5 were treated, a total of 14 (see table 2).

Carrying out the solar abundance, T=100,000K, log g=5 TLUSTY calculation with the new model atom generated a spectrum almost identical to that of PRO2 (figure 4). This is a very important result. We have demonstrated agreement between NLTE calculations performed by two completely independent programmes for the first time. Consequently, we can have high confidence that both codes are giving sensible results. We have also shown the necessity of including sufficient detail in the model atoms used by any model atmosphere calculation in order to establish the populations of the levels involved in important transitions correctly. However, it can also be seen that, at least when considering the continuum fluxes, there is a limit above which additional detail is unimportant. This could be of benefit in reducing computation times by lowering the number of independent quantities that a programme has to calculate.

5. NLTE analysis of the soft X-ray observations

Having established the reliability of the PRO2 and TLUSTY codes and determined the limitations of the ATMOS in the context of its assumption of LTE and the model O atom used, the second question that must be answered is how well the NLTE soft X-ray fluxes match the observations. We have used the same technique to analyse the data with NLTE models as for the LTE grid. However, the longer computing time required limits the range and number of different compositions presently available. Table 3 lists the models, with which programme they were calculated, their surface gravities and their abundances. All the models have a minimum temperature of 60,000K and span the temperature range up to $\approx 200,000$K in 20,000K steps.

The soft X-ray data for PG1159-035, H1504+65, PG1144+005 and KPD0005+5106 were compared with all the log g=7 models. In the case of PG1159-035 only the AO and NB grids fit the both the soft X-ray data and observed N_H simultaneously. The NB model which gives the best fit to the optical data gives a temperature of 129,000 ± 500K. Similarly, the AO and NB models can also account for the H1504+65 EXOSAT fluxes. However, the inclusion of the ROSAT WFC S1a count rate in the analysis (see Barstow et al. 1991) allows us to reject the AO model and restrict the temperature range allowed by the NB grid to 185,000±500K. All the models fit the PG1144+005 data. If it has the same abundance as PG1159-035 then its temperature will lie in the range 108,000-130,000K. The situation is similar for KPD0005+5106 but the IUE determined temperature provides an

Table 2. Number of NLTE levels considered for the Oxygen ions included in each model atmosphere programme.

Oxygen Ion	Husfeld et al.	PRO2	Red. PRO2 in TLUSTY
OIII	2	–	–
OVI	1	1	1
OV	3	6	6
OVI	2	36	14
OVII	1	1	1

Table 3. NLTE models used to analyse the soft X-ray data

Program	Model[1]	log g	He:H	C:H	N:H	O:H
PRO2	AO	7-8	0.1	4.84×10^{-4}	1.0×10^{-4}	8.39×10^{-4}
PRO2	NC	7-8	1.0	4.84×10^{-4}	1.0×10^{-4}	8.39×10^{-4}
PRO2	NB	7-8	4.75	3.33	4.75×10^{-3}	6.25×10^{-1}
TLUSTY	N000	5-6	0.1	3.0×10^{-4}	7.0×10^{-5}	6.0×10^{-4}
TLUSTY	N001	7	0.01	3.0×10^{-5}	7.0×10^{-6}	6.0×10^{-5}

1. Models AO and N000 are $\approx \odot$ abundance and model N001 is $0.1\odot$.

Figure 4. A comparison of NLTE models generated by Husfeld et al/TLUSTY+Husfeld et al Oxygen model atom (- - - -), PRO2 (- - —) and TLUSTY+reduced PRO2 Oxygen model atom (——).

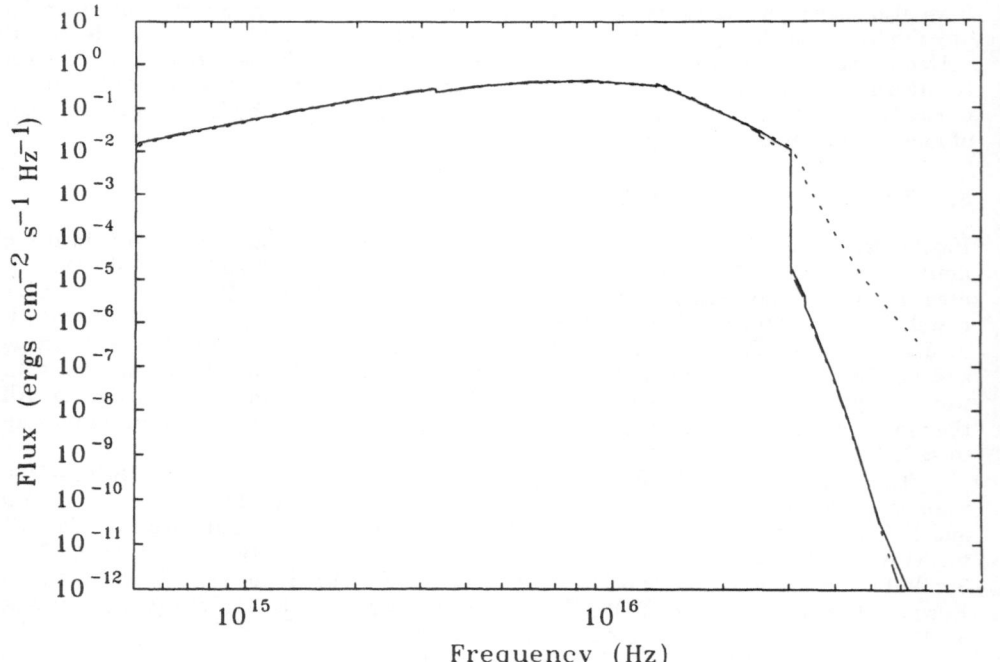

important constraint. The temperature allowed by the AO and NB grids exceeds the upper bound of 100,000K determined from the far-UV slope. However, the 1/10 solar abundance model gives a temperature range of 80,000-105,000K which is consistent with the IUE result.

We now consider NGC246 and K1-16. Although spectroscopically similar to the PG1159 objects NGC246 has a lower gravity (log g=6). The gravity has not been determined for K1-16 but as a near twin of NGC246 it is also likely to be low. In comparing the model predictions with the K1-16 observations we find that the AO and NB models are consistent but they require a value of N_H that is higher than observed. There is only a single NLTE grid (solar abundance) with log g=6 available. It fits the K1-16 EXOSAT data but only agrees with the measured N_H if T is $\approx 100,000$K. Similarly, this model fits the NGC246 data giving a temperature range of 110,000-120,000K. Such a range is in agreement with the work of Kudritzki and Mendez (1989). However, the He abundance that they determine is about 5-9 times the solar value. Clearly we must extend this work to model greater He and CNO abundances.

6. Discussion

Since we have only studied NLTE models with a few different compositions this work is not yet complete. However, a few general conclusions can be drawn from the results obtained thus far. The composition and temperature of PG1159-035 are now reasonably well established by the optical work of Werner et al (1991). We have found that the soft X-ray data is in agreement with this result, although a slightly lower temperature is obtained. Combining the soft X-ray analysis with the temperature range determined for KPD0005+5106 with the IUE data we find that its He and CNO abundance could be as low as 1/10 solar. However, optical data do not show the presence of hydrogen (Werner, private communication) and it may be that He-rich models containing much smaller traces of metals, which we have not yet tested, could also explain the EXOSAT fluxes.

The difference in the temperature of PG1159-035 estimated from the optical and soft X-ray data may be of importance. Since there are no visible H transitions in the optical data, only very coarse limits can be placed on its abundance relative to He. If we already know the abundance of CNO relative to He, this can be fixed in the soft X-ray analysis. Since, the soft X-ray fluxes are very sensitive to the presence of H, 'diluting' the He and metal opacities, it should be possible to measure the H abundance directly by varying it in the models to force the temperature allowed by the soft X-ray fluxes to be equal to that determined from the optical data.

7. Conclusion

In this paper we set out to discover why the abundances determined by LTE analysis of soft X-ray data and NLTE analysis of optical spectra differed by 2-3 orders of magnitude. We have established that the cause was a combination of the inadequacy of the LTE approximation and insufficient detail in the oxygen model atom used in the LTE calculations. In doing so we have also been able to demonstrate consistency between two entirely independent NLTE codes, capable of including C,N and O, for the first time. As a result of this we have found that the soft X-ray fluxes predicted by NLTE models with high He and CNO abundances, as favoured by the optical data, are consistent with the observations.

Using some new grids of NLTE model atmospheres we have estimated the temperatures and abundances of the PG1159 stars included in the earlier LTE analysis. It is clear from this work that it is necessary to compute more models with differing abundances to properly test the uniqueness of the results.

8. References

Apparao,K.M.V. and Tarafdar,S.P., 1989, *Ap.J.*, **344**, 826.
Barstow,M.A., 1990, *Mon.Not.R.astr.Soc.*, **243**, 182.
Barstow,M.A. and Holberg,J.B., 1990, *Mon.Not.R.astr.Soc.*, **245**, 370.
Barstow, M.A. and Tweedy,R.W., 1990, *Mon.Not.R.astr.Soc.*, **242**, 484.
Barstow,M.A., Holberg,J.B., Grauer,A.D. and Winget,D.E., 1986, *Ap.J.*, **306**, L25.
Barstow,M.A. et al., 1991, these proceedings.
Grauer,A.D., and Bond,H.E. 1984, *Ap.J.*, **277**, 211.
Hubeny,I., 1988, *Computer Physics Communications*, **52**, 103.
Husfeld,D., Kudritzki,R.P., Simon,K.P. and Clegg,R.E.S., 1984, *Astron.Astrophys.*,
     134, 139.
Koester,D., 1989, *Ap.J.*, **342**, 999.
Kudritzki,R.P. and Mendez,R.H., 1989, in *Planetary Nebulae*, ed. S.Torres-Peimbert, 273.
Nousek,J.A., Shipman,H.L., Holberg,J.B., Liebert,J., Pravdo,S.H., White,N.E. and
     Giommi,P., 1986, *Ap.J.*, **309**, 230.
Paerels,F.B.S. and Heise,J., 1989, *Ap.J.*, **339**, 1000.
Shipman,H.L., 1989, *proceedings of IAU Colloquium 114 - White Dwarfs*,
     ed. G.Wegner, Springer-Verlag, 220.
Sion, E.M., Liebert, J. and Starrfield, S.G., 1985, *Ap.J.*, **292**, 471.
Tarafdar,S.P. and Apparao,K.M.V., 1988, *Ap.J.*, **327**, 342.
Tweedy,R.W., Barstow,M.A. and Werner,K., 1991, these proceedings.
Werner,K., Heber,U. and Hunger,K., 1991, *Astron.Astrophys.* and these proceedings, in
press.

DISCUSSION

HEBER :

I like to make three comments :

i) there are no traces of N and Ne lines in our optical spectra, which gave an upper limit on the N abundance close to the solar value. For Ne we cannot set a limit yet.

ii) when we did the analysis of the PG1159 optical spectra we had to realize that five free parameters have to be determined. Besides T_{eff} and log g there are H/He, C/He and O/He that determine the atmospheric structure. You cannot group C, N and O into one group. Therefore, I do not see how you could derive 5 parameters from just 4 EUV-fluxes.

iii) Husfeld (1988, Tucson-conference Faint blue stars) already analysed NGC 246 and derived abundances for it that are far from solar.

BARSTOW :

i) Yes but NV features have been seen in the far-ultraviolet by IUE. However the abundance of N in the NLTE high metal abundance models reflects your comment

ii) I agree, we have always been aware that there are too many free parameters for them to be determined by broad band photometry. This is why we choose to hold our C : N : O abundances constant in the model grids. However, the EUV measurements are still important and complement the optical work at Kiel. For exemple, given knowledge of the C : N : O ratio it may be possible to place limits on traces of H or He at abundances (or temperatures) where the optical data are insensitive, since these species will dilute the EUV opacity if present.

iii) At the moment only a solar abundance log g = 6 NLTE model grid has been calculated. Certainly there is a need to investigate whether higher abundances are consistent with the EUV/Soft X-ray fluxes.

BUES :

You mention the metal abundances as "solar". Do you think of just the CNO group or of other species like Ne also? I temember old model atmospheres of central stars of PNs, where NeV and its absorption edges were important for the structure of the models.

ANALYSIS OF THE SOFT X-RAY DATA FROM THE CENTRAL STAR OF NGC 7293.

R.W.Tweedy & M.A.Barstow
Physics Department,
Leicester University,
University Road,
Leicester LE1 7RH,
U.K.

Abstract

Results from soft X-ray observations of NGC 7293 by the Einstein IPC and Exosat LE have been analysed using both LTE and non-LTE model atmospheres. Using the temperature obtained by Mendez and Kudritzki (1988) of 90,000K±10,000K as a guide, the best-fit model has a He:H ratio of 0.01, and CNO:H ratio of 0.1*solar. This model also produced the best fit for the DOZ white dwarf KPD0005+5106, and since the IUE spectra of the two objects have remarkably similar characteristics, the possibility that the central star of NGC 7293 may be a less evolved version of KPD0005+5106 is considered.

1   Introduction

Conventionally, there are two ways of obtaining elemental abundances in a white dwarf. One is to simulate the spectral line profiles by varying the temperature, gravity, and elemtental composition of a model stellar atmosphere. However, models that treat elements heavier than H and He in non-LTE are still under development, and thus are not generally available. The other method uses the high energy tail of the white dwarf spectrum, which is very sensitive to the presence of metals. An example of this is shown in figure 1, where two spectra - both non-LTE - are compared. The difference between the two is that the lower curve has ten times more CNO than the other, and thus the absorption edges are considerably deeper. Overlaid on this diagram are the effective areas of the Exosat low-energy telescope and the Einstein IPC, which operate at exactly these wavelengths. Because of this, soft X-ray count-rates - even broad-band ones - are a sensitive indicator of the temperature and elemental abundances of the stellar photosphere.

2   Current understanding of NGC 7293

Mendez and Kudritzki (1988) observed the central star of NGC 7293 with the Cassegrain

29

G. Vauclair and E. Sion (eds.), White Dwarfs, 29–38.
© 1991 *Kluwer Academic Publishers.*

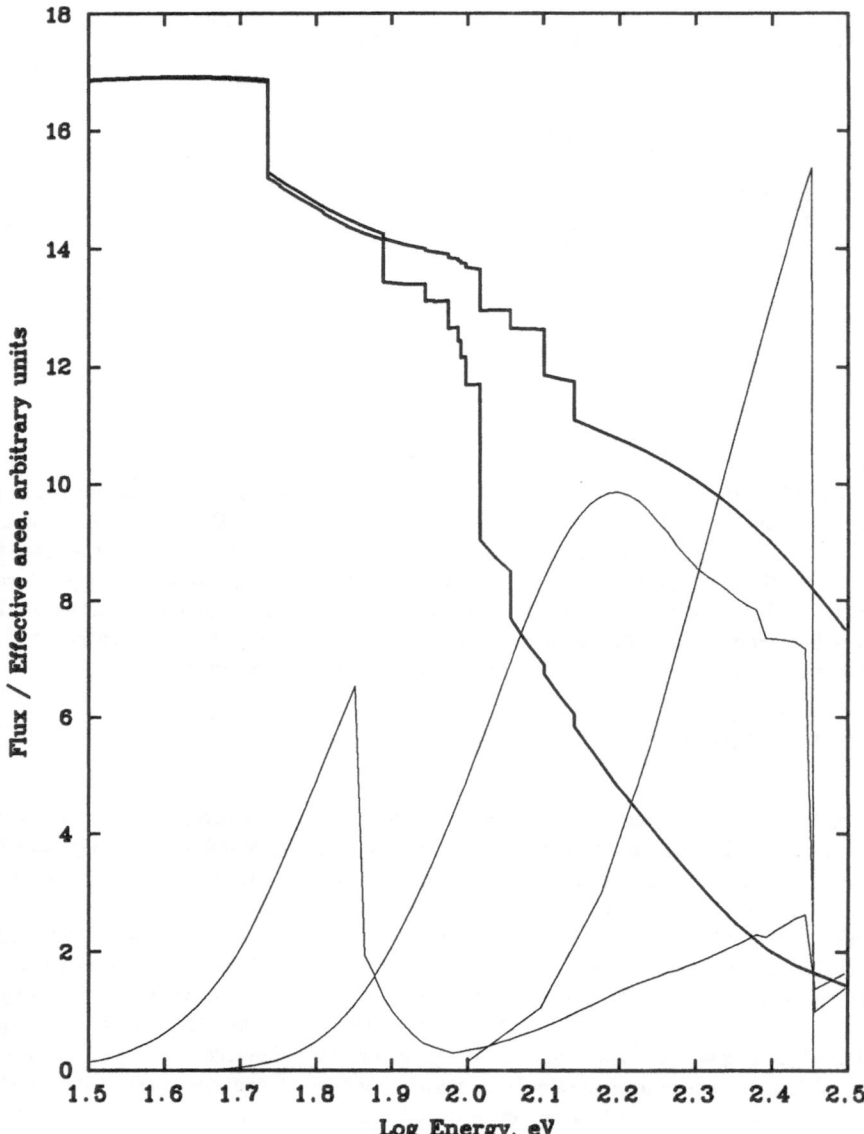

Figure 1 Effective areas of the soft X-ray instruments used on NGC 7293, overlaid on two spectra. Both spectra are H:He=100; the top one has H:He=0.1*solar, the bottom has H:He=0.01*solar. The three filters are, from left to right, the Exosat Aluminium/Parylene filter, the Exosat thick Lexan filter and the Einstein IPC (effective area shown at one tenth of the real value).

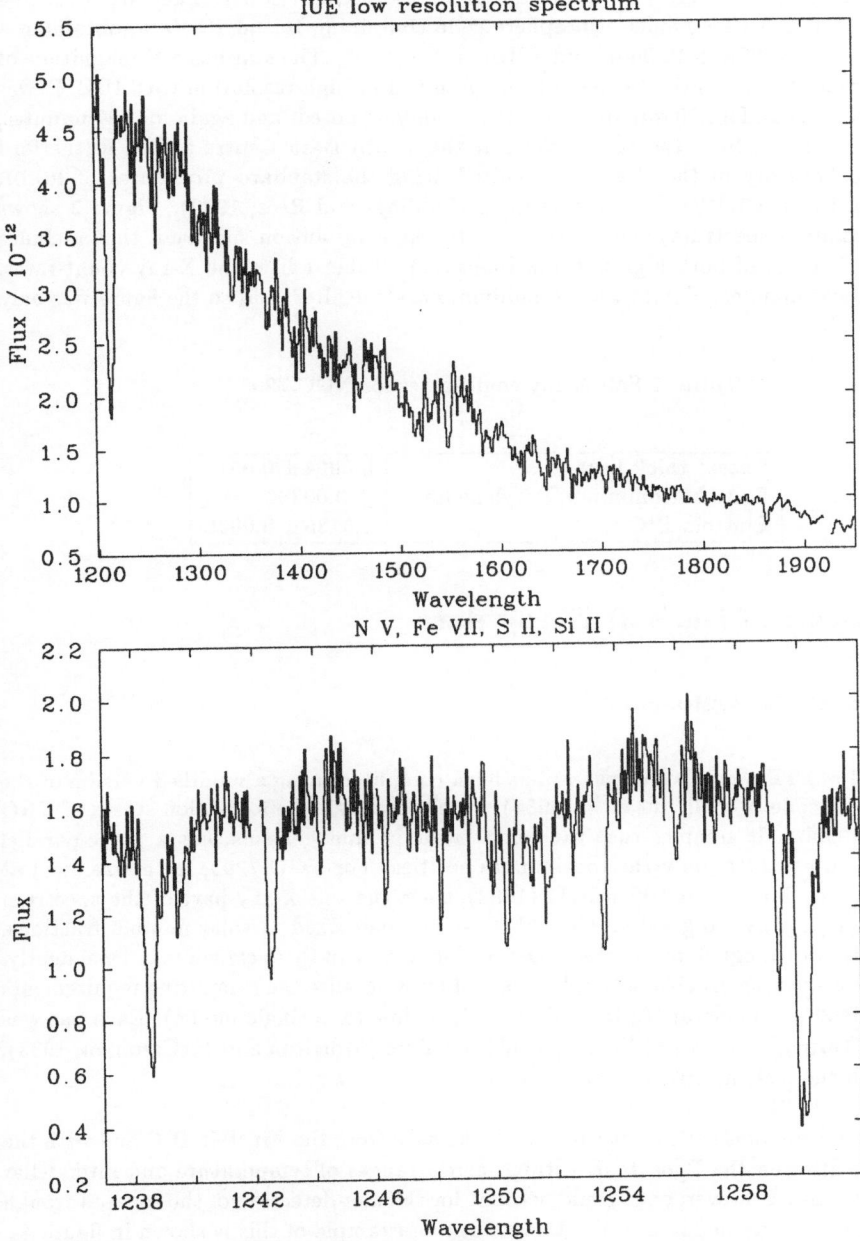

Figure 2 Low resolution spectrum of NGC 7293, and section of the high-resolution spectrum.

echelle spectrograph (CASPEC) at the European Southern Observatory. They fitted the line profiles using a non-LTE model atmosphere code containing H and He data, obtaining a temperature of 90,000K ± 10,000K and a H:He ratio of 100. The star has a V magnitude of 14.43, and was thus just bright enough to be observed at high-resolution with IUE. It was observed twice, once for 270 minutes when it was underexposed, and again for 720 minutes. The data was taken from the IUE archive at the World Data Centre at the Rutherford Appleton Laboratory in the U.K, and reduced using the standard procedures of IUEDR available on the STARLINK computer network (Giddings and Rees, 1989). Figure 3 shows the low resolution spectrum, and a section of the high resolution spectrum that includes absorption features of both high and low ionisation. Table 1 lists the X-ray count-rates. The source was undetected with the Aluminium/Parylene filter, and so the figure was only an upper limit.

Table 1 Soft X-ray count-rates for NGC 7293.

Exosat thick Lexan	0.0054 ± 0.0015
Exosat Aluminium/Parylene filter	≤ 0.00386
Einstein IPC	0.0135 ± 0.0039

3   Analysis of the soft X-ray data

3.1   LTE MODEL ATMOSPHERES

A large grid of LTE model atmospheres has been computed using a modified version of the self-consistent code of Williams et al (1987), which treats all the ionisation states of CNO (Barstow, 1990b). It assumes that the composition is homogeneous, has a plane-parallel geometry, and considers only the continuum opacities. For NGC 7293, log g was fixed at 7.0, obtained by Mendez and Kudritzki (1988), since the soft X-ray part of the spectrum is relatively insensitive to gravity. The C:N:O values were fixed at solar number fractions, since there is not enough data to determine the abundances of these elements independently. Extending the existing grid was a simple task, not least because the computing requirements on a VAX 3400 was moderate (3 to 4 minutes CPU time for a single model). Each one was then folded through a model of the interstellar medium (Morrison and McCammon, 1983), and through the instrumental response profiles.

All the H and He models were consistent with the data from the Einstein IPC and with the thick Lexan filter on the Exosat LE within narrow ranges of temperature and interstellar hydrogen column. However, none could account for the non-detection of the source through the Aluminium/Parylene filter on the Exosat LE; an example of this is shown in figure 4a.

In order for the three count rates to agree with an LTE model, a CNO:He abundance of at least 0.05 times solar was needed. However, it was not possible to obtain an upper limit;

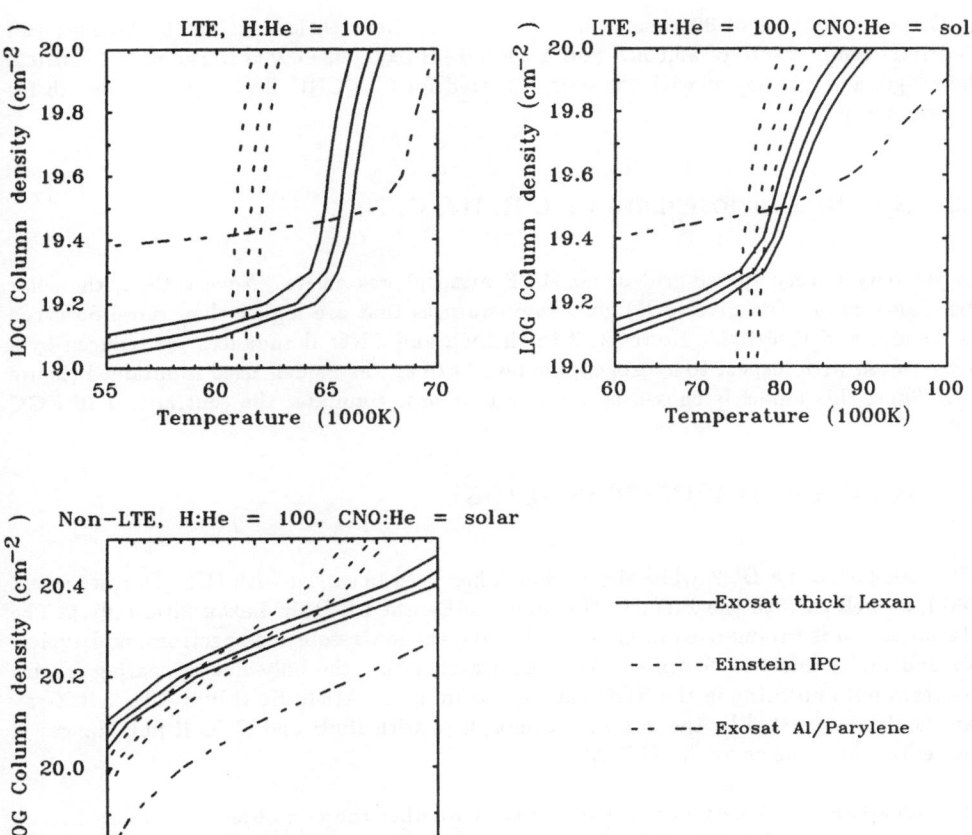

Figure 3 Results of the soft X-ray analysis of NGC 7293. The lines represent constant count-rate contours in the $T - N_H$ plane. A good fit occurs when the contour bands (taking into account one sigma errors) from all three observations overlap. (See Barstow 1990b, these proceedings).

CNO:He abundances of 80 times solar fit the data. Nevertheless, using the Mendez and Kudritzki temperature of 90,000K and $Y=H/(H+He)=0.009\pm0.005$ considerably restricts the range; and agreement with the data occurred for $0.5 \leq CNO:He \leq 2.0$ (see figure 4b for an example.)

3.2 Non-LTE atmospheres with H, He, C, N, O

As yet only a very sparse grid of non-LTE atmospheres exists. Nevertheless, the solar abundance model (relative to H) gives temperatures that are higher than those observed by Mendez and Kudritzki. However, if the helium and CNO abundances are reduced by a factor of ten with respect to hydrogen, the best fit of all the models used is obtained (figure 4c). Thus, this model is chosen as the one that best simulates the central star of NGC 7293.

4 Analysis of KPD0005+5106

KPD0005+5106 is a DOZ white dwarf, which has been observed with IUE (Downes et al., 1987), as well as with the Einstein IPC and the Exosat LE (thin Lexan filter only.). The IUE spectrum is summarised in figure 5, showing the low resolution spectrum, and typical low and high ionisation features. At these wavelengths, the only line appearing in this spectrum not appearing in the NGC 7293 spectrum is the $\lambda 1640$ He II line. The soft X-ray data is also best fitted by the non-LTE atmosphere with He:H and CNO:H abundances of one tenth, the same as for NGC 7293.

This remarkable similarity begs the question of whether the two objects are related in an evoultionary sense. NGC 7293 has been regarded as a typical DA progenitor (since the H abundance is high), and yet it shows a striking similarity to a DOZ. Thus, is there such a thing as a DA progenitor?

5 Summary

Model stellar atmospheres have been used to obtain temperatures and elemental abundances of the central star of NGC 7293. In addition to H and He, a CNO abundance of one twentieth * solar is needed for consistency between a model and all three soft X-ray observations with Exosat and Einstein. Nevertheless, these count-rates cannot constrain the abundances any further, unless other information is included. If the Mendez and Kudritzki (1988) temperature of 90000K±10000K is used, the model that provides the best agreement has H:He=100, and CNO:He=solar. This model is also consistent with the observations of KPD0005+5106, a DOZ white dwarf which has a UV spectrum remarkably similar to NGC 7293. This suggests that there may be an evolutionary link between the two objects.

35

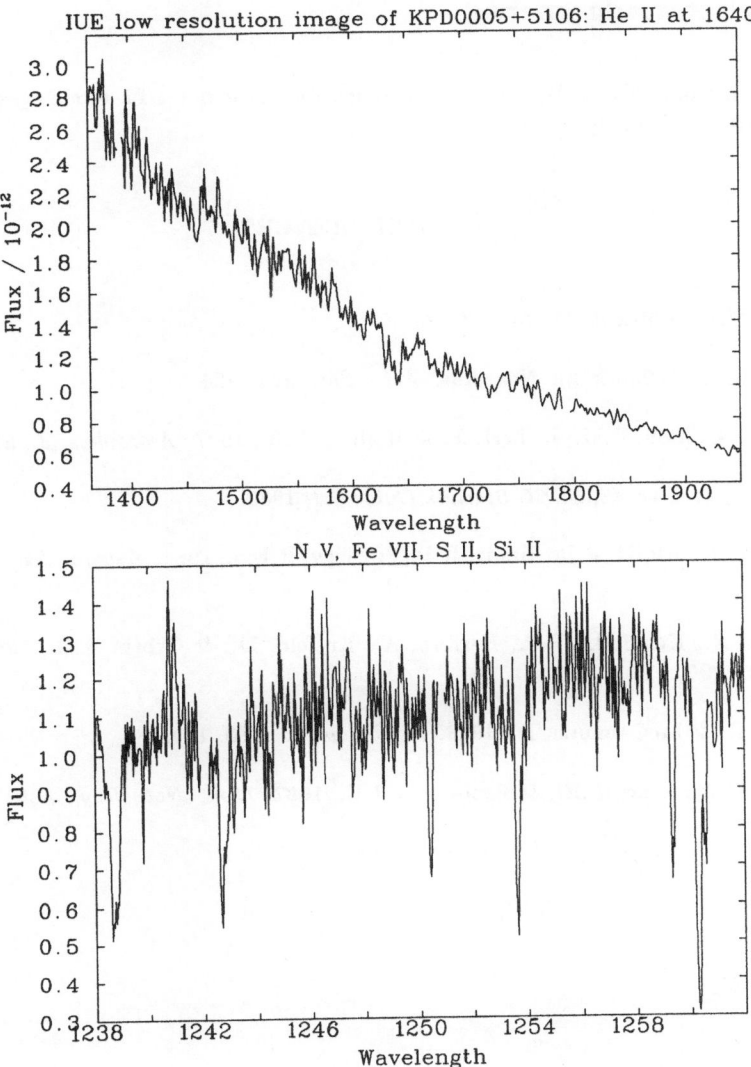

Figure 4 Low resolution spectrum of KPD0005+5106, and section of the high-resolution spectrum.

6 Acknowledgements

The authors thank Klaus Werner for providing some of the non-LTE models produced with PRO2 used in this analysis.

7 REFERENCES

Barstow, M.A., 1991a *these proceedings.*

Barstow, M.A., 1990b *Mon. Not. Roy. Astr. Soc.*, **242**, 484.

Downes, R.A., Sion, E.M., Liebert, J., & Holberg, J.B., 1987. *Astrophys. J.*, **321**, 943

Giddings, J. & Rees, P., *SERC Starlink Guide 3.1*, 1989.

Husfeld, D., Kudritzki, R.P., Simon, K.P., & Clegg, R.E.S., 1984. *Astron. Astrophys.*, **134**, 139

Mendez, R.H., Kudritzki, R.P., Herrero, A., Husfeld, D., & Groth, H.G., 1988. *Astron. Astrophys.*, **190**, 113

Morrison, R. & McCammon, D., 1983. *Astrophys. J.*, **270**, 119.

Williams, G.A., King, A.R., & Brooker, J.R.E., 1987, *Mon. Not. Roy. Astr. Soc.*, **266**, 725.

DISCUSSION

SION :

I would like to point out that the high ionization, far ultraviolet resonance lines in KPD0005 appear to arise in an expanding circumstellar halo. Their velocity shifts are shifted longward by a few km/s relative to the local interstellar medium in that line of sight.

TWEEDY :

Feibelmann and Bruhweiler (Astrophys. J., 1990, vol. 357, 548) recently showed a correlation between Fe VII and NI features in a large number of planetary nebula central stars, as well as DO white dwarfs such as KPD 0005+5106. The only reasonable explanation of this is that the NI -and thus other low-ionisation material -is circumstellar, rather than interstellar. Thus the velocity shift you observe is more likely to be due to motion in the (low-ionisation) circumstellar material. This is also consistent with the soft X-ray results which suggest the presence of CNO of around one-tenth the solar abundance, relative to hydrogen, and which would produce significant photospheric lines.

LIEBERT :

The assumption of H/He=100 for KPD 0005 is impossible given the appearance of the optical spectrum and the analysis of Wesemael, Green and Liebert on DAO and DO white dwarfs and, for example, that of NGC 7293 by Mendez and Kudritzki. Such a star at H/He $\sim$ 100, Te $\sim$90.000 would have a DAO spectrum with strong H lines and weaker HeII like the CSPN of N7293. Instead KPD 0005 has a DO spectrum with no evidence for H features **in absorption** (except for that explainable as nearly-coincident HeII). It does show unexplained **emission** lines of H, NV? ... and I suggest that your soft X-ray excess over a "He" model might be due to a nonthermal (mass loss ?) component.

TWEEDY :

First of all, unless the hydrogen is the dominant constituent of the atmosphere, the soft X-ray data cannot be explained. Given that the optical spectrum of KPD0005 is so peculiar, I suggest that this is what needs the explanation rather than the soft X-ray data. Recent work by Fritz, Leckenby and Sion shows that most of the IUE absorption features are circumstellar rather than photospheric, including the NV, which had hitherto been thought to be clearly photospheric. Thus the optical emission lines are more likely to be due to circumstellar material.

More work is therefore needed on the NGC7293 IUE data, in order to ascertain whether the high ionization absorption features are also circumstellar rather than photospheric. Given the striking similarity between the IUE spectra of both objects, this is a distinct possibility. Synthetic line profiles will also be generated to ascertain what features are expected for an atmosphere with the abundances favored for NGC 7293.

There remains the problem of the strange correlation between the appearance of Fe VII lines and the strength of the NI features, recently reported by Feibelman and Bruhweiler. Clearly the features that were thought to be purely interstellar such as NI and SiII, may have a large circumstellar component. High resolution spectroscopy (for example the Hubble Space Telescope) may be needed to sort out the components.

PLANETARY NEBULAE NUCLEI WITH WHITE DWARF SPECTRA

R. NAPIWOTZKI* AND D. SCHÖNBERNER
*Institut für Theoretische Physik und
Sternwarte der Universität Kiel
Olshausenstr. 40, D-2300 Kiel 1
Federal Republic of Germany*

ABSTRACT. We report here on the first results of an ongoing spectroscopic investigation of very old planetary nebulae together with their central stars, with the aim of a better understanding of the late planetary and early white dwarf evolution. We found that most of our observed planetaries have hot degenerates as their central stars, two of them being twins of "PG 1159" stars. The hydrogen-rich degenerates, except one which is a twin of the hot white dwarf EG 247 (G 191 B2B), show at least traces of helium, indicating that gravitational separation of hydrogen and helium occurs even at high gravities on a rather long timescale ($\geq 10^5$ yrs). For some of the central stars we were able to derive distances from the strengths of the interstellar sodium lines. In nearly all cases these distances turned out to be larger than earlier assumed. However, spectroscopic analyses of the hydrogen-rich degenerate central stars reveal a severe discrepancy: their position in the T_{eff}-g plane suggest evolutionary ages of about 10^6 yrs, ten times larger than the largest kinematical ages inferred from sizes and expansion velocities of their nebulae.

1. Introduction

The upper left part of the Hertzsprung-Russell diagram where white dwarfs begin their evolutionary path is currently a region of very active research. A variety of different kinds of stellar remnants, viz planetary nebulae, subdwarf O stars, and the so-called "PG 1159" stars are located there and about to evolve into one of the two main spectroscopic sequences of classical white dwarfs: the hydrogen-rich (DA) or helium-rich (non-DA) sequence. The largest, hence oldest, planetaries are especially important since they mark an important phase in the life of stars, i. e. the change-over from nuclear to gravitational and thermal energy release as the main contribution to the stellar luminosity. Thus we expect to see a substantial fraction of the hottest degenerates residing within the oldest planetaries. Unfortunately, these central stars are intrinsically very faint, and very little is known about their properties. So far, only four degenerates within planetaries have definitively been detected: three with hydrogen dominated surface layers, of spectral type DAO (A 7, NGC 7293, and EGB 6), one with a helium dominated surface, or spectral type DOZ corresponding to PG 1159 stars (VV 47).

In order to improve our knowledge of this important phase of stellar evolution we initiated an observing program for studying the properties of old central stars and their nebular

* Visiting astronomer, German-Spanish Astronomical Center, Calar Alto, operated by the Max-Planck-Institut für Astronomie Heidelberg jointly with the Spanish National Commission for Astronomy

39

G. Vauclair and E. Sion (eds.), White Dwarfs, 39–51.
© 1991 Kluwer Academic Publishers.

shells by long slit spectroscopy. Already the first observing run (4 nights) yielded a quite astonishing result: of 13 observed planetaries, 11 have degenerate central stars with spectral types ranging from DA over DAO to DOZ. The finding of two new DOZ central stars, viz those of Jn 1 and IW 1, has already been reported by Schönberner and Napiwotzki (1990). Here we report on the preliminary analyses of all so far gathered spectrograms.

2. Observations and Data Reduction

The observations were performed in October, 1989, with the 3.5 m telescope at the Calar Alto Observatory, equipped with the new Cassegrain Twin Spectrograph (CTS). As CCD detectors we selected a RCA chip ($15 \times 15 \, \mu m^2$ pixel size) for the blue ($\lambda \leq 5500 \, \text{Å}$) and a GEC chip ($22 \times 22 \, \mu m^2$ pixel size) for the red channel. The corresponding gratings were T 13 (300 grooves/mm, dispersion 144 Å/mm) for the blue and T 11 (270 grooves/mm, dispersion 160 Å/mm) for the red channel. The corresponding scales on the CCDs are 2.2 Å/pixel and 3.5 Å/pixel, respectively. The long slit (240" projected on the sky) was always oriented in the E-W direction. Because of the rather poor seeing we selected a slit width of 0.4 mm corresponding to 2.4" on the sky. The RCA chip was binned over 2×2 pixels, the GEC chip over 2 pixels perpendicular to the direction of dispersion. The resulting resolution is about 5.5 Å in the blue and 6.0 Å in the red. We took also several spectrograms of standard stars (BD +28 4211, EG 247) in order to convert our spectrograms to absolute fluxes. However, since all nights were of rather poor photometric quality we did not attempt to derive absolute fluxes. After each program-star exposure, a spectrogram of the comparison lamp (Th-Ar) was made for the wavelength calibration.

All the spectrograms were reduced in the usual manner by means of the software package IDAS developed by G. Jonas at Kiel. Special care was taken in subtracting the proper background including nebular emission lines. The useful wavelength range of the blue channel was limited owing to the low transmission of the dichroic beam splitter shortward of about 4000 Å. The red region extends from $\lambda = 5550 \, \text{Å}$ to 7050 Å.

Our program stars are given in Table 1, together with additional information like magnitudes and spectral types, as known so far. Nebulae radii are based on the assumption of constant ionized mass, $M_i = 0.2 M_\odot$. Except for Jn 1, the objects were taken from the list of old, close-by planetaries of Ishida and Weinberger (1987).

3. The Spectra and their Interpretation

As already mentioned in the introduction, the result of our first observing run came as a surprise: from the 13 objects given in Table 1, only one showed a typical sdO spectrum (K 2-2), and one seems to be intermediate between sdO and DAO (NGC 6853, see Fig. 1). The nucleus of HFG 1 appeared very unusual in its spectral appearance. It has recently been identified as an interacting binary (Acker and Stenholm, 1990, see also Bond, 1989). The Figs. 2, 3, and 4 show the blue spectrograms of the remaining objects, which indicate clearly high gravities for them. Especially interesting are the central stars of IW 1 and Jn 1 (Fig. 2). They appear essentially hydrogen-free and show strong features that are typical for the high gravity PG 1159 stars. Up to now only one DOZ central star has been known, viz that of VV 47 (NGC 2474-5, Liebert et al., 1988). The nucleus of IW 2 deserves special

Table 1: List of program stars

Name	V	Spectral type		R_{PN}/pc
		literature	this work	
S 176	≈ 18	—	DA	0.47
S 188	17.4	—	DAO	0.29
HFG 1	≈ 14	—	—	0.81
HW 4	17.1	—	DAO	0.48
IW 1	16.6	—	DOZ (PG 1159)	0.62
WDHS 1	17.4	—	DA	0.72
PW 1	15.4	sdO	DAO	0.70
K 2-2	15.0	—	sdO	0.48
NGC 6853	13.8	cont./O7	sdO/DAO	0.27
A 74	17.1	sdOp/DC	DAO	0.46
IW 2	17.7	—	DAOZ	0.57
DHW 5	≈ 15	—	DA	0.51
Jn 1	16.1	—	DOZ (PG 1159)	0.56

attention (Fig. 3). It shows a weak CIV/HeII absorption trough, but also hydrogen lines comparable in strength with those of EG 247. This object may well be the prototype of a new class of hybrid hot degenerates, which show hydrogen, helium, *and* carbon lines in the optical. The remaining objects all have hydrogen dominated spectra with varying strength of the He II line at 4686 Å. Except for one case, the hydrogen lines are of similar strength. The exception is the nucleus of WDHS 1, which shows shallow and very broad hydrogen lines, indicating higher effective temperature and larger gravity than is true for the rest of this group (Fig. 3).

A few important conclusions can already be drawn from inspection of Fig. 1 to 4: firstly, the large majority ($\approx 80\%$) of central stars residing within old planetaries have still hydrogen-dominated surfaces. Secondly, many of these nuclei are high-gravity objects already somewhat evolved along the white dwarf "cooling" sequence. The hydrogen-rich hot degenerates show in general helium, though with varying strength. This, if combined with the fact that also a few DAO white dwarfs without any planetary shell exist, indicates that the gravitational separation between hydrogen and helium occurs during the late planetary nebula phase and may even continue beyond. From the kinematical ages of our objects, $\frac{R_{PN}}{v_{exp}} \leq 10^5$ yrs, a minimum timescale for diffusion of the same order of magnitude may be estimated. Thirdly, there exists also for shell helium-burning central stars, which are completely deprived of hydrogen, an evolutionary sequence from the high luminous part of the HR diagram down to rather low luminoities ($\leq 100 L_\odot$), as is the case for nuclei with hydrogen-rich surfaces burning hydrogen in a shell (Schönberner und Napiwotzki, 1990).

The nucleus of K 2-2 (Fig. 1) deserves an extra comment. It has been identified photometrically by deep CCD exposures (Kwitter et al., 1988) and appears now confirmed by our spectroscopy. However, Kaler and Jacoby (1989) placed this object way down onto a position in the HR diagram ($L/L_\odot \approx 7, T_{eff} \approx 93000$ K) corresponding to a 0.8-0.9$M_\odot$ "cooling" track, and with gravities of log $g \approx 8.0 \ldots 8.5$. Our spectrogram in Fig. 1 is clearly at variance with such a high gravity. First comparisons with NLTE model atmospheres

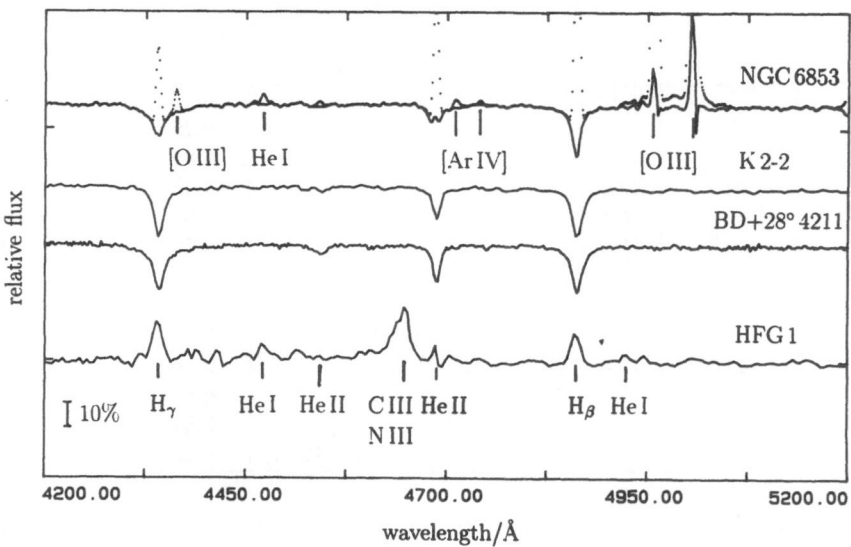

Figure 1: Blue spectrograms of NGC 6853, K 2-2, BD +28° 4211, and the interacting binary HFG 1. Important spectral lines are indicated. NGC 6853 is shown with (solid line) and without (dotted line) the nebular emission lines.

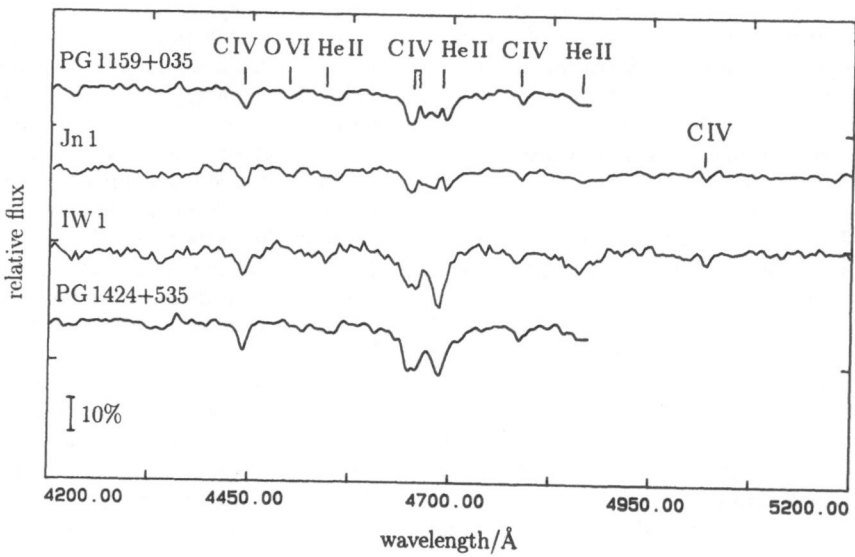

Figure 2: Blue spectrograms of Jn 1 and IW 1 together with those of the stars PG 1159+035 and PG 1424+535 analyzed by Werner et al. (1990), degraded to our resolution.

43

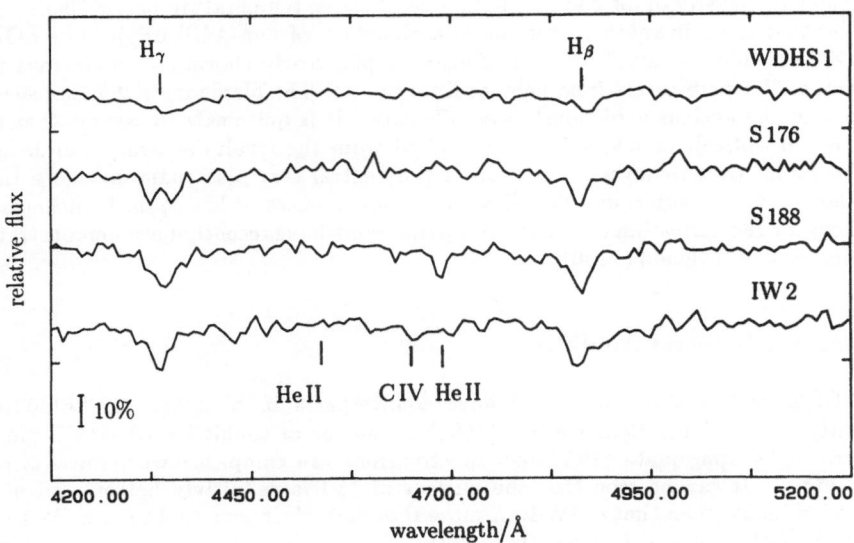

Figure 3: Blue spectrograms of WDHS 1, S 176, S 188, IW 2

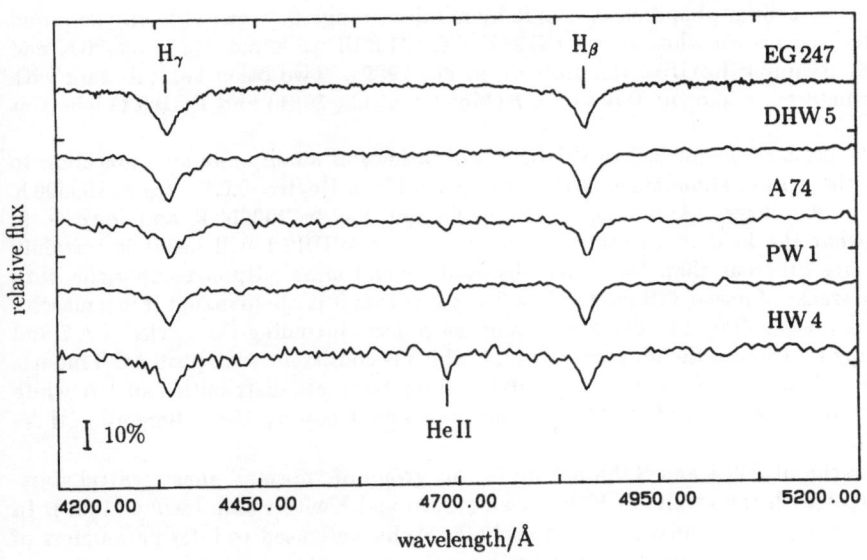

Figure 4: Blue spectrograms of EG 247, DHW 5, A 74, PW 1, and HW 4

indicate instead a gravity of $\log g \approx 5 \ldots 6$, and an effective temperature below 60000 K. It may well be that K 2-2 is another example of a planetary of non-AGB origin, like EGB 5 and PHL 932 (Mendez, et al., 1988a, b). This example clearly shows how dangerous the derivation of stellar parameters from nebular lines can be. The planetary K 2-2 is of sickle-like shape, and the nucleus is obviously well off-center. It is quite safe to assume that the planetary is not optically thick, a fact also evident from the results of Kaler and Jacoby (1989) who found the crossover magnitude of the central star discrepant by more than 2^m from the observed magnitude. The disastrous consequences of low optical thickness of nebular shells for the derivation of central-star parameters have recently been demonstrated by Schönberner and Tylenda (1990).

4. Preliminary Spectral Analysis

The two DOZ central stars of Fig. 2 have counterparts in a sample of "PG 1159" stars recently analyzed by Werner et al. (1990) by means of sophisticated NLTE model atmospheres. The appropiate "PG 1159" spectrograms are compared with those of our objects in Fig. 2. It can be seen that the nucleus of Jn 1 is definitivly hotter, and most likely more luminous, than that of IW 1. A more thorough discussion of Jn 1 and IW 1 can be found in Schönberner und Napiwotzki (1990).

The hydrogen-dominated spectrograms are investigated by Koester's updated model atmospheres (Koester, priv. comm.). The preliminary fit to the hydrogen line profiles and to the helium line strengths yielded the somewhat astonishing result that, with the exception of WDHS 1, all (7) central stars with strong hydrogen lines have rather low effective temperatures, ranging between 60000 and 70000 K, with gravities around $\log g \approx 7.5$. The helium abundances, He/H by number, range between virtually zero and 0.02. For the well-known white dwarf EG 247 = G 191 B2B we found $T_{\text{eff}} = 60000$ K and $\log g = 7.75$, assuming He/H=0 (cf. Holberg et al., 1989). Two other central stars with similar parameters are known: those of A 7 (Mendez et al., 1981) and EGB 6 (Liebert et al., 1989).

The spectrogram of the nucleus of WDHS 1 is of rather low quality and does not allow to precisely fix the helium abundance. With an upper limit of He/H= 0.001, $T_{\text{eff}} \approx 100000$ K and $\log g \approx 8.8$ follows. Assuming He/H=0, we got $T_{\text{eff}} \approx 200000$ K and $\log g \geq 9$. Regardless what the final parameters for the nucleus of WDHS 1 will be, it is certainly an object quite different than the other observed central stars. Upon comparison with evolutionary tracks of post-AGB models in a T_{eff}-g diagram it is obvious that it is a massive object of about $1 M_{\odot}$ (Fig. 5). The other analyzed objects including the nuclei of A 7 and EGB 6, occupy a region in the diagram of Fig. 5 which is consistent with post-AGB models of, say, 0.55 to 0.65 $M_{\odot}$. This is very gratifying since the mass distribution of DA white dwarfs peaks between 0.5 and 0.6$M_{\odot}$ (Weidemann and Koester, 1984, Bergeron, these proceedings).

Also the nuclei of PW 1 and S 188 belong to this group of "normal" mass central stars, which disagrees with the results of Kaler et al. (1990) and Kwitter and Jacoby (1989). In both of these analyses line fluxes from the nebular shells were used to infer parameters of the central stars and to derive their masses by comparison with post-AGB evolutionary tracks in the conventional HR diagram. The result is, in both cases, $M \approx 1 M_{\odot}$, in complete variance with our finding. A hot white dwarf of $1 M_{\odot}$ is expected to have a

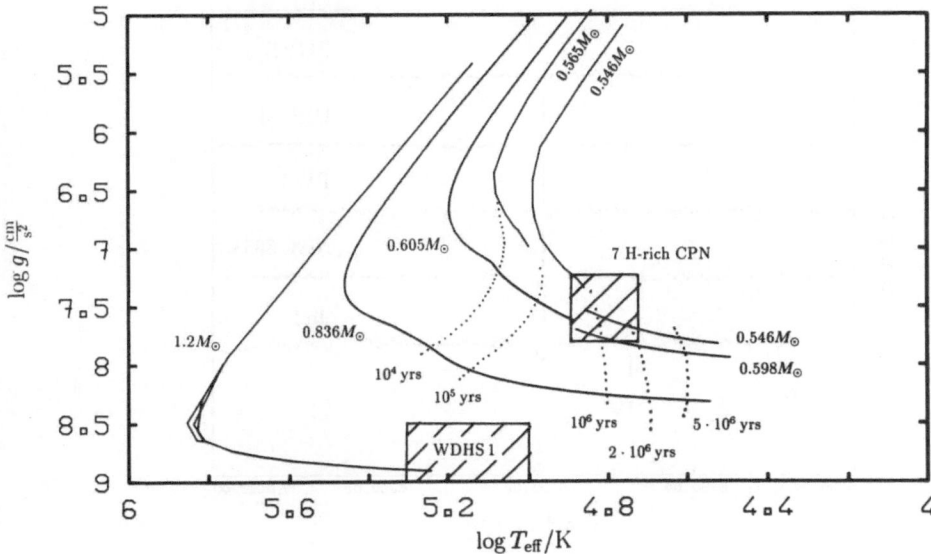

Figure 5: T_{eff}-g plane with post-AGB tracks of $0.546M_\odot$, $0.565M_\odot$ (Schönberner, 1983), $0.605M_\odot$, $0.836M_\odot$ (Blöcker, priv. comm.), $1.2M_\odot$ (Paczyński, 1971), and white dwarf tracks of $0.546M_\odot$ and $0.598M_\odot$ (Koester and Schönberner, 1986). Positions of PN nuclei with hydrogen rich white dwarf spectra are indicated by the hatched regions.

gravity $\log g \approx 8.5$, a value that is completely out of question for the spectrograms shown in Fig. 4. Thus we have another two examples where the use of insufficient or inaccurate data from the nebular shell lead to incorrect parameters for the central stars!

However, a severe discrepancy is evident from Fig. 5: The ages of the central stars, as they follow from evolutionary calculations ("cooling" of white dwarfs), are between 10^6 and $2 \cdot 10^6$ yrs, whereas none of the nebular shells has an kinematical age, $R_{\text{PN}}/v_{\text{exp}}$, exceeding 10^5 yrs! Such a discrepancy has already been noted by Liebert et al. (1989) in their investigation of EGB 6 (PG 0950+139).

5. Distances and the "Mystery" of White Dwarf Central Stars

In the last section, we mentioned already the discrepancy between evolutionary ages of some central stars with white dwarf-spectra and kinematical ages of their planetaries. One might think that this is only a problem of distances which are poorly known for planetaries. But this cannot be the explanation since we would need a distance increase by a factor of ten in order to reconcile both ages! Nevertheless, we tried to estimate distances for our program stars by using the interstellar Na D lines in our red spectrograms. Fig. 6 gives an illustration about the variation in strength of the Na D lines. Both components are, of course, not resolved at our spectral resolution.

Using an equivalent width-distance relation from the literature (actually Fig. 6 of

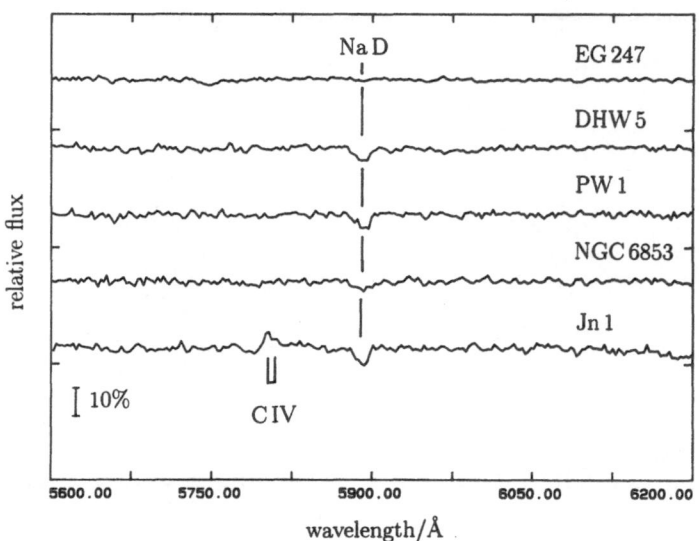

Figure 6: Na D lines from the red spectrograms of EG 247, DHW 5, PW 1, NGC 6853, and Jn 1

Binnendijk, 1952), we were able to derive distances for several of our objects. Table 2 presents the results, where the entries labelled "old" refer to the distances quoted in the literature, based on the standard assumption of constant ionized nebular mass, $M_i = 0.2 M_{\odot}$. The absolute magnitude, M_V, includes correction for interstellar absorption, and the kinematical ages are preferentially based on the H_{α} velocities. This choice takes into account that in these old nebulae with low luminosity nuclei the O III zone may not extend all the way to the nebular shells outer rim.

Upon comparison of the "old" with our new distances it can be seen that the latter are generally larger, on the average by a factor of roughly two! The only two exceptions are Jn 1 and NGC 6853. Because of these increased distances, the extremly low absolute brigthness of some objects is removed, and both the hydrogen-rich and helium-rich nuclei have the same mean absolute magnitude of $M_V \approx 7$, with a rather small scatter. Wesemael et al. (1985) derived $M_V = 6.7^{+1.3}_{-1.5}$ for the DOZ PG 2131+066 by estimating a spectroscopic parallax for its companion, in good agreement with our result. The increase of the distances leads, of course, also to larger kinematical ages, but the amount is insufficient to remove the discrepancy between evolutionary and kinematical ages! The ages listed in Table 2 are certainly upper limits, since one has to consider a deceleration of the nebular expansion due to the ambient stellar and interstellar matter (cf. Hippelein and Weinberger, 1990). The new distances lead to larger ionized shell masses: they range between a few tens up to several solar masses. Such high values are not unreasonable when one considers that the PN expansion velocity is larger than that of the AGB wind and that most of the stellar envelope is lost during the last 10^5 years on the AGB. Very old planetaries are thus expected to have swallowed most of the former stellar envelope by accretion of the AGB

Table 2: Objects within the galactic plane ($|z| \leq 250\,\mathrm{pc}$) with distances based on interstellar Na D lines

Objects		old		new		
		d/pc	M_V	d/pc	M_V	kin. age/yrs
DOZ	IW 1	330	8.4	630	7.0	$1 \cdot 10^5$
	Jn 1	700	6.9	500	7.6	$3 \cdot 10^4$
					7.3	
DA	DHW 5	400	7.0	500	6.5	$1 \cdot 10^5$
DAO	PW 1	240	8.7	630	6.6	$8 \cdot 10^4$
	A 74	230	9.7	510	8.0	$5 \cdot 10^4$
	S 188	220	9.8	680	7.3	$5 \cdot 10^4$
DAOZ	IW 2	260	10.0	1070	6.9	$1.5 \cdot 10^5$
					$7.1^{+0.9}_{-0.6}$	
sdO/DAO	NGC 6853	270	6.7	330	6.3	$1 \cdot 10^4$

material (cf. Schmidt-Voigt and Köppen, 1987).

With better absolute magnitudes for a number of our program stars we are in a position to construct an age-M_V diagram analogous to that of Schönberner (1981) but extending to larger ages. First we discuss the hydrogen-free objects (DOZ from Table 2) and compare them with evolutionary calculations of post-AGB models that experienced a late thermal pulse after thay have left the AGB which turned them into shell helium-burning objects, and lost the hydrogen-rich envelope by the action of a stellar wind (Iben, 1984, Blöcker, priv. comm.). Fig. 7 is supplemented by the white dwarf evolutionary paths of Koester and Schönberner (1986) without surface hydrogen. Also, some isothermes and the temperature at the turn-around points are given. In Fig. 7 the positions of the nuclei of Jn 1 and IW 1 are determined by their M_V and the kinematical ages of their planetaries from Table 2, with the implicit assumption that the latter ages equal the post-AGB evolutionary ages. A possible offset of both ages is certainly only of the order of 10^3 yrs and is completely unimportant for the very old planetaries considered here. Also errors of the kinematical ages generate only horizontal shifts. The nuclei of Jn 1 and IW 1 have positions that agree with the evolutionary models between 0.6 and $0.8 M_\odot$, and the latter even predict the observed effective temperatures of at least 100000 K.

Fig. 8 compares the hydrogen-rich central stars with the corresponding models. These models are taken form Schönberner (1983) and Blöcker (priv. comm.) and encompass the mass range 0.55 to $0.84 M_\odot$. During the shell hydrogen-burning phase ("plateau" evolution) the evolutionary speed is highly dependent on mass. When the envelope becomes too small, hydrogen burning ceases abruptly, and the release of gravitational-internal energy takes over as the main source for the surface luminosity. The evolution slows down, giving rise to the sharp "knee" typically for the evolutionary tracks of the more massive remnants. These knees can be considered to mark the beginning of the white dwarf "cooling" paths. Fig. 8 contains also two white dwarf evolutionary tracks of Koester and Schönberner (1986), but this time with "thick" hydrogen-rich envelopes ($M_H = 10^{-4} M_\odot$). "Turn-around"

48

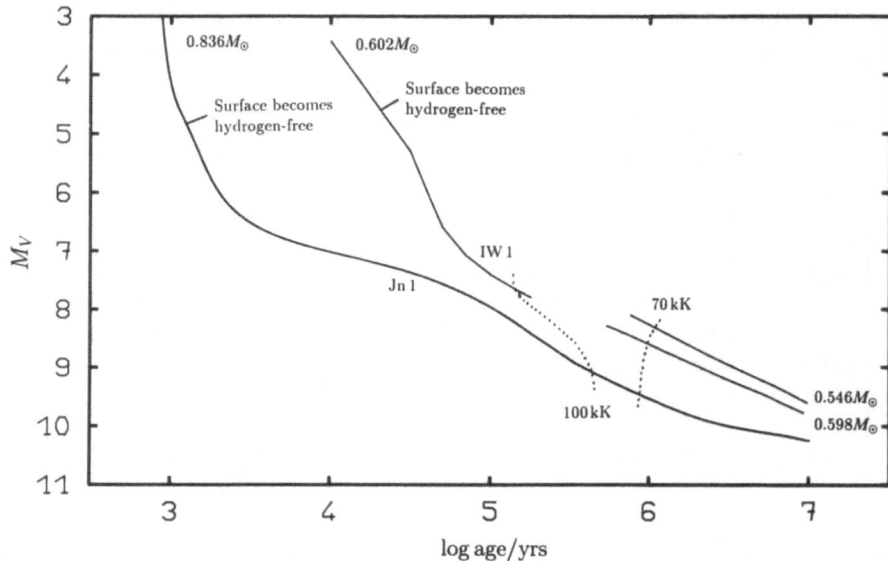

Figure 7: age-M_V diagram for the helium-rich central stars with post-AGB tracks of shell helium-burning central stars with $0.602M_\odot$ (Iben, 1984) and $0.836M_\odot$ (Blöcker, priv. comm.) and white dwarf tracks of Koester and Schönberner (1986) without hydrogen.

temperatures and some isothermes are also given. Comparison with Fig. 7 illustrates nicely the fact that shell hydrogen-burning post-AGB remnants "fade" about three times faster to $M_V \approx 6.5$ than corresponding models burning only helium in a shell.

We plotted in Fig. 8 two "test" objects: the first is the central star of NGC 6853 from Table 2, plotted by its absolute magnitude M_V and the kinematical age of its planetary. The other object is the hot, hydrogen-rich white dwarf EG 247 whose distance and effective temperature (≈ 60000 K) is well known. They are placed in the figure by their M_V and effective temperatures. The agreement with theory is encouraging: the nucleus of NGC 6853 has (Zanstra)-temperatures between 125000 and 140000 K and has just started its evolution as a degenerate star of about $0.6M_\odot$. HZ 43 and EG 247 are also degenerates of $\approx 0.6M_\odot$, but more than 10^6 yrs older.

The five hydrogen-rich nuclei from Table 2, which are substantially older than NGC 6853, seem to support this evolutionary picture at first glance: when plotted by M_V and age (from Table 2) their positions are consistent with masses of about $0.6M_\odot$ if allowance is made for possible distance errors. Closer inspection, however, reveals that the evolutionary models predict effective temperatures of about 100000K, whereas our photospheric analyses indicate temperatures well below that (see Sect. 4).

Thus we arrive at a situation which can be termed as the "mystery of degenerate central stars with hydrogen-rich surfaces". This "mystery" can be formulated as follows:

1. Reliable distances give $M_V \approx 7$ for the oldest central stars, and with the kinematical ages of their planetaries, $5 \cdot 10^4 \mathrm{yrs} \leq T \leq 10^5$ yrs, one finds close agreement, within the

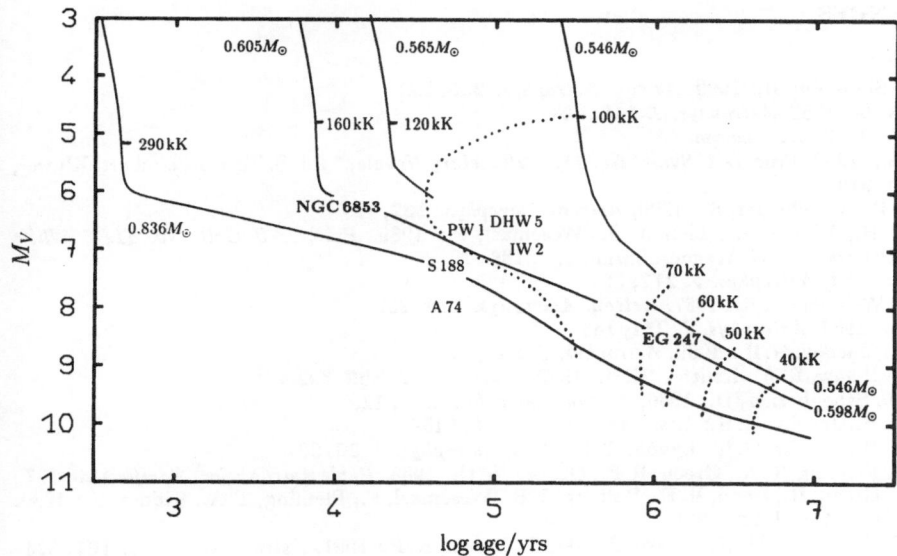

Figure 8: age-M_V diagram for the hydrogen-rich central stars with the evolutionary tracks from Fig. 5.

    errors, with appropiate post-AGB evolutionary tracks for about $0.6 M_\odot$ in the age-M_V plane. However the models predict effective temperatures of 10^5 K and more.

2. Spectroscopic analyses for the same central stars lead to $T_{\text{eff}} \approx 65000$ K, $\log g \approx 7.5$, consistent with post-AGB evolutionary tracks for about $0.55 M_\odot$ in the T_{eff}-g plane. However, these models predict $M_V \approx 8$ and an evolutionary age of $\approx 10^6$ yrs.

It should be repeated here that we did not encounter such a contradiction in our discussion of the two DOZ central stars.

6. Future Activities

We are clearly only in the beginning to understand the early white dwarf evolution, and more observations of intrinsically faint central stars are badly needed. But also their nebulae should be investigated carefully. This would allow the modelling of the whole stellar system and to put constraints on the temperature of the ionizing source. Also a independent check of the temperature scale of hot degenerates by careful intercomparison with IUE spectrograms seems to be in order (cf. Schönberner and Drilling, 1984).

ACKNOWLEDGEMENTS. The authors thank Dr. Weinberger for his help in preparing the finding charts and the DFG for a travel grant (Scho 394/2–1).

REFERENCES

Acker, A., Stenholm, B.: 1990, *Astron. Astrophys.* **233**, L21
Binnendijk, L.: 1952, *Astrophys. J.* **115**, 428
Blöcker, T.: 1990, *priv. comm.*
Bond, H. E.,: 1989, *Proc. IAU Symp. No. 131, "Planetary Nebulae"* Ed. S. Torres-Peimbert, Kluwer, p. 310
Hippelein, H., Weinberger, R.: 1990, *Astron. Astrophys.* **232**, 129
Holberg, J. B., Kidder, K., Liebert, J., Wesemael, F.: 1989, *Proc. IAU Coll. No. 114, "White Dwarfs"* Ed. G. Wegner, Springer, p. 188
Iben, I. Jr.: 1984, *Astrophys. J.* **277**, 333
Ishida, K., Weinberger, R.: 1987, *Astron. Astrophys.* **178**, 227
Kaler, J. B.: 1983, *Astrophys. J.* **271**, 188
Kaler, J. B., Jacoby, G. H.: 1989, *Astrophys. J.* **345**, 871
Kaler, J. B., Shaw, R. A., Kwitter, K. B.: 1990, *Astrophys. J.* **359**, 392
Koester, D., Schönberner, D.: 1986, *Astron. Astrophys.* **154**, 125
Kwitter, K. B., Jacoby, G. H.: 1989, *Astrophys. J.* **98**, 2159
Kwitter, K. B., Jacoby, G. H., Lydon, T. J.: 1988, *Astrophys. J.* **96**, 997
Liebert, J., Fleming, T. A., Green, R. F., Grauer, A. D.: 1988, *Publ. Astron. Soc. Pacific* **100**, 187
Liebert, J., Green, R., Bond, H. E., Holberg, J. B., Wesemael, F., Fleming, T. A., Kidder, K.: 1989, *Astrophys. J.* **346**, 251
Méndez, R. H., Kudritzki, R. P., Gruschinske, J., Simon, K. P.: 1981, *Astron. Astrophys.* **101**, 323
Méndez, R. H., Kudritzki, R. P., Herrero, A., Husfeld, D., Groth, H. G.: 1988a, *Astron. Astrophys.* **190**, 113
Méndez, R. H., Groth, H. G., Husfeld D., Kudritzki, R. P., Herrero, A.: 1988b, *Astron. Astrophys.* **197**, L25
Paczyński, B.: 1971, *Acta Astron.* **21**, 417
Schmidt-Voigt, M., Köppen, J.,: 1987, *Astron. Astrophys.* **174**, 223
Schönberner, D.: 1981, *Astron. Astrophys.* **103**, 119
Schönberner, D.: 1983, *Astrophys. J.* **272**, 708
Schönberner, D., Drilling, J. S.: 1984, *Astrophys. J.* **278**, 702
Schönberner, D., Napiwotzki, R.: 1990, *Astron. Astrophys.* **231**, L33
Schönberner, D., Tylenda, R.: 1990, *Astron. Astrophys.* **234**, 439
Weidemann, V., Koester, D.: 1984, *Astron. Astrophys.* **132**, 195
Werner, K., Heber, U., Hunger, K.: 1990, *Astron. Astrophys.* , submitted
Wesemael, F., Green, R. F., Liebert, J.: 1985, *Astrophys. J. Suppl.* **58**, 379

DISCUSSION

BARSTOW :

Can you comment on the absence of a Planetary Nebula around PG1159-035 in comparison with the CPN you have observed.

SCHÖNBERNER :

So far, only three PG 1159-like, low luminosity central stars are known which certainly are, or have been, shell helium-burning post-AGB stars. From evolutionary calculations it is known that the evolutionary time scale of shell helium-burning AGB remnants is about three times larger than that of corresponding shell hydrogen-burning remnants. This fact may explain the scarcity of planetaries around objects such as PG 1159-035.

SHIPMAN :

I am very interested in the star WDHS1, which has high temperature and high gravity. What is its evolutionary age? Is it quite young? Is the planetary nebula peculiar?

SCHÖNBERNER :

The 1.2 $M_\odot$ post-AGB model of Paczynski fades to the observed position of WDHS1 within about 10^4 yrs. More consistent calculations will lead, as I discussed earlier this morning, to larger ages. Thus I don't see age discrepancies for this particular object, which has a kinematical age of 35.10^4 yrs, deduced from its planetary. The latter is described in detail by Weinberger et al. (1983, Ap.J. **265**, 249). It appears to be one of the largest and faintest planetaries so far known. Other pecularities are not known.

LIEBERT :

How do you reconcile these results which require H-shell burning and $M_H \geq 10^{-4}$ $M_\odot$ with the strong evidence that the stars enter the DA white dwarf sequence with very thin hydrogen layer masses? Can there be a higher mass loss rate ($\sim 10^{-8}$ $M_\odot$ yr^{-1}) **after** your luminosity drop?

SCHÖNBERNER :

At present I don't see ways to reconcile the results of evolutionary calculations which predict "thick" residual hydrogen layer masses with observational evidences favoring "thin" hydrogen layers. Mass-loss rates of 10^{-8} $M_\odot$ yr^{-1} are typical along the plateau evolution of central stars, and such rates are also predicted by the theory of radiation-driven winds. I doubt that rates this large can also be sustained at luminosities below 100 $L_\odot$.

ATMOSPHERIC PARAMETERS OF SUBLUMINOUS B STARS

REX A. SAFFER, P. BERGERON, and JAMES LIEBERT
Steward Observatory
University of Arizona
Tucson, AZ 85721

D. KOESTER
Department of Physics and Astronomy
Louisiana State University
Baton Rouge, LA 70803

ABSTRACT. High signal-to-noise optical spectrophotometry of a sample of field subluminous B stars drawn largely from the Palomar Green UV excess survey is analyzed with a new grid of model atmospheres and synthetic spectra. The effective temperatures, surface gravities, and photospheric helium abundances are determined simultaneously from an analysis of hydrogen and helium absorption line profiles. The derived temperatures and gravities place the helium-deficient sdB stars in the H–R diagram along and just above theoretical sequences of the zero-age extended horizontal branch, lending strong support to the hypothesis that these stars are composed of helium-burning cores of $\sim 0.5 M_\odot$ overlain by very thin ($\lesssim 0.02 M_\odot$) layers of hydrogen. Various scenarios for their past evolutionary history are examined in the context of their probable future evolution into low-mass white dwarfs.

1. Introduction

Hot subdwarfs, stars with temperatures exceeding 20,000 K and surface gravities somewhat higher than main sequence stars of the same temperature, are thought to be the direct progenitors of the white dwarfs, although they likely comprise only a small fraction of all stars which evolve directly to that state. It is believed that most white dwarfs are formed by low and intermediate mass stars evolving toward their final states after the ejection of a planetary nebula at the tip of the asymptotic giant branch. Yet while the birthrates of planetary nebulae and white dwarfs are estimated to be comparable, the most common objects found in color-selected, magnitude limited samples such as the Palomar Green (Green, Schmidt and Liebert 1986) and Kitt Peak Downes Surveys (Downes 1986) are the hot subdwarfs. The subdwarf B (sdB) stars have hydrogen-dominated atmospheres, while the subdwarf O (sdO) stars have helium-dominated atmospheres. An intermediate class (sdOB) has been identified (Baschek and Norris 1975, Hunger *et al.* 1981); these stars have effective temperatures and photospheric helium abundances between those of the

G. Vauclair and E. Sion (eds.), White Dwarfs, 53–66.
© 1991 *Kluwer Academic Publishers.*

sdB and sdO stars. In the past two decades, a number of hypotheses on the origin of the hot subdwarfs have been advanced:

1) Greenstein (1971) and Greenstein and Sargent (1974) first suggested that the sdB stars are the field counterparts of the extreme blue extension of the horizontal branch (extended horizontal branch; EHB) of very metal-poor globular clusters. This interpretation is supported by results of the analysis of extreme blue horizontal branch stars in the globular cluster NGC 6752 by Heber *et al.* (1986). Some of those stars proved to have atmospheric parameters like those of field sdB and sdOB stars belonging to the disk population. Other globular clusters are known with blue horizontal branches; all have $[Fe/H] \lesssim -1.5$ and are thought to have ages comparable to that of the Galactic halo. Globular clusters with a metal content closer to the solar value and with estimated ages younger than that of the halo appear to lack blue horizontal branches (see, *e.g.*, Armandroff 1988).

2) The hottest and most luminous subdwarfs might be the descendents of AGB stars whose evolutionary times are much longer than the time required for the ejection of a planetary nebula. For example, Schönberner and Drilling (1984) and Mendez *et al.* (1986) discuss sdO stars of low surface gravity which lie in the same region as low mass post-AGB theoretical tracks. The problem remains that the vast majority of sdB stars are found in regions of the H–R diagram which do not overlap with theoretical post-AGB evolutionary tracks.

3) Close binary evolution could lead to post-RGB phases with effective temperatures and luminosities like hot subdwarfs. Iben and Tutukov (1986a,b) describe systems leading to the formation of helium degenerates in which the first mass transfer event occurs when the primary has developed a degenerate helium core with a mass less than required for core helium burning. Roche lobe detachment occurs when the primary's hydrogen-burning shell is quenched, and while the core is gravitationally contracting toward its final state as a helium degenerate, the star undergoes several short-lived phases passing through the region of the hot subdwarfs. Alternatively, Iben (1990) proposes that most subdwarfs could be the product of mergers of He–He cores in very close, short-period orbits, the bulk of the hydrogen-rich envelopes of both stars having been lost in common envelope events. The merger initiates helium burning in the surface layers of the merged product which propogates to the center in a series of helium flashes. The result is a core helium-burning star with little or no overlying hydrogen; an EHB star.

In this work, we establish the atmospheric parameters of a subsample of the sdB stars defined in the PG Survey. In Section 2 we describe the observations and present optical spectra of a subset of the program objects. In Section 3 we describe the model spectra and the procedure used to fit the observations. In Section 4 we present results of the analysis and discuss their distributions in the H–R diagram as diagnostics of the evolutionary status of the sdB stars, and we present a summary in Section 5.

2. Observations

The observations described here were obtained at the Steward Observatory Kitt Peak Station with the 2.3–m telescope equipped with the Boller & Chivens Cassegrain spectrograph and UV–flooded TI CCD. This instrumentation provided wavelength coverage $\lambda\lambda 3650$–5200 at a spectral resolution of ~6Å FWHM. This spectral coverage and resolution are ideal for observations of the Balmer series in subdwarf B stars, covering Hβ through the Balmer jump, while providing spectral resolution sufficient to resolve both the line cores of the Balmer absorption profiles and the narrower He I lines often present in the spectra. The spectra were extracted from the two-dimensional frames and reduced to linear wavelength and intensity scales using standard reduction packages in the Image Reduction and Analysis Facility (IRAF).

A subset of spectra obtained for the sample is presented in Figure 1.

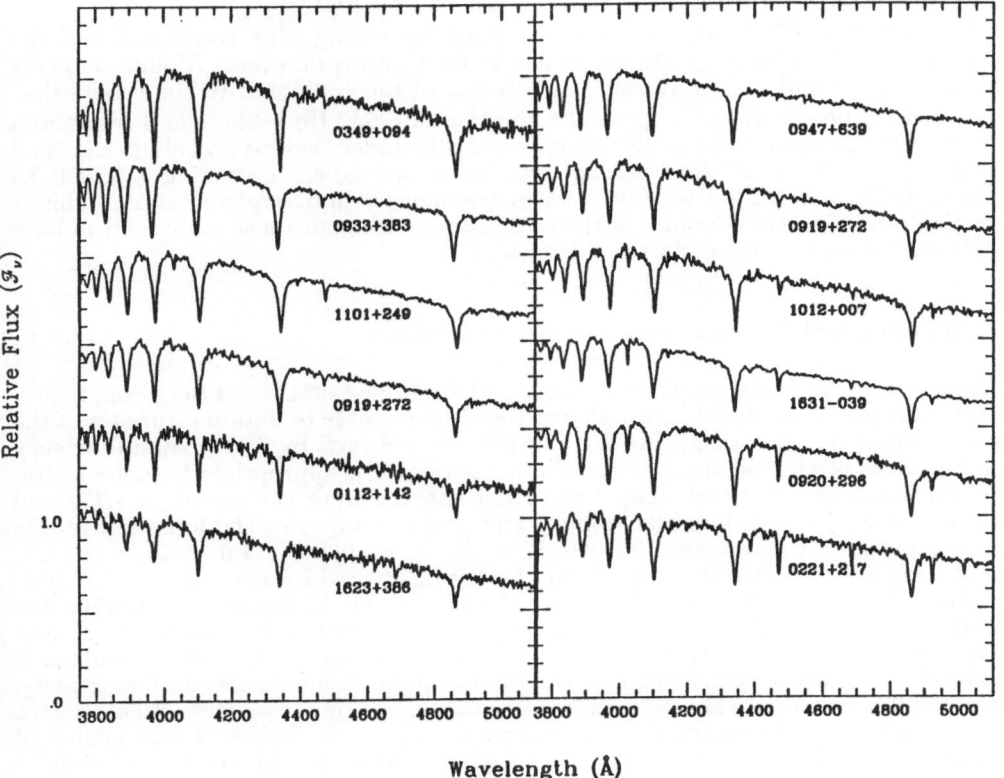

Wavelength (Å)

Figure 1. Left panel: Examples of confirmed normal sdB stars. The spectra are arranged in order of effective temperature, ranging from ~24,500 K at the top to ~40,000 K at the bottom in increments of ~3,000 K. Right panel: Helium abundance variations in the narrow ranges $30,000\ K < T_e < 34,000\ K$ and $5.6 < \log g < 6.1$. From top to bottom, the derived photospheric helium abundances in solar units are: 0, .05, .1, .25, .55, 1.1.

The analysis described in Section 3 identifies 57 of the 78 program objects as normal sdB stars; the atmospheres are helium-depleted, the effective temperatures lie between about 20,000 and 40,000 K, and the surface gravities exceed $\log g = 4.5$. An additional 8 program objects have surface gravities in excess of $\log g = 4.5$ but have helium abundances near or even larger than solar. Of the remaining

13 program objects, 11 have surface gravities less than log $g = 4.5$ and resemble main sequence B stars at the intermediate spectral resolution employed here. Alternatively, these stars could be contaminators from the blue horizontal branch (BHB). Spectral features which would distinguish between these two classes are not well-resolved. For example, at the spectral resolution of the observations shown here, Mg I λ4481 in main sequence B stars would be nearly invisible in the red wing of He I λ4471. The last 2 objects in the sample show very narrow Balmer absorption up to n = 16, characteristic of giant stars. A complete description of the sample will be presented elsewhere (Saffer *et al.* 1991).

The normal sdB spectra are characterized by strong blue continuua and the presence of moderately Stark–broadened Balmer absorption lines. Helium appears to be underabundant, on average, by a factor of ten compared to the solar value. In most of the spectra at a spectral resolution of $\sim$6Å, He I λ4471 and sometimes He I λ4026 are detected weakly, while in the remainder there is no helium detected whatsoever. In most stars having effective temperatures exceeding $\sim$32,000 K, He II λ4686 is detected weakly, even in the most helium-depleted stars. This is reminiscent of the appearance of the sdOB stars, although those stars tend to have helium abundances closer to the solar value.

3. Theoretical Spectra and Fitting Procedure

Daou *et al.* (1990), Bergeron, Wesemael, and Fontaine (1991), and Bergeron, Saffer, and Liebert (1991a, 1991b) have shown that it is possible to obtain estimates of the atmospheric parameters T_e and log g in DA white dwarfs by fitting simultaneously line profiles of the members of the Balmer series with appropriate model spectra. In the analysis described here, the model spectra were calculated in LTE and incorporate hydrogen-line blanketing. The grid encompasses the following ranges of atmospheric parameters: $20,000\ K < T_e < 40,000\ K$, $4.0 < \log g < 6.0$., and $0.00 < N(\mathrm{He})/N(\mathrm{H}) < 0.10$. In the fitting procedure used for the analysis of the white dwarf samples, the observed and theoretical Balmer line profiles are normalized to a continuum set to unity, and the atmospheric parameters T_e and log g are determined with least-squares fitting techniques. Here the technique is extended to three dimensions, and the photospheric temperature, surface gravity, and helium abundance are determined simultaneously by fitting theoretical spectra to a single optical spectrum of the program object. In Figure 2, the effects of variations in effective temperature and surface gravity on the theoretical Balmer line profiles are shown.

At the surface gravities characteristic of white dwarfs, a proper treatment of the quenching of the higher members of the Balmer series is crucial for the determination of the surface gravities of those stars. In the analyses cited above, the quenching formalism used is that described by Hummer and Mihalas (1988). At the lower surface gravities characteristic of sdB stars, quenching is less important, becoming significant only in stars having the highest surface gravities. This is demonstrated in the right panel of Figure 2, where the upper transitions Hϵ and H8 begin to show the effects of quenching in the saturation of the residual intensity near log $g = 6.0$. At even higher surface gravities, the quenching becomes more effective, and the residual intensities begin to decrease faster than can be compensated for by increased Stark broadening. For those upper transitions, equivalent widths then begin to decrease with increasing surface gravity. Despite these differences, the sensitivity to surface gravity achieved by fitting several Balmer line profiles simultaneously remains high, comparable to that achieved in the white dwarf regime.

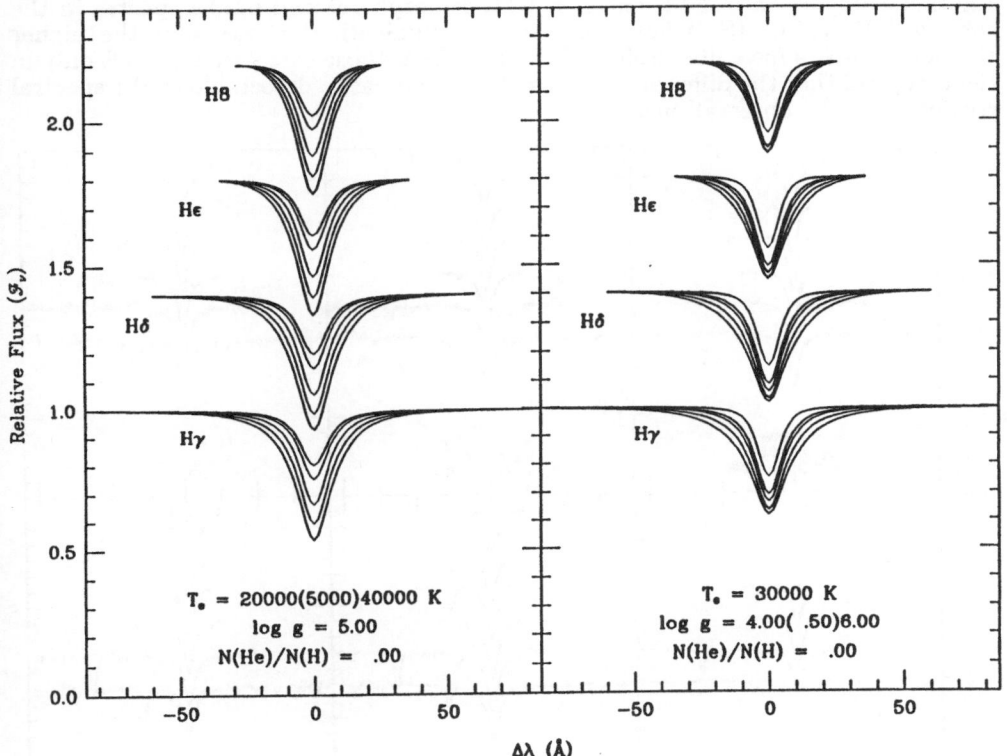

Figure 2. Left panel: The effect of temperature variations on the Balmer line profiles. In the temperature range of interest, the line equivalent widths decrease with increasing effective temperature. Right panel: The effect of variations in surface gravity on the Balmer line profiles. In the subdwarf regime, the equivalent widths of all Balmer lines up to H8 increase monotonically with increasing surface gravity, although in Hϵ and H8 the effects of quenching begin to become important as the surface gravity approaches $\log g = 6.0$.

In Figure 3 are shown model spectra fits to the stellar spectra displayed in the right panel of Figure 1. The figure shows that the model spectra fit the observed spectra very well; indeed, for many of the fits it is difficult to distinguish the model spectra from the data. Regarding the use of model spectra calculated in LTE model atmospheres for the analysis performed here, it is well-known that for hot stars the details of both absorption and emission line profiles are strongly affected by non-LTE processes (see, e.g., Kudritzki and Hummer 1990 for a review of quantitative spectroscopy of hot stars). However, some previous comparisons suggest that the use of LTE model atmospheres in the sdB regime is not seriously

58

inappropriate. Wesemael *et al.* (1980) find good agreement for continuum fluxes in comparisons of LTE and non-LTE calculations at subdwarf surface gravities, although an overpopulation of the n = 2 level strengthens the Balmer line cores. Another comparison of LTE and non-LTE atmospheres and model spectra in the subdwarf B regime (S. Vennes, private communication) shows that the higher Balmer members have line profiles that differ from those calculated in LTE only in the core, and that the differences are small and not easily detectable at the spectral resolution of the observations.

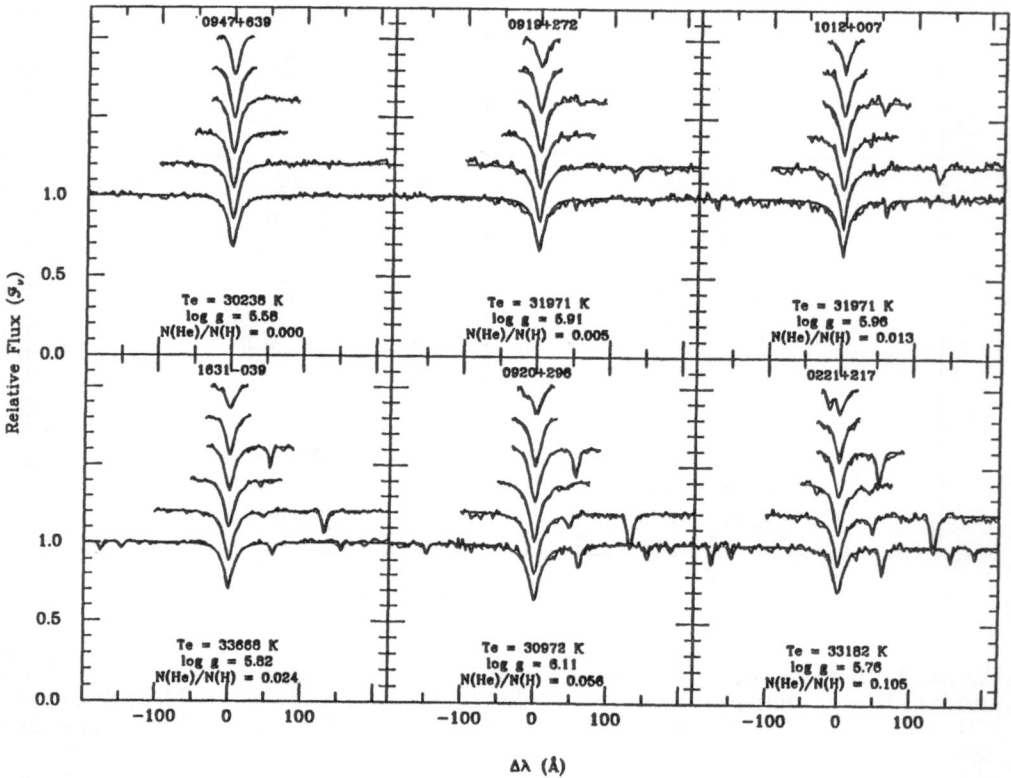

Figure 3. Model spectra fits to the sdB stars shown in the right panel of Figure 1, in order of increasing photospheric helium abundance. Spectrum segments are normalized to unity with a linear fit to normal points at the segment ends and are offset vertically.

It is worthwhile to compare the derived atmospheric parameters of the sample with previous determinations in the literature. Unfortunately, with the exception of the recent works of Moehler *et al.* (1990) and Moehler, Heber, and de Boer (1990), with which comparisons will be discussed elsewhere (Saffer *et al.* 1991), most previous analyses of stars analyzed here either have quite uncertain surface gravity determinations or lack them altogether. The only other extensive body of accurate surface gravity determinations (see, *e.g.*, Heber *et al.* 1984, Groth *et al.* 1985, and

Heber 1986) is for southern hemisphere stars for which there is little overlap with this sample. It is possible, however, to compare derived effective temperatures for a considerable portion of the sample using Strömgren photometry.

Fontaine *et al.* (1987) have graciously provided the authors with a machine-readable copy of the results of their Strömgren photometry program, in which nearly every PG survey sdB and sdOA star brighter than magnitude 15.5 was observed. Using the reddening-free index $Q' = (u - b) - 1.56(b - y)$ described by Bergeron *et al.* (1984) and their $Q'-T_e$ calibration, effective temperatures were calculated for 50 stars in this sample for which the requisite colors have been measured. There is good general agreement between the two methods, although there is a suggestion that temperatures derived from colors are somewhat hotter ($\lesssim 1500$ K) than those derived from line profile fitting. It should perhaps be noted that the relation between the reddening-free index Q' and effective temperature used here was derived by Bergeron *et al.* (1984) from model spectra calculated without the Hummer-Mihalas quenching formalism. The good general agreement between the derived temperatures is consistent with the previous assertion that level quenching in the subdwarf regime is less important than in the white dwarf regime, both in the absolute differences in the derived atmospheric parameters, and for the interpretation in the context of stellar evolution.

4. Results and Discussion

The derived effective temperatures and surface gravities for the entire sample of 78 sdB candidates are displayed in Figure 4. The open symbols distinguish ranges of photospheric helium abundance, and the connected filled symbols represent various theoretical loci, as described in the figure caption and legend. The stars represented in the figure can be divided into three broad classes: 1) Normal sdB and sdOB stars having effective temperatures between 20,000 and 40,000 K and surface gravities exceeding log $g = 4.5$, and having helium-deficient photospheres ($N(\text{He})/N(\text{H}) \lesssim 0.02$). 2) Stars with temperatures and surface gravities characteristic of normal sdB and sdOB stars but having normal or even enhanced helium abundances compared to solar. These stars tend to have the highest effective temperatures and lie near the helium main sequence and the helium-rich sdO stars. 3) Lower gravity main sequence B and blue horizontal branch stars having photospheric helium abundances ranging from normal to mildly depleted compared to solar. These stars are difficult to distinguish from each other at the intermediate spectral resolution of the observations.

The derived atmospheric parameters for the low gravity stars should be regarded with some reservation. In this regime of effective temperature and surface gravity, metals in the atmosphere are not depleted by diffusion, and the helium line cores are affected to some extent by departures from LTE. Moreover, at the spectral resolution of the observations, weaker lines useful as luminosity indicators are not detected. More accurate determinations of the atmospheric parameters await an analysis of higher resolution spectra with appropriate models. Even so, the analysis performed here serves to cull these low-gravity stars from the confirmed sdB sample and gives a rough estimate of their luminosities.

The confirmed sdB stars and their helium-normal counterparts cluster in a well-defined sequence lying parallel to theoretical sequences of the zero age extended horizontal branch (ZAEHB; Caloi 1972, 1989) and zero age horizontal branch

(ZAHB; Sweigart 1987). For all stars with log $g \gtrsim 4.5$, only three lie outside a sloping band with vertical cross-section ~0.5 dex and bounded below by the ZAEHB sequence with mass ~0.5 $M_\odot$. The sequence is parameterized by $q = M_c/M_*$, the ratio of the helium-burning core mass to total mass. For the $0.5 M_\odot$ sequence shown in Figure 4, $q = 0.92$ and 0.98 at $T_e = 20{,}000$ and 26,000 K, respectively. For $q \gtrsim 0.96$, corresponding to an envelope mass $M_e \lesssim 0.02 M_\odot$, it is not possible to sustain hydrogen shell burning, and the star is unable to ascend the asymptotic giant branch after the end of the core helium-burning phase. Its subsequent evolution should resemble that of a pure helium star of the same mass.

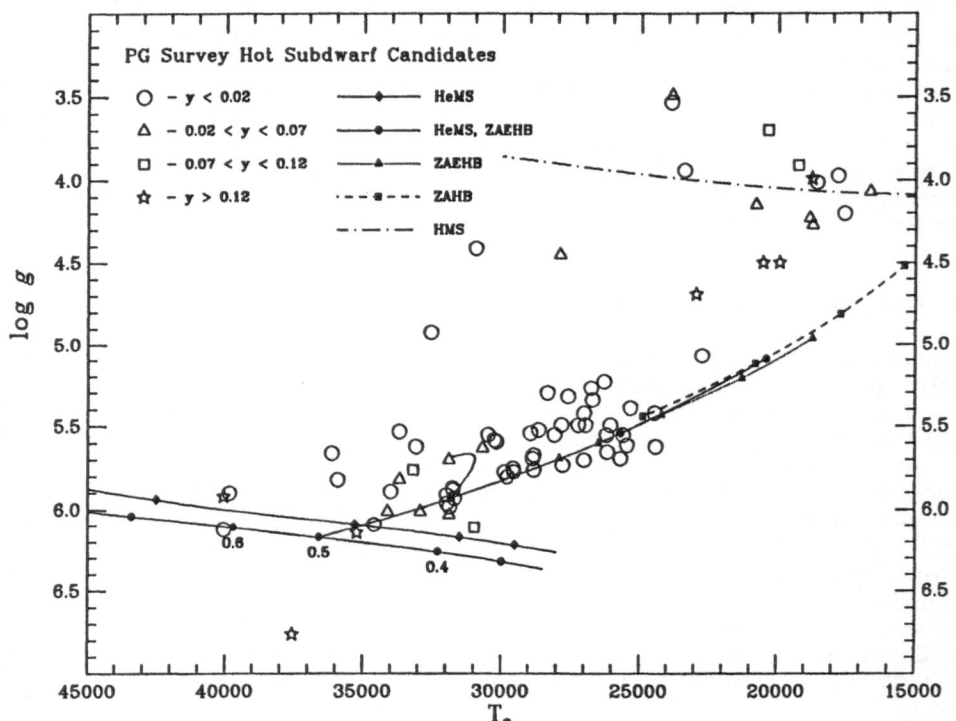

Figure 4. Derived effective temperatures and surface gravities for 78 sdB candidates (open symbols). Four ranges of helium abundance $y = N(\text{He})/N(\text{H})$ are distinguished as indicated in the legend. Connected filled symbols denote the helium-burning main sequence (HeMS; Paczyński 1971, Caloi 1972), zero-age extended horizontal branch (ZAEHB; Caloi 1972, 1989), zero-age horizontal branch (ZAHB; Sweigart 1987), and hydrogen-burning main sequence (HMS; Allen 1973) as indicated in the legend. The HeMS is labelled with the stellar mass in solar units. The hooked track on the lower part of the ZAEHB illustrates evolution of the stellar luminosity during the core helium-burning phase.

The sdB sequence is particularly well populated at surface gravities exceeding log $g = 5.0$ and effective temperatures exceeding 25,000 K. Near $T_e = 24{,}000$ K, where the overlying hydrogen-rich envelope of the model sequence has a mass

$\sim 0.02 M_{\odot}$ ($q \sim 0.96$), the sequence abruptly becomes sparse, and there are no sdB stars found cooler than 22,500 K. The sequence is populated by hydrogen-rich stars at temperatures as high as 40,000K, overlapping to some extent with the helium-rich sdO stars. The width of the sequence is nearly constant and is due largely to variations in derived surface gravity associated with observational error and to evolution in luminosity of a $\sim 0.5 M_{\odot}$ star during its $\sim 10^8$ yr core helium-burning lifetime. Assuming independent contributions from observational error, luminosity evolution, and core mass variations, a standard deviation of the core mass of at most $0.04 M_{\odot}$ suffices to explain the width of the observed distribution. The mean core mass $M_c = 0.5 M_{\odot}$ adopted here depends most strongly on the surface gravity zero point implicit in the models. As has been discussed elsewhere in the context of the mean mass of DA white dwarfs (Bergeron, Saffer and Liebert 1991b), uncertainties in level quenching might result in systematic shifts in derived surface gravity. If the same were true in the sdB regime, the mean derived core mass might differ, but the core mass distribution would remain very narrow.

The sdB sequence shown in Figure 4 strongly supports the core helium-burning interpretation of the EHB and is inconsistent with the hypothesis that the majority of these stars are in post-AGB evolutionary configurations. If the interpretation of the sequence as one of continuously decreasing envelope mass is correct, and if the narrow width of the distribution correctly implies a narrow range of core masses, evolution from RGB progenitors might account for the majority of the field sdB stars. The kinematics of the sample (Saffer and Liebert 1991) suggest membership in the intermediate to old disk population, and upon completing RGB evolution such a population would contribute large numbers of helium-burning stars with core masses near $0.5 M_{\odot}$ to the field horizontal branch population. Only a small fraction of these stars would have to lose the bulk of the hydrogen-rich envelope to account for the field sdB stars.

Alternate hypotheses for the origins of the sdB stars must confront the morphology of the observed sdB sequence. Iben and Tutukov (1986a) have calculated the evolution of a close binary system that first undergoes Roche lobe overflow when the primary has become a giant with a degenerate helium core with a mass of $0.3 M_{\odot}$. When the binary detaches, its subsequent evolution is driven by the contraction of the inert helium core. There are three stages of evolution with luminosities reaching or exceeding those of the sdB stars. The first and longest-lived of these stages lasts for about 5×10^6 yr, evolving at a nearly constant luminosity dominated by residual hydrogen burning at the base of the envelope. The other 2 stages last only for 3×10^5 yr and 5×10^3 yr, respectively. More massive remnants would evolve even more rapidly. If one takes the longest-lived phase for the $0.3 M_{\odot}$ core and uses the sdB space density of Downes (1986; 2×10^{-6} pc^{-3}), the resulting birthrate is dn/dt $\sim 2 \times 10^{-12}$ yr^{-1} pc^{-3}, or about the same as the birthrate of all white dwarfs. This is too high if all sdB stars evolve directly to the white dwarf cooling sequence.

There would be a spectrum of remnant core masses, depending on the initial mass of the primary, mass ratio, and orbital separation in the way described by Iben and Tutukov (1986b). This picture is inconsistent with the narrow width of the sdB distribution in surface gravity. At $T_e = 25,000$ K, the $0.3 M_{\odot}$ core would have log $g \sim 4.6$, with the higher mass cores having even lower surface gravity. Only for very low mass cores would the evolutionary tracks overlap the observed sdB surface gravities. These calculations also predict that the bulk of sdB stars should be radial velocity variables with amplitudes of many tens of $km \ s^{-1}$.

Saffer and Liebert (1991) have made multiple measurements of radial velocities of nearly 50 of the stars in the sample presented here. Only 5 of these stars have radial velocities measured at different epochs which differ by more than 3 standard deviations ($\sigma \sim 15$–$30\ km\ s^{-1}$), inconsistent with this close binary hypothesis for the origin of the sdB stars.

Iben (1990) calculates the evolution of close binary systems leading to mergers of He–He and CO–He cores. The He–He mergers result in core helium-burning stars with masses ranging from $0.3 M_\odot$ to $0.7 M_\odot$. Such a dispersion in core mass would result in a smearing of the distribution in the theoretical H–R diagram in the direction parallel to the HeMS, inconsistent with the observed distribution. However, as the somewhat hotter sdO stars exhibit a considerably larger range of surface gravities than the sdB stars analyzed here (see, e.g., Hunger et al. 1981, Schönberner and Drilling 1984, and Groth et al. 1985), it is possible that some fraction of those stars might have evolved from both types of mergers.

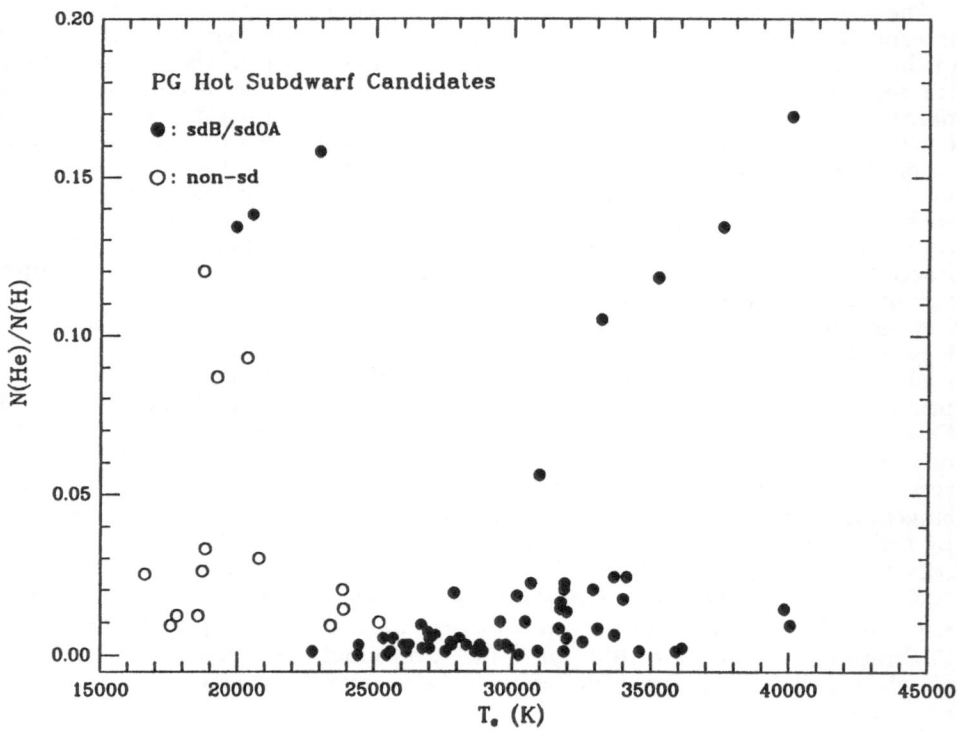

Figure 5. Helium abundance vs. effective temperature. Photospheric helium is depleted by a factor of 10 compared to the solar value at temperatures lower than $T_e \sim 33,000\ K$. At higher effective temperatures, both the mean value and the dispersion about the mean become larger, inconsistent with detailed diffusion calculations.

The distribution of photospheric helium abundances also helps to test the various origin hypotheses. In Figure 5, the derived photospheric helium abundances for all program objects are plotted against effective temperature. Michaud *et al.* (1989) have made detailed non–LTE calculations of radiative acceleration on helium in model atmospheres appropriate to sdOB stars. For $N(\text{He})/N(\text{H}) = 0.1$, the radiative accelerations always are at least 10 times smaller than the gravitational acceleration in the line-forming region, so that helium should indeed diffuse below the photosphere and become underabundant. However, even at the abundances derived here, $N(\text{He})/N(\text{H}) \sim 0.01$, the radiative acceleration remains less than the gravitational acceleration, and diffusion should lead to abundances even lower than are observed. The observed abundances thus cannot be explained by diffusion, and other processes must be invoked. Mass loss is one attractive possibility; Michaud *et al.* (1985) have shown that the equilibrium abundances of heavy elements in sdOB stars can be modified significantly in the presence of a mass loss rate of only $2 \times 10^{-15}\ M_\odot\ \text{yr}^{-1}$. Such calculations for helium have not yet been reported. Moreover, such low mass loss rates are completely undetectable observationally. The explanation for the observed helium underabundances in sdB and sdOB stars thus remains a completely open question.

5. Summary

In this paper, we have presented the results of a program to determine the atmospheric parameters of subdwarf B stars using a detailed analysis of the Balmer and helium line profiles. The method yields the most precise determinations to date of effective temperatures and surface gravities for the sdB sequence in the ranges $20,000\ K < T_e < 40,000\ K$ and $5.0 < \log g < 6.0$. The photospheric helium abundances, determined simultaneously with T_e and $\log g$ by fitting model spectra to optical spectrophotometry, range from completely depleted (formally, $N(\text{He})/N(\text{H}) \equiv 0$) to nearly twice the solar value. The determinations of T_e and $\log g$ beautifully fit the theoretical sequences of the extended horizontal branch of Caloi (1972, 1989) and Sweigart (1987) and give strong support to the interpretations of the EHB proposed by Greenstein and Sargent (1974) and Heber (1986) for a plausible helium-burning core mass of $\sim 0.5 M_\odot$. Taking into account estimates of the uncertainties in the derived atmospheric parameters, the width of the sequence in surface gravity, only 0.5 dex, is consistent with that predicted by the evolutionary increase in the stellar luminosity during the core helium-burning lifetime. The position and morphology of the sdB sequence in the theoretical H–R diagram are inconsistent with other proposed interpretations of the sdB stars. Photospheric helium is underabundant by a factor of ten, on average, for the majority of the sdB stars, although at effective temperatures higher than $\sim 33,000$ K helium abundances rise. The observed abundances are higher than expected from detailed diffusion calculations, and other processes, such as mass loss, must be invoked.

This work was supported in part by NSF grant AST 89–18471, by an NSERC Postdoctoral Fellowship to one of us (PB), and by NATO.

REFERENCES

Allen, C.W. 1973, *Astrophysical Quantities*, (London:Athlone).

64

Armandroff, T.E. 1988, *A. J.*, **96**, 588.
Baschek, B. and Norris, J. 1975, *Ap. J.*, **199**, 694.
Bergeron, P., Fontaine, G., Lacombe, P., and Wesemael, F. 1984, *A. J.*, **89**, 374.
Bergeron, P., Wesemael, F., and Fontaine, G. 1991, *Ap. J.*, in press.
Bergeron, P., Saffer, R.A., and Liebert, J. 1991a, in *Confrontation between Stellar Pulsation and Evolution*, ed. C. Cacciari, A. S. P. Conference Series (Astronomical Society of the Pacific : Provo, Utah), in press.
Bergeron, P., Saffer, R.A., and Liebert, J. 1991b, these proceedings.
Caloi, V. 1972, *Astr. Ap.*, **20**, 357.
Caloi, V. 1989, *Astr. Ap.*, **221**, 27.
Daou, D., Wesemael, F., Bergeron, P., Fontaine, G., and Holberg, J.B. 1990, *Ap. J.*, **364**, 242.
Downes, R.A. 1986, *Ap. J. Suppl.*, **61**, 569.
Fontaine, G., Wesemael, F., Lamontagne, R., Bergeron, P., and Green, R.F. 1987, *I.A.U. Colloquium No. 95, The Second Conference on Faint Blue Stars*, eds. Philip *et al.* (Schenectady:Davis), p. 615.
Green, R.F., Schmidt, M., and Liebert, J. 1986. *Ap. J. Suppl.*, **61**, 305.
Greenstein, J.L. 1971, *IAU Symposium No. 42, White Dwarfs*, ed. W.J. Luyten (Dordrecht: Reidel), p. 46.
Greenstein, J.L. and Sargent, A.I. 1974, *Ap. J. Suppl.*, **28**, 157.
Groth, H.G., Kudritzki, R.P., and Heber, U. 1985, *Astr. Ap.*, **152**, 107.
Heber, U. 1986, *Astr. Ap.*, **155**, 33.
Heber, U., Hunger, K., Jonas, G., Simon, K.P., 1984, *Astr. Ap.*, **130**, 119.
Heber, U., Kudritzki, R.P., Caloi, V., Castellani, V., Danziger, J., and Gilmozzi, R. 1986, *Astr. Ap.*, **162**, 171.
Hummer, D. G. and Mihalas, D. 1988, *Ap. J.*, **331**, 794.
Hunger, K., Gruschinske, J., Kudritzki, R.P., and Simon, K.P. 1981, *Astr. Ap.*, **95**, 244.
Iben, I. 1990, *Ap. J.*, **353**, 215.
Iben, I. and Tutukov, A.V. 1986a, *Ap. J.*, **311**, 742.
Iben, I. and Tutukov, A.V. 1986b, *Ap. J.*, **311**, 753.
Kudritzki, R. D. and Hummer, D. 1990, *Ann. Rev. Astr. Ap.*, **20**, 303.
Mendez, R.H., Miguel, C.H., Heber, U., and Kudritzki, R.P. 1986, *I.A.U. Colloquium No. 87, Hydrogen Deficient Stars and Related Objects*, eds. Hunger *et al.* (Dordrecht:Reidel), p. 323.
Michaud, G., Bergeron, P., Wesemael, F., and Fontaine, G. 1985, *Ap. J.*, **299**, 741.
Michaud, G., Bergeron, P., Heber, U., and Wesemael, F. 1989, *Ap. J.*, **338**, 417.
Moehler, S., Richtler, T., de Boer, K.S., Dettmar, R.J., and Heber, U. 1990, *Astr. Ap. Suppl.*, **86**, 53.
Moehler, S., Heber, U., and de Boer, K.S. 1990, *Astr. Ap.*, in press.
Paczyński, B. 1971, *Acta Astr.*, **21**, 1.
Saffer, R. A., Bergeron, P., Liebert, J., and Koester, D. 1991, in preparation.
Saffer, R.A. and Liebert, J. 1991, in preparation.
Schönberner, D. and Drilling, J.S. 1984, *Ap. J.*, **278**, 702.
Sweigart, A.B. 1987, *Ap. J. Suppl.*, **65**, 95.
Wesemael, F., Auer, L.H., Van Horn, H.M., and Savedoff, M.P. 1980, *Ap. J. Suppl.*, **43**, 159.

DISCUSSION

Van HORN :

I am impressed with the sharpness of the mass distribution you find for the sdB's. Do you understand why they seem to be so singular, cutting off at ~ 0.5 $M_\odot$?

SAFFER :

If the interpretation of these stars as core helium-burning stars is correct, then such a sharp cutoff would be consistent with evolution directly from the tip of the red giant branch, where disk stars would have grown helium cores of about 0.5 solar masses prior to igniting helium.

THEJLL :

How many of the sdB's in your sample have independent masses (from orbits for example), and what are they?

SAFFER :

Many sdB stars in the PG survey have been found to have companions, but I know of no case where followup observations have been obtained to derive orbital parameters for those systems. For sdO stars, the only orbital mass determination is by Howarth and Heber for HD 128220, a double-lined spectroscopic binary consisting of a subdwarf O and a G giant. This subdwarf has an estimated mass of 0.55 solar masses. This is an area which needs much more attention.

HEBER :

The existence of the sdB stars is a mystery. These objects belong to population I according to their scale height and kinematics. Hence their progenitors on the main sequence must have 1 $M_\odot$ or more and must have lost more than 0.5 $_\odot$ on the first giant branch to reach the EHB. Since is not reproduced by canonical stellar evolution theory. Therefore, Iben and Tutukov have suggested an alternative scenario, which is the merging of 2 low masss Helium white dwarfs. This scenario predicts a mass spectrum from about 0.35 $M_\odot$ to 0.7 $_\odot$. The low masses, however, are favoured in the merger scenario. So it predicts that a lot of stars should lie **below** the ZAEHB, which is opposite to your finding. Selection effects, of course, provide a bias against the faint low mass objects. But your distribution really looks like a cut off, i.e. there are **no** stars below the ZAEHB.

WEIDEMANN :

Concerning the mass of sdB's being ~ 0.5 $M_\odot$ at the extended horizontal branch by huge mass loss at the first giant stage (versus the merging hypothesis) : this is not impossible as demonstrated by the existence of RR Lyrae stars of population I, which also come from progenitors at ~ 1 $M_\odot$ and have lost half of their mass (but of course it is not yet understood).

WHITE DWARF SPACE DENSITIES
AND BIRTH RATES RECONSIDERED

V.WEIDEMANN
Institut für Theoretische Physik und
Sternwarte der Universität Kiel
Olshausenstraße 40, D-2300 Kiel 1
Federal Republic of Germany

ABSTRACT. The space density derived from a restricted 10 pc ensemble according to the method of Liebert (1978) but with a newly derived smaller velocity correction factor is estimated to be $N = 0.005$ WD/pc^3 down to M_{bol}. Comparison with predictions of synthetic models of galactic evolution by Yuan (1989) for single white dwarfs which are constrained to reproduce the total mass in the solar neighborhood with a reasonable binary factor (1.6 to 2) favors a present white dwarf birth rate of 1.4 resp. 2.3 $\cdot 10^{-12}$pc^{-3}yr^{-1} for single white dwarfs resp. for all white dwarfs including binaries.

1. Introduction

White dwarf space densities have been estimated during the last decades by attempts to establish the luminosity function, essentially by number counts down to certain limiting absolute magnitudes, until which the counts were assumed to be complete. Birth rates were then obtained by division through a theoretically derived cooling time down to the same limiting luminosity. Milestones on this way were the investigations by Weidemann (1967) who used the Eggen-Greenstein, Luyten, and 10 parsec ensembles in connection with the Mestel cooling theory, by Sion and Liebert (1977) who repeated the analysis with increased numbers and by Liebert (1978) who tried to predict the extension of the luminosity function (LF) towards cooler and fainter stars, using space densities by Green(1977, 1980) and different cooling laws, then the investigation by Fleming, Liebert and Green (1986) who evaluated the PG survey to derive space densities down to $M_V = 12.75$, and most recently by Liebert, Dahn and Monet (1988, 1989) (LDM) who constructed the luminosity function down to the low luminosity cut-off at about $\log L/L_\odot = -4.2$.

Synthetic calculations of models of galactic evolution by Yuan (1989) allowed to compare the observed white dwarf LFs with predictions, depending on a number of parameters like age of the galactic disk, initial mass function, star formation rate, initial to final mass relation, cooling laws and scale height evolution. They also predict space densities down to given luminosity limits and have been used to estimate the number and mass of white dwarfs below the observed cut-off luminosity (see Weidemann 1990). Her calculations were calibrated by the postulate that the present white dwarf birth rate, χ_0, should be equal to $1 \cdot 10^{-12}$pc^{-3}yr^{-1} but demonstrated that the birth rate is by no means constant but evolves even for a constant star formation rate, whereas the local space density decreases due to dilution by the increase of scale height. Therefore the birth rate determination by simple division of space density through cooling time is valid only approximately. For comparisons of birth rates one thus should refer to χ_0.

G. Vauclair and E. Sion (eds.), White Dwarfs, 67–73.
© 1991 *Kluwer Academic Publishers.*

In deriving space densities from observations one has to account for several incompleteness factors: sky coverage, selection effects by survey limitations e.g. by use of proper motion catalogues, or hidden white dwarfs in binaries. These correction factors have been dealt with in an exemplary fashion by Liebert (1978).

I shall here use his method with new data in order to determine a revised space density and to derive birth rates χ_0 which are consistent with Yuan models yielding the well-determined total stellar mass in the solar neighborhood column.

2. Space density in a restricted 10 parsec volume

We reconsider Table 2 of Liebert (1978) which compares expected and observed numbers of single white dwarfs within 10 pc. The sky factor, 1.79, corresponds to the coverage of the Luyten Palomar survey, the essential source for the faint and cool white dwarfs which determine the total space density (maximum of the LF below $M_V = 13$). Liebert calculates predicted numbers for different cooling theories below $M_V = 13$ but uses a birth rate based on the space densities derived by Green (1977,1980). He obtains for the bolometric intervals < 13, $13 - 14$, and $14 - 15$ magnitudes 4.2, 3.9 and 7.4 predicted stars (for Shaviv/Kovetz cooling) compared with 2, 3 and 4 observed white dwarfs. Since the fainter stars were found in proper motion surveys favoring larger tangential velocities he suggested a velocity correction factor of 2.6, derived from a comparison of the tangential velocity distribution for red degenerates and for the Gliese McCormick stars (M dwarfs) in the solar neighborhood. Whereas for the latter 62% have v_{tang} below 40 km/sec, all red degenerates have v_{tang} larger as demonstrated in Fig.1 of Liebert(1978).Application of a velocity correction factor of 2.6 would raise the observed numbers from 3 and 4 to 8 and 10 for the 13-14 and 14-15 mag intervals, in somewhat better agreement with the predictions.

However, in the meantime the space density and birth rate have been revised downward by Fleming et al (1986) since spectrocopic investigations had shown many of Greens hot blue stars not to be white dwarfs. If we correct the figures of Liebert (1978) accordingly, and also takes into account one newly found star for the observations, we now obtain a number ratio of predicted to observed stars of 1.3/4 or 2.5/4 for the bolometric magnitude intervals 13-14 or 14-15. This suggests that the Fleming et al birth rate, around $0.6 \cdot 10^{-12} \mathrm{pc}^{-3} \mathrm{yr}^{-1}$,is somewhat low. However,had a velocity factor of 2.6 been applied to the observed figures the discrepancy would be extreme. We therefore reconsider the determination of velocity correction factors in the next section.

3. A revised velocity correction factor

Liebert, Dahn and Monet (1989) have presented the distribution of tangential velocities against absolute magnitude for an increased ensemble of faint proper motion stars (with $\mu < 0.''8$ or $0.''7$ respectively) which shows the absence of stars with v_{tang} smaller than 40 km/sec for $M_V > 13$. For the brighter stars, with $M_V < 13$, the median v_{tang} is much smaller: Weidemann (1979) found 37 km/sec for 45 DA within 25 parsec, whereas from Sion et al (1988) we derive 39 km/sec for 78 DA within 70 pc (with $T_{eff} > 13000$ K).Even if one includes all white dwarfs from the McCook/Sion catalogue which is partly weighted towards higher proper motions, one obtains only 47 km/sec. This has to be compared with the median values of v_{tang} for the fainter white dwarfs in the LDM surveys, which are 74,

82, 98 or 116 km/sec for successive M_V intervals from $13 - 14, 14 - 15, 15 - 16$ or > 16 magnitude.

Of course one expects the increasingly older white dwarfs to have larger tangential velocities due to the increase of the velocity dispersion, a fact which is also reflected in the increased scale height for older stars.According to the diffusion theory for the dynamical evolution of the galactic disk as outlined by Wielen and Fuchs (1983) - see also Fuchs and Wielen (1987) - the W velocity dispersion increases as $\sigma_w^2 = \sigma_{w,0}^2 + \alpha D\tau$ (where α is a coefficient dependent on Oort constants, D is the diffusion coefficient, and τ the total age of the star) which for a Schwarzschild distribution and under neglegence of anisotropy also holds for the temporal evolution of σv_{tang} (I am indebted to B.Fuchs for a mathematical proof). After a billion years the first term is neglegible so that one expects $\langle v_{\text{tang}} \rangle = \sqrt{\pi/2} \sigma v_{\text{tang}}$ to increase - like the scale height - with the square root of τ.

Applying this square root law (with $\alpha = 0.8$ and $D = 2 \cdot 10^{-7}$) to the increase of the median value of v_{tang} and calibrating the relation to $v_{\text{tang}} = 47$ km/sec for $M_V < 13$ we obtain the data of Table 1.

Table 1: Median tangential velocities and velocity correction factor

M_V	$\log L/L_\odot$	$\langle \tau \rangle$	median v_{tang} pred.	obs.	$N(v_{\text{tang}})$ > pred.	< pred.	v correction calc.	adopt.
12-13	-3.0	$3.5 \cdot 10^9$	47	47	50%	50%	1.0	1.00
13-14	-3.5	$5.0 \cdot 10^9$	56	74	12	3	1.6	1.33
14-15	-3.9	$6.8 \cdot 10^9$	65	82	10	3	1.5	1.5
15-16	-4.2	$8.7 \cdot 10^9$	73	98	13	5	1.5	1.5
>16	-4.4	$10 \cdot 10^9$	78	136	3	-	2.0	

The total average age, $\langle \tau \rangle$, is taken from Yuan models (see Fig.2 in Weidemann 1990), it includes pre-white dwarf evolution. Comparison with the observed values from the LDM data in the fifth column allows us to estimate the number of missing low velocity stars in each interval and thus to determine the velocity correction factors given in the last columns. The first row of course has no velocity correction by definition. In order to keep the correction factor as small as possible, I have here used the highest value for the calibrating value, 47 km/sec, and changed the value for the interval 13-14 from 1.6 to 1.33 as an interpolation between 1.0 and 1.5 (last column).

4. Revised space density

We now apply the results to construct a revised Liebert table for the number of stars in the restricted 10 pc volume, as given in Table 2.

Since data below $M_{\text{bol}} = 15$ are more uncertain, I proceed for the estimate of the space density within 10 pc as follows: there are 9 observed stars with $M_{\text{bol}} < 15$ vs. 12 expected if the velocity corrections are applied, i.e. 3 lower velocity white dwarfs are still missing. For the full sky (factor 1.79, Liebert 1978) there should be 21 single white dwarfs down to $M_{\text{bol}} = 15$ or $\log L/L_\odot = -4.1$. This corresponds to a space density of 0.0050 WD/pc^3.

LDM (1988) by comparison obtained from integration over the total luminosity function - i.e. including the few stars with $\log L/L_\odot < -4.1$ - a space density of 0.0033 WD/pc^3,

Table 2: Expected number of WD in restricted 10 pc volume

M_{bol}	scale height	observed	$v_{corr.}$	expected	missing
<13	300-400	1	1	1	---
13-14	400-500	4	1.33	5	1
14-15	500-600	4	1.5	6	2
15-16	580-600	3	1.5	4.5	1
>16	>600	1	2	2	1

however they did not apply a velocity correction factor under the assumption that the V_{max} method automatically accounts for selection effects caused by the use of proper motion catalogues. Since the LF shows the sharp cut-off below $M_{bol} = 15$, probably caused by the finite age of the local disk (cf.Winget et al 1987 and M.Wood at this Conference), the total space density may not differ very much from the one down to $M_{bol} = 15$. However if one applies the synthetic models of Yuan (1989) the fraction of the cooled-down degenerates below this limit can be estimated: it is of the order of 15 to 25% for disk ages from 9 to 12 Gyr (see Yuan 1989 Fig.2 and Weidemann 1990).

5. The birth rate question

The synthetic galactic evolution models calculated by Yuan (1989) are specified by a given IMF, SFR, age of the Galaxy (more exactly: by the age of the local disk), by an initial-final mass relation, with nuclear burning and white dwarf cooling ages as function of initial and final mass, and by a scale height function; normalized to yield a present WD birth rate of $1 \cdot 10^{-12} WDpc^{-3}yr^{-1}$. They allow to determine WD space densities (per cubic as well as per square parsec - here enters the scale height function - down to arbitrary luminosity limits, but also to obtain several other figures of interest, e.g. the total stellar mass per square parsec .

As for the white dwarf birth rate, χ, the calculations demonstrate, as mentioned in the Introduction, that χ is a function of time even for constant SFR and that even for a constant birth rate the present space density is smaller than that obtained by multiplication with cooling time due to the the scale height inflation. One thus has to specify an evolutionary model and use χ_0 for comparisons.

In a first step we give an example: Yuan's standard model (IMF modified Salpeter, SFR constant, galactic age 12 Gyr, $\chi_0 = 1$ (in the units of $10^{-12} pc^{-3}yr^{-1}$)) predicts 0.0034 WD per cubic parsec down to $\log L / \log L_\odot = -4.1$. Since our observational value is 0.0050 WD/pc^3 we have to scale up the model to $\chi_0 = 0.0050/0.0034 = 1.47$ in order to produce the observed space density. This value is given for 6 different models in column 3 of Table 3 below (χ_0, single). Secondly we consider those χ_0 values which had to be used in order to scale the Yuan models to provide the — well-determined — total stellar mass in the local column, 28 $M_\odot$, which includes binaries. These are tabulated for 9 different models in Weidemann(1990) where the local mass budget is considered. We list the corresponding χ_0 values in column 4 (χ_0, $M^* = 28$). If we remember that the value of the total local mass includes binaries whereas the Yuan models consider only single star evolution, we can derive a binary factor which should reflect an increase of χ_0 to $\chi_0(M^* = 28)$. It is given in column 5 of Table 3.

Table 3: Comparison of birth rates

Yuan model	$N(< -4.1)$ pred. $\chi_0 = 1$	χ_0 (single)	χ_0 (M^*)	binary factor
Standard, disk age 12 Gyr	0.0034	1.47	1.65	1.12
Standard, disk age 9 Gyr	0.0032	1.56	2.00	1.28
Standard, but initial burst	0.0038	1.31	1.13	0.86
Standard, but SFR expon.decl.	0.0036	1.40	1.40	1.01
Standard, but IMF Miller/Scalo	0.0034	1.47	3.11	2.11
Standard, but IMF Scalo 1986	0.0034	1.40	2.13	1.52

Explanations to first column: The Standard model uses IMF Salpeter down to 0.25 $M_\odot$, then constant; SFR= const.; M(initial)/M(final) Weidemann (1987); WD cooling, Winget et al (1987); scale height function,Scalo (1986); main sequence lifetimes, Maeder/Mermillod (1981) for intermediate overshooting; disk age 12 Gyr. The model with exponential decline uses a SFR with $exp(-t/5)$ for the first 5 Gyr, then constant. The IMF Miller/Scalo and Scalo (1986) declines more strongly than Salpeters IMF for larger masses.

Empirically determined binary factors for stars in the solar neighborhood are estimated to be between 1.6 and 2.0 (e.g. Liebert 1978, Jahreiss 1987). The figures of Table 3 then favor models with a non-Salpeter IMF as derived by Scalo (1986). We thus obtain as best values for χ_0 based on $N = 0.005$WD/pc^3 down to $\log L/\log L_\odot = -4.1$

$$\chi_0(\text{single}) = 1.4 \cdot 10^{-12}\text{pc}^{-3}\text{yr}^{-1} \qquad \text{or} \qquad \chi_0(\text{all}) = 2.3 \cdot 10^{-12}\text{pc}^{-3}\text{yr}^{-1}.$$

Had we used the space density of LDM, 0.0033 WD/pc^3, we would have obtained $\chi_0(\text{single}) = 0.92$ or $\chi_0(\text{all}) = 1.5 \cdot 10^{-12}\text{pc}^{-3}\text{yr}^{-1}$ In the first case we expect a total of 21, in the second of 14 single WD within 10 pc.

Although the application of a velocity correction factor and the revised IMF favor the first alternative, even the second case predicts $\chi_0 > 0.9 \cdot 10^{-12}\text{pc}^{-3}\text{yr}^{-1}$ and thus a higher value than the range found by Fleming et al (1986), around $0.6 \cdot 10^{-12}\text{pc}^{-3}\text{yr}^{-1}$, or advanced by Liebert (at this Conference) who points out that the new WD mass scale (Bergeron, this Conference) lowers masses and therefore increases luminosities and reduces the derived space densities. However in this case the well known discrepancy with the birth rate for planetary nebulae (PN) the best value of which is estimated to be $2.4\pm0.3\cdot10^{-12}$PNpc^{-3}yr^{-1} by Phillips (1989) would become unsurmountable, whereas our preferred result, $\chi_0(\text{all}) = 2.3 \cdot 10^{-12}\text{pc}^{-3}\text{yr}^{-1}$ is in good agreement. The PN value depends strongly on the distance scale. For further discussion of the latter, and of the white dwarf fraction not going through the PN channel cf. Weidemann(1989): in any case χ_{PN} should be smaller than $\chi_0(\text{all})$. The comparatively higher value for $\chi_0(\text{single})$ has to stand future observational tests, e.g. by the ongoing surveys. It should be remarked in this context that the quasar survey by Boyle (1989) increased the space density of DA white dwarfs by a factor of 1.22 compared to Fleming et al (1986).

72

6. References

BOYLE, B.J. 1989: *Mon.Not.R.astr.Soc.* **240**, 533.
FLEMING, T.A., LIEBERT, J., GREEN, R.F. 1986: *Astrophys.J.* **308**, 176.
FUCHS, B., WIELEN, R. 1987: *The Galaxy* eds. Gilmore, G. and Carswell, B., Reidel, p. 375.
GREEN, R.F. 1977: *Thesis,* California Inst.Technology.
    1980Astrophys.J. 238685 JAHREISS, H. 1987: *Mem.S.A.It.* **58**, N° 1, 53.
LIEBERT, J. 1978: *Astron.Astrophys.* **70**, 125.
LIEBERT, J., DAHN, C.C., MONET, D.G. 1988: *Astrophys.J.* **332**, 891.
LIEBERT, J., DAHN, C.C., MONET, D.G. 1989: *IAU Coll.114 : White Dwarfs* ed. G.Wegner, Springer, p. 15.
MAEDER, A., MERMILLOD, J.C. 1981: *Astron.Astrophys.* **93**, 136.
PHILLIPS, P. 1989: *IAU Symp.131: Planetary Nebulae* ed S. Torres-Peimbert, Kluwer , p. 425.
SCALO, J.M. 1986: *Fund.Cosm.Phys.* **11**, 1.
SION, E.M., LIEBERT, J. 1977: *Astrophys.J.* **213**, 468.
SION, E.M., FRITZ, M.L., MCMULLIN, J.P., LALLO, M.D. 1988: *Astron.J.* **96**, 251.
WEIDEMANN, V. 1967: *Zeitschr.f.Astrophysik* **67**, 286.
WEIDEMANN V. 1979: *IAU Coll.53, White Dwarfs and Variable Degenerate Stars,*eds.Van Horn, H.M., Weidemann,V., Univ.Rochester NY, p. 206.
WEIDEMANN, V. 1987: *Astron.Astrophys.***188**, 74.
WEIDEMANN, V. 1989: *Astron.Astrophys.***213**, 155.
WEIDEMANN, V. 1990: *Baryonic Dark Matter,* eds. Lynden-Bell, D., Gilmore, G., NATO ASI Series C **306**, p. 87.
WIELEN, R., FUCHS, B. 1983: *Kinematics, Dynamics and Structure of the Milky Way,* ed. W.L.H.Shuter, Reidel , p. 81.
WINGET, D.E., HANSEN, C.J., LIEBERT, J., VAN HORN, H.M., FONTAINE, G., NATHER, R.E., KEPLER, S.O., LAMB, D.Q. 1987: *Astrophys.J.* **315**, L77.
YUAN, J.W. 1989: *Astron.Astrophys.* **224**, 116.

DISCUSSION

LIEBERT :

We have never claimed to have a complete luminosity function (LF) for the cool magnitudes drawn from a proper motion sample. The $1/V_{max}$ method does not correct for the "missed" low velocity stars. The apparent match of the kinematically-unbiased LF for the hotter stars with this determination at $M_v \sim +13$ suggested that we were not terribly incomplete. However, in Winget et al. (1987), we claimed only that the cutoff was real. I would not try to fit the "shape" of the LF. Therefore, I think your determination of moderate velocity correction factors is reasonable. (The Fuchs-Wielen relation for $\tau > 10^9$ years lacks for the most part a reliable empirical calibration, so I hadn't tried to use it).

A SPECTROSCOPIC DETERMINATION OF THE MASS DISTRIBUTION OF DA WHITE DWARFS

P. BERGERON, REX A. SAFFER, and JAMES LIEBERT
Steward Observatory
University of Arizona
Tucson, AZ 85721

ABSTRACT. Previous determinations of the mass distribution of DA white dwarfs are shown to be sensitive to both the input physics and the assumed atmospheric composition. A new mass distribution based on fitting observed hydrogen line profiles to theoretical model spectra is presented. A sample of 127 DA stars has been selected above $T_{\rm eff} \sim 15,000$ K where these theoretical uncertainties are minimal. The results of our analysis indicate that the mean mass of the distribution is somewhat lower – near 0.53 $M_\odot$ – than found previously. Extended tails which contain clear cases of low and high mass objects are also found.

1. Previous Determinations of Mass Distribution of DA White Dwarfs

The determination of the mass distribution of white dwarfs is crucial for our understanding of the evolution of these stars and is a critical boundary condition in understanding the mass-loss phenomenon in the red giant, asymptotic giant branch, and planetary nebula phases of evolution. Because of the well known relationship between mass and radius for white dwarfs, this problem also can be studied by measuring the radius, or equivalently, the surface gravity of the star. For DA white dwarfs, most of the work in this field has been accomplished through analyses of Strömgren and narrow-band colors. For example, Shipman (1979), and Koester, Schulz, and Weidemann (1979; hereafter KSW) have estimated independently, using either stellar parallax techniques or color–color diagrams, that the mean radius of DA white dwarfs is $\sim 0.012\ R_\odot$, with a corresponding mass of 0.58 $M_\odot$ (or log $g \sim 8.0$). The sample of white dwarfs used in these studies is mainly concentrated below $T_{\rm eff} \sim 15,000$ K, where the color indices are most sensitive to surface gravity, and also because the bulk of white dwarfs is found in this range of temperatures. However, as we argue below, the model atmospheres in this particular region are very sensitive to the input physics and the assumed atmospheric composition, and previous estimates of atmospheric parameters of white dwarfs may need to be reconsidered.

1.1 THE COOL DA WHITE DWARFS

Previous determinations of atmospheric parameters of cool DA white dwarfs have always been performed assuming a pure hydrogen atmospheric composition.

75

G. Vauclair and E. Sion (eds.), White Dwarfs, 75–87.
© 1991 *Kluwer Academic Publishers.*

However, Koester (1976), Vauclair and Reisse (1977), and D'Antona and Mazzitelli (1979) have shown that below $T_{eff} \sim 12,000\ K$, mixing between the thin superficial hydrogen layer and the more massive underlying helium layer can turn a hydrogen-rich star into a helium-rich star, provided the mass of the hydrogen layer is small enough ($M_H < 10^{-8}\ M_\odot$). It is generally assumed in these scenarios that convective mixing will turn DA stars into non-DA stars, since the helium convection zone is so much more massive than its hydrogen counterpart. But as discussed by Bergeron *et al.* (1990), the efficiency of the convective mixing process may be sufficiently poorly understood that the atmospheric composition expected once mixing occurs remains uncertain. Furthermore, below $T_{eff} \sim 13,000$ K, helium becomes spectroscopically invisible in a DA atmosphere, and it is then no longer possible to rely on the presence of helium lines to infer the atmospheric composition. The case where convective mixing would lead to an atmosphere simply enriched in helium *while preserving its DA character* has apparently never been granted much attention. Were the latter picture the correct one, however, past and current determination of atmospheric parameters of cool DA stars could be seriously affected, as they would then all assume an unrealistic chemical composition.

For example, we show in Figure 1 the Strömgren color–color diagrams of DA white dwarfs for pure hydrogen models with different surface gravities, and also for models with constant surface gravity but with different helium compositions.

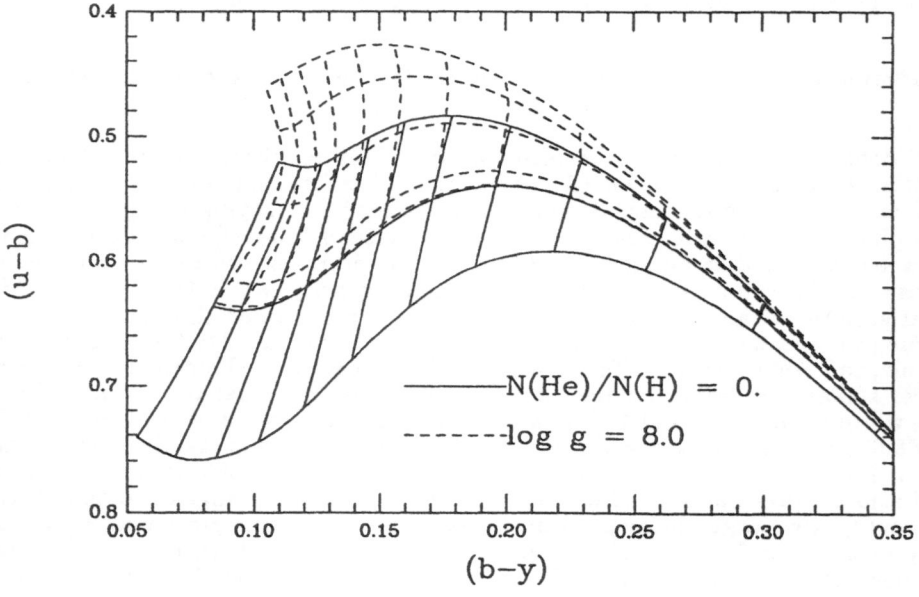

Figure 1. Strömgren color–color diagrams for DA white dwarfs. The effective temperature starts at 12,000 K and decreases by steps of 500 K. The solid lines are for pure hydrogen models with surface gravities of (from bottom to top) log $g = 7.5$, 8.0, and 8.5. The dashed lines are for log $g = 8.0$ models with helium abundances of (from bottom to top) $N(\text{He})/N(\text{H})= 0.01, 0.1, 1, 5$, and 10.

These results indicate that the effect on the predicted colors of an increase in surface gravity, as measured by the strength of the Balmer jump, can also be reproduced by increasing the helium abundance, as noticed also by Wegner and Schulz (1981). It is clear that in such diagrams, DA white dwarfs with pure hydrogen atmospheres could as well be interpreted as helium-enriched objects with lower surface gravities.

A detailed analysis of the high Balmer line profiles could potentially separate helium and gravity effects; the presence of helium increases the photospheric pressure, and thus produces a quenching of the upper levels of the hydrogen atom which, in turn, affects the line profiles. Bergeron, Wesemael, and Fontaine (1991) have shown, however, that an increase in surface gravity quenches the atomic levels in a similar fashion. Even by using line profiles, it is always possible to compensate almost exactly the effect of decreasing the surface gravity by increasing the helium abundance, provided that the effective temperature remains constant. An example of such an equivalence is presented in Figure 2. Finally, it is found that these spectroscopic equivalences correspond almost exactly to the those obtained from color–color diagrams.

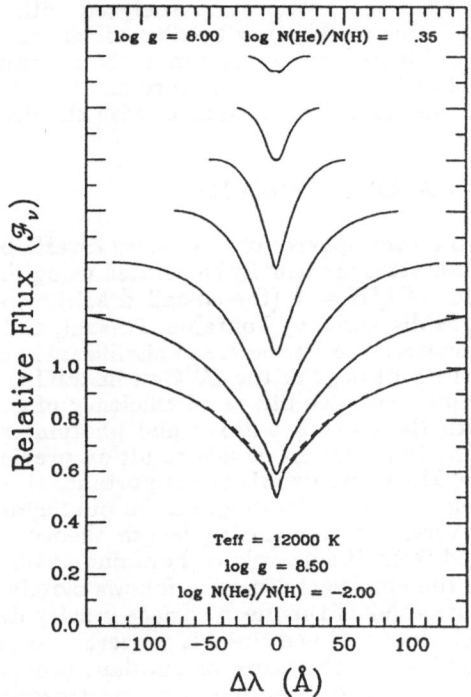

Figure 2. Equivalence in the line profiles between models at different helium abundances and surface gravities. The lines are (from bottom to top) Hα to H9, and have been normalized and offset vertically from each other. The solid line spectrum is calculated with the atmospheric parameters indicated at the bottom. The dashed line spectrum is calculated at the same effective temperature, but with the helium abundance and surface gravity indicated at the top.

Another possible way around this problem is to use the stellar parallax. Indeed, stellar parallax techniques directly measure the radius, and therefore the surface gravity for a given mass–radius relation. Although it is true that the monochromatic flux at V is not a very sensitive function of the atmospheric composition, the bolometric correction, however, depends on the helium abundance. Therefore, the two effects are still coupled, and it is not possible to measure directly the radius unless the atmospheric compositon is known *a priori*. A detailed discussion of this problem will be presented elsewhere.

Therefore, Bergeron, Wesemael, and Fontaine (1991) conclude that, on the basis of optical spectroscopy and photometry (and probably stellar parallax), *it is not possible to separate the pressure effects originating from an increased helium abundance from those stemming from an increased surface gravity.* These results bear profound implications for previous determinations of the mass (or surface gravity) distribution of cool DA white dwarfs. In particular, Bergeron *et al.* (1990) have shown that in order to fit the line profiles of their sample of 37 cool DA stars with pure hydrogen models, the mean surface gravity would have to be increased to log $g \sim 8.2$. Since there is no reason to believe that the mean log g of these stars should be different than the hotter stars analyzed with similar techniques (*e.g.* Daou *et al.* 1990 and this work), they attributed instead this high value of the mean surface gravity to the presence of helium in the atmospheres of most cool DA stars. The presence of helium has been interpreted as evidence for convective mixing between the thin, superficial hydrogen layer with the deeper, more massive helium layer.

1.2 THE CONVECTIVE DA ATMOSPHERES

Below $T_{\mathrm{eff}} \sim 15,000$ K, the atmospheres of DA white dwarfs become convective. Model atmospheres for these stars are usually calculated using the standard mixing length theory, with a value of $l/H = 1$ (the so-called ML1 version of the mixing length theory). However, as discussed by Fontaine, Tassoul, and Wesemael (1984), the efficiency of convection needs to be increased significantly in order to explain the observed location of the blue edge of the ZZ Ceti instability strip. When used in model atmosphere calculations, this increased efficiency of convection produces significant changes in both the emergent fluxes and photometric colors (see also Wesemael *et al.* 1991). The implications of this result on previous determinations of the mass distribution of DA white dwarfs are important.

Figure 3 displays Strömgren color–color diagrams for pure hydrogen models using both the ML1 and ML3 version of the mixing length theory. Also shown is the empirical sequence derived from the sample of Fontaine *et al.* (1985). Although with both sets of models the empirical sequence follows closely a line of constant surface gravity, the *absolute value* of the mean surface gravity depends strongly on the assumed convective efficiency. Since there is no reason *a priori* to adopt one parametrization of the mixing length theory or another, previous determinations of the mass distribution using either photometric, spectroscopic, or even stellar parallax techniques in this range of effective temperature should be considered model-dependent. In particular, had a more efficient convection theory been used in previous analyses, *the mean surface gravity of DA white dwarfs would have been lower than the canonic log $g = 8.0$ value.*

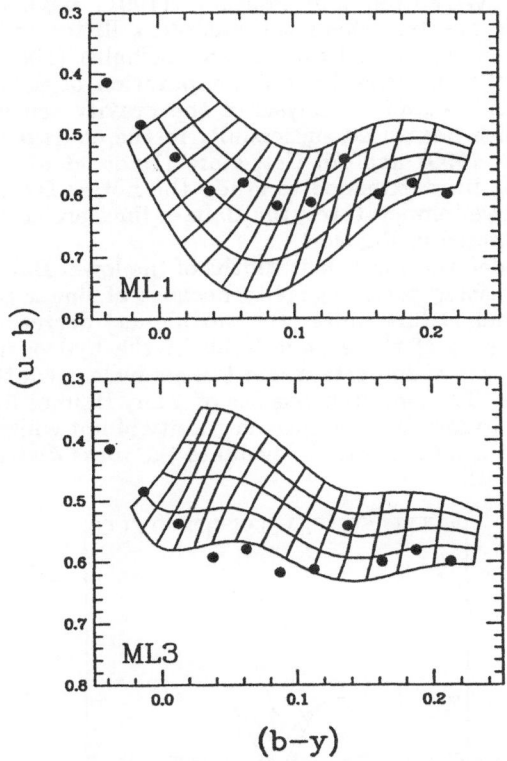

Figure 3. Influence of the convective mixing efficiency on the morphology of the Strömgren diagram of cool DA stars. The effective temperatures range (from left to right) from 8000 to 15,000 K by steps of 500 K, and the values of log g (from bottom to top) from 7.5 to 8.5 by steps of 0.25. The dots represent the empirical sequence derived from the sample of Fontaine *et al.* (1985).

1.3 HOTTER DA STARS

Above $T_{eff} \sim 15,000$ K, the atmospheres of DA white dwarfs are completely radiative and thus do not suffer from the uncertainties in convection theory discussed previously. Furthermore, the absence of convection ensures that the chemical separation in the atmospheric regions by gravitational settling is maintained, and the atmospheres are known to be hydrogen-rich. Therefore, the uncertainties in the input physics and/or atmospheric compositions are minimized. Unfortunately, because of the gradual disappearance of the Balmer jump at high effective temperature, photometric analyses become rapidly insensitive to surface gravity above $T_{eff} \sim 20,000$ (see e.g. KSW). Furthermore, the errors on the measured parallax are larger than those associated with the cooler stars since the hot white dwarfs are intrinsically brighter, and therefore farther away.

Recently, Bergeron, Wesemael, and Fontaine (1991) have developed a new generation of synthetic spectra where the emergent fluxes are calculated with the new occupation probability of Hummer and Mihalas (1988) which provides an improved theoretical framework for the computation of detailed atomic level populations and permits a careful analysis of the gravity sensitive high Balmer lines. These models have been used successfully in the context of cool DA white dwarfs (Bergeron *et al.* 1990) and ZZ Ceti stars (Daou *et al.* 1990). That grid of synthetic spectra has been extended to study the hotter DA stars. It is found that even at high effective temperatures, the Balmer lines are still very sensitive to surface gravity as illustrated in Figure 4.

In this particular model, the equivalent widths of the lower Balmer lines (Hβ and Hγ) increase with increasing surface gravity because of linear Stark broadening, while those of the higher Balmer lines (Hϵ and higher) decrease with increasing log g due to the quenching of the atomic levels of the hydrogen atom. For Hδ, both physical processes are as important and the equivalent width remains almost constant at all gravities. The simultaneous use of many Balmer lines thus provides us with a powerful tool to measure the surface gravity of hot white dwarfs, and it is arguably the most accurate technique to evaluate the mass distribution of a large sample of DA white dwarfs.

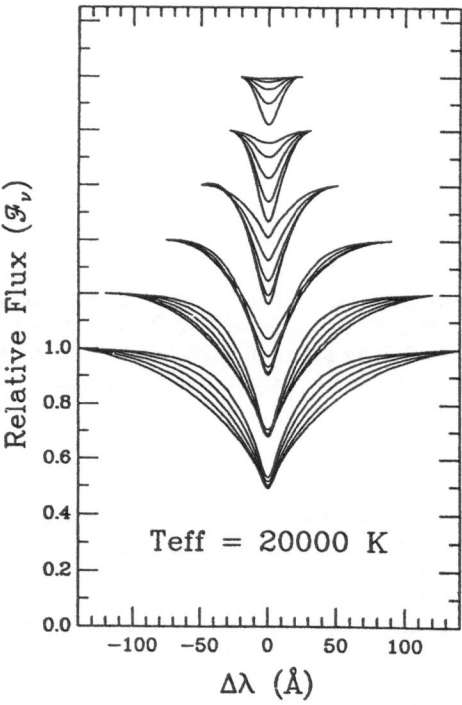

Figure 4. Theoretical line profiles of models at $T_{\mathrm{eff}} = 20,000$ K and log $g = 7.0(0.5)9.0$. The lines are (from bottom to top) Hβ to H9, and have been normalized and offset vertically from each other. The behavior of each line with surface gravity is discussed in the text.

2. The Spectroscopic Technique

2.1 THE DATA SAMPLE

We have selected from the catalog of McCook and Sion (1987) a complete sample of 127 DA white dwarfs brighter than $V = 15.6$ with spectral types DA2 and DA3. This corresponds roughly to effective temperatures in the range 15,000 K to 30,000 K. High signal-to-noise ($S/N > 80$) spectra have been obtained with the Steward 2.3-m reflector and ultraviolet–sensitive TI CCD detector. Such a high signal-to-noise is needed in order to make precise determination of atmospheric parameters. The spectra have 8 Å resolution and cover the range $\lambda\lambda3750–5100$. A subsample of our spectra is displayed in Figure 5.

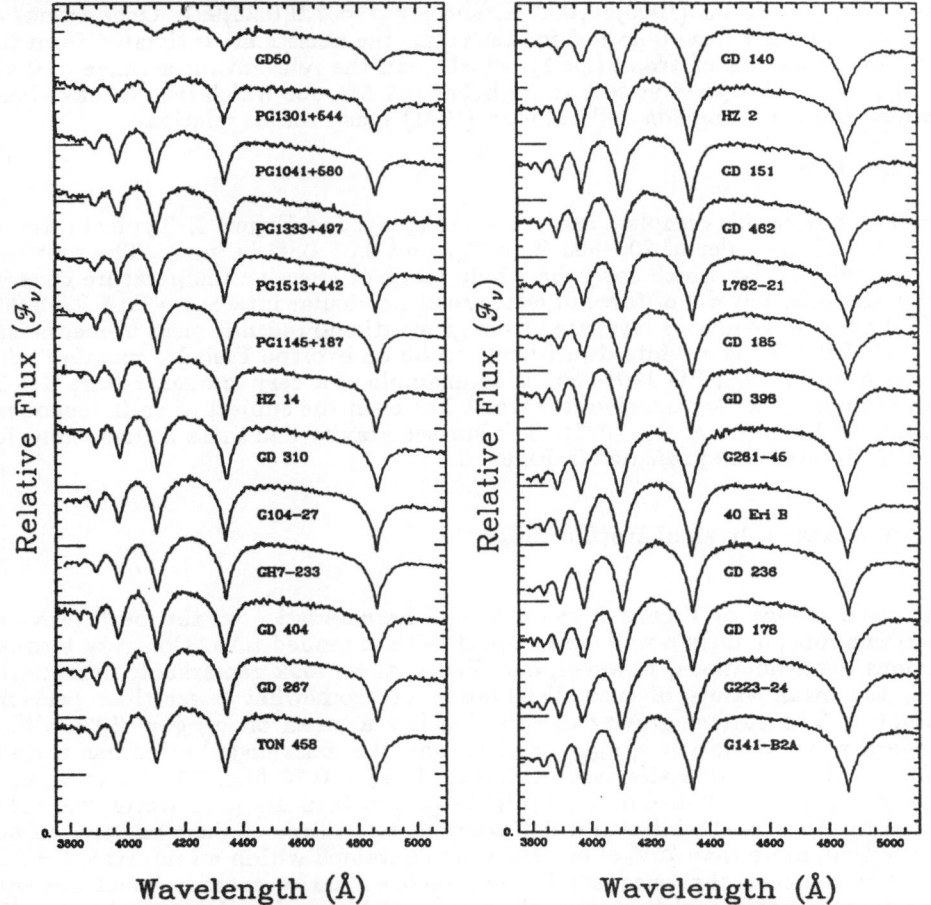

Figure 5. Optical spectra of a subsample of our objects. The spectra are normalized at 4500Å, and are shifted vertically; the various zero points are indicated by long tick marks. The effective temperatures decrease from upper right to bottom left.

2.2 THE FITTING TECHNIQUE

The technique used to derive the atmospheric parameters is similar to that used by Bergeron *et al.* (1990) and Daou *et al.* (1990) where all the line profiles are fitted simultaneously. The first step is to normalize the line flux, in both observed and model spectra, to a continuum set at a fixed distance from the line center. The comparison with model spectra, which are convolved with a Gaussian instrumental profile, is then carried out in terms of these line shapes only. The fitting technique we employ here relies on the nonlinear least-squares method of Levenberg-Marquardt (Press *et al.* 1986), which is based on a steepest descent method. The calculation of χ^2 in our case is carried out using the normalized lines profiles as defined above. The theoretical models are similar to those described in Bergeron, Wesemael, and Fontaine (1991). The grid covers a range of $T_{eff} = 12,000(1000)60,000$ K, and $\log g = 7.0(0.25)9.0$. Once values of T_{eff} and $\log g$ have been obtained for each star, the masses are estimated from the evolutionary models of Wood (1991), which span the relevant mass range of $0.4 - 1.2 M_{\odot}$. Several objects appear to lie below $0.4 M_{\odot}$ for which masses have been obtained from the Hamada and Salpeter (1961) mass–radius relation.

2.3 RESULTS

Sample fits from our complete analysis are displayed in Figure 6. Typical internal errors are of the order of 200–500 K in T_{eff} and 0.03–0.05 in $\log g$. The first row are examples of fits which span the whole range of effective temperature covered in our analysis. Several of these objects are of particular interest: VR16, LDS455, and 40 Eri B have masses estimated from gravitational redshift measurements (see also §4); LB 1497 is a white dwarf likely to be an evolved Pleiades member with a mass near $0.9 M_{\odot}$; PG 1101+364 is an example of a very low mass star; GD 50 is an example of a very massive star which has been the subject of an independent analysis by Bergeron *et al.* (1991). The surface gravity and mass distributions for our 127 DA stars are presented in Figure 7.

3. Some Astrophysical Implications

The distributions of Figure 7 share some characteristics of the best previous determinations, namely a well-defined peak with extended tails. Contrary to most previous determinations, however, our distributions look remarkably symmetric. Also, the mean values of each distribution are somewhat lower than previous estimates: The surface gravity distribution has a mean of $\log g = 7.85$ with a standard r.m.s. deviation of $\sigma_{\log g} = 0.24$, and the mass distribution has a mean of $M = 0.53 M_{\odot}$ with a standard deviation of $\sigma_M = 0.13 M_{\odot}$. The latter value is slightly larger than that obtained by KSW ($\sigma_M = 0.10 M_{\odot}$). However, the value quoted by KSW is the interval which contains two thirds of their sample. In our distribution, more than 75% of our stars are contained within an interval of ± 0.10 $M_{\odot}$. This indicates that our mass distribution has a narrower central peak but with more extended tails which contain clear cases of low and high mass objects. We believe that this is the result of the sensitivity of our technique to resolve with high accuracy the fine structure present in the mass distribution of DA white dwarfs.

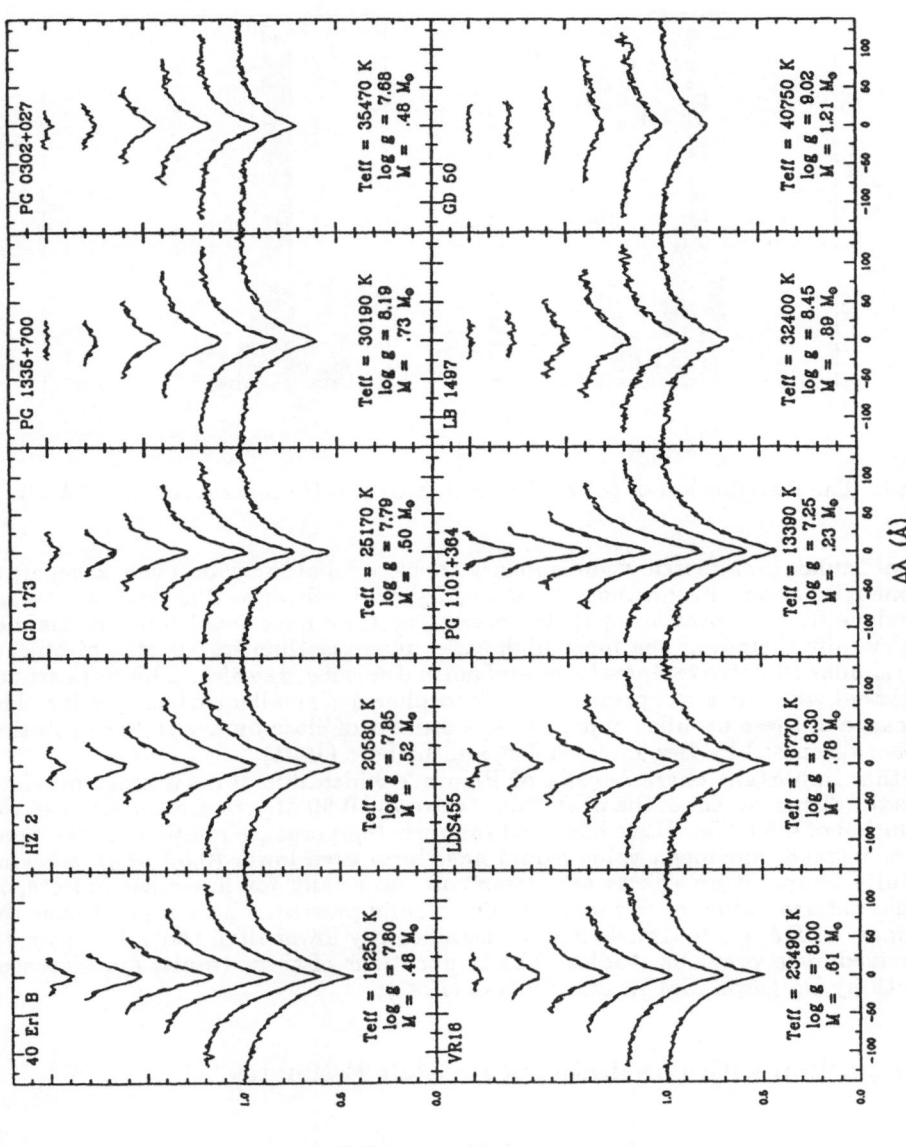

Figure 6. Fits to the individual Balmer lines for a subsample of our objects. The lines range from Hβ (bottom) to H9 (top), each offset vertically by a factor of 0.2.

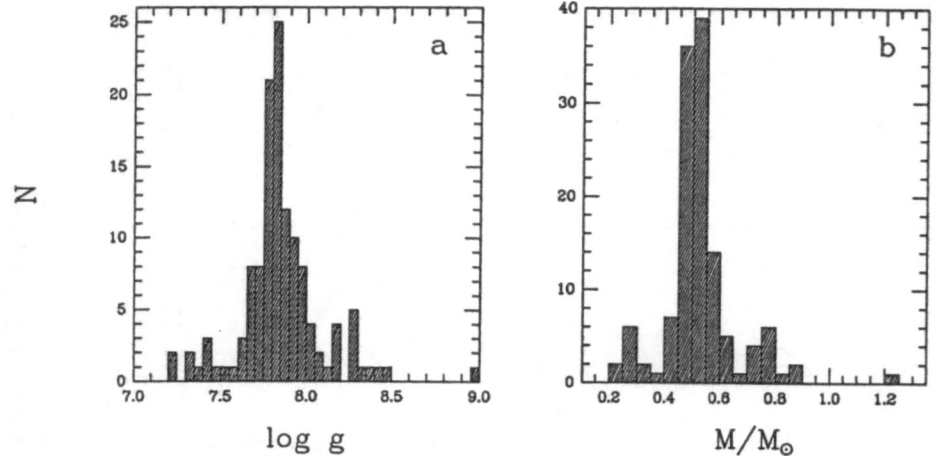

Figure 7. The distribution of (*a*) surface gravities and (*b*) masses for 127 DA stars.

In particular, our distribution appears to have what may be even a separate component of about 10 low mass (≤ 0.35 $M_\odot$) white dwarfs. The low mass stars are obvious from inspection of their spectra, as they have much sharper Balmer lines generally including one more high series member than for spectra of objects having similar effective temperatures and normal surface gravities. This appearance is consistent with the lower pressures in atmospheres of smaller surface gravity. The significance of these peculiar objects in the context of close binary stellar evolution is further discussed by Bergeron, Saffer, and Liebert (1991).

Another important characteristic of Figure 7 which differs from most previous work is the displacement of the peak value to around 0.50 $M_\odot$, or a mean value of the distribution of 0.53 $M_\odot$. If we had used the zero–temperature relation of Hamada–Salpeter instead, our mean value would have been even lower (0.50 $M_\odot$), showing that finite temperature effects are important, especially for low-mass objects, in the mass determination of our sample. The results presented here suggest that the mean mass of DA white dwarfs may be significantly lower than the value near 0.6 $M_\odot$ arrived at in previous studies. The implications of these results are discussed at length by Bergeron, Saffer, and Liebert (1991).

4. Comparison with Gravitational Redshift Estimates

As discussed by Bergeron, Saffer, and Liebert (1991), although we believe that our technique provides the most accurate determination of individual atmospheric parameters, it is also new and relies strongly on the accuracy and validity of the Hummer–Mihalas formalism used in our calculations. In particular, since the errors of our individual determinations are small, the *relative* distributions displayed in Figure 7 are probably reliable. The *absolute* values, however, may suffer from a zero-point calibration problem and it is therefore important to compare the results of individual objects with independent determinations. Here we make use of the masses determined from gravitational redshift measurements.

Table 1 presents the results of our comparison. The values obtained from gravitational redshifts, M_{GR}, are taken from the review paper by Wegner (1989). Although there are some discrepancies between a few objects, the general agreement is extremely good, *even for low or high mass stars*. In particular, the mean masses are different by only 0.01 $M_\odot$! The first four objects in Table 1 are members of the Hyades cluster. Since these stars have similar effective temperatures, they were formed at the same time and therefore should have similar masses. Indeed, the masses obtained from our analysis are clustered around 0.60 $M_\odot$ with a standard deviation of only 0.008 $M_\odot$. The corresponding values obtained from gravitational redshift determinations show a much larger scatter of 0.076 $M_\odot$. Finally, as noticed by Wegner, Reid, and McMahan (1989), the mean mass of the Hyades white dwarfs is measureably higher than for the field white dwarfs.

TABLE 1. Comparison with gravitational redshift measurements

Name	$T_{\rm eff}$ (K)	$M_{GR}/M_\odot$	$M_{BSL}/M_\odot$
HZ 14	27,050	0.47	0.59
VR16	23,490	0.60	0.61
HZ 7	21,020	0.65	0.59
VR7	18,720	0.66	0.60
LDS455	18,770	0.80	0.78
40 Eri B	16,250	0.50	0.48
L970-27	15,280	0.44	0.49
G148-7	15,120	0.54	0.53
Wolf 485A	14,040	0.52	0.50
G142-B2A	13,630	0.55	0.45
	$\overline{M} =$	0.573	0.563
	$\sigma =$	0.102	0.090

Although we are still exploring the systematic effects potentially present in our analysis, the results presented in Table 1 suggest than our current determinations are accurate even in an absolute sense. We are currently comparing our results with those obtained from stellar parallax techniques.

This work was supported in part by the NSF grant AST 89-18471, by a Postdoctoral Fellowship to one of us (PB), and by NATO.

REFERENCES

Bergeron, P., Kidder, K.M., Holberg, J.B., Liebert, J., Wesemael, F., and Saffer, R.A. 1991, *Ap. J.*, in press.

Bergeron, P., Saffer, R.A., and Liebert, J. 1991, in *Confrontation between Stellar Pulsation and Evolution*, ed. C. Cacciari, A. S. P. Conference Series (Astronomical Society of the Pacific : Provo, Utah), in press.

Bergeron, P., Wesemael, F., and Fontaine, G. 1991, *Ap. J.*, in press.

Bergeron, P., Wesemael, F., Fontaine, G., and Liebert, J. 1990, *Ap. J. (Letters)*, **351**, L21.

D'Antona, F., and Mazzitelli, I. 1979, *Astr. Ap.*, **74**, 161.

Daou, D., Wesemael, F., Bergeron, P., Fontaine, G., and Holberg, J.B. 1990, *Ap. J.*, in press.

Fontaine, G. Bergeron, P., Lacombe, P., Lamontagne, R., and Talon, A. 1985, *A. J.*, **90**, 1094.

Fontaine, G., Tassoul, M., and Wesemael, F. 1984, in *Proc. 25th Liège Astrophysical Colloquium: Theoretical Problems in Stellar Stability and Oscillations*, eds. A. Noels and M. Gabriel (Liège: Université de Liège), p. 328.

Hamada, T., and Salpeter, E. E. 1961, *Ap. J.*, **134**, 683.

Hummer, D.G., and Mihalas, D. 1988, *Ap. J.*, **331**, 794.

Koester, D. 1976, *Astr. Ap.*, **52**, 415.

Koester, D., Schulz, H., and Weidemann, V. 1979, *Astr. Ap.*, **76**, 262 (KSW).

McCook, G.P., and Sion, E.M. 1987, *Ap. J. Suppl.*, **65**, 603.

Press, W.H., Flannery, B.P., Teukolsky, S.A., and Vetterling, W.T. 1986, *Numerical Recipes* (Cambridge: Cambridge University Press).

Shipman, H.L. 1979, *Ap. J.*, **228**, 240.

Vauclair, G., and Reisse, C. 1977, *Astr. Ap.*, **61**, 415.

Wegner, G. 1989, in *IAU Colloquium 114, White Dwarfs*, ed. G. Wegner (New York: Springer), p. 401.

Wegner, G., Reid, I.N., and McMahan, R.K. 1989, in *IAU Colloquium 114, White Dwarfs*, ed. G. Wegner (New York: Springer), p. 378.

Wegner, G., and Schulz, H. 1981, *Astr. Ap. Suppl.*, **43**, 473.

Weidemann, V., and Koester, D. 1983, *Astr. Ap.*, **121**, 77.

Wesemael, F., Bergeron, P., Fontaine, G., and Lamontagne, R. 1991, these proceedings.

Wood, M.A. 1991, these proceedings.

DISCUSSION

THEJLL :

What is the importance on the average mass of the choice of the guenching formalism? Will the low $\overline{M}$ you get move up significantly if you dont use the above formalism?

BERGERON :

We are still investigating this issue but we believe that the real comparison has to be made with masses obtained with independent reliable techniques, e.g. gravitational redshift.

OSWALT :

Do the variation in atmospheric seeing impose an additional scatter in your mass distribution, i.e. is the true dispersion in mass somewhat less even than you show?

BERGERON :

No. The line cores are broad enough that this effect is not important.

THE AGE AND FORMATION OF THE GALAXY: CLUES FROM THE WHITE DWARF LUMINOSITY FUNCTION

M. A. WOOD
Département de Physique,
Université de Montréal
C.P. 6128, Succ. A
Montréal, QC, H3C 3J7
Canada

ABSTRACT. Since the onset of star formation in the local disk of the Galaxy, stars have been evolving into white dwarf stars, which have cooled with time to produce the white dwarf luminosity function that we observe today. Because the cooling timescale of a typical white dwarf is longer than the age of the local disk, even those white dwarfs formed from the earliest generation of stars are still visible. Using homogeneous sets of mass-dependent cooling curves and analytical approximations to the details of galactic evolution, it is possible to synthesize theoretical white dwarf luminosity functions which can then be compared directly with the observations to probe the age and star formation history of the local disk, and the sensitivity of these to the inputs. I have completed just such a parametric study, and report here in brief some of the major results and their bearing on our understanding of the process of spiral-galaxy formation.

1. Introduction

The turndown in the observed luminosity function (LF) reported by Liebert (1979; see also Liebert 1980) was discussed by Winget *et al.* (1987) in terms of the finite age of the Galaxy. They showed that by combining the single-sequence luminosity functions,

$$N = \frac{d\tau}{d\log(L/L_\odot)}, \tag{1}$$

obtained from pure-carbon models of masses 0.4, 0.6, 0.8, and 1.0 $M_\odot$, and weighted by the observed mass distribution of Weidemann and Koester (1984), it is possible to computed theoretical LFs which can be compared against the observations. Winget *et al.* computed LFs of ages 6, 9 and 12 Gyr, laid these over an improved determination of the observed LF turndown, and after adding 0.3 Gyr as an estimate of the mean white dwarf progenitor lifetime derived an estimated Galactic age of 9.3 ± 2.0 Gyr. In using this method, however, the authors implicitly assumed that the white dwarf birthrates of each of the four masses were constant over time, and further, neglected the mass dependence of the evolutionary timescales of the white dwarf progenitors.

Three followup studies were published in quick succession, all using the Winget *et al.* evolutionary sequences but including more sophisticated treatments of the

89

G. Vauclair and E. Sion (eds.), White Dwarfs, 89–97.
© 1991 *Kluwer Academic Publishers.*

galactic evolution (Iben and Laughlin 1989, Yuan 1989, and Noh and Scalo 1990). Because Winget *et al.* were correct in their assumption that the main sequence evolutionary timescales are short enough to be relatively unimportant, the ages that these three arrived at also clustered near 9 Gyr. However, they did find that the peaks of the Winget *et al.* LFs were probably too sharp as a result of the coarseness of the mass steps and the neglect of the main sequence evolutionary timescales.

2. Calculations

I have completed a fresh analysis of the problem, preliminary results of which can be found in Wood (1990b). I began by computing several homogeneous sets of white dwarf evolutionary sequences spanning the mass range 0.4–1.2 $M_\odot$ using the Rochester/Austin white dwarf evolution code (see Wood 1990a, and Lamb and Van Horn 1975). These include helium-envelope (DB) sequences with carbon cores, oxygen cores, and mixed C/O cores with the profile

$$X_O = \begin{cases} 0.8, & 0.0 \leq q \leq 0.5, \\ 0.8 - 2(q - 0.5) & 0.5 < q \leq 0.9, \\ 0. & 0.9 < q \leq 1.0, \end{cases} \qquad (2)$$

where $X_C = X_O - 1$ for all but the outermost helium layer. This C/O core profile is oxygen rich in the center and is patterned after C/O profiles predicted by MS→WD evolutionary calculations (*e.g.*, Mazzitelli and D'Antona 1986, D'Antona and Mazzitelli 1989) using the Caughlan and Fowler (1988) estimate of the $^{12}C(\alpha, \gamma)^{16}O$ reaction rate. The helium layer masses I used in most of the sequence calculations were $\log(M_{He}/M_*) = -3$ and -4, based on the observational/theoretical estimate of Pelletier *et al.* (1987) that $\log(M_{He}/M_*) = -3.5 \pm 0.5$. The ages of the sequences scale linearly with the logarithm of He-layer mass (see Figure 1), and so the LF ages below can be scaled quite simply to estimate the results for thicker He-layer masses. At $\log(L/L_\odot) = -4.2$, the scaling is approximately 0.5 Gyr per decade in $\log(M_{He}/M_*)$, and at $\log(L/L_\odot) = -4.6$, it is closer to 1 Gyr per decade. Note that I computed DB sequences because the large majority of DA stars have surface H layers thin enough $[\log(M_H/M_*) = -10]$ to mix convectively with the subsurface He layer at roughly $T_{eff} \approx 10,000$ K and characteristic evolutionary times of less than 1 Gyr.

The LFs are calculated by integrating

$$\Phi = \int_{M_L}^{M_U} \int_{L_L}^{L_U} \psi(t) \, \phi(M) \, \frac{dt_{cool}}{d\log(L/L_\odot)} \, \frac{dm}{dM} \, dL \, dM, \qquad (3)$$

where M_L and M_U, L_L and L_U are the lower and upper mass and luminosity limits to the integration, respectively. I typically take the upper mass limit to be 8 $M_\odot$— the calculations are relatively insensitive to the specific choice. The lower mass limit is the main-sequence turnoff mass for the input disk age (t_{disk}). The upper

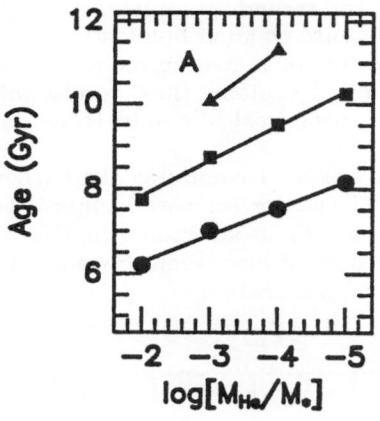 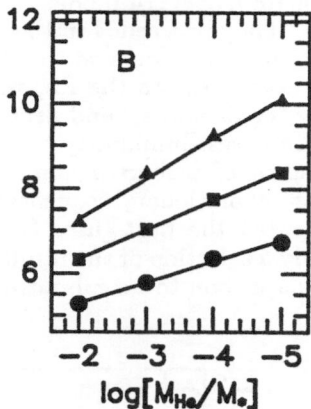

Figure 1. Age *vs.* He-layer mass at three selected luminosities for 0.6 M$_\odot$ C-core (*panel A*) and O-core (*panel B*) sequences. Because He has opacities in general smaller than either C or O, thinner He layers yield older ages at the luminosities $\log(L/L_\odot) = -4.2$ (*circles*), $\log(L/L_\odot) = -4.4$ (*squares*), and $\log(L/L_\odot) = -4.6$ (*triangles*). The lines in the figure result from linear least squares fits to the points.

luminosity limit is $10L_\odot$, and the lower luminosity is determined for each mass by interpolating between sequences for the luminosity corresponding to the age

$$t_{\rm cool}^{\rm max}[M_{\rm WD}(M_{\rm MS})] = t_{\rm disk} - t_{\rm MS}(M_{\rm MS}). \tag{4}$$

The other inputs to the model include the star formation rate as a function of time, [SFR $\equiv \psi(t)$], the initial mass function [IMF $\equiv \phi(M)$], the initial→final mass relation (dm/dM), and of course the mass-dependent WD cooling curves. For these parameters, I define my *standard model* to have a constant SFR, a Salpeter IMF $\phi(M) = (M/M_\odot)^{-2.35}$, a pre-WD lifetime $t_{MS} = 10 \, (M/M_\odot)^{-2.5}$, an initial→final mass relation $M_{\rm WD} = 0.49 \cdot \exp(0.09 \cdot M_{\rm MS})$, and no scale height inflation with time. The sequences I use here all have He-layer masses of $\log(M_{\rm He}/M_\star) = -4$.

3. Results

I present the results over the observed LF of Liebert, Dahn, and Monet (1988). These authors computed their results in two ways, first by applying bolometric corrections from cool DA model atmospheres, and second by applying no bolometric correction. They proceeded in this manner because no good bolometric corrections existed for cool DB stars, and the DA atmosphere bolometric corrections were likely to be overestimates. In the Figures that follow, I combine these results into low-luminosity "error boxes," and define plausible theoretical LFs to be those that pass through the lowest-luminosity box.

In the spirit of Liebert, Dahn, and Monet (1988), I computed the carbon- and oxygen-core evolutionary sequences to bracket the correct core composition, and then computed the best-guess C/O-core sequences in addition. In this work, I present only a selection of the results, both because of space limitations and because the details are soon to be submitted for journal publication.

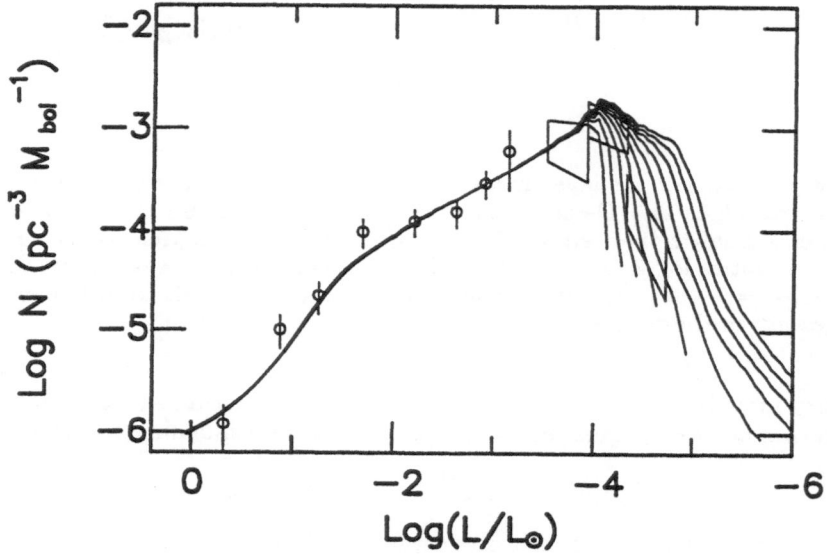

Figure 2. Integrated disk luminosity functions for Carbon-core sequences using the standard model parameters. Luminosity functions with input disk ages of 7–16 Gyr (at intervals of 1 Gyr) are shown over the observed luminosity function of Liebert, Dahn, and Monet (1988). Those which pass through the lowest luminosity box have ages 9-12 Gyr.

Figure 2 shows a series of LFs computed with the C-core model sequences and the standard model parameters. The LFs range in age from 7 to 16 Gyr at intervals of 1 Gyr. Those that pass through the low-luminosity box have ages ranging from 9 to 12 Gyr. Note that the original paper by Winget *et al.* (1987) showed data that had had the bolometric corrections applied, and so corresponded to the bright edge of the low-luminosity box, and that their 9 Gyr LF and ours are therefore in good agreement. The bright ends of the LFs are all nearly identical, as we would expect, and the curves diverge from each other at approximately $\log(L/L_\odot) = -3.7$, and peak shortly after this. The turndown of a given LF is given approximately by the age–luminosity relation for the peak of the mass distribution, and the contributions beyond this are from higher-mass sequences, which evolve more quickly (and whose progenitors also evolved quickly).

Figure 3 shows a series of LFs computed with the O-core model sequences and the standard model parameters. Here the LFs range in age from 6 to 12 Gyr at intervals of 1 Gyr, and those that pass through the low-luminosity box have ages ranging from roughly 7 to 10 Gyr — again, the uncertainty from the observations alone is roughly 3 Gyr. The age range suggested by these models is shifted roughly 2 Gyr lower than that of the C-core results, reflecting both the lower integrated heat capacity of the O-core models, and the fact that the O-core models crystallize at higher luminosities and so their evolutionary timescales are less affected by the release of latent heat of crystallization. The O-core models diverge at a luminosity of $\log(L/L_\odot) = -3.4$, and the luminosity functions are flatter at the peak than the C-core models.

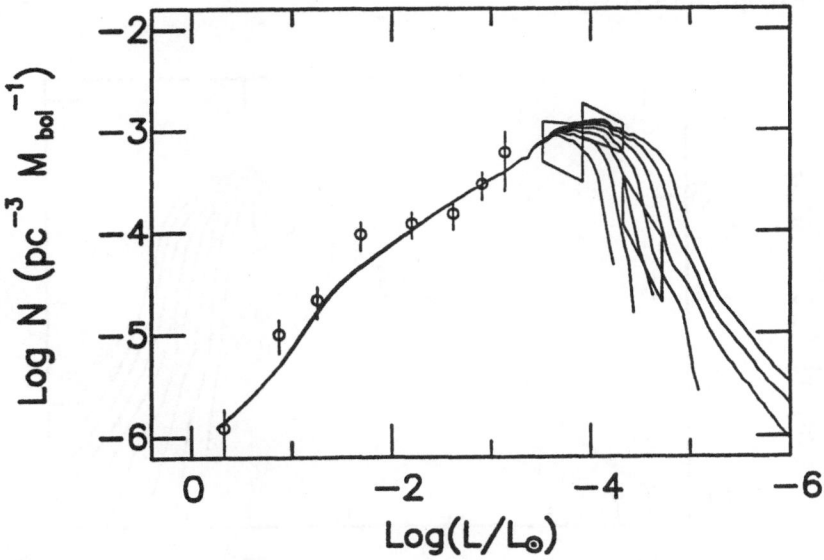

Figure 3. Integrated disk luminosity functions for O-core sequences using the standard model parameters. Luminosity functions with input disk ages of 6–12 Gyr (at intervals of 1 Gyr). Those which pass through the lowest-luminosity box have ages of approximately 7-10 Gyr.

Finally, Figure 4 shows the best-guess model results, computed with the C/O-core model sequences. Here, I have replaced the constant SFR with a more physical SFR based on the Clayton (1988) infall model, with a gas consumption timescale of 2 Gyr and a timescale of infall from the halo of 3 Gyr. This results in $\Psi(t)$ peaking roughly 2 Gyr after $t = 0$, but the rise time to 50% of the peak SFR is less than 0.5 Gyr, and the effect on the derived disk age is correspondingly small. The initial→final mass relation I used is $M_{WD} = 0.40\exp(0.125M_{MS})$, which results in a mean mass close to 0.5 $M_\odot$ compared to the roughly 0.6 $M_\odot$ mean mass of the standard-model initial→final mass relation. I chose this relation because the mass function derived by Bergeron, Saffer, and Liebert (these proceedings) suggests that the mean mass of the youngest ($T_{eff} \gtrsim 15{,}000$ K; $\tau_{WD} \lesssim 0.5$ Gyr) white dwarfs is $\langle M/M_\odot \rangle \approx 0.53$. Finally, I included modest scale height inflation at about half the Twarog (1980) estimate — *i.e.*, here I use a factor of 3 inflation over 12 Gyr. The LFs shown in Figure 4 have input ages ranging from 7 to 14 Gyr, and those that pass through the low-luminosity box suggest that the range which best estimates the disk age is from approximately 8 to 11 Gyr. Remembering that the work of Pelletier *et al.* (1987) suggests that there is a roughly one decade uncertainty in the He-layer mass, the total range for the best estimate to the local disk age is approximately 7.5 to 11 Gyr.

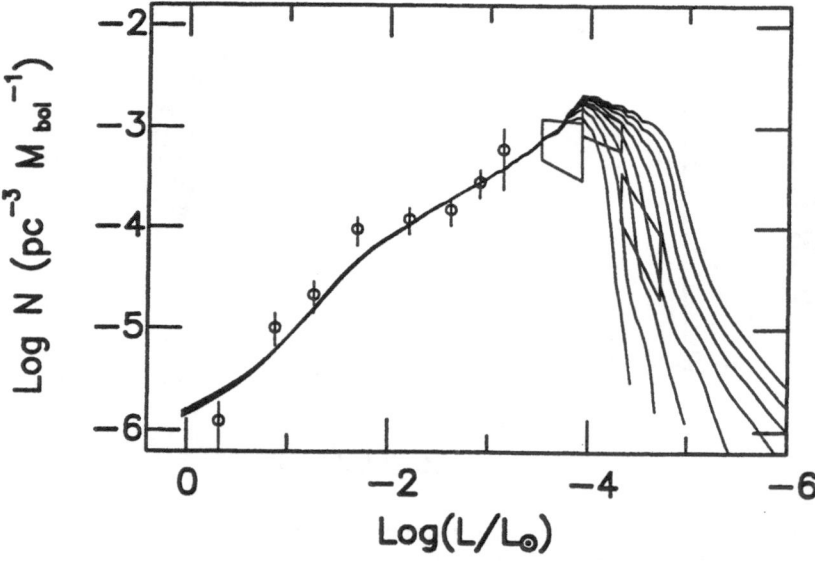

Figure 4. Luminosity functions using the C/O-core sequences and best-guess parameter set. These curves combined with the scaling for the uncertainty in the He-layer mass suggest that our current best estimate for the age of the local disk lies within the approximate range 7.5 to less than 11 Gyr.

4. The Emerging Picture of Early Galactic Evolution

The results from the white dwarf luminosity function are particularly exciting because they put constraints on the models of the evolution of the early Galaxy. Specifically, these models fall broadly into two classes, the rapid-collapse models (*e.g.*, Eggen, Lynden-Bell, and Sandage 1962), and the pressure-supported collapse (or thick disk) models (Larson 1976; Gilmore, Wyse, and Kuijken 1989). The former suggests a collapse of the proto-Galaxy on a timescale of a few 10^8 yr, and the onset of global star formation on a comparable timescale. The latter suggests that the collapse occurred over a much longer timescale, that star formation began first in the bulge and halo of the Galaxy, and that it spread out through the spiral arms on a timescale of a few billion years. Through the comparison of the white dwarf cosmochronological results with the results of other, independent methods, a unified picture begins to emerge, and the pressure-supported collapse models are clearly preferred.

Estimates of the Galactic age have traditionally been determined from isochrone fitting of the halo globular clusters, where the isochrones were calculated using the evolutionary timescales suggested by standard main-sequence evolutionary calculations. These idealized calculations included in general no rotation, diffusion, or magnetic fields, and treated convection using the known-to-be-flawed mixing-length theory. Using these model results, the globular cluster ages fall into the approximate range 13 to 18 Gyr — *i.e.*, up to a factor of two larger than the white dwarf LF results. Recent refinements to the main-sequence evolutionary calculations, however, suggest that the original timescales are too long by several billion years. First, Proffit and Michaud (1991) have recently investigated the effects on evolutionary timescales of gravitational settling of He and Li, and found that the settling of He *decreases* the turnoff ages of low-mass, metal-poor stars by about 20%. Second, Canuto and Mazzitelli (1991) have developed a new theory of convection which surpasses the standard mixing length theory by implicitly taking into account convective eddies of all sizes, instead of the unphysical single eddie size of the MLT. This new theory of convection is more efficient than the standard MLT, and when included in stellar evolution calculations requires for the observational turnoff a larger mass and thus a younger age — again suggesting that the original globular cluster age estimates are too large. Thus, it appears likely that the pressure-supported collapse models of Galactic formation are essentially correct, and that star formation began in the bulge and halo some ~10-12 Gyr ago, and that the onset of star formation at our galactocentric radius occurred some 7.5-11 Gyr ago.

In the next few years, the picture of the evolution of the early Galaxy should come into increasingly sharp focus. Improvements in the bolometric corrections applicable to cool DB white dwarfs and improved distance determinations will greatly reduce the uncertainties in the cool end of the observed luminosity function, which now account for fully one half of the total uncertainty in the white dwarf disk age determination. Given these improvements, the white dwarf stars should soon provide the most reliable estimates of the age of the local disk of the Galaxy, and in doing so will provide tight constraints for any model of the age and evolution of spiral galaxies.

This material is based upon work supported in part by the North Alatlantic Treat Organization under a Grant awarded in 1989, and in part by the NASA Graduate Student Researchers Program.

References

Canuto, V. M., and Mazzitelli, I., March 20, 1991, *Ap. J.*, in press.

Caughlan, G. R., and Fowler, W. A. 1988, *Atomic Data and Nuclear Data Tables*, **40**, 334.

Clayton, D. D. 1988, *M.N.R.A.S.*, **234**, 1.

D'Antona, F., and Mazzitelli, I. 1989, *Ap. J.*, **347**, 934.

Eggen, O. J., Lynden-Bell, D., and Sandage, A. R. 1962, *Ap. J.*, **163**, 748.

Gilmore, G., Wyse, R. F. G., and Kuijken, K. 1989, *Ann. Rev. Astron. Astrophys.*, **27**, 555.

Iben, I., Jr., and Laughlin, G. 1989, *Ap. J.*, **341**, 312.

Lamb, D. Q., and Van Horn, H. M. 1975, *Ap. J.*, **200**, 306.

Larson 1976, *M.N.R.A.S.*, **176**, 31.

Liebert, J., 1979, in *IAU Coll. #53: White Dwarfs and Variable Degenerate Stars*, ed. H. M. Van Horn and V. Weidemann (Rochester: University of Rochester Press), p. 146.

Liebert, J. 1980, *Ann. Rev. Astron. Astrophys.*, **18**, 363.

Liebert, J., Dahn, C. C., and Monet, D. G. 1988, *Ap. J.*, **332**, 891.

Mazzitelli and D'Antona 1986, *Ap. J.*, **308**, 706.

Noh, H.-R., and Scalo, J. 1990, *Ap. J.*, **352**, 605.

Pelletier, C., Fontaine, G., Wesemael, F., Michaud, G., and Wegner, G. 1986, *Ap. J.*, **307**, 242.

Proffit, C., and Michaud, G., 1991, *Ap. J.*, in press.

Twarog, B. A. 1980, *Ap. J.*, **242**, 242.

Weidemann, V., and Koester, D. 1984, *Astron. Astrophys.*, **132**, 195.

Winget, D. E., Hansen, C. J., Liebert, J., Van Horn, H. M., Fontaine, G., Nather, R. E., Kepler, S. O., and Lamb, D. Q. 1987, *Ap. J. (Letters)*, **315**, L77.

Wood, M. A. 1990a, Ph.D. thesis, The University of Texas at Austin (photocopies available by request from author).

Wood, M. A. 1990b, *J. Roy. Astron. Soc. Can.*, **84**, 150.

Wyse, R. F. G., and Gilmore, G. 1988, *Astron. J.*, **95**, 1404.

Yuan, J. W. 1989, *Astron. Astrophys.*, **224**, 108.

DISCUSSION

VAN HORN :

Last week I learned about a new calculation of galactic "chemodynamics" by Burkert and Truran. They have included detailed models for the heating and cooling of the infalling, pre-galactic gas, and they find that the model first halts with a scale height $\sim$ 1kpc, where it "cooks" for $\sim$ 5-6 Gyr. Then it collapses to an "old thin disk" with a scale height $\sim$ 250 pc. This and Larson's calculations suggest that may be both the globular cluster ages **and** the white dwarf cooling ages may be right.

I'd also like to follow up Prof. Schatzman's question to you this morning. We've seen lots of indications about the importance of mixing, diffusion, and accretion in modifying the surface abundances of the white dwarfs. That will affect the opacities and the cooling ages. Have you done any model calculations with H surface layers or with different heavy element abundances that can illustrate the magnitude of this effect?

WOOD :

I had not heard of the Burkert and Truran calculations, but am encouraged that the apparent age differences between the globular clusters and the coolest white dwarfs are perhaps understandable within the context of increasingly sophisticated theoretical treatments. As for the sensitivity of the white dwarf ages to the surface abundances of the models, I do not find a strong dependence in the numerical experiments I have conducted to date. I believe that this is because the surface layers are convective, and so the temperature gradient is adiabatic; however, I am still looking into this question.

EARLY RESULTS FROM THE ROSAT WIDE FIELD CAMERA

M.A.BARSTOW, A.F.ABBEY, R.E.COLE, M.DENBY, C.G PAGE,
G.S.PANKIEWICZ*, K.A.POUNDS, J.P.PYE, A.E.SANSOM, M.R.SIMS,
J.E.SPRAGG, D.J.WATSON, A.A.WELLS and R.WILLINGALE
Physics and Astronomy Department
University of Leicester
University Road
Leicester, LE1 7RH, UK

G.M.COURTIER, J.A.GOURLAY, A.W.HARRIS*, B.J.KENT,
D.H.READING, A.G.RICHARDS, B.M.SWINYARD and J.S.WRIGHT
Rutherford Appleton Laboratory.

C.V.GOODALL
School of Physics and Space Science, University of Birmingham.

R.D.BENTLEY, E.R.BREEVELD, P.R.GUTTRIDGE, H.E.HUCKLE,
and A.J.McCALDEN
Mullard Space Science Laboratory, University College, London.

A.BEWICK, G.K.ROCHESTER and T.J.SUMNER
Imperial College of Science, Technology and Medicine, London.

* *Currently at the WFC Quick Look Facility, MPE, Garching.*

ABSTRACT. We present early results from observations made by the UK's imaging EUV telescope, the Wide Field Camera (WFC). This telescope is being flown aboard the West German satellite ROSAT. In conjunction with a large X-ray telescope, built by the West Germans, the WFC is conducting a survey of the whole sky. This is the first in the EUV waveband. Afterwards ROSAT will be devoted to pointed observations of specific objects. We use results from the earliest phases of the ROSAT mission to highlight the importance of WFC observations for the study of white dwarfs and related objects such as the central stars of planetary nebulae.

1. Introduction

On June 1 1990 the ROSAT mission was launched from Kennedy Space Centre by a Delta II rocket. ROSAT is an international mission led by West Germany with US and UK partners. The spacecraft carries two coaligned imaging telescopes. An X-ray telescope (XRT), developed by the Max Planck Institut Fur Extraterrestriche Physik, covers the energy range 0.1-2keV and has a 2° field of view. The second telescope, which has a 5° field of view, is an extreme ultraviolet instrument spanning a total energy range 20-200eV and was constructed by a consortium of UK space research groups. The primary aim of ROSAT is to perform an imaging survey of the whole sky in both EUV and soft X-ray bands to search for point and extended sources of emission. After this phase of the mission, which lasts for six months, ROSAT will be used for pointed observations of

99

G. Vauclair and E. Sion (eds.), White Dwarfs, 99–108.
© 1991 *Kluwer Academic Publishers.*

specific objects.

In this paper we will review the initial phases of ROSAT operations. Particular attention will be paid to the results of observations of white dwarfs and the related objects such as the nuclei of planetary nebulae (PNN) but other work of more general interest will also be covered. Since the XRT is the responsibility of the MPE, it is not appropriate for us to report here on observations made with it and this paper will concentrate mainly on results from the EUV telescope, the wide field camera (WFC). However, for reference, we include a brief description of the XRT and briefly allude to its scientific potential with regard to white dwarfs.

2. The ROSAT telescopes

2.1 THE X-RAY TELESCOPE

The ROSAT XRT has been described several times before, most recently by Briel et al (1990). The telescope is a 4-fold nest of Wolter I grazing incidence mirrors with an aperture of 80cm and a 2° field of view. These mirrors have the smoothest optical surface ever produced resulting in a focussed image having a very small scattered light halo. Three detectors are mounted on a carousel in the focal plane. Two position sensitive proportional counters (PSPCs), a redundant pair, are available for both the survey and pointed phases of the mission. An additional detector utilising a pair of microchannel plates is also available for pointed observations. The high resolution imager (HRI, Zombeck et al, 1990) has higher spatial resolution than the PSPC (1.7 cf. 25arcsec at 1keV) but smaller field of view (38arcmin) and limited energy resolution. Essentially, the instrument is a twin of the Einstein HRI but achieves a higher quantum efficiency by the use of a CsI photocathode.

The energy resolution of the PSPC is dependent upon the energy of the incident photon spectrum, ranging from 130% FWHM at 0.1keV to 30% at 2keV. At low energies, where the resolution is poorest, additional spectral information can be obtained with the use of a Boron filter. This divides the lowest energy channel into two. The HRI has a high intrinsic efficiency at energies well below the design range of the XRT extending into the EUV and far-UV bands. Hence, a fixed UV blocking filter is mounted in the HRI assembly to define the low energy cutoff of the instrument ($\approx$0.1 keV).

2.2 THE WIDE FIELD CAMERA

The WFC (figure 1) is also a grazing incidence telescope of Wolter I design but incorporates the Schwarzchild modification to the optical figure to reduce off-axis aberrations. A nest of three mirrors yields a total collecting area of 456cm^2 from a maximum aperture of 57.6cm. The nest has a focal length of 52.5cm and a field of view of 5°. The design and performance of the mirrors has been discussed in detail by Willingale (1988). A redundant pair of microchannel plate detectors (Barstow and Sansom, 1990) are mounted in the focal plane and can be deployed into the telescope focus by a turret mechanism. A unique feature of these detectors is that the microchannel plates in each detector are curved, like a watch glass, to match the optimum focal surface of the telescope, ensuring that the best resolution possible is achieved across the whole field of view. This is of particular importance during the survey where individual sources are scanned across the entire field of view. As the edges of the field make the largest contribution to the overall exposure during the survey, improved off-axis spatial resolution will increase the signal-background ratio for any source and, as a result, lower the detection limit for a given exposure time.

Figure 1. The ROSAT Wide Field Camera

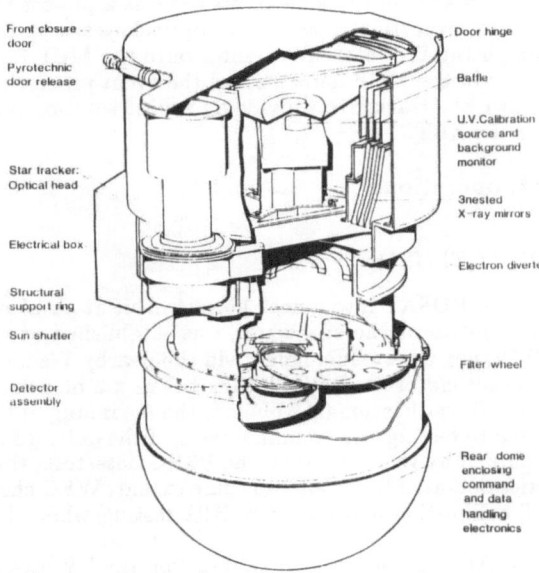

In the WFC spectral information is obtained by inserting filters into the field of view, restricting the detected flux to a narrow energy range. This is the analogue of optical UBVRI photometry. In addition, the filters must perform the functions of rejecting geocoronal background emission at energies outside the filter bandpasses and suppression of the UV flux from bright stars which could give rise to ambiguous source detections. Filters covering 4 bandpasses are available in the WFC (Kent et al., 1990) and the details of these are summarised in Table 1. Only the two highest energy bands, designated S1 and S2, are used during the survey and these filters cover the entire field of view. Two additional filters (P1 and P2), which occupy approximately half the field of view, can be used for pointed phase observations.

Table 1. ROSAT WFC Science Filters

Filter type	Materials	Survey(S)/Pointed(P)	Bandpass (eV)
S1 (a)	$B_4C/C/Lexan$	S + P	90-190
S1 (b)	C/Lexan	S + P	90-210
S2 (a)	Be/Lexan	S + P	62-111
S2 (b)	Be/Lexan	S + P	62-111
P1	Al/Lexan	P	56-83
P2	Sn/Al	P	17-24

In addition to the major telescope components, the WFC has several other important subsystems on which the operation of the instrument depends. The protection and calibration system (PCS) has two functions. Two detectors (a Geiger tube and channel electron multiplier, CEM) mounted forward of the telescope baffles, looking directly into space, monitor the background arising from high energy particles and soft electrons. When the observed count rates in either of these detectors exceeds a pre-set threshold, the MCP detector is switched off as a protection measure. An optical system in the PCS can project a pattern of spots through the UV interference filter onto the MCP detector. This serves to monitor the calibration of the MCP detector and the event processing electronic chain. The WFC also has its own startracker to provide an aspect solution independently of the ROSAT spacecraft and the XRT.

3. The early in-orbit operations

3.1 PROGRAMME OF ACTIVITIES

The Delta II launch placed ROSAT into a near perfect orbit at an altitude of 575km and an inclination of 53°. Control of the spacecraft was established at the German Space Operations Centre (GSOC) at Oberpfaffenhofen, via the nearby Weilheim ground station, near the beginning of its second orbit. ROSAT then began a 2 month period of checkout, calibration and performance verification (PV) before the beginning of the sky survey. The first 19 days were devoted to testing and commissioning of the onboard systems, beginning with the spacecraft, followed by the XRT with the PSPC detectors, the WFC and finally the HRI. These activities operated in parallel to some extent, WFC checkout taking place during the initial XRT calibration and similarly HRI testing while the WFC was being calibrated.

After completion of the main instrument calibration, the PV programme began with a 5 day long 'mini-survey' of the sky as a test of the spacecraft operations for the main survey. Since, the orbit of ROSAT precesses by about 1° per day a swath of sky 10 degrees across was covered during the 5 days. Some results from the analysis of these data will be presented in this paper. Further PV observations up to the beginning of the main survey were all in pointed mode and consisted of parts of the approved AO-1 programme. Some indication of the potential scientific return from the ROSAT WFC will be illustrated with reference to some early analysis of these data.

3.2 WFC CHECKOUT

The technical aspects of the early checkout and calibration of the WFC have been covered in some detail by Wells et al. (1990). As the aim of this paper is to highlight the early scientific return we will only note the most important details here. All the main WFC subsystems were checked out between June 6, when the telescope door was opened, and June 16 with the exception of the science filters. No problems were found with any of the units. Both detectors operated well and one of these (detector 2) was deployed into the field of view. Initial observations on 15 and 16 June showed an anomalously high background in the UV calibration and so called 'opaque' filters, indicating an unexpected UV or EUV light leak. However, on June 17, when the first survey filter (S1a) was deployed the background count rates were close to prelaunch predictions. With the WFC and XRT pointing at the cluster of galaxies Abell 2256 a new EUV source was discovered during a brief 300s exposure in the S1a filter. An observation of the source with the S2a filter is shown in figure 2. Palomar and HST guide star fields reveal a single candidate for the source with a V magnitude of ≈ 13.0. Subsequently, the object has been identified as a previously uncatalogued DA white dwarf with a dM star companion, rather like Feige 24. A detailed paper concerning this object is in preparation.

Figure 2. Image of the first new EUV source detected by the WFC.

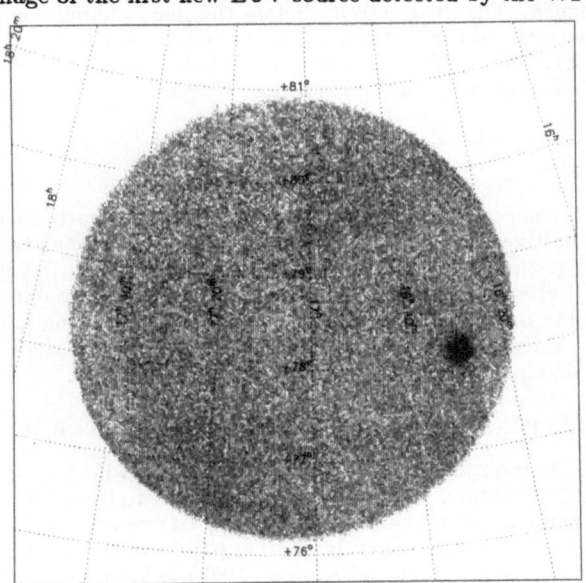

Figure 3. WFC telescope point spread function constructed from an observation of the DA white dwarf HZ43. The X-axis scale is in pixels (1 pixel ≈0.7 arcmin.).

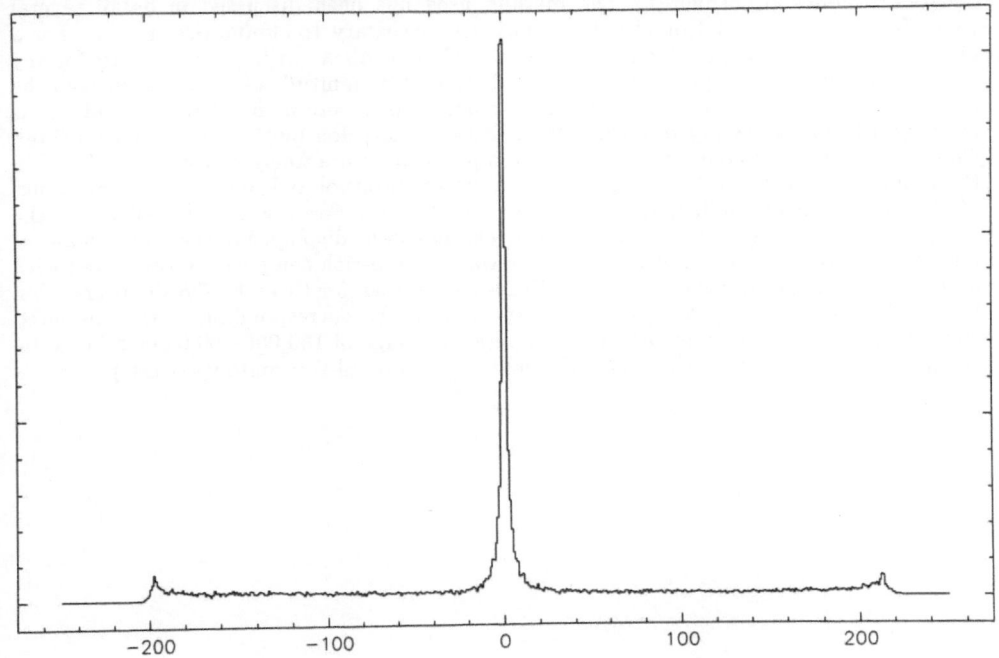

All the survey and pointed observation filters were tested during this period and were found to be undamaged. The final activity before the main calibration programme commenced was a measurement of the instrument focus made by observing the currently brightest known EUV source, the hot DA white dwarf HZ43. The resulting telescope point spread function, illustrated in figure 3, has a FWHM of 70arcsec showing the instrument to be in focus.

4. Early scientific results

At the time of writing this paper (October 1990), it is still a little early in the mission to be able to include definitive scientific results. For example, uncertainties still exist regarding the instrument performance, because calibration data is still being analysed, and the final results could alter conclusions drawn now. Also, not all the data pertaining to an individual observation may yet be available in processed form. The aim here is to use a few examples from the early phases of the mission to illustrate the scope of the science that the WFC is doing and the likely impact of the final results.

4.1 PV PHASE OBSERVATIONS OF THE HOT DEGENERATE STAR H1504+65

The star H1504+65 is the hottest known white dwarf. It seems to be related to the hot helium-rich subgroup of PG1159 stars. These objects are thought to lie on an evolutionary path linking the OVI central stars of planetary nebulae and DO white dwarfs. H1504+65 and the PG1159 objects are discussed extensively in other papers in this volume (Barstow, Tweedy and Werner, 1991; Werner, Heber and Hunger, 1991). This star has already been observed by EXOSAT (Nousek et al, 1986; Barstow and Tweedy, 1989) and these data have been used to constrain its temperature and the composition of its atmosphere by comparing the observed count rates in several filters with the predictions of computer generated model atmospheres. The method used has been discussed in detail several times (eg. Barstow and Holberg, 1990) but it is necessary to summarise it here. For a given temperature it is possible to calculate a predicted soft X-ray/EUV count rate for any star by folding a model spectrum, normalised to the V magnitude of the star, through the instrument response of each band in which observations were made. Unless an object is close by it is also necessary to include the effect of absorption by the interstellar medium. Count rates are predicted for each model temperature and a finely spaced grid of N_H in the range $10^{18} - 10^{22}$ cm^{-2}. The resulting tables are interpolated to determine the values of T and N_H corresponding to the observed count rates. For any model (defined by the composition), a family of constant count rate curves can be displayed in the T/N_H plane as depicted in figure 4. The model chosen here was a metal-rich composition computed with a non-LTE code (Barstow et al, 1990). Curves are shown for three EXOSAT filters (thin lexan - 3Lx, aluminium/parylene - Al/P and boron - B) corresponding to the measured count rates and statistical errors. The temperature range of 150,000-190,000K allowed by the model and the data is the region of the figure where all the contours overlap.

Figure 4. Curves of constant count rate, calculated for a metal-rich model atmosphere spectrum, corresponding to the range of observed H1504+65 fluxes in three EXOSAT bands (3Lx, AlP and B) and in the S1a filter of the WFC.

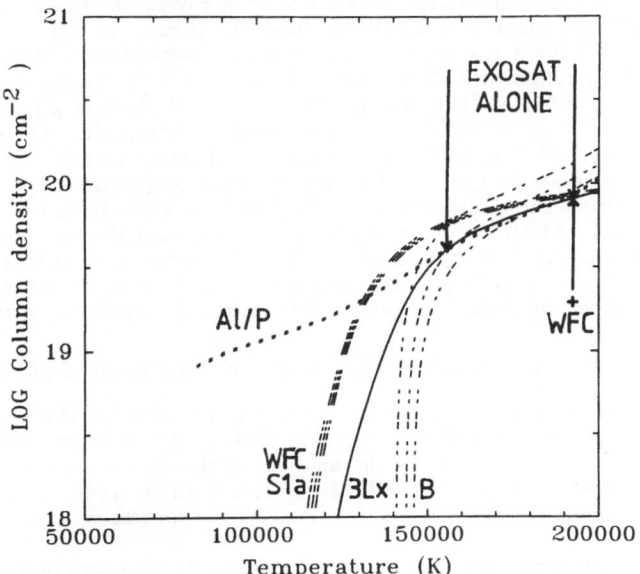

H1504+65 was observed by the WFC as part of the PV programme. All the different filter types (S1, S2, P1 and P2) were used but only the S1a data have been processed. A set of constant count rate curves were generated for the S1a filter as described above and have been included in figure 4. The spread of the S1a curves is defined not by the statistical errors but more realistically by an estimate of the current uncertainty in the absolute calibration of the WFC. Adding this single additional data set (note, there will be others when the data have been processed) has a dramatic effect, restricting the temperature of the star to a narrow range ($\approx 2,000$K) around 190,000K. The main reason for the improvement in the temperature constraint is that the WFC filters have much narrower (ie. spectrally pure) passbands than those of EXOSAT.

4.2 OBSERVATIONS OF V471 TAURI DURING THE MAIN SURVEY

The consortium of WFC institutions maintains a Quick Look Facility (QLF) at the MPE in Garching. During each daily cycle of Weilheim ground contacts, data from three orbits of WFC observations are processed immediately by GSOC and copied to the QLF over the computer network. From these data we can monitor the health of the instrument and deal rapidly with any instrument anomalies. In addition the data can be used for science analysis. During the survey the three orbits of data amount to 1/5 of the full sky exposure.

Any individual source in the region of sky currently being surveyed by the WFC is observed once every orbit. The exposure time depends upon the off-axis position of the source as it passes through the telescope field of view. This is a maximum of 80s for an object travelling through the centre of the field, decreasing to zero at the very edge. The precession of the satellite's orbit drifts the scan direction across the celestial sphere at a rate $\approx 1°$ per day. The ecliptic poles are mapped all the time and a low latitude source, such as V471 Tauri, will be scanned for about 5 days.

V471 Tauri lies in the Hyades cluster and is an eclipsing binary system comprising a dwarf K star and a DA white dwarf. It is a known X-ray source having been observed extensively by EXOSAT (Jensen et al, 1986). Both the X-ray and optical fluxes exhibit pulsations, and dips in the X-ray light curve have been observed outside the white dwarf eclipse. The pulsation periods of 555s are not well matched to the survey observing times but the white dwarf eclipses are 1 hour long. Figures 5a and 5b show a pair of exposures in the region of V471 Tau made with the S2a filter on successive orbits. These images were drawn from the QLF 3-orbit data. Fortuitously, the first image coincides with an eclipse of the white dwarf while in the second the white dwarf has reappeared from behind the K star. These observations show that almost all the EUV radiation from the system comes from the white dwarf. At optical wavelengths the K star contributes 75% of the total luminosity. Consequently, longer term study of the EUV emission from this system (as is planned for AO-1) will be very important since it is possible to observe the behaviour of the white dwarf in isolation from the K star unlike optical or soft X-ray studies.

4.3 THE MINI-SURVEY AND QLF DETECTIONS IN THE MAIN SURVEY

The mini-survey occupied a period of 5 days during the PV phase of the mission. A strip of sky some 10° wide was scanned but only the central degree or two was exposed to the full depth. A detailed discussion of the analysis of the mini-survey results has been presented by Pounds et al (1991). We will briefly review these results here. In addition we will also include some results from the 'S1' survey, the sources detected in the QLF's 3-orbit data.

Images created for each filter from the mini-survey data were automatically searched for point sources. The resulting list was then cross-correlated with several stellar and galaxy catalogues to determine if an obvious candidate for the source of the EUV emission exists within the errors of the position determination. The S1 point source analysis at the QLF is a manual task and sources are picked out 'by eye'. Consequently, the sensitivity of this survey is not necessarily uniform and the low exposure means that only the brightest sources are detected.

The analysis of the mini-survey data has at this stage resulted in 33 reliable point source detections. An additional 64 sources have been found by the QLF from 45% of the sky up to 20th October 1990. Zombeck's handbook of 'Astronomy and Space Science' (1990) lists 7 previously identified EUV sources, a number that we have now increased by an order of magnitude. The distribution of these sources on the sky is shown in figure 6. The earlier detections are noted, only the brightest of these, the DA white dwarfs HZ43, G191-B2B and Sirius B, occurring in both samples at this stage. All of the identified sources are either late-type stars or white dwarf related objects.

About a third of the source detections in the mini-survey and 25% of those in the S1 list have no obvious identifications. In these cases discovery of the emitting object requires a more detailed optical search. First, the Palomar or ESO/UK Schmidt plates will be scanned, within the detection position errors, for candidate objects. Secondly, these objects will be observed spectroscopically at the telescope to determine their nature and ascertain the likelihood that they are responsible for the EUV emission. Prior to the launch of ROSAT, several estimates were made of the likely rates of detection of objects of particular types (eg. Barstow, 1989 and Pye and McHardy, 1988). These calculations suggest that most of the EUV sources discovered will be late type stars and white dwarfs, in roughly equal quantities, with a handful of other objects such as AGN or CVs. Consequently, on a percentage basis, about 50% of the unidentified sources should

Figure 5. Images of V471 Tau from the sky-survey, recorded on successive orbits with a) the white dwarf in eclipse and b) the white dwarf visible.

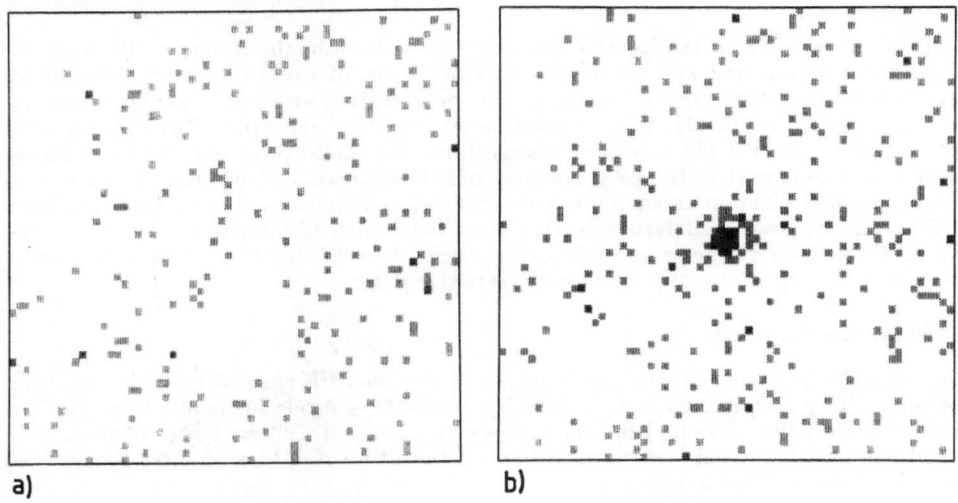

a)   b)

Figure 6. An all-sky plot of the locations of EUV sources discovered during the mini-survey and in the main survey '3-orbit' data.

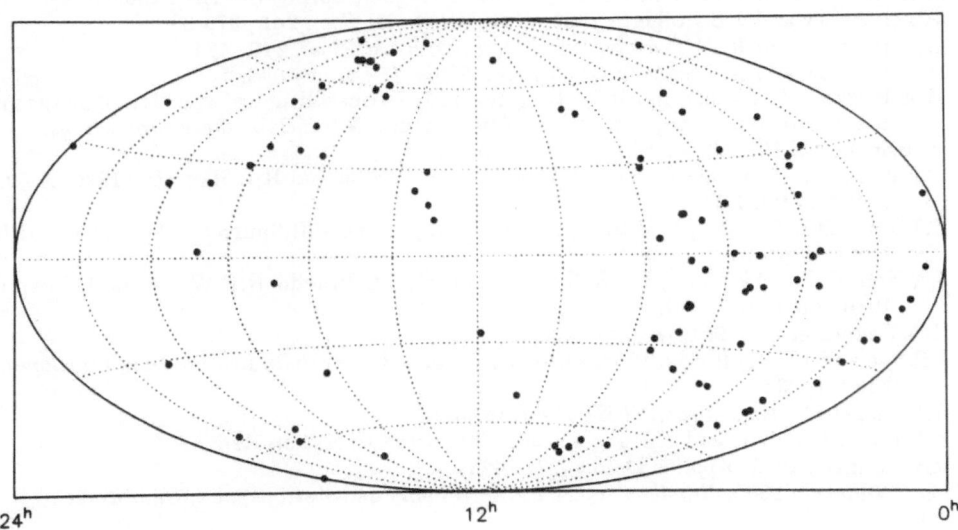

24^h 12^h 0^h

turn out to be previously undiscovered hot white dwarfs.

5. Conclusion

Early observations with the ROSAT telescopes show considerable promise. Both instruments are working well and the initial mission phases of calibration and performance verification have been very successful. In this presentation we have concentrated on observations made by the UK Wide Field Camera, an EUV telescope. Drawing on some examples from the PV phase pointed observations, the mini-survey and the main survey we have been able to show the great potential of EUV observations, particularly for studies of white dwarfs. The main survey is now over 25% complete and a considerable body of data already awaits analysis. Furthermore, the health of both instruments and the spacecraft is stable giving us an optimistic outlook of continuing observations through to the end of the sky survey and for several years thereafter.

6. Acknowledgments

The Wide Field Camera has been built by a consortium of UK space research groups from Leicester University, Birmingham University, Rutherford Appleton Laboratory, Mullard Space Science Laboratory and Imperial College of Science Technology and Medicine. We would like to acknowledge the contributions of the many members of these groups who have also been involved in this project. The success of the WFC has also been dependent upon the efforts of the West German teams, for building ROSAT and operating the ground system, and the USAF and NASA, for a perfect Delta II launch. We are grateful to all those involved.

7. References

M.A.Barstow, 1989, in 'White Dwarfs', ed. G.Wegner, Springer-Verlag, 156.

M.A.Barstow and J.B.Holberg, 1990, *Mon.Not.R.astr.Soc.*, **245**, 370-383.

M.A.Barstow and R.W.Tweedy, 1990, *Mon.Not.R.astr.Soc.*, **242**, 484.

M.A.Barstow and A.E.Sansom, 1990,*Proc. SPIE*, **1344**, in press.

M.A.Barstow, K.Werner and R.W.Tweedy, 1991, **Proceedings of the 7th European Regional Workshop on White Dwarfs**, ed. G.Vauclair, *these proceedings*.

U. Briel et al., 1990 *Proc. SPIE*, **1344** in press.

K.A.Jensen, J.H.Swank, R.Petre, E.F.Guinan, E.M.Sion and H.L.Shipman, 1986, *Ap.J. Letters*, **309**, L27.

B.J.Kent, D.H.Reading, B.M.Swinyard, E.B.Graper and P.H.Spurrett, 1990, *Proc. SPIE*, **1344**, in press.

J.A.Nousek, H.L.Shipman, J.B.Holberg, J.Liebert, S.H.Pravdo, N.E.White and P.Giommi, 1986, *Ap.J*, **309**, 230.

K.A.Pounds et al., 1991, in preparation.

J.P.Pye and I.M.McHardy, 1988, *O.Havnes et al. eds., Activity in Cool Star Envelopes*, Kluwer, 231.

A.Wells, et al., 1990, *Proc. SPIE*, **1344**, in press.

K.Werner, U.Heber, and K.Hunger, 1991, submitted to *Astron.Astr.*

R.Willingale, 1988, *Applied Optics*, **27**, 1423.

M.V.Zombeck, 1990, *Handbook of Space Astronomy and Astrophysics*, Cambridge University Press.

M.V.Zombeck, F.R.Harnden and A.Roy, 1990, *Proc. SPIE*, **1344**, in press.

THE STELLAR COMPONENT OF THE HAMBURG SCHMIDT SURVEY

U. HEBER*, S. JORDAN*, V. WEIDEMANN
Institut für Theoretische Physik und
Sternwarte der Universität Kiel
Olshausenstr. 40, D-2300 Kiel 1
Federal Republic of Germany

ABSTRACT. We report on follow-up spectroscopy of stellar candidates from the Hamburg Schmidt survey, an objective prism survey which primarily aims at quasars in the northern sky. In a pilot project 59 stellar candidates with magnitudes ranging from $13^m 5$ to $18^m 5$ have been observed at a spectral resolution of 4Å. 54 of of these were indeed stars of either of four spectral classes: white dwarfs, subluminous B- or O stars, horizontal branch A stars or sdF stars. Only two objects turned out to be non-stellar but bright low-redshift quasars. A spectroscopic binary star, an UX UMa star and an unclassified star with an almost featureless spectrum were also discovered.

1. Introduction

The success of the Palomar Green survey (Green *et al.*, 1986) has led to a "boom" for UV excess object surveys. Therefore many colour selected or objective prism surveys have been initiated. One such objective prism survey is the Hamburg Schmidt (HS) survey carried out with the 80cm Calar Alto Schmidt telescope (see Engels *et al.*, 1988). It is primarily a survey for quasars in the northern sky. The specifications of the Hamburg Schmidt survey, listed in Table 1, are very similar to that of the Case Low Dispersion survey (Pesch and Sanduleak, 1983) which is carried out with the Burrell Schmidt telescope at Kitt Peak. The most important difference between the two surveys concerns the candidate selection procedure. While in the Case Low Dispersion survey the selection is done by visual inspection, for the HS survey the plates are digitized in Hamburg with a PDS 1010G microdensitometer. About 30-50,000 spectra in the magnitude range $13^m 5$ to $18^m 5$ are found on each plates. From such samples, lists containing quasar candidates as well as faint blue star candidates are extracted by means of an automated search software on the base of blue continua and/or emission lines (cf. Hagen, 1987, Hagen *et al.*, 1987). At present the discrimination between quasar and stellar candidates is made by our colleagues in Bergedorf by visual inspection of the Schmidt spectra. In the near future this step will also be automized with the aid of our follow-up observations which will help to improve selection criteria.

Since complete samples of white dwarf and subluminous stars can be drawn from the plates to a certain magnitude limit, a collaboration between the Hamburg and Kiel institutes has been established with the following objectives (see also Jordan *et al.*, this

---

*Visiting Astronomer, German-Spanish Astronomical Center, Calar Alto, operated by the Max-Planck- Institut für Astronomie Heidelberg jointly with the Spanish National Commission for Astronomy

G. Vauclair and E. Sion (eds.), White Dwarfs, 109–119.
© 1991 *Kluwer Academic Publishers.*

conference):

(i) to determine the space densities and scale heights of white dwarfs and hot subdwarfs.

(ii) to enlarge the number of objects in those subclasses where up to now only very few members are known.

(iii) to search for new "exotic" classes of stars such as the PG1159 stars found by the Palomar-Green survey.

(iv) to check the completeness of the quasar sample. The stellar list might be contaminated by some quasars which escaped the visual inspection of the plates (and vice versa).

Table 1: Specification of the Hamburg Schmidt survey in comparison with the Case Low Dispersion survey

	Case Low Dispersion	Hamburg Schmidt
dispersion ($Å\,mm^{-1}$)	1500	1390
size of field	5° * 5°	5.5° * 5.5°
wavelength coverage	3300-5350Å	3300-5300Å
photographic plates	Kodak IIIa-J	Kodak IIIa-J
limiting B magnitude	18	18.5
candidate selection	visual inspection	computer search

The results of the PG survey, which is limited to objects brighter than $B = 16\overset{m}{.} 2$, can be used to predict cumulative surface densities for objects of interest in the HS survey ($B = 18\overset{m}{.} 5$) since both surveys are carried out at high galactic latitudes. In the PG survey the dominating class of objects are the hot subdwarfs (sdB, sdO) comprising 53% of the 1715 objects in the catalogue and outweighing the white dwarfs (26%) and the quasars (5.4%). Extrapolating from Fig. 2 of Green *et al.* (1986) a completely different picture emerges for a limiting magnitude of $18\overset{m}{.} 5$: Now quasars and white dwarfs should be equally frequent (about 1 per square degree) and should outweigh the hot subdwarfs (0.3 per square degree). Moreover, the HS survey will also detect objects redwards of the PG colour cut-off (U-B=-0.4). Hence the ratio of white dwarfs to hot subdwarfs in the HS survey should be even larger than predicted above. For the same reason, we expect to find additional classes of stars in the HS survey. In particular, the horizontal branch A (HBA) stars and the old main sequence halo stars (sdF/sdG) will be encountered.

First results from the deepest HS plates indicate that the above estimates are roughly correct (Engels, priv. com.). Hence, as far as the stars are concerned, the HS survey is primarily a survey for white dwarfs.

2. Follow-up spectroscopy

In a pilot project follow-up optical spectroscopy of stellar candidates has been carried out in January 1989 and 1990 at the DSAZ (Calar Alto, Spain) using the 3.5m telescope. A Boller and Chivens (B&C) spectrograph was used in 1989 equipped with a RCA CCD detector. In 1990 we used the Cassegrain Twin Spectrograph (CTS), which includes two separate spectroscopic channels behind the common slit aperture. The spectral range from 3850Å to 5550Å (and additionally 5600Å to 7000Å with the CTS) was recorded at a spectral resolution of about 4Å (B&C) and 4.5Å (CTS), respectively. 48 candidates selected from six Schmidt plates were observed in 1989 whereas weather permitted only eleven spectra in 1990.

The data were reduced in Kiel using the program package IDAS written by G. Jonas. The spectra were flat-field corrected by selecting suitable dome flats and converted to absolute fluxes using observations of Feige 34. Visual magnitudes were estimated from the absolute fluxes at 5500Å. However, most of our observations were obtained under non-photometric weather and variable seeing conditions.

3. Results

The Schmidt spectra of stellar candidates are classified into four categories, which are thought to represent either of four spectral classes: DA white dwarfs, subluminous O- or B-stars, horizontal branch A-stars or sdF-stars. However, due to the low spectral resolution such a classification is difficult, especially for the faintest spectra. Our first aim was to check the classification of the Schmidt spectra from our follow-up spectroscopy. Therefore, in the first observing run we observed stellar candidates whose Schmidt spectra were classified as HBA or sdF although these objects are not relevant to our programme. It turned out that the classification of DA and HBA stars is highly reliable because of their large Balmer absorption line strenghts, whereas a few misclassifications occured among the other subclasses. Outstanding mismatches are two bright objects, HS0624+6907 (V=14^{m}2, see Fig. 1) and HS1227+4530 (V=16^{m}3), which are low-redshift quasars (Groote *et al.*, 1989). Their objective-prism spectra are almost identical to those of sdF stars since no prominent emission features fall into the region of photographic wavelengths due to their low redshifts. HS0624+6907, actually, is the brightest quasar discovered by optical selection. These discoveries are used to fine-tune the selection criteria in order to ensure that low-redshift quasars do not escape detection.

Amongst the 59 objects observed, four have already been observed by the PG survey since both surveys overlap. Interestingly, two of them have been misclassified previously. PG1234+4811 (=HS1234+4811) was classified as sdB by Green *et al.* (1986) but turned out to be one of the rare hot DA stars (see Jordan *et al.*, this conference for a detailed discussion). Also, we reclassify PG1236+4754 (=HS1236+4754=TON096) as sdB while Green *et al.* list it as DA.

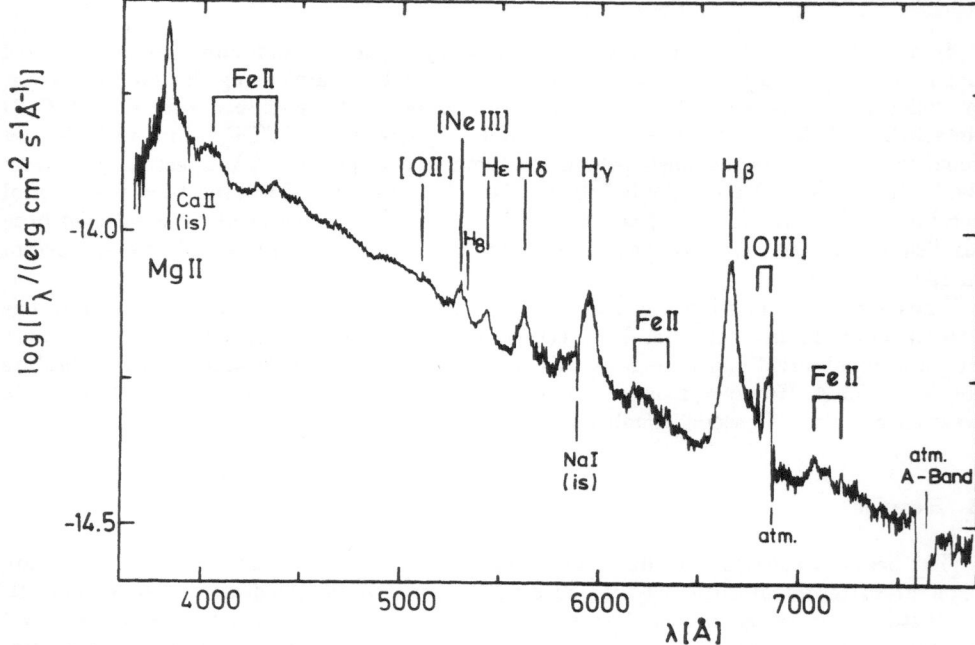

Figure 1: Spectrum of the bright low-redshift quasar HS0624+6907, the brightest quasar discovered by optical selection.

In Fig. 2 to 10 we display a gallery of spectra that might illustrate the variety of stellar spectra we have obtained. The fluxes are plotted on a logarithmic scale. In Fig. 2 to 7 the objects are arranged according to increasing temperature. The numbers of objects observed for each class are listed in Table 2.

For the sake of completeness Fig. 2 and 3 show spectra of the non-programme stars (sdF and HBA classes). It is worthwhile to note that complete samples of such stars could be drawn from the HS survey as well. Although most of the HBA stars are found amongst the brightest stars observed, some of them are as faint as $V=18^m 0$. Since these stars have relatively large absolute visual magnitudes, they must be quite far away. This is in accordance with the finding of Sommer-Larsen *et al.* (1989) who derived distances for HBA stars up to 30 kpc from the galactic plane.

Let us now turn to the true programme stars: the white dwarfs and hot subdwarfs. In Fig. 4 we display typical spectra of three DA white dwarfs whose effective temperatures are close to that for maximum Balmer line strength (see Jordan *et al.*, this conference). Fig. 5 presents the spectra of a sdB star and a sdOB star in comparison to the well studied sdB HD 4539 (Baschek *et al.*, 1972). The Balmer lines are much weaker than in the DA spectra and weak He I lines are detectable in the sdB spectrum, while the sdOB HS0938+4807 shows He II 4686Å in addition. While the sdB stars form a homogenous spectroscopic class (see e.g. Heber, 1990 for a review), the hotter sdO stars display a variety of different spectra as illustrated in Fig. 6 and 7. The spectra of two helium weak-lined sdO stars are presented in Fig. 6 which are very similar to the standard star Feige 34. In contrast the spectrum of the sdO HS1000+4704 is He strong-lined and shows great resemblence to that of HZ44 (Greenstein, 1966). Although HS is amongst the faintest stars observed, the quality of the spectrum is sufficient for quantitative work.

Three stars did not fit in the classification scheme. HS0942+4608 (Fig. 8) displays a composite spectrum and probably is a sdB star with a G-type compagnion. HS0139+0559 (Fig. 9) displays broad Balmer lines filled in by emission, especially in Hβ, and a broad and shallow He I 4471Å line. Comparison with PHL227 and V3885 Sgr (see Hunger et al., 1985) suggest that HS0139+0559 like the latter is an UX UMa star, a subclass of cataclysmic variables not showing pronounced emission lines. One object, HS0247+0537, defied classification. Its blue spectrum is nearly featureless (see Fig. 10): the only absorption features visible are broad and very shallow lines identified as Hβ and He I 4471Å. The star has been reobserved recently (not shown) at better spectral resolution with the CTS which allows the Hα-region and the blue spectrum to be recorded simultaneously. Surprisingly, Hα absorption was detected with an equivalent widths almost ten times as strong as Hβ. Also simultaneous ground based photometry and IUE spectrophotometry have been obtained. These data will hopefully give a clue to the nature of the star.

Table 2: Summary of results by spectral type and comparison to the Case Blue stars (CBS) studied by Wagner et al. (1988).

spectral type	number of objects (this paper)	CBS
sdF/sdG	8	7
HBA	18	8
DA	16	$\widehat{42}$
DB	1	11
DZ		1
mag. WD		1
sdB/sdO	11	35
binaries	1	1
CV	1	5
unclassified	1	
quasars/AGN	2	20

114

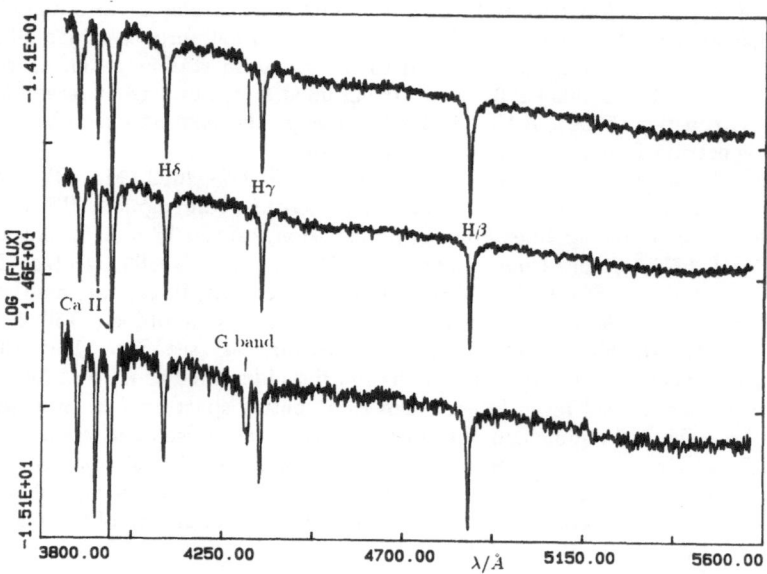

Figure 2: Spectra of HS1002+4552 (V=13.^{m}8, top) and HS0938+4807 (V=14.^{m}5, bottom) compared to the well studied sdF star BD+26° 2606, Hartmann and Gehren, 1988).

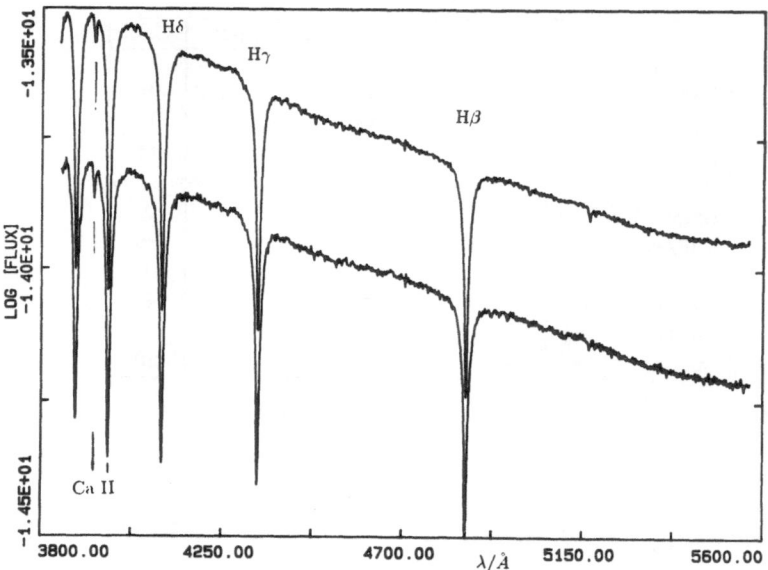

Figure 3: Spectrum of HS1002+4636 (V=13.^{m}8, bottom) compared to the field horizontal branch A star HD100995.

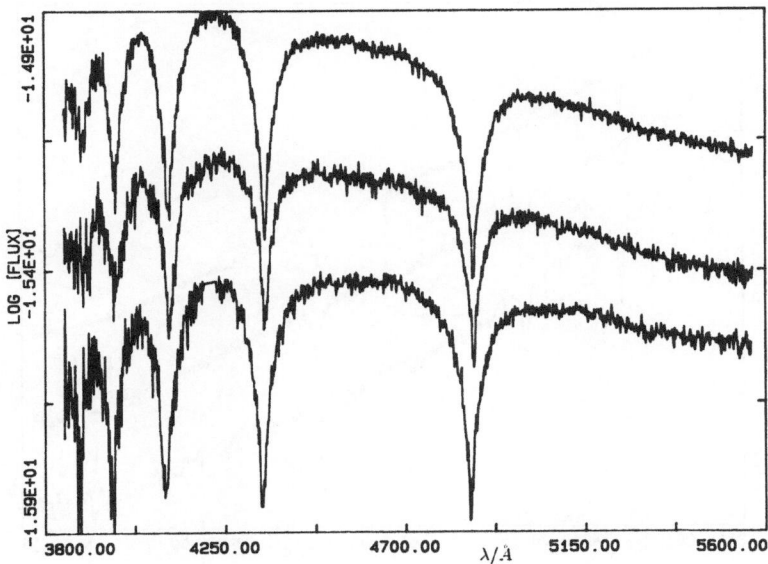

Figure 4: Spectra of HS1232+4754 (V=14.^{m}6, top), HS1231+4631 (V=16.^{m}1, middle) and HS0130+0800 (V=17.^{m}0, bottom), three typical DA white dwarfs.

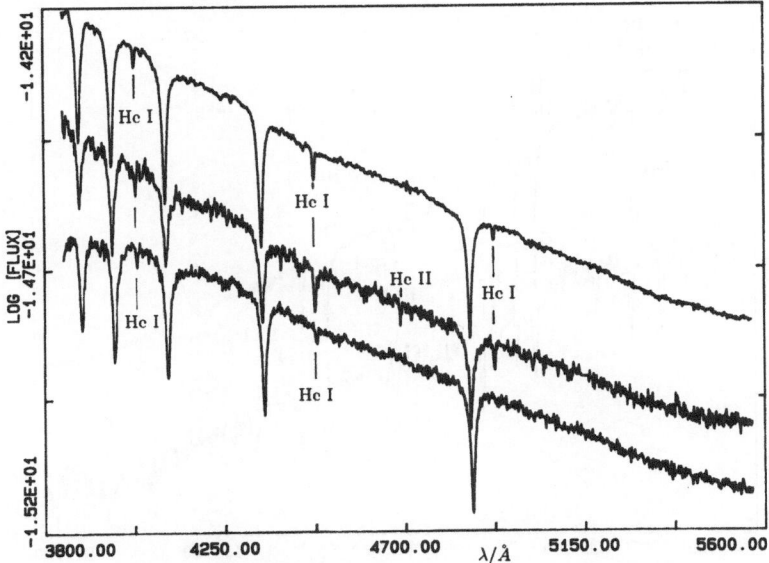

Figure 5: Spectrum of the sdB star HS1236+4754=PG1236+4754 (V=15.^{m}6, bottom) and the sdOB star HS0600+6602 (V=16.^{m}4, middle) compared with the well known sdB star HD4539 (top).

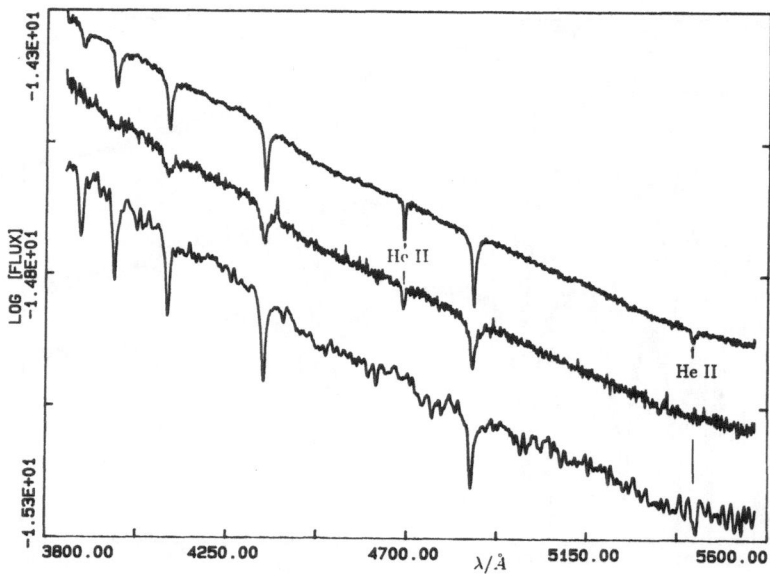

Figure 6: Spectra of two helium weak-lined sdO stars HS0231+0505 (V=16.^{m}5, middle) and HS0941+4649 (V=16.^{m}8, bottom) compared to Feige 34 (top).

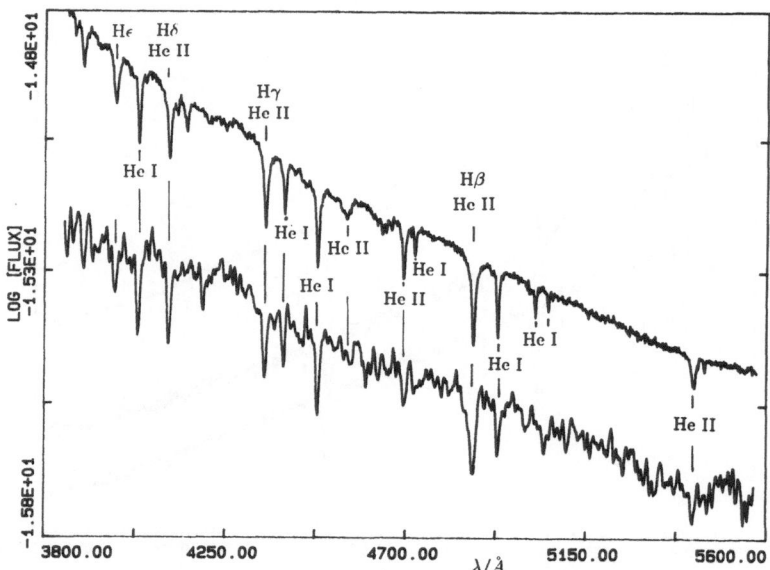

Figure 7: Spectrum of the helium strong-lined sdO star HS1000+4704 (V=18.^{m}2, bottom) compared the sdO star HZ44.

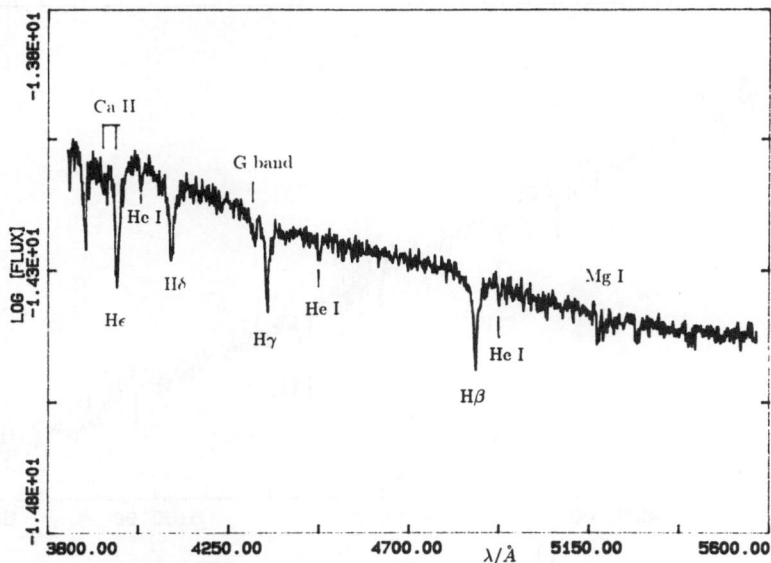

Figure 8: Spectrum of HS0942+4608

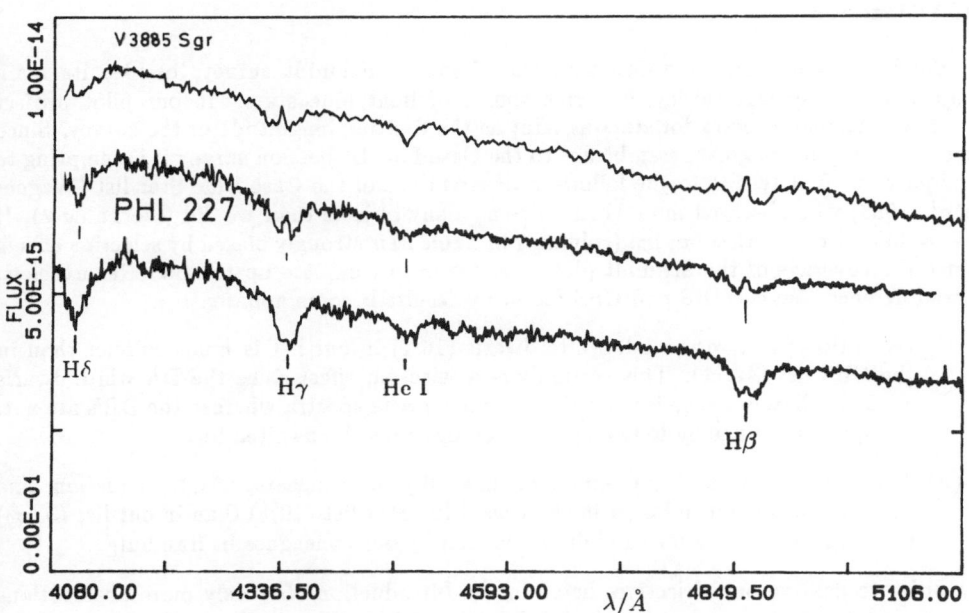

Figure 9: Spectrum of the UX UMa star HS0139+0559 (V=15^m 3, bottom) compared to two well known UX UMa stars V3885 Sgr (top) and PHL227 (middle).

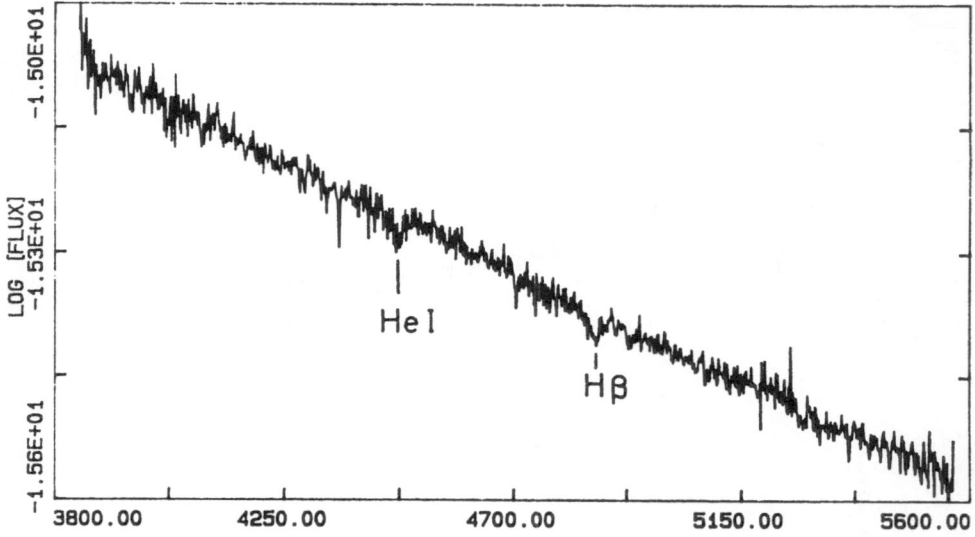

Figure 10: Spectrum of the unclassified star HS0247+0537 (V=17.^{m} 7)

4. Discussion

We have demonstrated above that the Hamburg Schmidt survey, besides its main objective as a quasar survey, is a rich source of faint blue stars. In our pilot project we have obtained spectra for stars as faint as the limiting magnitude of the survey. Since the HS survey bears great resemblence to the Case Low Dispersion survey it is tempting to compare our first results to the follow-up observations of the Case Blue Star list (Wagner *et al.*, 1988) who observed more than twice as many objects than we did (see Table 2). It has to be mentioned that our list (column 1 in Table 2) is strongly biased by selection effects since the coverage of the Schmidt plates are far from complete up to now. Nevertheless, two differences between HS and CBS follow-up results become apparent:

(i) the ratio of DA versus DB white dwarfs (16:1) in our list is much smaller than in the CBS list (42:11). This certainly is a selection effect since the DA white dwarfs are immediately recognized on the objective prism spectra whereas the DB's are not. Complete coverage by follow-up spectroscopy must be awaited for.

(ii) The "contamination" by emission line objects (quasars, AGN, emission line cataclysmics) is much larger in the Case Blue Star list (19%) than in our list (3.5%) pointing to the superior candidate selection by our colleagues in Hamburg.

In order to achieve our objectives listed in the Introduction obviously more observations are required, which are already scheduled for late 1990. These spectra shall be used for quantitative spectroscopic analyses based on model atmospheres. First preliminary results are reported by Jordan *et al.* at this conference. This will allow spectroscopic distances of the objects to be determined.

ACKNOWLEDGEMENTS. We thank our colleaques in Hamburg, Prof. D. Reimers, who initiated this collaboration, and Dieter Engels, Detlef Groote, Hans Hagen and Frank Toussaint, who provided the selection and classification of the candidate samples. The authors are grateful to the Calar Alto staff, especially to Dr. U.Hopp, for the assistence at the observatory.

REFERENCES

Baschek, B., Sargent, W.L.W., Searle, L. 1972: *Astrophys. J.* **173**, 611.
Engels, D., Groote, D., Hagen, H.J., Reimers, D. 1988: *Publ. Astron. Soc. Pacific Conf. Ser.***2**, 143.
Groote, D., Heber, U., Jordan, S. 1989: *Astron. Astrophys.* **223**, L1.
Green, R.F., Schmidt, M., Liebert, J. 1986: *Astrophys. J. Suppl.* **61**, 305.
Greenstein, J.L. 1966: *Astrophys. J.* **144**, 496.
Hagen, H.J. 1987: *Ph. D. thesis Hamburg University*, .
Hagen, H.J., Groote, D., Engels, D., Haug, U., Reimers, D. 1986: *Mitt. Astron. Gesell.***67**, 184.
Hartmann, K., Gehren, T. 1988: *Astron. Astrophys.* **199**, 269.
Heber, U. 1990: *Proceedings of IAU symposium No. 145: Stellar Evolution: The photospheric abundance connection*, eds. G. Michaud and A. Tutukov, Kluwer Academic Publisher, Dordrecht , in press.
Hunger, K., Heber, U., Koester, D. 1985: *Astron. Astrophys.* **149**, L4.
Pesch, P., Sanduleak, N. 1983: *Astrophys. J. Suppl.* **51**, 171.
Sommer-Larsen, J., Christensen, P.R., Carter, D. 1989: *Monthly Notices Roy. Astron. Soc.* **238**, 225.
Wagner, R.M., Sion, E.M., Liebert, J., Starrfield, S.G 1988: *Astrophys. J.* **328**, 213.

WHITE DWARFS IN THE HAMBURG SCHMIDT SURVEY

S. JORDAN*, U. HEBER*, V. WEIDEMANN
*Institut für Theoretische Physik und
Sternwarte der Universität Kiel
Olshausenstr. 40, D-2300 Kiel 1
Federal Republic of Germany*

ABSTRACT. With follow up-observations and subsequent analysis of all white dwarf candidates in some of the fields of the Hamburg Schmidt Survey we hope to obtain important statistical material for the clarification of many white dwarf problems. Two observation runs are completed and preliminary results of the analysis for 13 white dwarfs observed during the first run are reported. Effective temperatures range from 11000 K to 60000 K. HS 1234+4811=PG 1234+4811, previously misclassified as a sdB star, turned out to be one of the rare DA's with $T_{eff} \gtrsim 60000$ K.

1. Introduction

Many important questions concerning white dwarfs can not be settled with sufficient certainty since the ensemble of the known white dwarfs is strongly biased by selection effects. Even at a distance of only 20 pc there is a significant deficit of observed white dwarfs (Jahreiß 1987). The Hamburg Schmidt Survey is able to identify all white dwarfs down to a visual magnitude of about $18^m.5$. Depending on the effective temperature this means that we will obtain white dwarf samples which are complete within $\approx$ 160 pc and $\approx$ 1000 pc for 10000 K and 70000 K, respectively.

At the German-Spanish Astronomical Center (DSAZ) on the Calar Alto in Spain we will make follow up observations of all white dwarf candidates identified from the objective prism spectra in several 5.5° × 5.5° test fields at high galactic latitudes ($\approx$ 25 white dwarfs candidates per field).

Up to now we have obtained spectra of 18 new and 3 known white dwarfs. 13 of them have already been analyzed with the newest version of the Koester model atmospheres (see Koester et al. 1979 for a discription of the basic methods) which e.g. now contain the Hummer and Mihalas (1988) occupation probability formalism mentioned several times at this conference.

During our next observating runs in October and November 1990 we hope to complete our first test field down to the magnitude limit. This and some additional complete samples will broaden the observational basis for several important problems concerning white dwarf stars.

---

*Visiting astronomers, German-Spanish Astronomical Center, Calar Alto, operated by the Max-Planck-Institut für Astronomie Heidelberg jointly with the Spanish National Commission for Astronomy

G. Vauclair and E. Sion (eds.), White Dwarfs, 121–127.
© 1991 *Kluwer Academic Publishers.*

2. Motivation

2.1 DISTRIBUTION OF SPECTRAL TYPES WITH THE TEMPERATURE

One of the most challenging problems is the fact that the number ratio between DA and non-DA stars changes dramatically with the effective temperature (see e.g. Liebert 1986). Especially the complete lack of DB white dwarfs between 30000 K and 45000 K and the absence of DA's with temperatures higher than 75000 K can not be explained by theories which predict a white dwarf atmosphere to remain hydrogen rich or helium dominated during the whole cooling sequence. An alternative explanation are very thin outer layers of hydrogen ($\approx 10^{-13}\,M_{\odot}$) with diffusion and mixing beeing responsible for the transitions of DA's into non-DA's and vice versa (see e.g. Shipman 1989 for a discussion of the various scenarios). The Hamburg Survey may help to construct a more reliable distribution of the spectral types with the temperature.

2.2 THE WHITE DWARF LUMINOSITY FUNCTION

The observed luminosity function of the white dwarfs is an important indicator for the evolution of the stars in our galaxy. It depends on several parameters: the initial mass function, the time dependent star formation rate, the initial-final mass relation (and therefore on mass loss rates and overshooting), the age of our galaxy, the time scales of the various stages of stellar evolution and the cooling times of white dwarfs (e.g. Yuan 1989). With the Hamburg Survey a luminosity function for $L/L_{\odot} \gtrsim 10^{-3}$ can be constructed.

2.3 THE MASS DISTRIBUTION OF THE WHITE DWARFS

Masses can only be measured for the brighter white dwarfs in the Hamburg Survey since a high signal to noise ratio is necessary for a good $\log g$ determination. In these cases temperature dependent mass-radius relations (Koester and Schönberner 1986) can be used.

2.4 THE GALACTIC SCALE HEIGHT OF THE WHITE DWARFS

Two fundamantal quantities for the understanding of galactic evolution are the space density of the white dwarfs and their galactic scale height. White dwarfs are old stars and not necessarily close to the location of their origin. Pertubations by large molecular clouds with masses up to 10000 $M_{\odot}$ or black holes (Fuchs and Wielen 1987) can lead to considerable components of the white dwarf velocities perpendicular to the galactic disk. The scale height resulting from this diffusion and oscillation around the center of the disk is about 275 pc (Boyle 1989). The exact determinaton of the value is very difficult because — as mentioned above — the white dwarf ensemble is very incomplete for distances greater than 20 pc. With the evolutionary models of Koester and Schönberner (1986) and the effective temperatures from the model atmosphere analysis of our spectra with model atmospheres distances can be obtained by assuming a mass of 0.6 $M_{\odot}$.

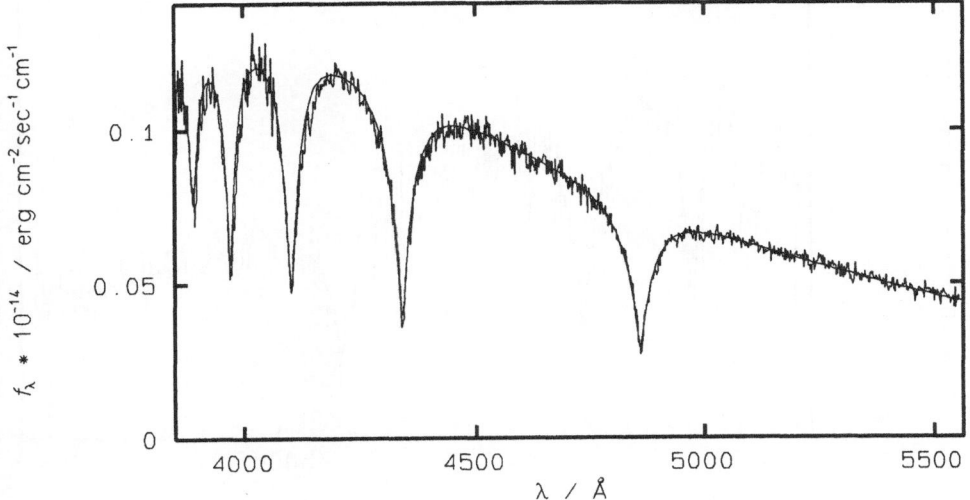

Figure 1: Spectrum of the cool DA white dwarf HS0235+0655 compared to a model with $T_{\text{eff}} = 11000$ K and $\log g = 8.00$.

3. First Results

Follow-up observations have been made in the Cassegrain focus of the 3.5 m telescope of the DSAZ using a Boller and Chivens spectrograph with a dispersion of 120 Åmm^{-1} and a RCA CCD detector in January 1989 and a twin spectrograph with 144 Åmm^{-1} in January 1990. The nine spectra of the first observation run have been reduced and flux calibrated as described by Heber, Jordan and Weidemann (these proceedings). These spectra were analyzed by fitting synthetic spectra using the latest version of the Koester model atmosphere code. A model grid was set up which covers the range from 10000 K to 65000 K. For this preliminary analysis the temperature steps were 1000 K. This will be refined in a later version.

The spectra have not been normalized because the Balmer lines are so strong that no true continuum points can be seen (Figure 1 and 2). Therefore it was not intended to fit the profiles of individual lines, but to match the entire observed spectrum by a synthetic one. Although the observed spectra were flux calibrated and corrected for the airmass, the slope of the continuum can not be used as a reliable T_{eff} indicator because most of the observations were carried out under non-photometric weather and variable seeing conditions using a small slit ($\approx 2''$ wide). Therefore the following procedure was adapted: During the process a logarithmic flux scale was used because for these hot stars the optical continuum flux can be approximately described by the Rayleigh-Jeans law which leads to a straight line in the $\lambda - \log f_\lambda$ plane.

124

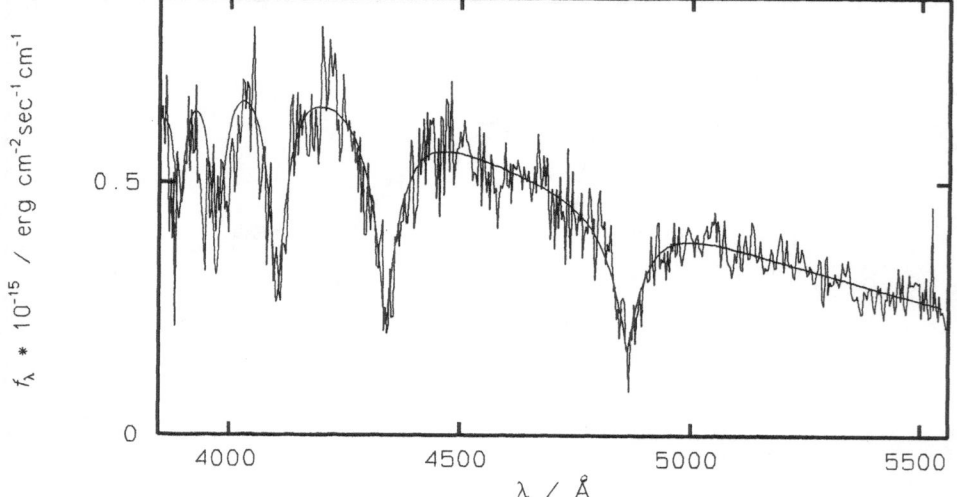

Figure 2: The faint DA white dwarf HS0949+4935 ($m_V = 18\overset{m}{.}5$); $T_{\text{eff}} = 12000$ K and $\log g = 8.00$.

In the first step wavelength intervals in the "pseudo continuum" were choosen well between the spectral lines. Subsequently the difference between the observed and the synthetic logarithmic "pseudo continuum" fluxes were calculated. In the case of photometric conditions and a perfect fit, this difference is a constant. In order to allow for small deviations from a constant the differences were fitted to a third order polynomial which we subtracted from the $\log f_{\lambda,\text{theory}}$ of the whole spectrum including the line intervals. Now the "continua" of the observed and the synthetic spectra match and the line profiles can be compared. T_{eff} and $\log g$ can then be obtained by a least square fit to the model grid taking only the spectral line intervals into account.

Two examples of spectral fits are displayed in Figure 1 and 2. Both objects have temperatures close to that of the maximum Balmer line strength ($T_{\text{eff}} = 12000$ K). Note that even for the faint object HS0949+4935 ($V = 18\overset{m}{.}5$) a satisfactory fit can be achieved (Figure 2).

An alternative to our fitting algorithm is to define the fluxes at certain wavelengths in the line wings to be unity in the observed and the theoretical spectrum. If the line profiles are divided by a suitable function, e.g. a straight line going through the the flux values at these wavelengths, one gets a "normalized" spectrum. The resulting best fit for the atmospheric parameters is very similar but since our fitting procedure outlined above needs much smaller "corrections" to the original spectrum we regard it to be superior.

The six 1990 twin spectra are not yet reduced. Five additional spectra have been taken by the Hamburg group with a Boller and Chivens spectrograph at 240 Åmm^{-1} dispersion. They were only preliminarily reduced and analyzed. Table 1 summarizes our results for 18 DA white dwarfs and 2 DB white dwarfs. For all objects except for HS1234+4811 $\log g = 8$ was assumed. In a forthcoming paper we will try to get a best fit for the gravity as well. It has been reported in this conference (e.g. by Dolez, Vauclair and Koester) that near $T_{\text{eff}} = 12000$ K two "best fit" solutions (T_{eff}, $\log g$) exist. This will be tested for our sample

Table 1: Preliminary atmospheric parameters of the white dwarfs in the Hamburg Survey

object	T_{eff}	$\log g$	m_V	spectral type	comment
HS 1234+4811	60000 ±4000 K	7.75 ±0.25	14^{m}5	DA	(1)
HS 0946+5009	30000 ±1000 K	8.00	17^{m}5	DA	(2)
HS 1241+4821	13000 ± 500 K	8.00	16^{m}9	DA	(2)
HS 0943+4724	11000 ±1500 K	8.00	18^{m}5	DA	(2)
HS 0949+4935	12500 ±1000 K	8.00	18^{m}5	DA	(2)
HS 0130+0800	12500 ± 500 K	8.00	16^{m}4	DA	(2)
HS 1232+4752	15000 ± 500 K	8.00	14^{m}6	DA	(3)
HS 0235+0655	11000 ± 500 K	8.00	16^{m}5	DA	(2)
HS 1231+4631	22000 ±1000 K	8.00	16^{m}1	DA	(4)
HS 0943+4852	18000 ±2000 K	8.00		DA	(5)
HS 0146+0723	19000 ±2000 K	8.00	16^{m}5	DA	(5)
HS 0235+0514	15500 ±2000 K	8.00	17^{m}8	DA	(5)
HS 0946+4848	18000 ±2000 K	8.00	17^{m}8	DA	(5)
HS 1002+4518				DA	(6)
HS 0949+4508				DA	(6)
HS 1003+4852				DA	(6)
HS 1001+4651				DA	(6)
HS 0727+6915				DA	(6)
HS 0700+6853				DB	(6)
HS 0841+2613	16000 K		14^{m}0	DB	(7)

(1) HS1234+4811=PG1234+4811; was misclassified as sdB; Jan 1989.
(2) $\log g = 8$ was assumed for this first analysis; Jan 1989
(3) HS1232+4752=GD148; Jan 1989; $\log g = 8$ assumed
(4) HS1231+4631=TON 82; Jan 1989; $\log g = 8$ assumed
(5) only crudely calibrated spectra with 240 Å/ mm; Hamburg group
(6) not yet calibrated; Jan 1990
(7) HS0841+2613=TON10; T_{eff} from Wegner and Nelan 1987; Hamburg group

as well as the question if small amounts of helium may disturb the $\log g$ determination.

The reason that only two follow-up spectra of DB's have been taken is probably caused by the fact that the broad Balmer lines of the DA's are much easier recognized among the objective prism spectra. But we expect a higher portion of DB's if spectra of all white dwarf candidates in a test fields are taken.

3.1 HE HOT WHITE DWARF HS1234+4811

We discovered HS1234+4811 to be a hot DA white dwarf very similar to G191B2B (T_{eff}=62250 K, Holberg et al. 1986). It had previously been detected in the PG survey. Green (1980) classified it as a white dwarf without giving a spectral type. Later on, Green et al. (1986) reclassified it as a sdB star. Therefore PG1234+4811 is not listed in the catalogue of spectroscopically identified white dwarfs (McCook and Sion, 1987).

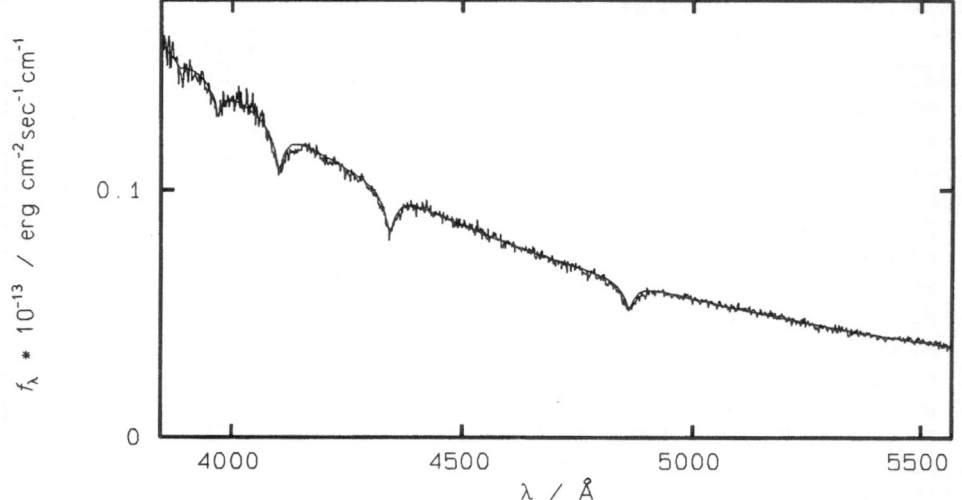

Figure 3: The hot DA white dwarf HS1234+4811; $T_{\text{eff}} = 60000$ K and $\log g = 7.75$.

With $T_{\text{eff}} = 60000$ K and $\log g = 7.75$ HS1234+4811 (see Fig. 3) is one of the hottest DA white dwarfs (only six with $T_{\text{eff}} \gtrsim 60000$ K are known) and relatively close to the 75000 K limit above which no DA's are found.

Recently, IUE spectra were taken in low and high resolution (Jordan, Koester and Heber). These spectra are consistent with our temperature determination and show no traces of helium. A complete analysis including the UV will be published in a forthcoming paper.

3.2 THE ZZ CETI STARS

Some of the DA white dwarfs in Table 1 are in the temperature region of the ZZ Ceti instability strip. Since the exact temperature interval is still uncertain (see e.g. Dolez, Vauclair and Koester in these proceedings) an accurate temperature determination and high-speed photometry of these objects might give further constraints.

3.3 MAGNETIC WHITE DWARFS

One of the first interesting stellar discoveries of the Hamburg Survey was the magnetic white dwarf HS1254+3430 (Hagen et al. 1987). Its nature was already visible in the objective prism spectrum so that it can be expected that more bright magnetic white dwarfs with strong Zeeman features can be identified just from the prism spectra.

The object was analyzed by Jordan (1989). A newer analysis, which will be presented in a forthcoming paper, gave a lower effective temperature of about 9500 K. The configuration of the dipole field (polar field strength of 18 MG, a viewing angle with respect to the dipole axis of 80°) did not change significantly compared to the first preliminary result.

3.4 THE HAMBURG ESO SURVEY

The success of the HS on the northern sky led to a similar ESO key project for the southern sky (magnitude limit $\approx 18^m$, Reimers 1990). With 480 Åmm^{-1} the ESO Schmidt prism has a much higher dispersion than the one at Calar Alto (1390 Åmm^{-1}). This will enable us to identify the principal characteristics of an object directly in the prism spectra (at least for the brighter ones). Hence it is not in any case necessary to make follow-up observations for the determination of the spectral type (DA vs. non-DA).

ACKNOWLEDGEMENTS. We thank our colleaques in Hamburg, especially Dieter Engels, Hans Hagen and Frank Toussaint for the object selection and for providing us with CCD spectra of five white dwarfs from their observing run. We are also indebted to Detlev Koester for providing us with the newest version of his model atmosphere code.

REFERENCES

Boyle, B.J. 1989: *Mon.Not.R.astr.Soc.* **240**, 533.
Fuchs, B., Wielen, R. 1987: *The Galaxy,* eds. Gilmore, G., Carswell, B., Reidel, p.375.
Green, R.F. 1980: *Astrophys.J.* **238**, 685.
Green, R.F., Schmidt, M., Liebert, J. 1986: *Astrophys.J.* **61**, 305.
Hagen, H.-J., Groote, D., Engels, D., Haug, U., Toussaint, F., Reimers, D. 1987: *Astron.Astrophys.* **183**, L7.
Holberg, J.B., Wesemael, F., Basile, J. 1986: *Astrophys.J.* **306**, 629.
Hummer, D.G., Mihalas, D. 1988: *Astrophys.J.* **331**, 794.
Jahreiß, H. 1987: *Mem.S.A.It.* **58**, N° **1**, 53.
Jordan, S. 1989: *Proceedings of IAU Coll. No. 114: White Dwarfs,* ed. G. Wegner, Springer Verlag, p.333.
Koester, D., Schönberner, D. 1986: *Astron.Astrophys.* **154**, 125.
Koester, D., Schulz, H., Weidemann, V. 1979: *Astron.Astrophys.* **76**, 262.
Liebert, J. 1986: *Proceedings of IAU Coll. No. 87: Hydrogen-Deficient Stars and Related Objects,* eds. K. Hunger, D Schönberner, and N.K. Rao, Dordrecht:Reidel, p. 367.
McCook, G.P., Sion, E.M. 1987: *Astrophys.J.Suppl.* **65**, 603.
Reimers, D. 1990: *The Messenger* ESO 60, 15.
Shipman, H.L. 1989: *Proceedings of IAU Coll. No. 114: White Dwarfs,* ed. G. Wegner, Springer Verlag, p. 220.
Wegner, G., Nelan, E.P. 1987: *Astrophys.J.* **281**, 916.
Yuan, J.W. 1989: *Astron.Astrophys.* **224**, 116.

Asteroseismology of White Dwarf Stars
With the Whole Earth Telescope

D. E. Winget

Department of Astronomy and McDonald Observatory

The University of Texas at Austin

ABSTRACT

The Whole Earth Telescope (WET) project is a coöperative effort to obtain global time-series photometry of compact variable stars. The aim of WET is to obtain essentially continuous light curves, from which we compute high-resolution, high signal-to-noise power-spectra. These enable us to apply the powerful theoretical machinery of asteroseismology to determine many of the fundamental parameters of white dwarf stars: rotation rates, magnetic field strengths, total mass, compositional stratification of the envelope, core composition, and more.

The purpose of this review is to provide a brief update on the WET project. I present a list of principle targets for which extensive data sets currently exist, or are planned (as of May 1991). All of the data on each target is in the public domain 18 months after a run, and possibly earlier (at the discretion of the principle investigator for the specific target). I discuss the results for the DOV star PG 1159-035 as a case study, emphasizing the science that has just begun to come pouring out, and finally attempt to chart some of the future directions we can expect the WET project to take.

1. Physical Context and an Introduction to Asteroseismology

Many diverse avenues of research — ranging from condensed matter physics to cosmochronology — intersect in the field of the white dwarf stars; this has motivated considerable activity in the study of these faint, relatively nearby stars. Historically the bulk of the observational work has been limited to proper motion studies, or spectroscopic and multi-color photometric investigations of the photospheres. Much of the corresponding theoretical work has aimed at using this information to infer the basic physical parameters of the white dwarf stars (T_e, M_*, R_*, L_*), as well as the conditions below the photosphere: envelope chemical stratification, and core composition. The discovery in the 1970's and 1980's of multi-periodic nonradial pulsators led to a new and complementary way of studying the white dwarf stars, and particularly, for probing their interiors: asteroseismology.

The principle behind asteroseismology is simple, and is the same for many different kind of physical systems: we learn about the structure of the underlying system by determining the frequencies of its normal modes of oscillation. As R.E. Nather is fond of pointing out, this is precisely the way in which we discovered the underlying structure of the atom at the beginning of this century, and continue to study it this way today—atomic seismology so-to-speak. The pulsation modes and their corresponding frequencies are the analogs of the atomic orbitals, and their energies or photon frequencies; if we know the frequencies of all the normal modes, we can derive the internal structure of the system.

129

G. Vauclair and E. Sion (eds.), White Dwarfs, 129–141.

© 1991 *Kluwer Academic Publishers.*

The similarity between the normal mode, or eigenvalue, problem for stars and atoms is more than superficial; the mathematical machinery for the two problems is nearly identical—reflecting the underlying spherical symmetry inherent to both. Both are cast as eigenvalue problems. The spherical symmetries are exploited by expanding the eigenfunctions in terms of spherical harmonics, reducing the system of equations to a purely radial problem. Thus each eigenfunction and its eigenvalue are associated with three specific "quantum numbers": the two angular quantum numbers corresponding to the indices of the spherical harmonic, ℓ, and m; and the radial quantum number, k.

Turning to the specific problem at hand, we indicate the dependence of the frequencies of the normal modes by labelling each frequency with its corresponding quantum numbers: $\sigma_{k,l,m}$. The quantum numbers represent the number of nodes in the eigenfunction.

If we consider the general case of the nonradial spheroidal modes (appropriate for a spherical star) then for each degree, ℓ, there exists a fundamental mode (except for $\ell = 1$, which implies movement of the center of mass) with no radial nodes. This mode serves as the central point for two "branches" of normal modes representing increasing radial overtones: the p-modes and the g-modes.

For the p-mode branch, pressure provides the restoring force. The first overtone p-mode (one radial node) occurs at a higher frequency than the fundamental mode, or f-mode. Each additional radial node in the eigenfunction results in a higher frequency still, and in the limit of high radial overtone, the consecutive radial overtones are separated by a constant difference in frequency. The theoretical accumulation point of this series of p-modes is infinity; a more practical upper limit occurs when the system can no longer respond mechanically and the surface reflection condition is violated. The displacements of these modes are primarily radial in direction. This may be the reason no p-modes have yet been observed in white dwarf stars: they require substantial motion across the equipotential surfaces of very strong gravitational fields (e.g. $\log g \sim 8$).

The other branch of modes, the g-modes, have predominantly horizontal displacements (along equipotential surfaces), and are thus easier to excite to observable amplitude in a high gravity star. These are the modes with periods which match those observed in the white dwarf stars. The lowest radial overtone g-mode of a given degree has a single radial node and a frequency lower than the the f-mode, each successive node gives rise to an even lower frequency, forming a series with an accumulation point at zero frequency; in this case a practical *lower* limit is set by the violation of the surface reflection condition. Note that the exact location of this critical frequency contains important information about the surface physical condition. In the limit of high radial overtones, these modes are roughly evenly spaced in period. You can see that as we add radial nodes, we increase the period in a regular way, until we reach a point where the new node would fall near a boundary in the star—such as a composition boundary in a white dwarf. Then the existence of the boundary affects the position of the node, and therefore the structure of the eigenfunction, and ultimately the frequency of the mode. This effect is known as resonant mode-trapping, and can be detected by irregularities in the period spacings.

Because there is no preferred axis of symmetry in a non–rotating, non–magnetic (non–physical!) stellar model, the fequencies are independent of the azimuthal quantum number m. On the other hand, in a case closer to reality, slow rotation, or modest magnetic fields (relative to the strength of the surface gravity) breaks the degeneracy. In the case of slow uniform rotation, for example, the frequency splitting produces $2\ell + 1$ modes for each radial overtone of degree ℓ. These modes are uniformly split by an amount which is a function only of ℓ and the rotation frequency, Ω. Magnetic fields, on the other hand, also break the degeneracy which depends only on the absolute value of m, but produce only $\ell + 1$ modes, so the effects can be distinguished from each other.

The amount of asteroseismological information is a sensitive function of the number of modes present. The dense pulsation mode spectra of the white dwarf pulsators makes them a theorist's dream but an observer's nightmare. Disentangling the closely spaced modes requires maximum run lengths with minimum internal gaps in the coverage.

The importance of eliminating these gaps led R.E. Nather to conceive the ultimate terrestrial solution: the Whole Earth Telescope. WET is a coöperative network of optical observatories, distributed in longitude, coördinated from a single site, and used as a single instrument. The development of this concept was traced in the review by Nather (1989), and the instrumental properties of the WET are described in detail in the paper by Nather et al. (1990). In the following I will comment briefly on the operation of WET and emphasize the changes since Nather et al. (1990).

2. The Whole Earth Telescope

The Whole Earth Telescope grew out of the necessity to reduce gaps in time-series data. The problems introduced by gaps are discussed and illustrated in detail by Nather (1989) and Nather et al. (1990). WET blossomed and developed in a natural way out of the co–operative spirit in the white dwarf community. It is this spirit that has caused it to succeed where other projects of similar scope have failed. Indeed, the WET is more a collection of people than of telescopes.

All of which brings me to a comment on the authorship of this paper. My name appears as the sole author because this is the custom for review papers. It does not imply that I am the leader of this project; I am not. The closest to the description of leader would be R.E. Nather, but to call him that would be misleading. He is more of a coördinator in an egalatarian coöperative project (in point of fact he describes himself as "the WET nurse", an apt description). I am only the "point man" for purposes of this overview. The mistakes in this paper are entirely mine, but the project itself and the science it produces belong to everyone listed in Table 1, and to a few whose names have no doubt been inadvertently left out.

The unique organization of WET has accomodated the growing interest and translated this into greater scientific productivity. The growing interest is documented by the increase in the number of sites participating, and so I include an updated map in Figure 1 to compare with Figure 1 in Nather (1989): 6 sites have been added. There is a straightforward way to measure the scientific success as well. The main goal of the project is to unambiguously resolve complex light curves. The spectral window of the data set—a single sinusoid sampled in exactly the same way

Table 1. A PARTIAL LIST OF PARTICIPANTS IN THE WET

France	G. Vauclair, N. Dolez, M. Chevreton
U.K.	M. Barstow, R. Tweedy, A.E. Sansom
Norway	J.-E. Solheim, A. Ulla, P.-I. Emmanuelsen
Brazil	S. O. Kepler, A. Kanaan
Canada	G. Fontaine, F. Wesemael, M.A. Wood
USA	A.D. Grauer
	S.D. Kawaler
	R.E. Nather, D.E. Winget, J.C. Clemens, J. Provencal,
	S.J. Kleinman, P.A. Bradley, C.F. Claver
	C.J. Hansen
	P. Bergeron
	B.P. Hine, III
Australia	N. Achilleos, D. Wickramasinghe
India	T.M.K. Marar, S. Seetha, B.N. Ashoka
Israel	T. Mazeh, E. Liebowitz
Poland	W. Dziembowski, J. Smak, P. Muskolic
South Africa	B. Warner, D. O'Donoghue, D.A. Buckley, A.J. Brickhill
	D. Kurtz, P. Martinez
Netherlands	T. Augustjein, Van Paradijs

Table 2. WET PRIMARY TARGET LIST

RUN	OBJECT	PI	TYPE	
XCOV I	PG 1346+082	Provencal	IBWD	Provencal et al.(1989)
(Mar 88)	V803 Cen	O'Donoghue	IBWD	O'Donoghue et al.(1990)
XCOV II	G29-38	Winget, Kepler	DAV	Winget et al.(1990)
(Nov 88)	V471 Tau	Clemens	Binary	
XCOV III	PG 1159-035	Winget	DOV	Winget et al.(1991)
(Mar 89)	G117-B15A	Kepler	DAV	Kepler et al.(1990)
XCOV IV	HZ-29	Solheim	IBWD	Solheim et al.(these Proceedings)
(Mar 90)	G117-B15A	Kepler	DAV	Kepler et al.(these Proceedings)
XCOV V	GD 358	Winget	DBV	
(May 90)	GD 165	Bergeron	DAV	
Planned:				
XCOV VI	PG 1707+427	Grauer	DOV	
(May 91)	GD 154	Vauclair	DAV	
	PG 1351+489	Hansen	DBV	

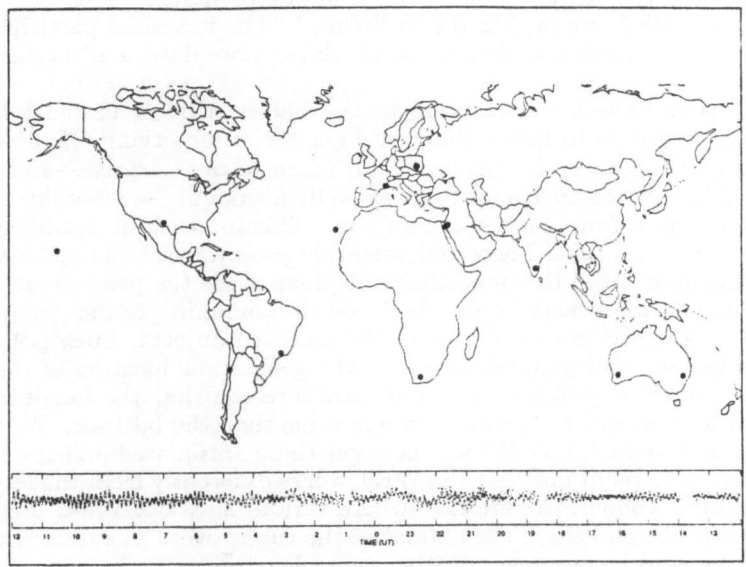

Figure 1. The Locations of the 12 sites participating in the Whole Earth Telescope project (upper panel). A 24 hr segment of the PG 1159-035 light curve (lower panel).

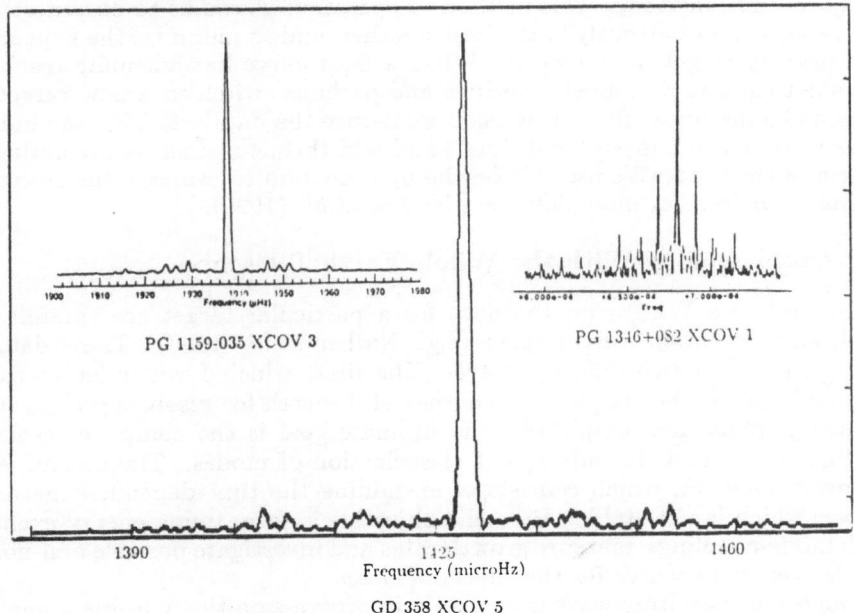

Figure 2. A comparison of Windows. Transform of a single sine curve sampled exactly as the data for the best data sets obtained in XCOV 1 (upper right), 3 (upperleft), and 5 (center)

as the data—tells the story. Witness the windows of XCOV (short for extended coverage run, or WET run) 1, 3 and 5 in Figure 1. The increased participation has led directly to more sites and thus to better data, more data, and better spectral windows.

There is another effect contributing to the better window in the last run as well. Weather forced us to take a desperate gamble on observing GD 358 (at +35 degrees) from the southern hemisphere and, much to our surprise—and southern hemisphere time allocation committees as well, no doubt—we obtained 6 hours of good quality data from each southern site. This increase in flexibility greatly improved our coverage, and hence our scientific productivity. There is no doubt that this "poaching" from the opposite hemisphere made the project successful.

This kind of situation casts serious doubt on the soundness of the policy at some southern sites of insisting on observing only southern objects. Such policies were put in place to take optimum advantage of the geographic location of the facility. The designers of these policies could not have foreseen that the longitude of the facility could have as much, or more, unique value than the latitude. We hope our documented success with GD 358 will help get these antiquated policies changed.

It is important to point out that the WET works differently from the less complicated "campaign" coöperative efforts—where various sites observe a single object as much as possible, and share the data after the run is over. The difference is that with WET, the sites communicate with a central coördinating location during the run. Multiple targets are observed during the run, optimizing the target assignment at each site to take best advantage of the weather circumstances at that site and sites adjacent in longitude. This method of operation allows us to obtain data on several targets simultaneously in the best weather, and to minimize the gaps in our highest priority target in the worst. When a light curve has adequate resolution and signal–to–noise, we adjust priorities and perhaps switch to a new target. In order to make the most efficient decision we reduce the data sets from the individual sites as they come in, in "real-time", and add them together, constructing the transform of the data. We use this on-the-fly reduction to evaluate the resolution and signal-to-noise. For more detail see Nather et al. (1990).

3. Asteroseismology With the Whole Earth Telescope

At the end of a WET run, the data for a particular target are carefully re–reduced, and we obtain a light curve (e.g. Nather et al. 1990). These data are then approached in two different ways. The first, which I will refer to as the "linear analysis", is the simplest. It consists of a search for resolved power, stable in frequency, phase and amplitude; the ultimate goal is the complete resolution of the light curve and the subsequent classification of modes. The second is the "non-linear" analysis, which consists of examining the time-dependent nature of the power which is not stable. Here the objective is to examine energy exchange through mode-couplings, measure growth rates and investigate possible non-normal modes, as well as to search for the effects of chaos.

Although much exciting work is currently in progress on the non-linear analysis, existing observational and theoretical tools for the analysis of the stable power are much more highly developed. Correspondingly, I will emphasize the linear analysis in most of the following discussion, and refer the reader to the papers in these proceedings by Serre and Goupil *et al.* for more insight into non-linear processes.

3.1 EXTRACTING THE PHYSICAL INFORMATION CONTENT OF THE LIGHT CURVE

The linear analysis begins with a search for individual frequencies which are stable in frequency, phase, and amplitude. Our most powerful tool for this search is the Fourier transform (FT). The high density of WET data resulting from the lack of significant gaps, combines with the total length of the data set to produce high-resolution, high signal-to-noise transforms. Ideally, the light curve is completely resolved in this way. If there exist sufficient numbers of stable frequencies, it is possible to classify the modes based on the distributions of the stable power. This is where the techniques of asteroseismology become exceptionally powerful.

In the limit of frequencies higher than $1,000\mu Hz$, we expect the low ℓ, normal modes to manifest themselves as a set of closely spaced multiplets, with like multiplets relatively evenly distributed in period. Recall that the fine structure comes about by the breaking of the spherical symmetry. Thus the spacing of the modes contains information about properties related to the breaking of this symmetry: the rotation properties $\Omega(r, \Theta, \phi)$, the magnetic field strength and structure, and the relative inclination of the pulsation and rotation axes.

The spacing of like multiplets (same ℓ), is roughly uniform in period; each multiplet represents a nonradial g-mode of different radial overtone number, k. The average separation is a very sensitive measure of the total stellar mass. The individual spacings, and their deviation from the average values, yield information about the chemical composition, masses of individual compositional layers, and transition zone structure. Thus the distribution of modes can give a relatively complete picture of the overall structure of the star. Finally, evolution produces a slow secular change in the periods of even the most stable modes. This change can be measured and used to study the evolution of the pulsating stars directly.

It is worth pointing out that much of the above theoretical interpretation has been developed by Kawaler and collaborators, and has been recently described in the review by Kawaler and Hansen (1989) and references therein.

As noted above, this is not a paper dealing with promises and futures, so I want to emphasize not what we *can* learn in principle, but what we *have* learned in practice. In Table 2 I present a complete list of all WET runs to–date, as well as the planned run in May 1991, and a list of principal targets (I do not include the secondary targets observed during each campaign), Principal Investigators (PI's), and publications. Below, I will illustrate what we have learned through a brief examination of a specific example: PG 1159-035.

3.2 PG 1159-035: A CASE STUDY

We completed the preliminary investigation of this object last spring (1990). The light curve and the data reduction techniques are presented in Nather *et al.* (1990) and the analysis and interpretation of the results have been submitted to the Astrophysical Journal (Winget *et al.* 1991). The reader is referred to these papers for figures and further details.

The FT of the light curve of the DOV star PG 1159-035 revealed 122 well-resolved pulsation modes which where stable on the timescale of the run (about 11 days). This was roughly an order of magnitude more modes than we had expected, and more than we needed to examine the distribution of modes in detail.

There are two distinct types of fine-structure multiplets in the transform. The first type include by far the largest amplitude modes, consisting of uniformly split triplets. The second type of fine structure multiplets have 3-5 components arranged in a pattern of uniformly split quintuplets, with typically 1 or 2 missing or low amplitude components in each multiplet. The absence of some of the components in the quintuplets underscores the need for caution in assigning ℓ values at this point based on a simple counting of components. The distributions, both within a given multiplet and between multiplets, hold the key to correct modal identification.

We first examine the spacings within a given multiplet. Averaging over the observed triplets and quintuplets separately, we obtain $4.22 \pm 0.04 \mu Hz$, and $6.92 \pm 0.07 \mu Hz$, respectively, for the splittings. The uniformity of the splittings tells us that magnetic-fields are not responsible; indeed the level of uniformity in the splittings can be used to calculate an upper limit to the magnetic field of $B < 6000G$ (see Jones et al. 1989). In contrast, we expect rotation to produce uniform splitting, and if the pulsation and rotation axes are aligned, to produce up to $2\ell+1$ component modes in each multiplet. The sizes of the splittings imply slow rotation, and in this limit we expect the ratio of the splittings to be a function only of their relative values of ℓ. For the lowest degree modes, pulsation theory predicts $\delta \nu_{\ell=1}/\delta \nu_{\ell=2} = 0.60$ (Hansen, Cox, and Van Horn 1977). The ratio we observe is 0.61. This indicates that the triplets are rotationally split modes of degree $\ell = 1$, and the quintuplets are rotationally split modes of degree $\ell = 2$. We note that we find no compelling evidence for splittings corresponding to modes of degree 3 or higher.

The above implies that the various sets of like multiplets are simply different radial overtone modes of the the same degree, ℓ. The period range observed further implies that these are modes of relatively high radial overtone (e.g. 15-50). We observe that the separations of like multiplets are roughly uniform in period, as we would expect for nonradial g-modes in the high overtone limit (Tassoul 1981). Theoretical calculations suggest that, in this limit, the ratio between the average period spacings for different degree modes depends only on the ratio of $\sqrt{\ell(\ell + 1)}$ (Kawaler 1986, 1987a,b, 1988). Using a variety of techniques we demonstrated that the mean period spacing from the observations is $21.51 \pm 0.03s$ for the triplets, and 12.67 ± 0.03 for the quintuplets. The theoretical ratio for $\ell = 1$ to $\ell = 2$ period spacings is $\sqrt{3} \approx 1.73$. The ratio of the observed mean period spacings is 1.72. This independently confirms the degrees of the observed multiplets to be 1 for the triplets and 2 for the quintuplets; thus we regard these identifications as secure.

With these identifications in hand, we can use the mean period spacings to determine the total stellar mass (Kawaler 1986, 1987a,b, 1988). The value we derive (only slightly dependent on the specific theoretical model) is $M/M_{\odot} = 0.59$; the values determined independently from the two sets of modes of different degree are consistent to within $0.005 M/M_{\odot}$.

In the realistic theoretical models the period spacing between individual modes is not constant. The deviations in the uniform period spacing are caused by resonant mode trapping (Kawaler 1987a,b, 1988); the resonance is between the radial structure of a particular eigenfunction and the chemical stratification. Put another way, resonance occurs when the node of a particular eigenfunction occurs in the region of rapid change in the chemical composition—within the composition transition zone. As shown by C.J. Hansen (private communication 1984) it is the node in the horizontal displacement eigenfunction that is most important—a sensible result given the dominance of horizontal over vertical displacements for non-radial g-modes. The alteration of the location of the node, forced by the composition transition zone, causes an alteration in the eigenfunction of the trapped mode and thereby a change in the frequency: the period spacing between a trapped mode and an adjacent mode is altered. Thus, as Kawaler and Hansen (1989) put it in their review in Hannover: "In the theorist's dream case, where each mode in a series were present, we could use the departures from uniform spacings to diagnose the compositional structure of the outer layers of WDV's."

As it turns out, we have exactly the theorists "dream case". The observed period spacing between consecutive radial overtones of the same degree is only roughly constant. We observe significant deviations from the mean value. The observed period spacings for PG 1159-035 are displayed in Figure 2. Comparison of this result with the handful of existing theoretical calculations of period distributions in compostionally stratified DOV models (Kawaler 1987a,b, Kawaler and Hansen 1989, and Cox 1989) indicates that PG 1159-035 is indeed compostionally stratified. This is sufficient to underscore the value of seismological techniques, and underscores the need to model the observed period structure sytematically. Given the number of observed frequencies, uniquely determining the compositional structure of the star in this way is a realistic goal.

The above is only a brief summary of the results we have already obtained on PG 1159-035; much more can be found in the aforementioned papers. I have, I hope, at least given you a taste of the power in the combination of the theoretical tools of asteroseismology and Whole Earth Telescope data.

3.3 UPDATES ON WET TARGETS FOR XCOV IV AND XCOV V

Our primary targets for the March WET run, XCOV IV, were the variable AM CVn, also called HZ-29—presumed to be an interacting twin degenerate system, and the DAV star G117-B15A.

(1) AM CVn (Principal Investigator J.-E. Solheim):
Our goal was to determine the nature and properties of the photometric variability. As is the case for all of the objects we have observed this past spring (1990), the analysis is in progress. During the run we roughly doubled the total amount of time series data obtained on the object in the past quarter century.

A preliminary first-look analysis of the is presented by the PI in these proceedings; I summarize only a few of the results here.

Surprisingly, we did not detect the period historically ascribed to the system: 1051 s. We did, on the other hand, find significant power at six consecutive harmonics of 1051 s.

The largest amplitude power is near the first harmonic of 1051 s, at 525.62 s. It is *not* a single peak, but has two higher-frequency companions split by $21\mu Hz$. The other harmonics (of the invisible 1051 s period) typically also have higher frequency, low amplitude companions also spaced at $21\mu Hz$. There is very little evidence of other significant power in the light curve.

The extreme simplicity of the power spectrum cries out for a correspondingly simple explanation; so far this eludes us. So is it pulsation, rotation, orbital motion, or what?

(2) G117-B15A (Principal Investigator S.O. Kepler):

With the density and timing accuracy obtained from the WET data, combined with single-site data from past observations, we were able to measure the time rate of change of the 215 s pulsation. The value we obtained is $dP/dt = (+12.0 \pm 3.5) \times 10^{-15}$ s/s. This is only the second pulsating degenerate with a measured rate of period change, and is the first such measurement for a DAV star. This result is a factor of 2 to 5 larger in magnitude than expected from simple-minded theoretical models based on C/O cores. At first glance, it looks like a heavier core composition model will be needed to agree with this rate of change if it is due entirely to cooling, but the situation is fairly complicated—other effects may be present—and the jury is still out (see Fontaine *et al.*, these proceedings, for a discussion).

Currently, Kepler is investigating the structure of the other 5 periodicities in the star. These were previously reported to be unstable in amplitude. He is attempting to determine if these periods were resolved into multiplets, or are intrinsically unstable. No definitive answer is available yet.

The principle targets for XCOV V in May 1990 were both pulsating degenerates, a DBV and a DAV.

(1) GD358 DBV (Principal Investigator D.E. Winget):

We obtained 170 hours of usable coverage during the WET campaign, and followed up with additional single-site data in June and July. The spectral window on this star, which indicates the amount of spectral leakage arising from gaps in the data, is the best we have done so far with WET (see Figure 1). There are no significant side-lobe aliases present. This is particularly important as we find in excess of 180 individual frequencies in the light curve. These are distributed in fine-structure multiplets. A search for period spacing patterns indicates four significant independent spacings. These correspond extremely well with the theoretically expected values for the asymptotic period splitting for orders $\ell = 1, 2, 3$ and 4.

The fine structure of the multiplet structure is puzzling, however. The splitting is in general not uniform in frequency, and all of the multiplets seem to be triplets (with several different average spacings). This last may be an inclination effect if we are observing the star at a small angle relative to its pulsation axis. The non-uniform splitting may be comprehesible in terms of an oblique pulsator model similar to Kurtz's (e.g. Kurtz 1990) models for the Ap stars, particularly since we also see some evidence for the "hyper-fine" structure that can arise from the presence of a significant magnetic field.

The interpretation of these data is clearly going to be complex but promises to be very exciting. The scientific yield should be comparable to or greater than that for our previous run on PG 1159-035.

(2) GD 165 DAV (Principal Investigator Pierre Bergeron):

We resolved the two principal regions of power, although some additional regions of power may not be resolved in the WET data set. This is the first DAV in which we have found the characteristic multiplet structure present in PG 1159-035 and GD 358. This bodes well for the asteroseismoligical potential of this data. As was the case for GD 358, however, the multiplets do not seem to have a simple uniform splitting structure. Our coverage was not so complete as on GD 358 but should be adequate for analysis.

4. Futures

The near future includes observations of PG 1707+427 (Principal Investigator A. D. Grauer), and GD 154 (Principal Investigator G. Vauclair). In the first case, CFHT observations by Fontaine, Vauclair and Bergeron during XCOV IV reveal that it has a rich multiplet structure qualitatively similar to PG 1159-035. We can therfore realistically expect, for example, to obtain an accurate mass on this object, as well as rotation rates, magnetic field strengths and perhaps most importantly, we should learn about the degree and kind of chemical stratification. The two dominant pairs of modes have been the subject of long-term investigations by Grauer (see Grauer 1989 and references therein), and with the WET data we will also be able to refine the accuracy of the measured rates of period change for these modes.

Observations of GD 154 promise a very different kind of scientific return. It appears to be a large amplitude pulsator with a simple kind of period structure dominated by a single frequency. The archival data on this object indicates that it has a harmonic structure similar to several other WDV stars (especially the DBV PG 1351+489) (see Winget 1988a,b and Kawaler and Hansen 1989 for a discussion of this point), including the presence of significant power near (but not exactly at) 3/2 and 5/2 the fundamental frequency. These objects have been posited as candidates for study of the development of chaos (e.g. Goupil *et al.* 1988, and these proceedings). WET data will provide the kind of data necessary for this analysis: many cycles, continuous coverage, and high signal-to-noise.

The rapid progress to date has been possible because of the existence of a theoretical framework. This rapid pace of progress will continue in the future only if theoretical work is pushed forward at the same enthusiastic pace with which the observations are pursued. Much remains to be done in the way of simple linear theory, but we should not get too carried away with our success here. Clearly, our biggest theoretical weakness is in the area of non-linear processes. Every one of the pulsators observed so far has displayed some type of non-linear behavior which has defied understanding or interpretation. Some of the results, such as those for G29-38, have been dominated by it.

On the instrumental side, we need to obtain more accurate amplitude information. This is a major weakness because only the Chevreton 3-channel photometer in France, and the Montreal 3-channel photometer currently have the capability for continuous sky monitoring; this is critical for accurate amplitude information. We need 3-channel capabilites at all the sites. We are actively investigating the prospect of adding a sky channel to our existing two-channel photometers, in a form which allows a retrofit.

In addition, the existence of low frequency power ($\sim 100\mu Hz$) is clear from the data in hand on PG 1159-035 and G29-38, and possibly other objects. There is a great deal of physics contained in the low-frequency power, particularly information about conditions near the photosphere. This information would provide the perfect opportunity to tie together the results of more traditional techniques of stellar spectroscopy, multi-color photometry, and atmosphere theory with the seismological results—giving a common ground for intercomparison and calibrations.

Although WET is extremely efficient in the domain of 1000 to 50,000 μHz, differential photometry is a superior technique below $\sim 100\mu Hz$. We are currently investigating the possibility of interlacing our global network with that of M. Breger and collaborators (e.g. Breger 1990)for the purpose of cross-calibrating, and extending the frequency domain of our investigations.

Finally, it is important to note that the results of the analysis of the PG 1159-035 WET data mark a turning point in the WET project. The scientific return exceeded even our most optimistic projections. Some of the information we obtained exceeded the accuracy possible using other techniques, but much of it cannot be obtained in *any* other way. This has made the potential of asteroseismology of compact pulsators a reality; it is currently our most powerful tool for looking below the photosphere. We need no longer point to the future, saying "we think we might be able to this...", or "it might be possible to do that...". On the contrary, that future is now!

I would very much like to thank the organizers of this meeting, and in particular Gerard Vauclair. The mutual respect and the friendships forged out of the free exchange of ideas in forums such as this enable us to reach around beauracratic entanglements and politics and just do science. International meetings of this sort are the very reason projects such as the WET can exist.

REFERENCES

Breger M. 1990 *Communications In Asteroseismology*, Number 19.

Cox, A.N. 1987, in IAU Colloq. # 95, *The Second Conference on Faint Blue Stars*, eds. A.G.D. Philip, D.S. Hayes, and J. Liebert (Schenectady: Davis Press), p. 631.

Goupil, M.J., Auvergne, M., and Baglin A. 1988 *Astr. Ap.*, **196**, L13.

Grauer, A.D., Liebert, J., and Green R. 1989 in IAU Colloq. # 114, *White Dwarfs*, ed. G. Wegner, (Springer-Verlag: Berlin), 119.

Hansen, C. J., Cox, J. P. and van Horn, H. M. 1977, *Ap. J.* **217**, 151.

Jones, P.W., Pesnell, W.D., Hansen, C.J., and Kawaler, S.D. 1989, *Ap. J.*, **336**, 403.

Kawaler, S.D. 1986, Ph. D. Thesis, University of Texas.

Kawaler, S.D. 1987a, in Stellar Pulsation, eds. A.N. Cox, W.M. Sparks, and S.G. Starrfield (Berlin: Springer-Verlag), p. 367.

Kawaler, S.D. 1987b, in IAU Colloq. # 95, *The Second Conference on Faint Blue Stars*, eds. A.G.D. Philip, D.S. Hayes, and J. Liebert (Schenectady: Davis Press), p. 297.

Kawaler, S.D. 1988, in IAU Symposium # 123, *Advances in Helio- and Asteroseismology*, ed. J. Christiansen-Dalsgaard and S. Fransden, (Dodrecht: Reidel), p. 329.

Kawaler, S.D., and Hansen, C.J. 1989, in IAU Colloq. # 114, *White Dwarfs*, ed. G. Wegner, (Springer-Verlag: Berlin), 97.

Kepler, S.O., *et al.* 1990, *Ap. J.*, **357**, 204.

Kurtz, D. W. 1990, *Ann. Rev. Atron, Astr.*.

Nather, R.E. 1989, in IAU Colloq. # 114, *White Dwarfs*, ed. G. Wegner, (Springer-Verlag: Berlin), p. 109.

Nather, R.E., Winget, D.E., Clemens, J.C., Hansen, C.J., and Hine, B.P. 1990, *Ap. J.*, **361**, 309.

O'Donoghue, D. *et al.* 1990, *M.N.R.A.S.*, **245**, 140.

Provencal, J.L. *et al.* 1989 in IAU Colloq. # 114, *White Dwarfs*, ed. G. Wegner, (Springer-Verlag: Berlin), 296.

Winget, D.E. 1988a, in IAU Symposium #123, *Advances in Helio- and Asteroseismology*, eds. J. Christensen-Dalsgaard and S. Frandsen (Dordrecht: Reidel), p. 305.

Winget, D.E. 1988b, in Workshop Proceedings, in Workshop Proceedings *Multimode Stellar Pulsations*, eds. G. Kovács, L. Szabados, and B. Szeidl (Budapest: Konkoly Observatory), p. 181.

Winget, D.E. *et al.* 1990, *Ap. J.*, **357**, 630.

Winget, D. E. *et al.* 1991, *Ap. J.*, submitted.

A MEASUREMENT OF THE EVOLUTIONARY TIMESCALE OF THE COOL WHITE DWARF G117-B15A WITH WET

S.O. KEPLER[1,2], A. KANAAN[1,*], D.E. WINGET[3], R.E. NATHER[3],
P.A. BRADLEY[3], J.C. CLEMENS[3], S.J. KLEINMAN[3], C.F. CLAVER[3],
J.L. PROVENCAL[3], A.D. GRAUER[4], G. FONTAINE[2,*], P. BERGERON[2,*],
F. WESEMAEL[2], M.A. WOOD[2], G. VAUCLAIR[5,*], T.M.K. MARAR[6],
S. SEETHA[6], B.N. ASHOKA[6], T. MAZEH[7], E. LEIBOWITZ[7],
N. DOLEZ[8], M. CHEVRETON[9], M. A. BARSTOW[10], A.E. SANSOM[10],
R.W. TWEEDY[10], B.P. HINE[11], J.-E. SOLHEIM[12], and P.-I. EMANUELSEN[12]

[1] Instituto de Fisica, Universidade Federal do Rio Grande do Sul, Brazil
[2] Département de Physique, Université de Montréal, Canada
[3] McDonald Observatory and Department of Astronomy, The University of Texas at Austin, USA
[4] Department of Physics and Astronomy, University of Arkansas, USA
[5] Observatoire Midi-Pyrenees, France
[6] Technical Physics Division, ISRO Satellite Centre, India
[7] Wise Observatory, The Sackler Faculty of Exact Sciences, Tel Aviv University, Israel.
[8] C.E.R.F.A.C.S, Toulouse, France
[9] Observatoire de Paris-Meudon, France
[10] Department of Physics and Astronomy, University of Leicester, UK
[11] NASA Ames Research Center, USA
[12] Institute of Mathematical and Physical Sciences, University of Tromsø, Norway
* Visiting astronomer at LNA-Laboratorio Nacional de Astrofisica, Brazil.
⋆ Visiting astronomer at CFHT-Canada-France-Hawaii Telescope.

ABSTRACT. We have measured the cooling rate of the 13,000K DA white dwarf G117-B15A by measuring the rate of period change with time for the main pulsation of this ZZ Ceti star, using the Whole Earth Telescope. The observed rate of period change is *larger* than the predictions of g-mode pulsation theory applied to C/O core white dwarf models.

1. Introduction

G117-B15A is a DAV white dwarf — a pulsating ZZ Ceti star — discovered to pulsate by McGraw and Robinson (1976). The DAV stars are single white dwarfs with pure hydrogen spectra which show luminosity variations. These pulsating stars are normal except for their luminosity variations, and are all inside a narrow instability strip around $T_{eff} \sim 12,000K$. Most, if not all, stars inside the instability strip are variable (Fontaine *et al.* 1982, Greenstein 1982), indicating that pulsation is a normal phase of the white dwarf evolution, i.e., the star pulsates as it cools

143

G. Vauclair and E. Sion (eds.), White Dwarfs, 143–151.

down to the temperature range of the instability strip. At this temperature a zone of partial ionization of H develops, due to the recombination of H at the surface. All ZZ Ceti stars show multiperiodic light variations, with main periods ranging from 109 sec to 1186 sec, and peak-to-peak amplitudes ranging from 2 to 34%.

Of the 21 known ZZ Ceti stars, G117-B15A is one of the hottest, at $T_{eff} = 13,200K$ (Daou *et al.* 1990). We have been measuring the light curve of G117-B15A (RY LMi), since 1975, to obtain its rate of change of period with time, which is a measure of the cooling rate of the white dwarf. The only measurement to this date of a stellar evolutionary timescale is for the hot pre-white dwarf PG1159-035 (GW Vir), by Winget *et al.* (1985). The detection we report in this paper is the first measurement of an evolutionary timescale for a *cool* white dwarf.

The observation of white dwarfs in open clusters suggest that they are the end results of the evolution of stars with masses less than 7-8 solar masses. The theory of stellar evolution applied to these intermediate mass stars says that the white dwarfs must have undergone hydrogen and helium core and shell-burning stages, leaving cores typically composed of the end products of He burning: carbon and oxygen. The exact ratio of these elements is still unknown due to uncertainties in the rates of He burning and in the amount of mass loss during the evolution. The rate of period change of a g-mode nonradial pulsation in a white dwarf reflects directly the cooling timescale of the star, which in turn reflects the core composition through the heat capacity. If we can measure the rate of period change we can then obtain a direct measurement of the core composition, which will help us to test our theories of stellar evolution and even the rates of the nuclear reactions.

Kepler *et al.* (1990) reported the study of all the photometric data on G117-B15A obtained during the last 14 years, up to 1989, and obtained a limit to the rate of change of period with time of $\dot{P} = (8.4 \pm 5.0) \times 10^{-15}$ s/s. That study suggested that it would take at least 4 years to confirm the measurement, if we continued to observe the star from a single site each night, because the uncertainty of the measurement decreases proportionally to the time span observed squared. It is also true that the precision of the measurement is proportional to the square root of the number of photons observed, so a large improvement in the photon statistics could also decrease the time necessary to obtain a 3σ measurement.

Considering the measurement of the evolutionary timescale to be extremely important, we decided to include G117-B15A in the 1990 target list for the Whole Earth Telescope (WET, Nather *et al.* 1990), to use the long span of the observations as a method for acquiring enough photons in a single season so that the improvement in the photon statistics would be enough to circumvent the small increase in the total baseline of the data.

2. Observations

We obtained 86.4 hr of high speed photometry in March 1990, using either two-star or three-star photometers on telescopes with sizes up to 3.6m in diameter (CFHT). Since G117-B15A is faint (V=15.52, Eggen and Greenstein 1965), and the light variations are confirmed to be g-mode nonradial pulsations — and therefore all light variation at differing wavelengths are in phase (Kepler 1984) — all the observations were obtained with unfiltered light, improving further our photon statistics.

The data were reduced and analyzed as described in Kepler *et al.* (1982), and the observed times were reduced to the Barycentric Julian Dynamical Date (BJDD), former Barycentric Julian Ephemeris Date, from the UTC received at each observatory, including the leap second corrections and the light travel time correction to the barycenter of the solar system due to the motion of the Earth. After obtaining a Fourier transform of each light curve individually to make sure that the main pulsation at 215 sec was still dominating the light curve and showed no evidence of amplitude or abrupt period changes, we obtained a Fourier transform of the whole data set at once, which showed that there is no contamination of the main pulsation by any other periodicity, since we see only one periodicity within the detection limit around 1 mmag.

After concluding that the 215 sec pulsation has remained stable in both frequency and amplitude since our first observation in 1975, we calculated the time of maximum for each observation separately, and included the new measurements in the (O-C) diagram with the data sets reported in Kepler *et al.* (1990), which comprised 196 hr of high speed photometry obtained from 1975 to 1989.

The rate of period change for the pulsation, as well as a correction to the period and epoch, was then obtained by fitting a parabola to the (O-C); *i.e.* we assume a fit of the form $C = E_0 + P \cdot E$ and what we obtain from the fit are corrections to the initial values of P and E_0, which we call ΔE_0 and ΔP:

$$(O - C) = \Delta E_0 + \Delta P \cdot E + \frac{1}{2} P \cdot \dot{P} \cdot E^2 \tag{1}$$

where ΔE_0 is the correction to the epoch of observation, ΔP is the correction to the period, E is the number of cycles elapsed since E_0, and $\dot{P}$ is the rate of change of period with time, dP/dt. Note the factor of $\frac{1}{2}$ in the above definition of $\dot{P}$, that comes in because $\dot{P}$ is in the second derivative with respect to time term, as you can see from the following derivation.

2.1 DERIVATION OF $\dot{P}$

The time of maximum of the pulsation T_{max} is supposed to follow the relation:

$$T_{max} = E_0 + P \cdot E, \tag{2}$$

and therefore

$$\frac{dT_{max}}{dE} = P. \tag{3}$$

If we now suppose the period is changing with time, we can expand T_{max} by a Taylor series:

$$T_{max} = Tmax\Big|_{E_0} + \frac{dT_{max}}{dE}\Big|_{E_0}(E - E_0) + \frac{1}{2}\frac{d^2T_{max}}{dE^2}\Big|_{E_0}(E - E_0)^2. \tag{4}$$

Writting

$$\frac{d^2T_{max}}{dE^2} = \frac{dP}{dE} = \frac{dt}{dE}\frac{dP}{dt} = P\frac{dP}{dt}, \tag{5}$$

we get

$$T_{max} = T_{max}\Big|_{E_0} + P(E - E_0) + \frac{1}{2}P \cdot \dot{P}(E - E_0)^2,$$

$$= \left(T_{max}\Big|_{E_0} - P \cdot E_0 + \frac{1}{2}P \cdot \dot{P} \cdot E_0^2\right) + P \cdot E + \frac{1}{2}P \cdot \dot{P}\left(E^2 - 2E_0 \cdot E\right). \quad (6)$$

Assuming $2E_0 \ll E$, we get:

$$T_{max} = T_{max}^0 + P \cdot E + \frac{1}{2}P \cdot \dot{P} \cdot E^2. \quad (7)$$

If we suppose: $O \equiv T_{max}^{obs} = T_{max}$, and $C \equiv T_{max}^1 + P_1 \cdot E$, we get:

$$(O - C) = (T_{max}^0 - T_{max}^1) + (P - P_1)E + \frac{1}{2}P \cdot \dot{P} \cdot E^2,$$

$$(O - C) = \Delta E_0 + \Delta P + \frac{1}{2}P \cdot \dot{P} \cdot E^2,$$

where $\Delta E_0 = (T_{max}^0 - T_{max}^1)$, and $\Delta P = (P - P_1)$, which is our equation (1).

By fitting equation (1) to the (O-C), we obtained a new value for the epoch of maximum, $E_0 = 244, 2397.917521\,\mathrm{BJDD} \pm 0.6\,\mathrm{s}$, a new value for the period, $P = 215.197387 \pm 0.000001\,\mathrm{s}$, and most importantly, a rate of period change of:

$$\dot{P} = (12.0 \pm 3.5) \times 10^{-15}\,\mathrm{s/s},$$

a 3.4σ measurement. The observed rate of period change implies a timescale of period change of:

$$P/\dot{P} = (5.7 \pm 1.7) \times 10^8\,\mathrm{yr},$$

which is directly related to the evolutionary timescale of the white dwarf.

3. Discussion

From the theory of stellar pulsations, we know that the period of an oscillation changes in response to the changes in the structure of the star, caused by its evolution. The theory of g-mode non-radial pulsations in white dwarfs predicts that the timescale of period change is related to changes in the radius and temperature of the star, as

$$\frac{P}{\dot{P}} = a\frac{T}{\dot{T}} - b\frac{R}{\dot{R}}, \quad (8)$$

with the a and b constants of the order of unity (see Winget, Hansen, and Van Horn 1983). For the DAVs, $\dot{R}/R \ll \dot{T}/T$, because white dwarfs at this temperature cool nearly at constant radius. Therefore, the timescale of period change is essentially a measurement of the cooling timescale of the white dwarf.

There are, however, other possibilities that can affect the rate of period change; the layer trapping the mode might be changing with time, affecting the period. It is well established that the selection mechanism for the modes present in the pulsating

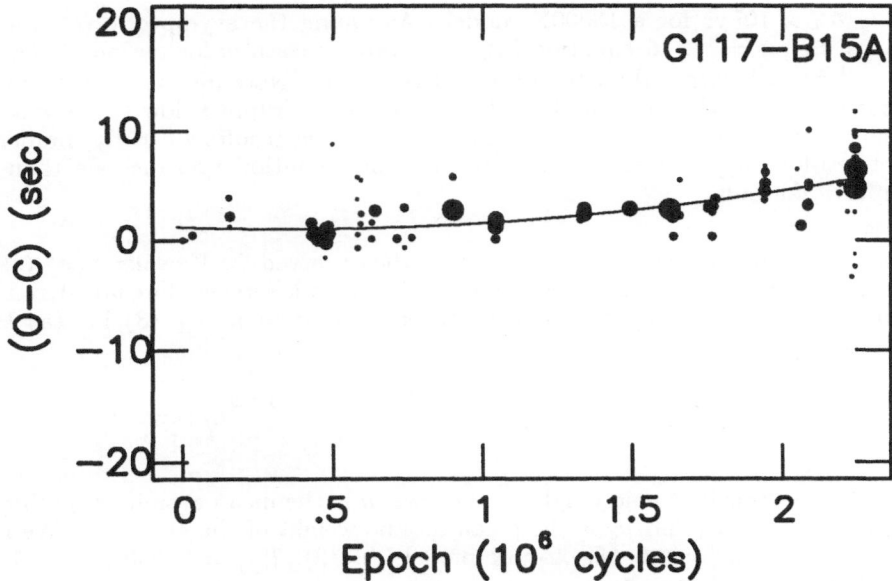

Figure 1. The (O-C) diagram for the 215 s pulsation of G117-B15A. The size of points is inversely proportional to the error of that point, so the larger the point the higher its weight. For the largest points, the error bars are smaller than the size of the point itself. The parabola is the best fit $\dot{P} = 12 \times 10^{-15}$ s/s.

white dwarfs is the trapping of the oscillations by the layered structure of the star (Winget 1981, Dolez and Vauclair 1981, Winget *et al.* 1982, Wood and Winget 1988). If, by effects of evolution, the layer width is changing, or the width of the transition zone between layers is changing, then the pulsation period might change accordingly. It is still an evolutionary effect, but not a direct measurement of the cooling of the white dwarf. Bradley and Winget (1990) have demonstrated however that this phenomenon, seen as avoided crossings, only decreases the observed rate of period change, but does not increase the theoretical rate.

3.1 THEORETICAL $\dot{P}$

It is important to compare the observed value for the timescale of period change and those predicted by the theory of g-mode pulsation applied to evolutionary models of white dwarfs in the temperature range of G117-B15A. The pulsation models calculated by Winget (1981), Bradley, Winget and Wood (1989), and Bradley and Winget (1990), all predict rates of period changes $\frac{\Delta P}{\Delta t} \simeq 3 - 4 \times 10^{-15}$ s/s, for the low ℓ, low k oscillations observed.

D'Antona and Mazzitelli (1990, private communication) calculated evolutionary models for a $10^{-8} M_H/M_*$ surface layer, 0.564 $M_\odot$, C/O core DA white dwarf, similar to their C/O-core DB models (D'Antona and Mazzitelli 1989), obtaining

$T/\dot{T} \simeq 4.7 \times 10^8$ yr for a 13000K model. Assuming the asymptotic relation for g-modes, *i.e.*, $a = 1/2$ in equation (3), we obtain timescales for period changes of 9.4×10^8 yr, still larger than the observed timescale. Note also that this value is uncertain because the 215 s periodicity is probably a trapped, low ℓ, low k mode, and so the asymptotic relation may not give reliable results. Finally, numerical model results suggest even larger values for the evolutionary timescale than the asymptotic formula.

3.2 TOTAL MASS

Let us now study the effect of total mass on the observed $\dot{P}$. Kawaler *et al.* (1986) derived a theoretical rate of period change assuming a Mestel cooling law, Kramer's opacity law, and the asymptotic value for the a constant on our eq. (3), i.e. (a=1/2). Their eq. (3) is

$$\frac{d\ln P}{dt} = 2 \times 10^{-30} A \left(\frac{\mu}{\mu_e^2}\right)^{0.286} \left(\frac{M}{M_\odot}\right)^{-1.190} T_{eff}^{2.857}, \tag{9}$$

where A is the mean atomic weight in the core, μ is the mean atomic weight of the ions in the envelope, and μ_e is the mean atomic weight of the electrons. We note that when applied to: 0.6 $M_\odot$, C-core model (A=12), $T_{eff} = 13000K$, $\mu = 1$, we get $\dot{P} = 4.6 \times 10^{-15}$ s/s, which compares well with the model calculations referred above. Here we are interested only on the mass dependence of $\dot{P}$, so the value of a is not important in this derivation; we assume the $\dot{P} = 4 \times 10^{-15}$ s/s of the 0.6 $M_\odot$ numerical models above, and derive a mass of 0.24 $M_\odot$ for a $\dot{P} = 12 \times 10^{-15}$ s/s like the one observed. Even if we use the upper value for $\dot{P}$ obtained from the models for 0.6 $M_\odot$, *i.e.*, 7×10^{-15}, the mass which would give the observed $\dot{P}$ is 0.38 $M_\odot$, still in disagreement with the value of $\log g = 7.81 \pm 0.06$ derived form the observed spectra (Daou *et al.* 1990), which, when compared to the evolutionary models of Wood (1990) gives a mass of $M = 0.49 \pm 0.03$ $M_\odot$. Therefore, we cannot explain the observed $\dot{P}$ from mass effects.

3.3 CORE COMPOSITION

Since the heavier the particles that compose the nucleus of the white dwarf, the faster it cools, the observed timescale implies a composition heavier than the C/O or pure C models calculated. A Ne-Mg core, which would give the observed $\dot{P}$, is an interesting possibility because traces of Ne and Mg have been detected in the spectra of novae (Ferland and Shields 1978, Williams *et al.* 1983), implying that these elements are synthesized in the progenitors of white dwarfs.

As stated above, the observed spectra of G117-B15A is inconsistent with a heavier than normal white dwarf. Why should then a normal (0.5 $M_\odot$) have such a heavy core composition, when the theory of stellar evolution says it should have a C/O core? Carbon-core models have been used most extensively because the currently accepted values for the nuclear reaction have suggested that the progenitors of white dwarf with masses less than 1 $M_\odot$, progenitors (main-sequence stars with masses of up to 6–7 $M_\odot$) build carbon/oxygen cores during the asymptotic giant branch double-shell-burning phase, but do not burn to heavier elements. The results presented here suggest that the core composition of the ~0.5 $M_\odot$ G117-B15A should have a mean atomic weight greater than carbon or oxygen.

4. Conclusion

The observed $\dot{P} = (12.0 \pm 3.5) \times 10^{-15}$ s/s suggests a core composition with a mean atomic weight A larger than 16, *i.e.*, a heavy core composition white dwarf.

This work was partially supported by grants from CNPq (Brazil), FAPERGS (Brazil), FINEP (Brazil), CAPES (Brazil), NSF (USA), NSF-NATO (USA), NASA (USA), CNRS (France), SERC (UK), NSERC (Canada), FCAR (Canada), and the Basic Research Branch of the Israeli Academy of Sciences.

REFERENCES

Bradley, P.A., and Winget, D.E., 1990, *Ap.J.*, in press.

Bradley, P.A., Winget, D.E., and Wood, M.A., 1989, in *Proceedings of the IAU Colloquium 114, White Dwarfs*, ed. G. Wegner, (Springer-Verlag, Berlin), p.286.

Caughlan, G. R., and Fowler, W. A. 1988, *Atomic Data and Nuclear Data Tables*, **40**, 334.

D'Antona, F., and Mazzitelli, I. 1989, *Ap.J.*, **347**, 934.

Daou, D., Wesemael, F., Bergeron, P., Fontaine, G., and Holberg, J.B., 1990, *Ap.J*, in press.

Dolez, N. and Vauclair, G. 1981, *Astron. Astrophys.*, **102**, 375.

Eggen, O.J., and Greenstein, J.L. 1965, *Ap. J.*, **141**, 183.

Ferland, G.J., and Shields, G.A. 1978, *Ap. J.*, **226**, 172.

Fontaine, G., McGraw, J.T., Dearborn, D.S.P., Gustafson, J., and Lacombe, P. 1982, *Ap. J.*, **258**, 651.

Greenstein, J.L. 1982, *Ap. J.*, **258**, 661.

Kawaler, S.D., Winget, D.E., Iben, I., Jr., and Hansen, C.J. 1986, *Ap.J.*, **302**, 530.

Kepler, S.O. 1984, *Ap. J.*, **286**, 314.

Kepler, S.O., Robinson, E.L., Nather, R.E., and McGraw, J.T. 1982, *Ap. J.*, **254**, 676.

Kepler, S.O., Vauclair, G., Dolez, N., and Chevreton, M., Barstow, M.A., Nather, R.E., Winget, D.E., Provencal, J.L, Clemens, J.C., and Fontaine, G. 1990, *Ap.J.*, **357**, 204.

McGraw, J.T., and Robinson, E.L. 1976, *Ap. J. (Letters)*, **205**, L155.

Nather, R.E., 1989, in *Proceedings of the IAU Colloquium 114, White Dwarfs*, ed. G. Wegner, (Springer-Verlag, Berlin), p. 109.

Nather, R.E., Winget, D.E., Clemens, J.C., Hansen, C.J., and Hine, B.P., 1990, *Ap.J.*, in press.

Williams, R.E., Ney, E.P., Sparks, W.M., Starrfield, S., and Truran, J.W. 1985, *MNRAS*, **212**, 753.

Winget, D.E., 1981, *Ph.D. Thesis*, University of Rochester.

Winget, D.E., Hansen, C.J., Liebert, J., Van Horn, H.M., Fontaine, G., Nather, R.E., Kepler, S.O., and Lamb, D.Q. 1987, *Ap. J.*, **315**, L77.

Winget, D. E., Hansen, C. J., and Van Horn, H. M. 1983, *Nature*, **303**, 781.

Winget, D.E., Kepler, S.O., Robinson, E.L., Nather. R.E., and O'Donoghue, D. 1985, *Ap. J.*, **292**, 606.

Winget, D.E., Van Horn, H.M., Tassoul, M., Hansen, C.J., Fontaine, G., and Carrol, B.W. 1982, *Ap. J.*, **252**, L65.

Wood, M.A., and Winget, D.E., 1988, in *Multimode Stellar Pulsations*, ed. G. Kovacz, L. Szabados, and B. Szeidl, (Konkoly Observatory-Kultura:Budapest), p. 199.

Wood, M.A., 1990, *Ph.D. Thesis*, The University of Texas at Austin.

Wood, M.A., and Winget, D.E., 1989, in *Proceedings of the IAU Colloquium 114, White Dwarfs*, ed. G. Wegner, (Springer-Verlag, Berlin), p. 282.

DISCUSSION

VAN HORN :

Could the star be in Debye cooling, with a smaller A?

KEPLER :

No, Ne/Mg starts to crystallize at 13.000 K, so it cannot be in Debye cooling.

ON THE INTERPRETATION OF THE dP/dt MEASUREMENT IN G117-B15A

G. FONTAINE, P. BRASSARD, F. WESEMAEL, S.O. KEPLER[1],
and M.A. WOOD
Département de Physique, Université de Montréal

The measurement of a rate of period change in the pulsating DA white dwarf G117-B15A reported by Kepler at this Workshop (Kepler *et al.* 1991) is both an exciting and important result since it potentially allows us to infer, for the very first time, the core composition of a white dwarf. This is because the period evolution of a ZZ Ceti star is directly tied to its cooling time scale which, in turn, depends sensitively on the core composition.

Using the reported value $\dot{P} \simeq 12.0 \times 10^{-15} s/s$ (Kepler *et al.* 1991) for the observed $215s$ pulsation mode, one obtains an estimate of the time scale for period change of the order $P/\dot{P} \simeq 5.7 \times 10^8 yrs$. According to Winget, Hansen, and Van Horn (1983), this also gives a rough estimate of the cooling time scale of G117-B15A. However, to make inferences about the core composition of the star, this estimate is too crude since it totally ignores the dependence on the actual mode being observed and on the stellar parameters. Instead, to use all the information, it is necessary to compare the observed rate of period change with values computed from detailed pulsation analyses of appropriate evolutionary models of G117-B15A.

In our comparisons, we have adopted a mass $M \simeq 0.5 M_\odot$ and an effective temperature $T_e \simeq 13,000K$ for G117-B15A. The temperature comes from the *IUE* analyses of Wesemael, Lamontagne, and Fontaine (1986; see also Lamontagne, Wesemael, and Fontaine 1987), the Strömgren photometry of Fontaine *et al.* (1985), and the recent spectroscopic study of Daou *et al.* (1990). The paper by Wesemael *et al.* (1991), presented at this Workshop, gives further details. The mass determination is based on the gravity estimates derived in Fontaine *et al.* (1985) and Daou *et al.* (1990) in conjunction with the evolutionary models of Wood (1990).

Our theoretical $\dot{P}$ values are based on the extensive adiabatic survey of Brassard *et al.* (1991*a,b*) which has examined some 80,000 modes for the evolutionary models of Tassoul, Fontaine, and Winget (1990; hereafter TFW). These equilibrium models are chemically layered and consist of a pure carbon core surrounded by a pure helium mantle, itself enveloped by a pure hydrogen layer. For each evolutionary sequence considered, we have compiled the expected $\dot{P}$ values for the particular overtone which has in the model nearest to $13,000K$, a period closest to the observed value of $215s$.

---

[1] Permanent address: Instituto de Fisica da UFRGS, Brazil

G. Vauclair and E. Sion (eds.), White Dwarfs, 153–158.
© 1991 *Kluwer Academic Publishers.*

In practice, the periods of the selected overtones fell in the range $190 - 240s$. Because we do not yet know the actual ℓ index of the observed g-mode in G117-B15A, we have repeated this exercise for modes with $\ell = 1, 2$, and 3, those being the most likely values in a ZZ Ceti star. We have compiled the results from some 28 evolutionary sequences which differ in total mass and in H/He layering. These results indicate that the most likely $\dot{P}$ value for a mode with a period around $P \simeq 215s$ in a $0.5M_\odot$, $13,000K$ DA white dwarf model is,

$$\dot{P}(\text{pure C core; TFW}) \simeq 4.3 \times 10^{-15} s/s \quad , \qquad (1)$$

with a rms deviation of $0.5 \times 10^{-15} s/s$. The most extreme individual $\dot{P}$ values that we found were,

$$\dot{P}_{\max}(\text{pure C core; TFW}) \simeq 5.3 \times 10^{-15} s/s \quad , \qquad (2a)$$

and

$$\dot{P}_{\min}(\text{pure C core; TFW}) \simeq 2.6 \times 10^{-15} s/s \quad . \qquad (2b)$$

To our knowledge, theoretical $\dot{P}$ values for ZZ Ceti stars are only published for models with pure carbon cores. Since our goal is to infer the chemical composition of the core of G117-B15A, we have to devise a scaling argument for compositions other than pure carbon. Fortunately, for pulsating DA white dwarfs, we can write,

$$\dot{P} = \frac{dP}{dT_e} \frac{dT_e}{dt} \quad . \qquad (3)$$

Here, the first term on the right hand side is a mechanical structure term which to a very good approximation (a few percent) does *not* depend on the internal composition for material with $\mu_e = 2$. This is because the period of a given mode for a model with known M and T_e depends essentially only on the degenerate electron pressure (the same for all material with $\mu_e = 2$) and (to a lesser extent) on the stratification of the outer layers.

The second term in equation (3) is the thermal structure term which, by contrast to the first one, depends almost exclusively on the core composition (for given M, T_e, and H/He layering) in the temperature range characteristic of cool white dwarfs.

The scaling strategy is therefore quite clear: use dP/dT_e as obtained in detailed pulsation studies such as that of Brassard *et al.* (1991a,b) for pure carbon-core models. Next, scale $\dot{P}$ as given by equation (3) using dT_e/dt obtained from various evolutionary calculations of white dwarfs with different core compositions and/or physics.

In this spirit, we have used the detailed evolutionary models of Wood (1990) to obtain improved estimates of theoretical $\dot{P}$ values for G117-B15A. These models differ from those of TFW in that they use an improved interior equation of state as well as more modern conductive opacities. As a result, they provide better estimates of the cooling time scales than the TFW models. Using the scaling relation given by equation (3), the results of Brassard *et al.* (1991a,b) for dP/dT_e, and the cooling sequences of Wood (1990) and TFW, we find,

$$\dot{P}(\text{pure C core; Wood}) \simeq 4.8 \times 10^{-15} s/s \quad , \qquad (4a)$$

and

$$\dot{P}(\text{pure O core; Wood}) \simeq 5.6 \times 10^{-15} s/s \quad . \tag{4b}$$

These values are, again, appropriate for pulsations around $\sim 215s$.

For core computations heavier than oxygen, detailed evolutionary calculations are unfortunately not available. Because of this, our scaling relation (equation [3]) cannot be used, and we must rely on the much cruder approximation provided by Mestel's theory of cooling (Mestel 1952), which suggests that,

$$\frac{dT_e}{dt} \propto A \quad , \tag{5}$$

where A is the mean atomic weight of the core constituent. Combining equations (3), (4b), and (5), we find the following approximate scaling law,

$$\dot{P} \simeq \frac{A}{16} 5.6 \times 10^{-15} s/s \quad , \quad A \geq 16 \quad , \tag{6}$$

a relation applicable to a mode around $215s$ in a $0.5 M_\odot$ star at $T_e \sim 13,000 K$.

The observed rate of period change is $\dot{P}_{\text{obs}} = (12.0 \pm 3.5) \times 10^{-15} s/s$ (Kepler et al. 1991), a value much larger than the predictions for pure carbon- or oxygen-core models (see our equations [4a] and [4b]). This immediately suggests core compositions heavier than oxygen. Indeed, using equation (6), we find

$$A \simeq 16 \frac{(12.0 \pm 3.5) \times 10^{-15}}{5.6 \times 10^{-15}} \simeq 34 \pm 10 \quad . \tag{7}$$

Taken at face value, this result indicates that the core of G117-B15A must at least be made of ions as heavy as Mg at the 1σ level. This is completely at odds with the predictions of standard evolution theory which suggest instead C/O cores for ordinary white dwarfs. Note that the maximum theoretical $\dot{P}$ value which we found in the TFW models is $\dot{P}_{\max}(\text{pure C core; TFW}) \simeq 5.3 \times 10^{-15} s/s$. Using this maximal value instead of the most likely value of $\dot{P}(\text{pure C core; TFW}) \simeq 4.3 \times 10^{-15} s/s$, and repeating our scaling procedure, the range of possible core compositions is then given by $A = 28 \pm 8$ (1σ). Thus, even for this extreme estimate, the core of G117-B15A consists of elements still at least as heavy as Ne, a result which remains in strong conflict with the predictions of evolution theory. There is thus no obvious way to reconcile the relatively large observed $\dot{P}$ in G117-B15A with the expectations of stellar evolution, unless the observed $\dot{P}$ value is pushed down to $5.0 \times 10^{-15} s/s$, 2σ away from the most probable value (10% probability).

The heavy core compositions inferred for G117-B15A certainly add to the list of "mysteries" discussed at this Workshop. Because ordinary white dwarfs are supposed to be made of C/O cores, we may ask the question: are we unlucky enough that G117-B15A (the very first ZZ Ceti star for which a $\dot{P}$ measurement has been made) seems to be a freak? On the other hand, we can turn this around and ask: are we lucky enough to have uncovered a star which poses a real challenge to stellar evolution theory? Indeed, how does one make a $0.5 M_\odot$ white dwarf with a core composition made of Ne or heavier material? Of course, it is always possible that we are not measuring what we think we are. There could be, for example, some (non-cooling) contribution to $\dot{P}_{\text{obs}}$ due to some as yet unknown mechanism. This possibility is further explored in a more detailed forthcoming publication (Fontaine et al. 1991).

156

This work has been supported in part by the NSERC Canada, by the Fund FCAR (Québec), and by NATO. G. Fontaine acknowledges additional financial support through a Killam Fellowship. S.O. Kepler acknowledges additional financial support by CNPq-Brazil, and M.A. Wood acknowledges support from the NATO Postdoctoral Fellowship Program.

REFERENCES

Brassard, P., Fontaine, G., Pelletier, C., and Wesemael, F. 1991*a*, these Proceedings.
Brassard, P., Fontaine, G., Wesemael, F., and Tassoul, M. 1991*b*, *Ap. J.*, in press.
Daou, D., Wesemael, F., Bergeron, P., Fontaine, G., and Holdberg, J.B. 1990, *Ap. J.*, in press.
Fontaine, G., Bergeron, P., Lacombe, P., Lamontagne, R., and Talon, A. 1985, *Ap. J.*, **90**, 1094.
Fontaine, G., Brassard, P., Wesemael, F., Kepler, S.O., Wood, M.A., Bergeron, P., and McGraw, J.T. 1991, *Ap. J.*, in press.
Kepler, S.O., *et al.* 1991, these Proceedings.
Lamontagne, R., Wesemael, F., and Fontaine, G. 1987, in *IAU Colloquium 95, The Second Conference on Faint Blue Stars*, ed. A.G.D. Philip, D.S. Hayes, and J. Liebert (Schenectady: L. Davis Press), p. 677.
Mestel, L. 1952, *M. N. R. A. S.*, **112**, 583.
Tassoul, M., Fontaine, G., and Winget, D.E. 1990, *Ap. J. Suppl.*, **72**, 335.
Wesemael, F., Lamontagne, R., and Fontaine, G. 1986, *A. J.*, **91**, 1376.
Wesemael, F., Bergeron, P., Fontaine, G., and Lamontagne, R. 1991, these Proceedings.
Winget, D.E., Hansen, C.J., and Van Horn, H.M. 1983, *Nature,* **303**, 781.
Wood, M.A. 1990, Ph. D. thesis, University of Texas at Austin.

DISCUSSION

VAN HORN :

We know that gravitational settling gradually changes the composition profile in the surface layers of the white dwarfs. In particular, calculations by your group show that the heavy elements sink increasing slowly into the core, building up a layer that is slightly enriched in metals. This will change the thermal structure of the surface region. Since the g-mode periods are very sensitive to the thermal structure, is there any chance that the $\dot{P}$ could be measuring this effect instead of the cooling time of the core?

FONTAINE :

This is a very interesting point. You have shown yourself that what $\dot{P}$ measures is the local cooling time scale in the region of period formation. In our case, this region is located in the envelope where indeed settling of heavy elements does occur. I do not know how large an effect the settling of heavy elements (and the associated change in the opacity) can have on $\dot{P}$. My guess is that it will fall short of the factor of 2 needed to explain the observed $\dot{P}$ in terms of 2 C/O core. But I may be quite wrong on that, we need to do the calculations. Also, the possible binary nature of G117-B15A may be a factor here, but this also needs to be investigated.

VAN HORN :

If the core composition is heavier, as you suggest, the star will reach Debye cooling faster, is this a possible explanation for the larger $\dot{P}$?

FONTAINE :

We have checked that and we find that crystallization effects are essentially negligible for core composition as heavy as calcium for a 0.5 $M_\odot$ at $T_e \simeq 13,000K$. So the answen is no.

WOOD :

Although the hydrogen/helium composition-transition profile reaches diffusive equilibrium well before reaching the ZZ Ceti instability strip, it seems plausible to me that the dynamics of the pulsations (i.e., the shear in the fluid layers) may act to increase the breadth of the transition zone. Recalling a figure of Winget and Fontaine (1982; in "Pulsations in classical and cataclysmic variable stars" eds. J.P. Cox and C.J. Hansen (Boulder; Univ. of Colorado Press) p 46) which shows the oscillation spectra obtained for 3 models differing only in transition-zone thickness, there is a trend for increasing period (for a given mode) for increasing transition-zone thickness. Because G117-B15A is at the blue edge of the instability strip, and unless the diffusion timescale is much less than 215 s at a depth of $\sim 10^{-10}$ M$_*$, this mechanism could account for some fraction of the observed P.

FONTAINE :

I am not sure of the real effects of pulsation dynamics on the transition zone structure. My guess is that the composition profile would react on a time scale smaller than the pulsation periods, so it would not be very much disturbed In any case, the increase at $\dot{P}$ due to a broader transition zone is not going to be very large. We have a sequence at models with an artificially reduced transition zone; the $\dot{P}$ differences with normal models are small.

THE BOUNDARIES OF THE ZZ CETI INSTABILITY STRIP

F. WESEMAEL
Département de Physique, Université de Montréal

P. BERGERON
Steward Observatory, University of Arizona

G. FONTAINE and R. LAMONTAGNE
Département de Physique, Université de Montréal

Several investigations over the past ten years have shown that the existence of pulsating, DA white dwarfs is restricted to a rather narrow instability strip. And indeed, fast-photometric searches for pulsating stars in that class, coupled to statistical studies of the resulting frequency of variable stars, provide strong evidence that the ZZ Ceti stars represent an evolutionary phase through which most, if not all, hydrogen-atmosphere white dwarfs are expected to cool. This result provides an obvious incentive to study these stars in greater detail, as seismological studies of ZZ Ceti stars might then eventually provide constraints on the properties not only of the two dozen or so known variable stars, but on those of the *whole* sample of DA stars as well.

The boundaries of the instability strip afford us with the opportunity of learning much about the structure of the outer layers of DA white dwarfs. Specifically, the comprehensive investigation of Winget *et al.* (1982) has demonstrated that:

i) The location of the blue edge is sensitive to the structure of the hydrogen convection zone, as parametrized by the convective mixing efficiency. This is because the blue edge can be identified as the effective temperature at which the thermal timescale at the bottom of the convection zone first becomes comparable to the shortest observable period associated with a particular model (conventionnally that of the $l = 3$, $k = 1$ mode).

ii) The location of the blue edge is sensitive to the thickness of the outer hydrogen layer in DA stars. The non-adiabatic calculations of Winget *et al.* (1982) demonstrate that, for fractional hydrogen layer masses exceeding 10^{-8}, the strong damping associated with the radiative zone below the driving region ensures stability until the effective temperature falls below 8000 K. For fractional layer masses below 10^{-12}, on the other hand, the hydrogen layer is too thin to support hydrogen-driving and the blue edge, now associated with helium-driving, is shifted to high effective temperatures. Thus the fractional hydrogen layer mass in ZZ Ceti stars appears effectively constrained, at least in this particular set of calculations.

iii) On the opposite side, it is generally agreed that the red edge is related to the mixing process expected when the hydrogen-rich surface layers of ZZ Ceti stars are mixed into the underlying helium envelope.

159

G. Vauclair and E. Sion (eds.), White Dwarfs, 159–165.
© 1991 *Kluwer Academic Publishers.*

We might thus be, eventually, in a position to use the observationally-determined boundaries of the instability strip to yield important constraints on the prior evolution of DA white dwarfs, as well as on the hydrodynamic properties of their outer layers.

1. A Review of Past Studies

The problem of determining the boundaries of the instability strip has been attacked with a variety of observational techniques. The initial attempts were made with broad-band colors, but Strömgren colors were later measured for a large number of ZZ Ceti and other DA stars, and reduced significantly the scatter in the location of the variable stars in two-color diagrams. In parallel, Greenstein measured a large number of those stars with the multichannel spectrophotometer, and that data base was used for several independent analyses of the instability strip. More recently, ultraviolet energy distributions obtained with the *I.U.E.* satellite have served as the basis of a new determination of the boundaries of the instability strip. The latest study relevant to this problem is the use of a spectroscopic technique which consists in fitting as much as possible of the blue optical spectrum of ZZ Ceti stars, all the way to $H9$. Slightly updated results from these various investigations are displayed, in effective temperature space, in Figure 1. Three bright ZZ Ceti stars are common to all the analyses mentioned above.

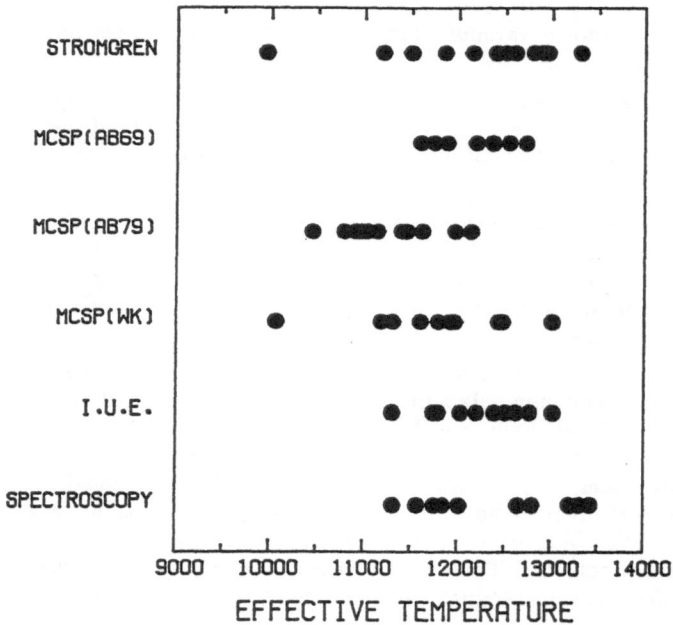

Fig. 1: Boundaries of the instability strip determined in various investigations.

Strömgren photometry of stars near the instability strip has been carried out by McGraw (1979) and Fontaine *et al.* (1985). While there have been other observations of cool DA white dwarfs on the Strömgren system, these studies are the only two which have explicitly emphasized the measurement of meaningful time-averaged colors. Both data sets are merged in Figure 1; when a star was observed in both studies, the colors were averaged. Uniform effective temperatures are obtained from Koester's models (see Weidemann and Koester 1984). The coolest variable in that sample is BPM 30551, a southern ZZ Ceti star with colors based on a single measurement by Wegner, which may not be a true time-averaged value. Because of this, its effective temperature near 10,000 K should be considered uncertain; to our knowledge, it has not been reobserved since. The next coolest star is R808, at 11,200 K. The blue edge is defined by G117-B15A, which is assigned an effective temperature near 13,300 K on the basis of its averaged colors; McGraw (1979) had fitted 13,640 K to that star on the basis of his observations and earlier models.

The most thorough studies of the atmospheric parameters of ZZ Ceti stars are based on the analysis of multichannel spectrophotometer data. Three such analyses were performed. That of Fontaine *et al.* (1982) is based on the colors of Greenstein (1976), which were on the so-called AB69 system of Oke and Schild (1970). We redetermine temperatures here from the $(G - R)$ color only, and from Shipman's (1979) models. The strip is bounded by G117-B15A, at 12,700 K, on the blue side, and by R808, at 11,600 K, on the red side. Greenstein's (1982) nearly simultaneous analysis was carried out with MCSP data on the AB79 system of Oke and Gunn (1983). Greenstein's instability strip is shifted to cooler temperatures, and is bounded by GD 154, at 10,450 K, and G117-B15A, at 12,130 K. The data base for the Weidemann and Koester (1984) analysis is identical to that of Greenstein (1982). However, their analysis makes use of improved colors, based on broader and optimally-defined bandpasses, and is based on the absolute flux calibration of Hayes and Latham (1975), rather than on the AB79 scale. Note that there is a shift of ~ 500 K between these scales, in the sense that temperatures based on the Oke-Gunn results are *cooler* than values based on the Hayes-Latham work. The coolest variable in the Weidemann and Koester analysis, the faint ZZ Ceti star G255-2, is assigned an uncertain temperature of 10,060 K and defines the red edge (this temperature is nearly 900 K *cooler* than that derived for this object by Greenstein 1982 on the AB79 scale). The next coolest variable, G38-29, is at 11,190 K. Their blue edge, set by G117-B15A, is at 13,010 K.

Wesemael, Lamontagne, and Fontaine (1986) and Lamontagne, Wesemael, and Fontaine (1987) discuss ultraviolet spectrophotometric observations in the 1200–1900 Å range of a subsample of ten ZZ Ceti stars. The temperatures presented in Figure 1 are based on updated models of Nelan and Wegner, and there now appears to be a satisfactory agreement between the various ultraviolet temperature scales (see Lamontagne *et al.* 1989 for a discussion of this point). In our updated determination, based on an analysis of an expanded sample of 11 ZZ Ceti variables, the instability strip extends from 11,310 K to 13,020 K.

Finally, the spectroscopic analysis of Daou *et al.* (1990) indicates that the ZZ Ceti stars studied have effective temperatures between 11,320 K and 13,420 K, with again GD 154 defining the red edge, and a clump of 3 stars (GD165, L19-2, and G117-B15A), all with temperatures within 220 K, defining the blue edge.

2. The Edges of the Strip

We summarize below the various determinations of the location of the blue edge of the instability strip found in the literature: McGraw (1979): 13,640 K and Fontaine *et al.* (1985): 13,000 K (averaged colors and improved models now yield 13,300 K, as shown in Fig. 1); Fontaine *et al.* (1982): 12,700 K; Greenstein (1982): 12,130 K; Weidemann and Koester (1984): 13,010 K; Wesemael, Lamontagne, and Fontaine (1986): 13,000 K, now updated to 13,020 K; and Daou *et al.* (1990): 13,420 K. While each one of these studies clearly suffers from its own uncertainties, we believe it significant that most of these analyses — based on distinct observational material often interpreted in terms of independent model atmosphere calculations — place the blue edge at or near 13,000 K. We believe that this result must be considered a serious constraint on the modeling of the pulsation properties of ZZ Ceti stars.

The work of Tassoul, Fontaine, and Winget (1990) makes it clear that there is no fundamental problem, in principle, in accommodating such a hot blue edge in current pulsation calculations. The location of the blue edge is related to the depth of the hydrogen convection zone, which itself is a function of the efficiency of convective energy transport in the envelope (Winget *et al.* 1982; Fontaine, Tassoul, and Wesemael 1984; Tassoul, Fontaine, and Winget 1990). The more efficient the convection, the hotter the blue edge. There is, of course, a limit to the efficiency of convective energy transport: it is reached when the temperature stratification in the convection zone becomes adiabatic. This specific case was considerered by Tassoul, Fontaine, and Winget (1990), who derive a limiting effective temperature for the blue edge of 14,000 K. As long as the observed blue edge is below that value, as is observed, there is no fundamental discrepancy between theory and observations, at least within this given set of constitutive physics. Note, however, that the recent nonadiabatic calculations of Cox *et al.* (1987) set the blue edge between 11,000 K and 11,500 K. The only determination consistent with the Cox *et al.* result is that of Greenstein (1982). According to Cox *et al.*, no conceivable change in the constitutive physics or convective efficiency can bring their blue edge near 13,000 K. Figure 1 suggests that these calculations appear at odds with the bulk of the observational results currently available on the location of the instability strip of ZZ Ceti stars.

As far as the red edge is concerned, the temperatures determined range from about 10,000 K to 11,600 K, with the average near 10,780 K. The uncertainty is currently large, but is due, in part, to the fact that *four different stars* currently define the edge in various analyses. Our knowledge of the cool boundary of the instability strip will undoubtedly improve when objects like G255-2 and BPM 30551, which define the red edge in some studies, are observed spectroscopically.

3. The Role of Convective Efficiency

It has been repeatedly emphasized in the past that a match to the observed blue edge required that a very effective convection be used in the calculation of the envelope structure (Winget *et al.* 1982; Fontaine, Tassoul, and Wesemael 1984). However, little attention has been paid to this requirement when time has come to determine observationally the location of the blue edge: model atmosphere calculations used in the interpretation of the data (whether narrow-band photometry, ultraviolet and optical spectrophotometry, or spectroscopy) have consistently made use of what one could term a conventional convective efficiency, akin to the ML1 theory (see Tassoul, Fontaine, and Winget 1990). We believe that this inconsistency is potentially an important, yet much overlooked, problem in the interpretation of ZZ Ceti stars. To investigate this problem further, we have carried out series of model atmosphere calculations for cool DA stars which incorporate different efficiencies for the convective energy transport. Figure 2 shows the influence of three different convective mixing efficiencies on the high Balmer series for several temperatures of interest. It is

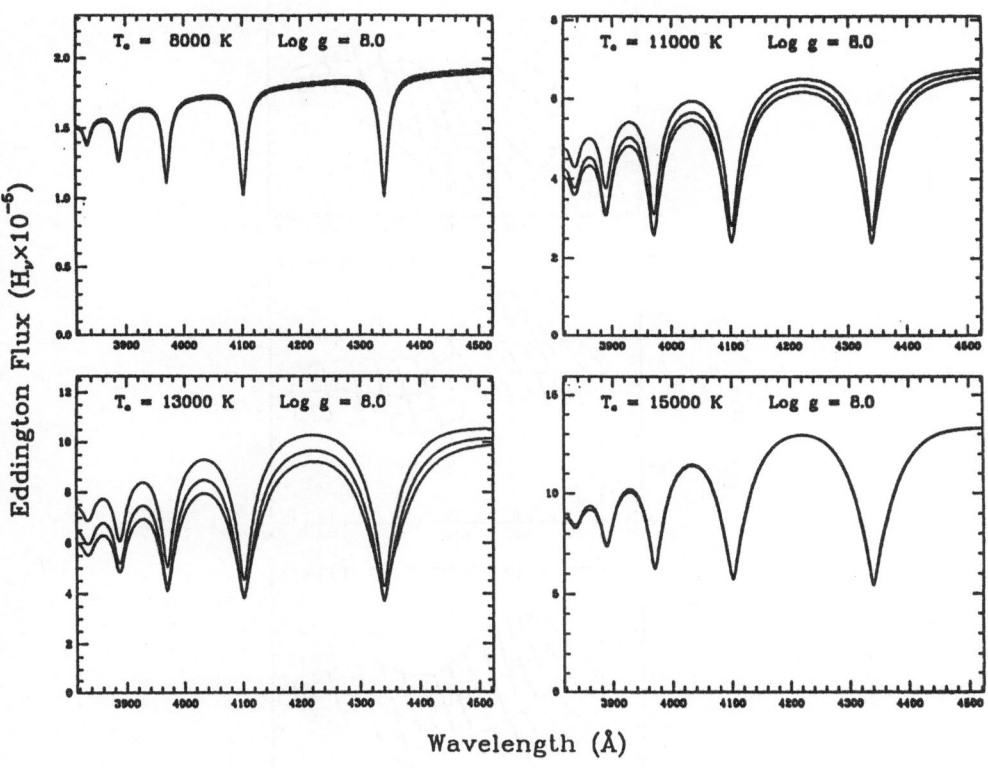

Fig. 2: Influence of the convective mixing efficiency on the optical spectra of cool DA stars for various effective temperatures. The efficiencies used are, from top to bottom, ML3, BS3, and ML1.

clear that the adopted efficiency is quite critical in the region of the ZZ Ceti stars, but unimportant at higher effective temperatures, where there is little convective energy transport, or at lower effective temperatures, where convection becomes adiabatic. Figure 3 shows the influence of the three same efficiencies (termed ML1, ML3, and BS3) on a standard Strömgren two-color diagram. Our preliminary analysis of the situation suggests that the results of our spectroscopic and photometric fits cannot be reconciled if ML3 is adopted as the convective prescription. However, the BS3 prescription, which incorporates the high efficiency of the ML3 theory with a Böhm-Stückl cutoff near the edges of the convection zone, could well be reconciliable with the empirical data. Are ZZ Ceti stars telling us that the deep envelope of cool DA stars is best described by very efficient convection, while the top of the convection zone, near the photosphere, is best described by less efficient convective transport?

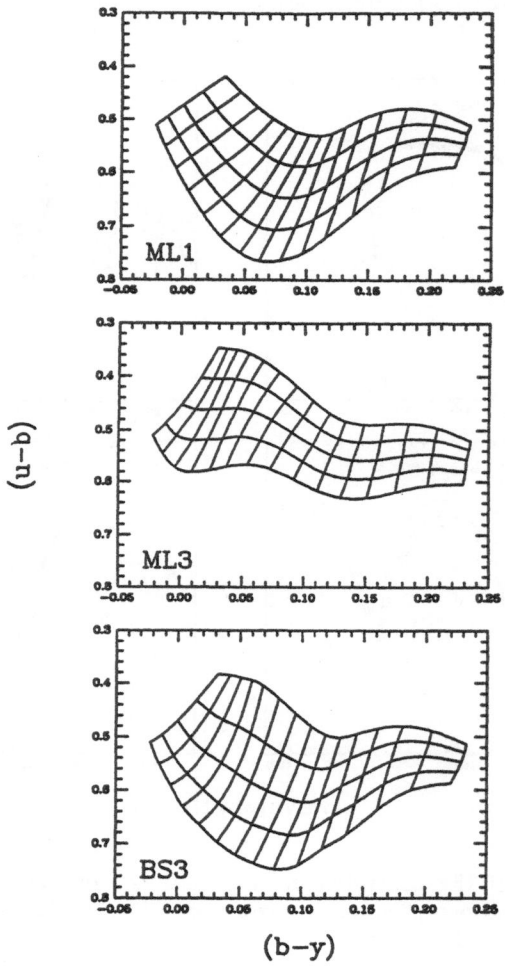

Fig. 3: Influence of the convective mixing efficiency on the morphology of the Strömgren diagram of cool DA stars. The effective temperatures range from 8000 to 15,000 K by steps of 500 K, and the values of log g from 7.5 to 8.5 by steps of 0.25

This work was supported in part by the NSERC Canada, by the Fund FCAR (Québec), by the NSF grant AST 88-18069, and by NATO. P. Bergeron and G. Fontaine acknowledge additional financial support through a NSERC Postdoctoral Fellowship and a Killam Fellowship, respectively.

REFERENCES

Cox, A.N., Starrfield, S.G., Kidman, R.B., and Pesnell, W.D. 1987, *Ap. J.*, **317**, 303.

Daou, D., Wesemael, F, Bergeron, P., Fontaine, G., and Holberg, J.B. 1990, *Ap. J.*, **364**, in press.

Fontaine, G., McGraw, J.T., Dearborn, D.S.P., Gustafson, J., and Lacombe, P. 1982, *Ap. J.*, **258**, 651.

Fontaine, G., Bergeron, P., Lacombe, P., Lamontagne, R., and Talon, A. 1985, *A. J.*, **90**, 1094.

Fontaine, G., Tassoul, M., and Wesemael, F. 1984, in *Proc. 25th Liège Astrophysical Colloquium: Theoretical Problems in Stellar Stability and Oscillations*, eds. A. Noels and M. Gabriel (Liège: Université de Liège), p. 328.

Greenstein, J.L. 1976, *A. J.*, **81**, 323.

———. 1982, *Ap. J.*, **258**, 661.

———. 1984, *Ap. J.*, **276**, 602.

Hayes, D.S., and Latham, D.W. 1975, *Ap. J.*, **197**, 593.

Lamontagne, R., Wesemael, F., and Fontaine, G. 1987, in *IAU Colloquium 95, The Second Conference on Faint Blue Stars*, eds. A.G. Davis Philip, D.S. Hayes, and J. Liebert (Schenectady: L. Davis Press), p. 677.

Lamontagne, R., Wesemael, F., Fontaine, G., Wegner, G., and Nelan, E.P. 1989, in *IAU Colloquium 114, White Dwarfs*, ed. G. Wegner (New York: Springer), p. 240.

McGraw, J.T. 1979, *Ap. J.*, **229**, 203.

Oke, J.B., and Gunn, J.E. 1983, *Ap. J.*, **266**, 713.

Oke, J.B., and Schild, R.E. 1970, *Ap. J.*, **161**, 1015.

Shipman, H.L. 1979, *Ap. J.*, **228**, 240.

Tassoul, M., Fontaine, G., and Winget, D.E. 1990, *Ap. J. Suppl.*, **72**, 335.

Weidemann, V., and Koester, D. 1980, *Astr. Ap.*, **132**, 195.

Wesemael, F., Lamontagne, R., and Fontaine, G. 1986, *A. J.*, **91**, 1376.

Winget, D.E., Van Horn, H.M., Tassoul, M., Hansen, C.J., Fontaine, G., and Carroll, B.W. 1982, *Ap. J. (Letters)*, **252**, L65.

LONG TERM VARIATIONS IN ZZ CETIS: G191-16 AND HL TAU-76

M. AUVERGNE[1], M. CHEVRETON[2], J.A. BELMONTE[3],
G. VAUCLAIR[4], N. DOLEZ[4], M.J. GOUPIL[1].

[1] Observatoire de Paris. D.A.S.G.A.L. URA CNRS n° 335.
92195 MEUDON FRANCE.
[2] Observatoire de PARIS. D.A.E.C. URA CNRS n° 173.
92195 MEUDON FRANCE
[3] Instituto de Astrophysica de CANARIAS.
38071 LA LAGUNA. TENERIFE SPAIN.
[4] Observatoire Midi-Pyrenées. URA CNRS n° 285.
14 Av. E. Belin. TOULOUSE FRANCE.

ABSTRACT. We report on new observations of two ZZ Ceti stars GD 191-16 and HL Tau-76. Variations of the dominant oscillation both in amplitude and period are found, over a time scale of several months. Previous interpretations of the observed oscillation in terms of chaotic dynamic are not in contradiction with present data, but no more confirmed.

1. Introduction.

Seismology studies of white dwarfs is of a great interest to constrain internal structure models. Large amplitude stars are generally considered quite stable with simple spectra in contrast with the general amplitude-complexity correlation suggested for the majority of other ZZ Ceti stars (McGraw, 1980). In some cases the apparent complexity have been solved thanks to long observation runs obtained by multisite campaigns of the WET network (Winget 1990). Up to one hundred frequencies have been detected on .

It has been also suggested that the complexity of the light curves could be due, in some cases to a choatic dynamic (Goupil et al. 1988). Subharmonic frequencies have been found in previous observations of G191-16 (Vauclair et al. 1989). For HL Tauri-76, C.G. Page (1972) and MC-Graw (1976) have reported the existence of a frequency at 3/2 times the main frequency which suggest also the existence of subharmonics. To confirm those results, new observations have been performed over the last two years. In the next sections we present these new observations and their Fourier analyses. Sections 2 and 3 are devoted successively to G191-16 and HL Tauri-76

2. Observations of G191-16

The DA white dwarf G191-16 was considered as a large amplitude pulsator, until 1988 (McGraw et al. 1981) with a main period of 885 seconds. Table 1 described our journal of observations of G191-16 during the last three years.

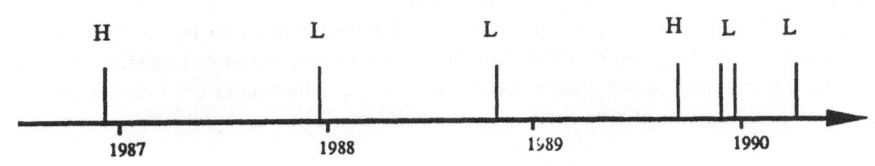

figure 1.

G. Vauclair and E. Sion (eds.), White Dwarfs, 167–174.

Our first run in 1988 has shown an unexpectedly low amplitude (reduced by a factor five) oscillation with a main period of 507 seconds. Such a low amplitude regime has been reobserved several times as shown on figure 1, where all runs are sketched with their low and large amplitude characteristic. The high amplitude regime is recurrent, and a switching time scale between the two regimes can be estimated to a value less or equal to 2 months. But it is not yet possible to give a significant value of this recurrence time.

Table 1: Journal of observations.

Star	Run number	Observatory	Date	Length of the run
G191-16	1	Tenerife	1989-10-26	4 h. 20
	2	Pic du Midi	1989-12-01	4 h.
	3	O.H.P.	1989-12-24	7 h. 50
	4		1989-12-25	8 h. 09
	5		1989-12-28	5 h. 20
	6	Calar Alto	1988-12-11	8 h. 10
	7		1990-02-17	4 h. 30
	8		1990-02-19	4 h. 30
HL Tauri	1	Tenerife	1989-11-04	6 h.
	2		1989-11-06	4 h.
	3	O.H.P.	1989-12-26	8 h. 10
	4		1989-12-27	7 h.

The Fourier analysis of the data has been done with a "clean" algorithm, described in Roberts et al. (1987), and applied on runs 3 4 and 5 on one hand, and on runs 7 and 8, on the other hand. Rough and residual spectra are shown on figures 2 and 3, and the detected frequencies on table 2 and figure 4. The same frequencies are found in december and february, exept one at 1.53 mHz. Four frequencies are identified at 1.67, 1.97, 2.52 and 3.57 mHz, with a high confidence level. In both spectra the peak at 1.37 mHz. has an amplitude of 2.10^{-3} which is of the same order than the accuracy of the data.

The observations performed in Tenerife (run 1) show a large amplitude oscillation with a period of the main peak equal to the value given by Vauclair et al. (1989). But the poor quality of these data does not permit to extract any other reliable frequency.

Only two frequencies (1.67 and 1.97 mHz) are the same in the two regimes. The identification of the peak at 1.36 with the peak at 1.41 is dubious because their distance 0.05 is slightly larger than the resolution.

We have, with G191-16 one example of a star having important amplitude and frequency content variations. This result shows one more time that continuous observations are fundamental to obtain a reliable description of the dynamic. Unfortunately the time scale involved, in this case, is so long that multisites observations, as they are done by the WET network will never give a complete description of this type of behaviour.

In the linear framework, a two or three months time scale can be seen as due to two close eigenmodes. But they are so close that they are probably resonnant modes and it is well know that such resonnance gives chaotic behaviour. On the other hand the fact that subharmonics are only seen on large amplitude runs, reinforces the nonlinear interpretation.

3. Observations of HL Tauri 76

The journal of observations of HL Tauri is given in Table 1. The rough and residual spectra of runs 3 and 4 are shown on figures 5 and 6. Table 3 and figure 4b present the detected frequencies which are exactly the same from OHP and Tenerife data. Our main peak at 1.85 mHz is new whereas the main component at 1.34 mHz observed in 1982 data is no longer seen. We have, in this case, a shortening of the main period on a very long time scale.

Frequencies larger than 3 mHz are linear combinations of those labelled 0, 1, 2 and 3 in table 3. Frequencies smaller than 3 mHz are probably independant modes. The frequencies labelled 0, 1 and 3 are equally spaced with a $\Delta\nu = 0.18$ mHz. It is also the case for 1.25 and 1.43 frequencies. If this splitting is a rotationnal one, it implies a rotationnal period of 1.5 hours. An adhoc model of nonlinearly coupled modes can gives spectra with equally spaced frequencies. In this case $\Delta\nu$ is the distance in the frequency of the modes. (Goupil et al. 1990)

Table 2: Identified frequencies in the power spectra of G191-16. Unit is mHz

Calar Alto (Feb. 90)	O.H.P (Dec. 89)	High Amplitude Frequencies
		0.52 ± 0.025
		0.60 ± 0.02
		0.84 ± 0.02
		1.12 ± 0.03
		1.20 ± 0.025
1.36 ± 0.02	1.37 ± 0.02	1.41 ± 0.02
	1.53 ± 0.02	
1.67 ± 0.02	1.66 ± 0.02	1.68 ± 0.045
1.97 ± 0.02	1.97 ± 0.02	1.96 ± 0.03
		2.23 ± 0.025
		2.32 ± 0.025
2.52 ± 0.02	2.53 ± 0.02	
		3.36 ± 0.025
		3.44 ± 0.02
3.57 ± 0.02	3.58 ± 0.02	

4. Conclusion

The new observations analysed in this paper show clearly variations on a long time scale of the spectra and amplitude for HL Tauri and G191-16. The interpretation of such a time scale can be done in a linear or nonlinear framework, and nothing in those data allow to make the distinction. Another DA white dwarf, GD 154 has a behaviour similar to G191-16. It is known that GD 154 has amplitude variations (Robinson et al. 1978). When its amplitude is large, frequencies at $3\nu/2$ and $5\nu/2$ are found (ν is the frequency of the main oscillation). In the small amplitude regime the $3\nu/2$ frequency dominates the light curve. When the amplitude of G191-16 is small the dominant frequency is the one identified as $7\nu/4$ in the large amplitude regime. A last remark can be done. G191-16 show a strong correlation between its dominant period and amplitude. This is qualitatively in agreement with the general period amplitude relation for DA white dwarf stars (Unno et al. 1989).

170

Table 3: Identified frequencies in the power spectra of HL Tauri-76. Unit is mHz.
⋆ indicates frequencies which can be considered as a linear combination of frequencies numbered 0, 1, 2 and 3. The resolution is ±0.025 as defined by the duration of the runs. Values in the third column comes from Warner and Nather 1972, Page 1972 and McGraw 1980.

	This work	Previous works
		1.00
	1.25	
		1.34
	1.43	
	1.50	1.51
		1.60
	1.53	
3	1.67	
0	1.85	
1	2.02	2.01
2	2.61	2.60
	3.35⋆	
	3.51⋆	
	3.69⋆	
	3.87⋆	
	4.29⋆	
	4.46⋆	
	5.54⋆	
	6.48	

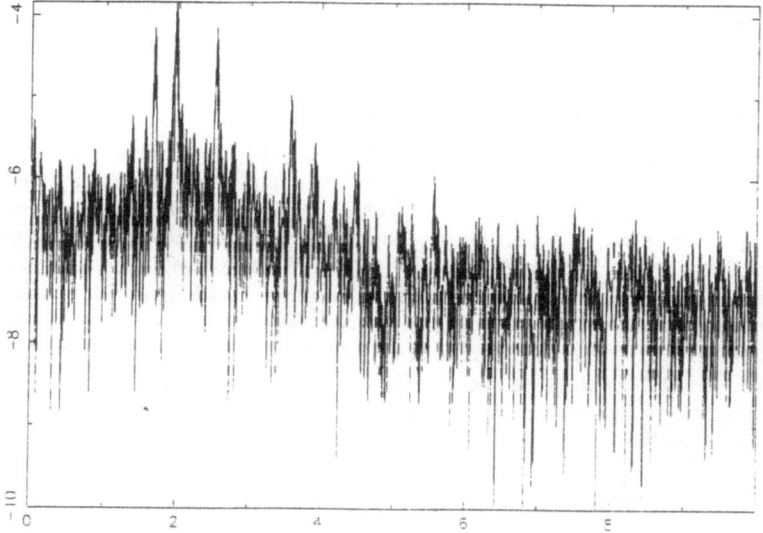

figure 2. Rough spectrum of G191-16 for runs 3, 4 and 5. Ordinate scale is the $\log_{10}$ of the power spectrum. Abscissae are in mHz

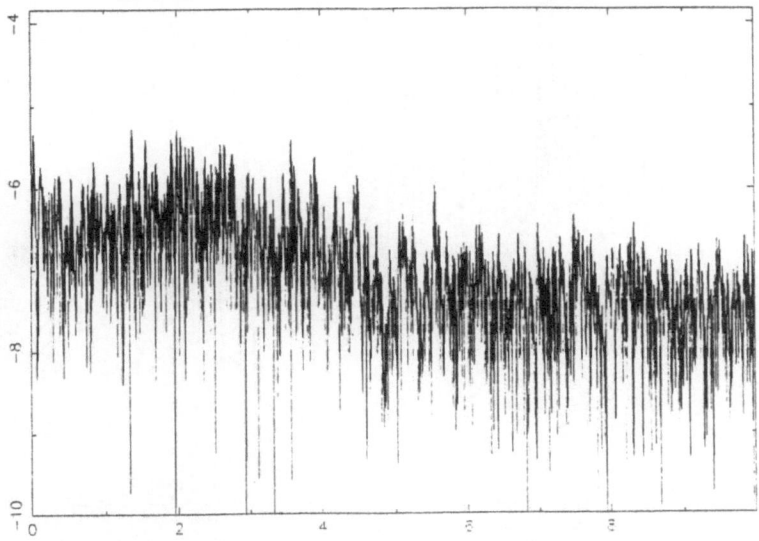

figure 3. Residual spectrum after the "cleanning" of the spectrum of figure 2.

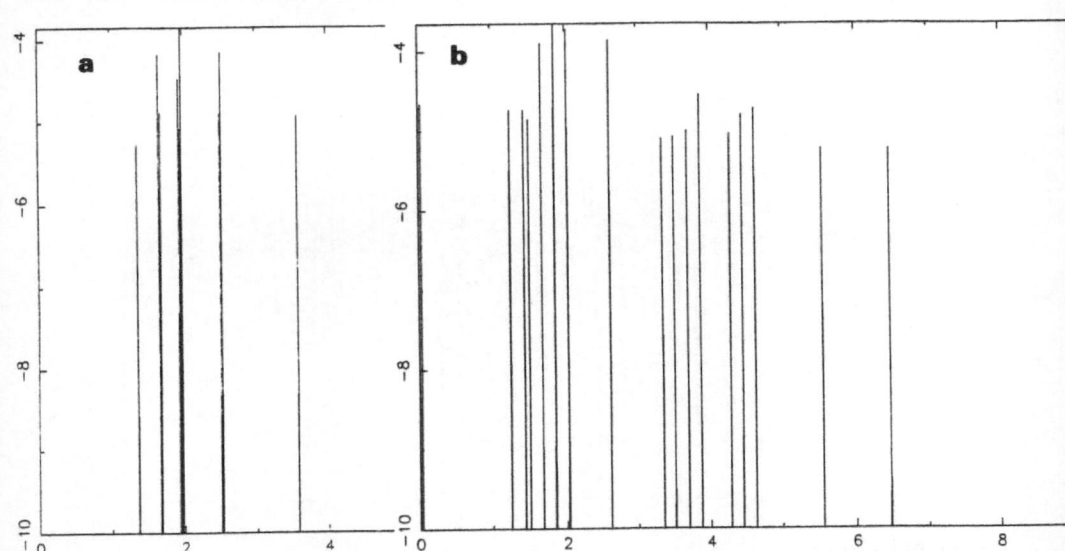

figure 4. Frequencies extracted from spectrum of G191-16 (a) and HL Tauri-76 (b) with the "clean algorithm". Axis scale are the same as on figure 2.

172

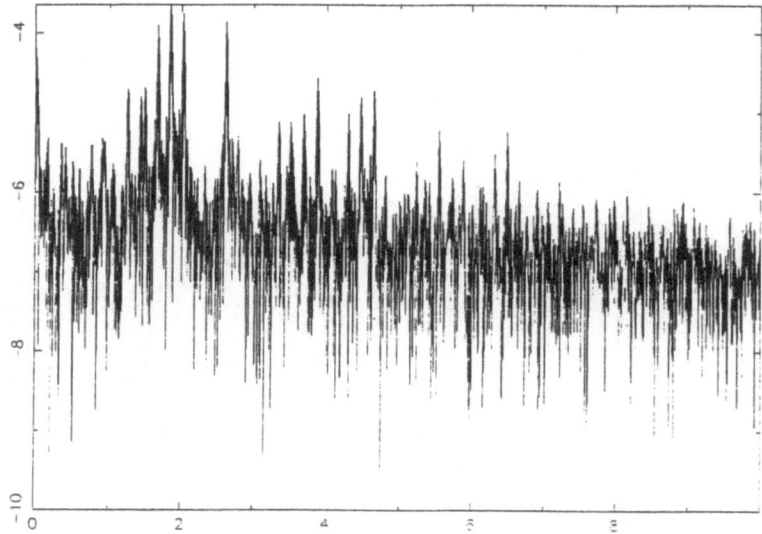

figure 5. Rough spectrum of HL Tauri for runs 3 and 4.

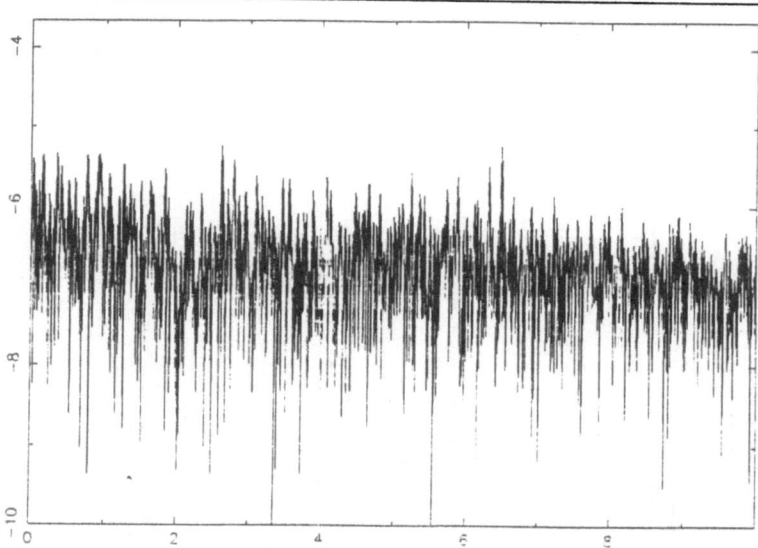

figure 6.Residual spectrum after the "cleanning" of the spectrum of figure 5.

References

Goupil M.J., Auvergne M., Baglin A.1988, *Astron. Astrophys.* **196**, L13.

Goupil M.J., Auvergne M., Baglin A. 1991, "Wavelet analysis of variable white dwarf stars", this meeting.

Page C.G. 1972, *Mont. Not. R. astr. Soc.* **159**,25P.

McGraw J.T. 1980, "Current problems in stellar Pulsation Instabilities" NASA Technical Memorandum 80625. D. Fischel, J. Rountree-Lesh, W.M. Sparks Eds.

McGraw J.T., Fontaine G., Dearborn D.P.S., Gustafson J., Lacombe P., Starrfield S.G. 1981, *Astrophys. J.* **250**,349.

Robinson E.L., Stover R.J., Nather R.E., McGraw J.T. 1978 *Astrophys. J.* **220**,614

Unno W., Osaki Y., Ando H., Saio H., Shibahashi H. 1989, "Non radial oscillations of stars". University of Tokyo Press Ed.

Vauclair G., Goupil M.J., Baglin A., Auvergne M., Chevreton M. 1989, *Astron. Astrophys.* **215**,L17.

Warner B., Nather R.E. 1972, *Mont. Not. R. astr. Soc.* **156**,1.

Winget D. 1991, "Results of the Whole Earth Telescope campaigns: a rewiev" this meeting.

DISCUSSION

BARSTOW :

It is interesting to note similar behaviour (compared to G191-16) in the DBV PG1115+158, observed from La Palma by the WET in March 1989. In this case the observed pulsation amplitudes were much reduced by comparison with the original observations of this star. Secondly, identified periods in one night's data were not repeated on the following nights.

WINGET :

Also with WET, we have seen a number of other objects with complicated non-linear behavior, such as G29-38. This star, during the WET run, was a large amplitude pulsator with a complicated light curve dominated by one frequency with an amplitude at 6×10^{-2}. The following year that mode went away, and the power disappeared-the star became a low amplitude pulsator. Currently large amplitude pulsations are reappearing at other frequencies.

PREDICTING WHITE DWARF LIGHT CURVES

T.SERRE[1], J.R.BUCHLER[2], M.J.GOUPIL[1]

(1) *Observatory of Meudon, DASGAL*
5, place Jules Janssen, 92195 Meudon Principal Cedex, France
(2) *Physics Department, University of Florida*
Gainesville, 32611 Florida USA

ABSTRACT. A predictive method for the evolution of a dynamical system is introduced. This method is based on the existence of a low dimensional attractor for the system and it is illustrated with an application to a well known chaotic attractor, *viz.* the Rössler attractor. Tests for predicting an observed light curve of a pulsating white dwarf are next carried out. Results show that if sufficiently long time-sequences are used so that the attractor is faithfully reconstructed, an irregular light curve can be predicted over several times the main 'period'.

I. Introduction

Predicting the light curves of pulsating stars presents some interest, in particular it offers the possibilities of filling gaps in observational data and of replacing or cleaning up noisy sections. For irregular light curves, this is not an easy task, in general. However, a pulsating star is a dynamical system and assuming the existence of an underlying attractor, one can take advantage of methods of prediction developed in the framework of dynamical system theory. In section II, we give a brief summary of the method of prediction we have chosen and illustrate the method with a simple dynamical system, *viz.* the Rössler system. In section III, we test the efficiency of the method on part of the observed light curve of a pulsating white dwarf PG1351+489 and we conclude in section IV.

II. Description of the method

We give here a brief summary of the prediction method that we have chosen. For a more detailed description we refer to Farmer and Sidorovitch (1987) (for a survey of other work also see Casdagli 1989). By means of the time delay method (Takens 1981), the trajectory

175

G. Vauclair and E. Sion (eds.), White Dwarfs, 175–183.
© 1991 *Kluwer Academic Publishers.*

176

$$\mathbf{X}(t) = (S(t), S(t - \tau), \ldots, S(t - (m-1)\tau)) \tag{1}$$

is constructed in an $m-$ dimensional embedding phase space from a single observable $S(t)$, with a time delay τ, for instance an observed pulsating light curve. The resulting reconstructed attractor is topologically equivalent to the real attractor of the system, *i.e.* the dynamical properties of the real dynamical system are preserved.

For two consecutive points $\mathbf{X}(j)$ and $\mathbf{X}(j+1)$ of the trajectory in this embedding phase space, $\mathbf{X}(j+1)$ is the known future of $\mathbf{X}(j)$ and both are thus linked by the flow $\mathbf{F}$

$$\mathbf{X}(j+1) = \mathbf{F}(\mathbf{X}(j)) \tag{2}$$

If these two points are close enough, we can approximate $\mathbf{F}$ by a linear or a quadratic expansion in the $\mathbf{X}$'s coordinates

$$\mathbf{X}(j+1) = \mathbf{F_{app}}(\mathbf{X}(j)) \tag{3}$$

The unknown coefficients of the expansion $\mathbf{F_{app}}$ are determined by a least squares method. The set of n_v neighbouring points $\mathbf{X}(j)$ to a point of interest $\mathbf{X}(i)$, *i.e.* the learning set in the neighbourhood of $\mathbf{X}(i)$, is used to perform an average of the n_v relations $\mathbf{F_{app}}$. This weighted relation is then applied to $\mathbf{X}(i)$ to produce the predicted future point $\mathbf{X}(i+1)$.

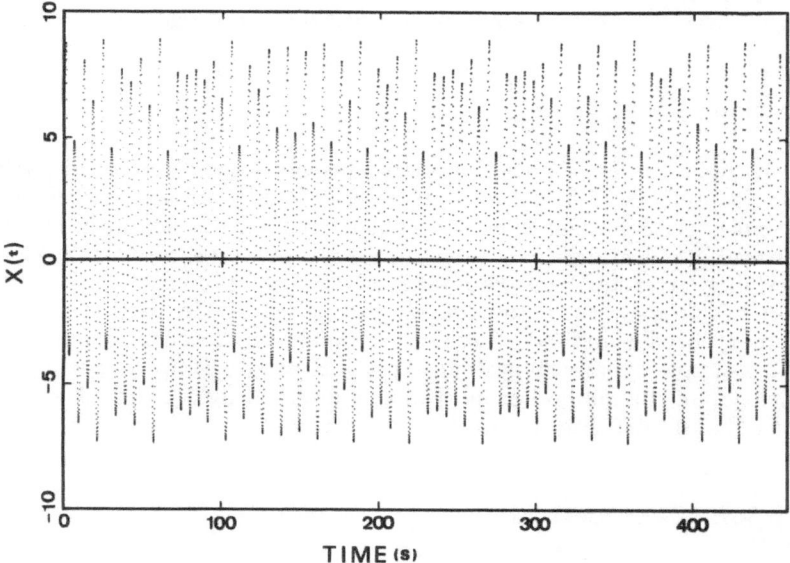

Fig(1): A part of $x(t)$ chaotic time serie.

The precision of the prediction can be evaluated by making first a prediction of a known part of the light curve, *i.e.* n_p points, and by calculating the difference between the actual value $\mathbf{X}_{true}$ and the predicted value $\mathbf{X}_{pred}$, and then comparing them to the variance of $\mathbf{X}_{true}$, $i.e.$

$$\sigma(\mathbf{F}_{app}) = \frac{1}{n_p} \frac{\sum_{i=1}^{n_p} |\mathbf{X}_{true}(i) - \mathbf{X}_{pred}(i)|^2}{VAR(\mathbf{X}_{true})} \tag{4}$$

With this method several parameters must be adjusted, among which the dimension m of the embedding phase space. As shown by Farmer (1987), the lowest value of m which minimizes the quantity (4), $i.e.$ which gives the best prediction, yields an approximate value of the minimal dimension of the embedding space necessary to reconstruct the real attractor. The departure from the exact minimal dimension mainly depends on the size of the learning set and on the degree of approximation of the flow $\mathbf{F}$ by $\mathbf{F}_{app}$. In the remaining of this paper, we choose for $\mathbf{F}_{app}$ the simplest linear approximation of $\mathbf{F}$. To check the efficiency of this prediction method we apply it first to the well-known chaotic Rössler attractor which is described by the three differential equations

$$\frac{dx}{dt} = -y - z$$
$$\frac{dy}{dt} = x + \delta y \tag{5}$$
$$\frac{dz}{dt} = \alpha + xz - \mu z$$

with parameters $\mu = 4.5$, $\delta = \alpha = 0.2$. We place ourselves under observational conditions in assuming that the temporal behavior of only one variable, here $x(t)$,

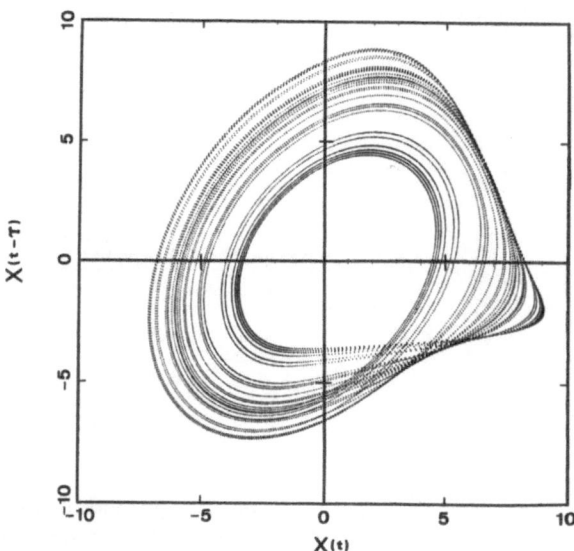

Fig(2): Projection on plane $(x(t), x(t - \tau))$ of the time delay reconstruction of Roessler chaotic attractor from $x(t)$ with time delay $\tau = 1.5s$ and $m = 3$.

is known (see fig.1). The reconstructed chaotic attractor is displayed in figure 2. Although the chaotic nature of the attractor (rapid divergence of nearby trajectories) puts a natural limit on the time length which is predictable, one is able to correctly predict, for instance, 3.5 periods with a learning set of 200 periods for determining $\mathbf{F_{app}}$ (fig.3). An embedding $m = 5$ gives the best prediction (see fig.4). The actual embedding dimension $m = 3$ could be better approched with a longer time-sequence (a larger learning set) and with a higher degree expansion of $\mathbf{F_{app}}$.

III. Application to a pulsating white dwarf : PG 1351+489

We now turn to observational data and consider a part of the light curve of the DB variable white dwarf PG 1351+489, which shows evidence of nonlinearities in its temporal behavior. In particular, the presence of alternatively large and small peaks, that gives rise to a subharmonic $\nu_o/2$ in the Fourier spectrum, suggests period doubling behavior, a well known scenario of systems on the route to chaos (Goupil et al. 1988). Thus a chaotic attractor is suspected to underly the pulsation of this star.

179

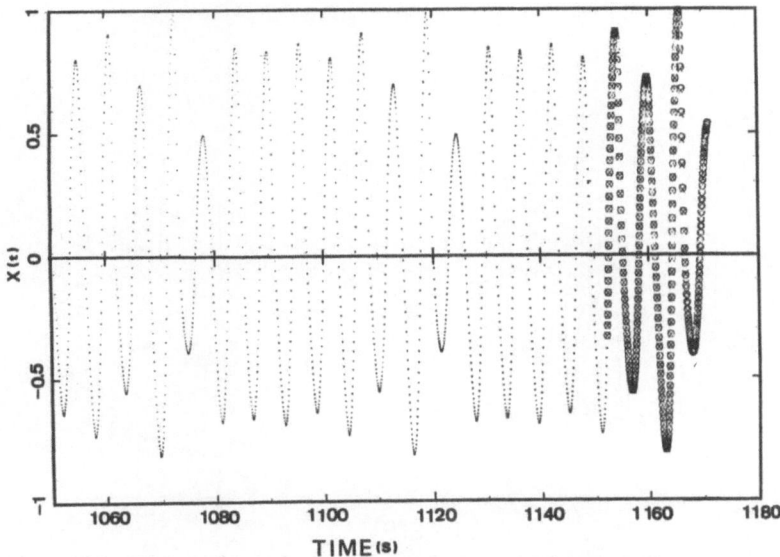

Fig(3): A prediction on 3 periods for a learning set of 200 periods. the last
boldface 3.5 periods are prediction perfectly matching the real signal.

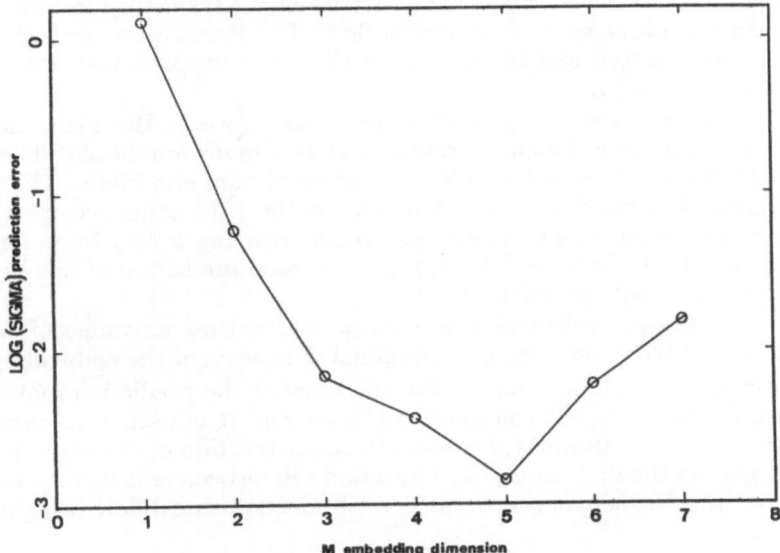

Fig(4): Prediction error following embedding
dimension m for Roessler attractor.

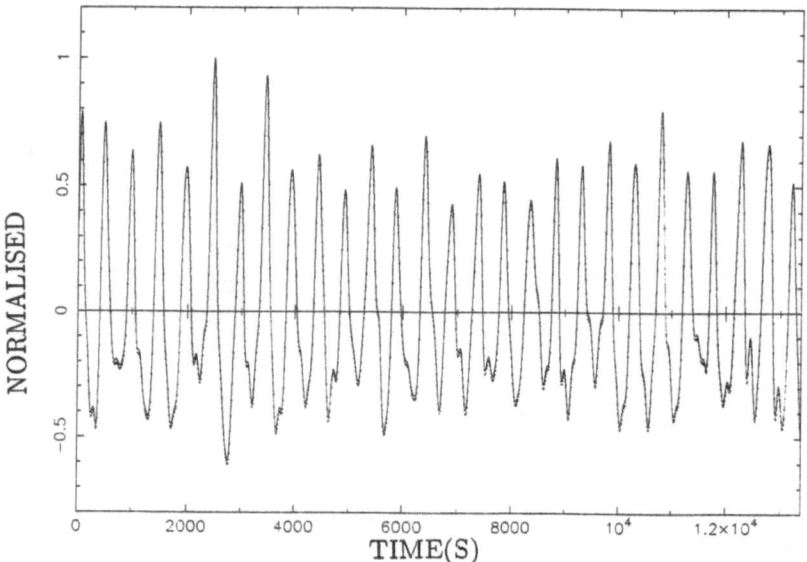

Fig(5): One nigth smoothed run of pulsating white dwarf PG 1351+489.

In order to check the efficiency of the prediction method we use an observational time-sequence of 28 periods (fig.5). The corresponding trajectory in a 3— dimensional embedding phase space is shown in fig.6. The learning set consists of 90% of the time series of fig.5 and the remaining 10% are compared with the resulting prediction shown in fig.7.

We see from fig.7 that the prediction correctly captures the main oscillation and the small amplitude features occurring at minimum amplitude. The largest departures from the real signal mainly occur at maximum amplitude. The primary cause of these discrepancies is the shortness of the time sequence used for the learning set. For example, the paucity of events involving a very large amplitude followed by an intermediate amplitude, prevent a good prediction of this behavior : the attractor is not fully reconstructed.

For the same reasons, calculations are not possible for higher values of m, and it has not been possible to determine the minimal dimension of the embedding space.

As mentioned above limitations to the efficiency of the prediction mainly come from the short length of the considered data set and it is useless to carry out a real prediction from it. However, because the reconstruction of the attractor is not vitiated by gaps in the time series, this limitation can be overcome by reconstructing the attractor with the help of time-sequences obtained during different nights (work in progress).

IV. Conclusion

We have tested a method which allows a prediction of the time behavior of pulsating

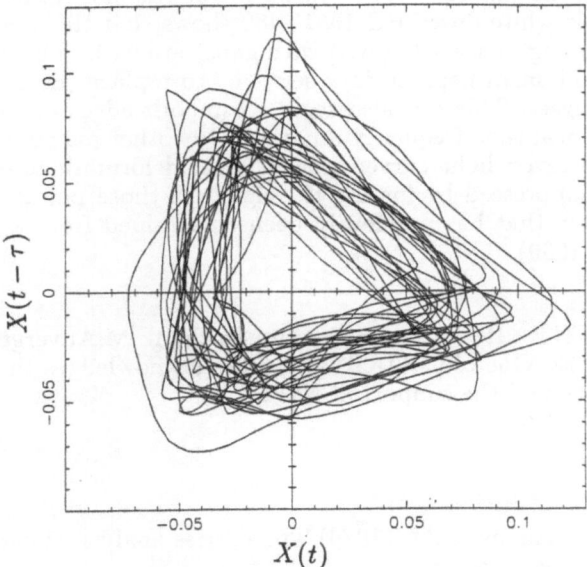

Fig(6): Projection on $(x(t), x(t - \tau))$ of Time delay reconstruction of PG 1351 with $\tau = 120s$ and $m = 3$.

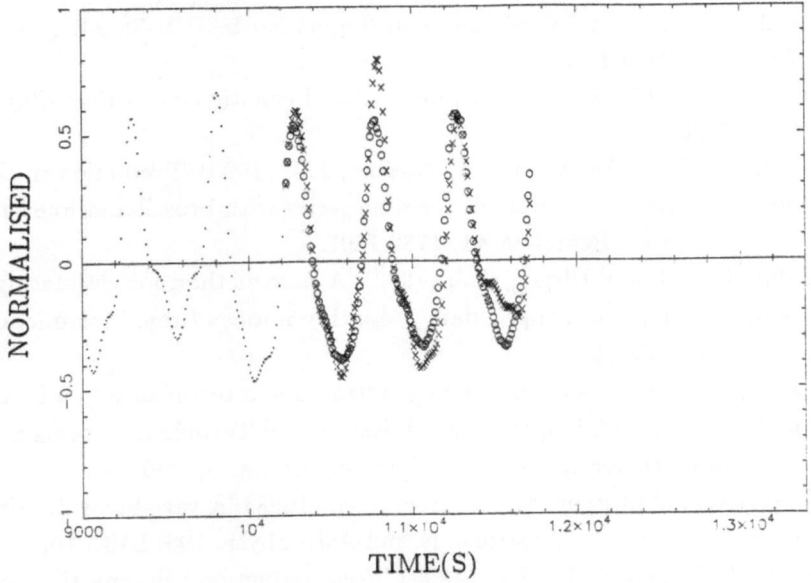

Fig(7): Prediction on 3 periods of time serie in Fig(6) with a past signal of 19 periods 'O' predicted 'x' to predict.

light curves (assuming the existence of an underlying attractor). Results on a short data set of the variable white dwarf PG 1351+489 shows that this method can be very efficient if long enough data sets (even with gaps) are used. This should allow one, for instance, to fill small gaps in data series or to replace too noisy parts of them, for further analyses. This can also serve to smooth edges of time intervals to improve results of local time frequency analyses. Our other purpose is to reduce noise effects in an observed light curve. Indeed, local deformations of the signal due to noise can be suppressed by forcing the signal at those points to obey the approximated F_{app} law that has previously been determined from a learning set (Kostelich and Yorke 1990).

ACKNOWLEDGEMENT : It is a great pleasure to thank M.Auvergne for useful suggestions on this work. One of us (JRB) gratefully acknowledges the hospitality of Meudon Observatory and the support of NSF.

REFERENCES

Box, G.E.P. and Jenkins, G.M. (1970) Time series analysis, forecasting and control, Holden day, San Franscisco.

Eckman, J.P. and Ruelle, D. (1985) 'Ergodic theory of chaos and strange attractors', Review of Modern Physic 57, 617-656.

Farmer, J.D. and Sidorovitch, J. (1987) 'Predicting chaotic time series', Phys ical Review Letter 59, 845-848 and Report No LA-UR-88-901 (1988) Los Alamos National Laboratory.

Casdagli, M. (1989) 'Nonlinear prediction of chaotic time series', Physica D 35, 335-356.

Abarbanel, H.D., Brown, R. and Kadtke, J.B. (1990) 'Prediction in chaotic nonlinear systems : Methods for time series with broadband fourier spec trum', Physical Review A 41, 1782-1807.

Fahlman, G.G. and Ulrych, T.J. (1982) 'A new method for estimating the power spectrum of gapped data', Monthly notice of royal astronomical society 199, 53-65.

Takens, F. (1981) ' Detecting strange attractors in turbulence ', in D.A. Rand and L.S.Young (Eds.), Dynamical System and Turbulence, springer lecture notes in mathematics, Vol 898, Springer, Berlin, pp.366-381.

Goupil, M.J., Auvergne, M. and Baglin, A. (1988) 'A variable white dwarf on its route to chaos? ', Astronomy and Astrophysic 196, L13-L16.

Kostelich, E. and Yorke, J.A. (1990) 'Noise reduction : finding the simplest dynamical system consistent with the data', Physica D 41, 183-196.

DISCUSSION

KEPLER :

1) Will the form (equation used) of the attractor change the predicted light curve?

2) Is there any physical reason to select one form, like the Rossler chaotic attractor, against another?

SERRE :

I answer to the two questions together. The answer is no, because we don't assume a form of attractor before prediction, in contrary we search for it. We construct the attractor directly from the observational signal : there isn't a predetermined form or model of the attractor which would be an a priori knowledge on dynamical properties.

A WAVELET ANAL'SIS OF THE ZZ CETI STAR G191 16

M.J. GOUPIL, M. AUVERGNE, A. BAGLIN
Observatoire de Paris Meudon, DASGAL
92195 Meudon Principal Cedex, France

ABSTRACT. In order to investigate nonlinear effects at work in the pulsations of a ZZ Ceti star, G191 16, a wavelet analysis is carried out on a part of its light curve. Results show that its pulsations are strongly nonlinear and cannot be interpreted in a linear framework, in which nonlinearities, apart from generating finite oscillation amplitudes, only act as small perturbations.

1. Introduction

Pulsations of most pulsating stars are satisfactorily interpreted, in a linear framework, i.e. as a linear superposition of pulsation modes. However, some variable white dwarfs show evidences of nonlinear effects as cross frequencies, harmonics and subharmonics in their Fourier spectra. The question is, then, the origin of these nonlinearities: are they simply due to observational effects because of the nonlinear dependence of the luminosity on temperature and radius or do we see nonlinearities of the pulsation itself? In this last case, nonlinearities of temperature and radius variations have to be taken into account, thereby giving access to additional informations on the pulsation mechanism. Another question that we must answer is whether these nonlinearities are small enough to be considered as perturbative effects, and hence studied by means of an amplitude equations formalism or they are strong and must be investigated with the help of the dynamical system theory applied to either nonlinear, simplified models or complete hydrodynamical models when they exist.

Elements of answer from observational data can be brought up by a local time frequency analysis, such as the wavelet analysis, which decomposes the signal in its different oscillations and allow to study separetely the temporal behavior of their amplitudes. It is then possible to detect nonstationary behaviors and to determine the possible existence of amplitude and phase correlations between the different oscillations of a light curve which are indicative of nonlinear processes.

Among the possible interpretations of the complex light curve of the ZZ Ceti star, G191 16, a chaotic (i.e. strongly nonlinear) pulsation has been suggested (Vauclair *et al.*, 1989). To investigate further its nonlinearities, we have carried out a wavelet analysis of a part of its light curve. In section 2, we first recall the definition of the wavelet transform. Sections 3 and 4 gives results of this analysis for G191 16 which are discussed in section 5.

G. Vauclair and E. Sion (eds.), White Dwarfs, 185–192.

2. The wavelet transform

The wavelet coefficients of a temporal signal $f(t)$ are defined as:

$$C(a, b) = \frac{1}{\sqrt{a}} \int f(t) g^*(\frac{t - b}{a}) dt \tag{1}$$

where a, b respectively are a scaling and a shift parameters. The analysing wavelet $g(t)$ is localized both in time and in frequency so that, for a given (a,b), the wavelet coefficient (1) is only sensitive to that part of the signal in a small time interval about b and to the energy present at characteristic time scales close to a.

Our choice of an analysing wavelet $g(t)$ is the wavelet of Morlet:

$$g(t) = e^{-t^2/2} e^{i\omega_m t} \tag{2}$$

where ω_m is a parameter greater than 5 (Grossmann et al., 1989). The wavelet of Morlet being complex, it is more convenient to work with the modulus $M(a,b)$ and phase $P(a,b)$ of the resulting complex wavelet coefficients (1).

Some insights into the wavelet transform can be gained with the signal $f(t) = A \cos(\omega_0 t + \phi)$, for which the wavelet modulus and phase are approximately given by:

$$M \sim A \sqrt{\frac{a\pi}{2}} e^{-x^2/2} \tag{3a}$$

$$P \sim b\,\omega_0 + \phi \tag{3b}$$

with $x = a\omega_0 - \omega_m$. In this case, the wavelet modulus (3a) is independent of b. The amplitude of oscillation is given by $A = M(a_0)/\sqrt{a_0 \pi/2}$ where $a_0 = \omega_m/\omega_0$ (i.e. for $\omega_m = 2\pi$, a_0 thus gives the period of the oscillation). The wavelet phase (3b) is independent of the scaling parameter a, of the amplitude of the oscillation A, and is periodic with the periodicity of the signal.

The wavelet analysis is not designed to resolve oscillations with close frequencies as does the Fourier analysis. For a signal $f(t) = A_0 \cos(\omega_0 t + \phi_0) + A_1 \cos(\omega_1 t + \phi_1)$, one actually has:

$$\frac{M^2}{a\pi/2}(a, b) = A_0^2 e^{-x_0^2} + A_1^2 e^{-x_1^2} + 2A_0 A_1 e^{-(x_0^2 + x_1^2)/2} \cos(\delta\omega b + \delta\phi) \tag{4}$$

with $x_0 = a\omega_0 - \omega_m$; $x_1 = a\omega_1 - \omega_m$; $\delta\omega = \omega_0 - \omega_1$ and $\delta\phi = \phi_0 - \phi_1$.
For $\delta\omega > \sqrt{8 \text{Ln} 2}/a$, one detects two oscillations at $a = 2\pi/\omega_0$ and $a = 2\pi/\omega_1$ (with $\omega_m = 2\pi$) while for $\delta\omega \leq \sqrt{8 \text{Ln} 2}/a$, one only detects one single oscillation at the mean period $< a_0 >$ which is a mean value of $2\pi/\omega_0$ and $2\pi/\omega_1$, weighted by the amplitudes A_0 and A_1. The amplitude of this single oscillation,

$$A(b) = \frac{M(< a_0 >, b)}{\sqrt{< a_0 > \pi/2}} \tag{5}$$

oscillates with the beat period $2\pi/\delta\omega$.

For more details about the wavelet analysis, see Grossmann *et al.* (1989) and its application to light curves of pulsating stars, Goupil *et al.* (1990).

3. Global wavelet representation of the light curve of G191 16

The light curve of the ZZ Ceti star G191 16 shows small and large amplitude regimes of oscillation with different Fourier frequency contents, alternating over several months (Vauclair *et al.*, 1989, Auvergne *et al.*, 1990). We analyse, hereafter, one run recorded during the large amplitude regime. In calculating (1) with $\omega_m = 2\pi$, values of a are chosen as $p_0/4 < a < 3p_0$ where $p_0 = 892.8s$ is the main period of oscillation during this run, as inferred from a Fourier analysis. The range of values of b is the time interval of the data set.

Resulting wavelet modulae $M(a, b)$ and phases $P(a, b)$ are shown on fig 1. Oscillations with mean periods $< a_0 >, < a_0/2 >, < a_0/3 >, < 2a_0 >$ are detected through their wavelet modulae which are non zero at these time scales and their periodic associated wavelet phases. Because a wavelet phase is only weakly sensitive to the amplitude of the oscillation, the three last of the above oscillations, of much smaller amplitudes than the first one, are best detected through the periodicities of their respective wavelet phases.

The wavelet modulae at the above characteristic time scales are seen to vary with time, indicating temporal variations of the amplitudes of the corresponding oscillations. It is particularly visible for the amplitude of the $< 2a_0 >$ oscillation.

4. Individual oscillations of G191 16

Once the different oscillations, that are simultaneously present in the signal, have been detected from fig.1, one takes advantage of the localization properties of the wavelet transform (1) in studying, separately, the temporal behavior of their amplitudes $A(b)$ as defined by (5).

4.1. THE MAIN OSCILLATION AND ITS HARMONICS

On figure 2, are plotted the amplitudes $A(b)$ of the oscillations with periods $< a_0 >, < a_0/2 >$, $< a_0/3 >$ versus time (i.e. b), calculated with $\omega_m = 6, 10, 12$ respectively. The reason for selecting different values of the free wavelet parameter, ω_m, comes from the fact that, for each investigated oscillation of a given frequency ω, it is possible to choose ω_m (and adjust $a = \omega_m/\omega$) in order to take two effects into account: first, the resolution depends on a (as seen at the end of section 2) i.e. it increases with decreasing ω (at fixed ω_m) and with increasing ω_m (at fixed ω); second, perturbations due to the finite length of the signal (window effects), which manifest themselves at the edges of the (b) interval , also increase with a. Therefore, at small frequencies, one chooses a small value of ω_m to limit window perturbations and for which the resolution still remains satisfying whereas, at large frequencies, at which window effects are small, one chooses a larger value of ω_m to increase the resolution.

As it is easily seen on fig.2, all three amplitudes oscillate about a mean value which, as expected, is smaller for the harmonics. They oscillate with a same period $P_b \sim 14 < a_0 >$, a little smaller than the time interval of the run. Thus, all three oscillations are amplitude modulated with the same period P_b. In addition, these modulations are on phase. This is emphasized by a dotted curve which represents the temporal variations of the wavelet modulus (5) of a signal of

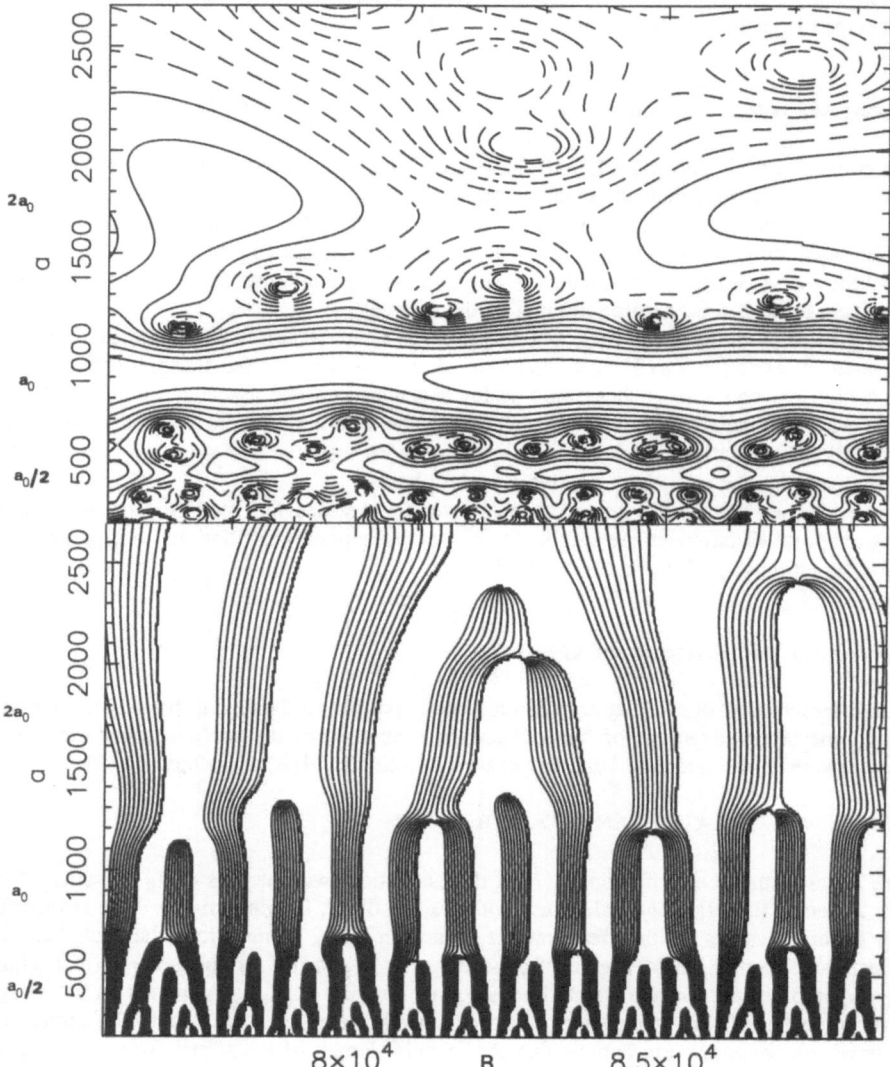

Figure 1. Wavelet modulae in logarithmic values (top) and phases (bottom) for one run of G191 16 light curve are represented by level curves in the (a,b) plane. The horizontal and vertical axes respectively represent the time interval (or b) and characteristic time scales a. The dotted lines are negative values. The wavelet phases are not plotted between $-\pi$ and 0, so that periodicities are more visible. The mean period of each detected oscillation is shown on the a-axis. Departures from horizontality of modulus level curves indicate a temporal variation of the amplitude of the oscillation detected at the corresponding time scale.

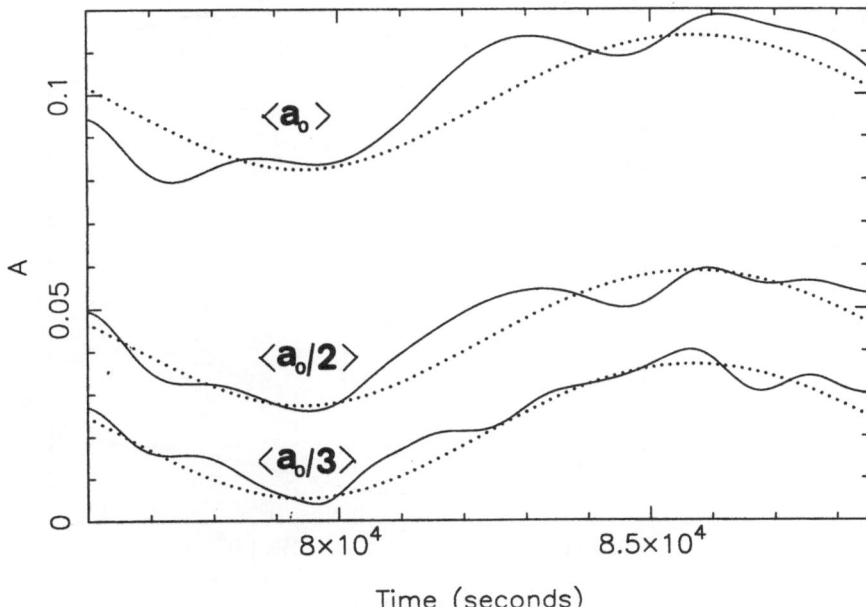

Figure 2. Amplitudes of the oscillations with mean periods $< a_0 >, < a_0/2 >, < a_0/3 >$ (full curves) versus time. They oscillate in phase with the same period as emphasized by the dotted curve (see 4.1)

two cosines with frequencies ω_0 and ω_1 (see section 5), calculated at the time scale $< a_0 >$. This dotted curve coincides with the amplitude variations of the $< a_0 >$ oscillation. Departures from this dotted curve comes from a lack of resolution. As mentionned above, a compromise must be made in choosing ω_m. Once shifted down to account for smaller mean amplitude values, the dotted curve is seen to also match amplitude variations of the harmonic oscillations. We checked that a different choice of ω_m does not affect periods or phases of modulation and only slightly decreases the amplitude of a modulation when increasing ω_m

4.2. THE MAIN OSCILLATION AND ITS SUBHARMONIC 1/2

Figure 3 displays the amplitudes of the $< 2a_0 >$ oscillation and again of the $< a_0 >$ oscillation for comparison ($\omega_m = 6$ in both cases). The subharmonic oscillation is also amplitude modulated with the modulation period P_b but its modulation is phase shifted with respect to the modulation of the $< a_0 >$ oscillation. Again, this is pointed out by the same dotted curve than in section 4.1, shifted down and scaled by a reducing factor because the $< 2a_0 >$ oscillation mean amplitude and amplitude of modulation are smaller than those of the $< a_0 >$ oscillation. Once shifted to the right to account for the phase shift between the two modulations, the dotted curve matches the $< 2a_0 >$, except at the edges of the time interval where window perturbations occur. The resolution is higher at larger time scales so that departures from the dotted curve are smaller for

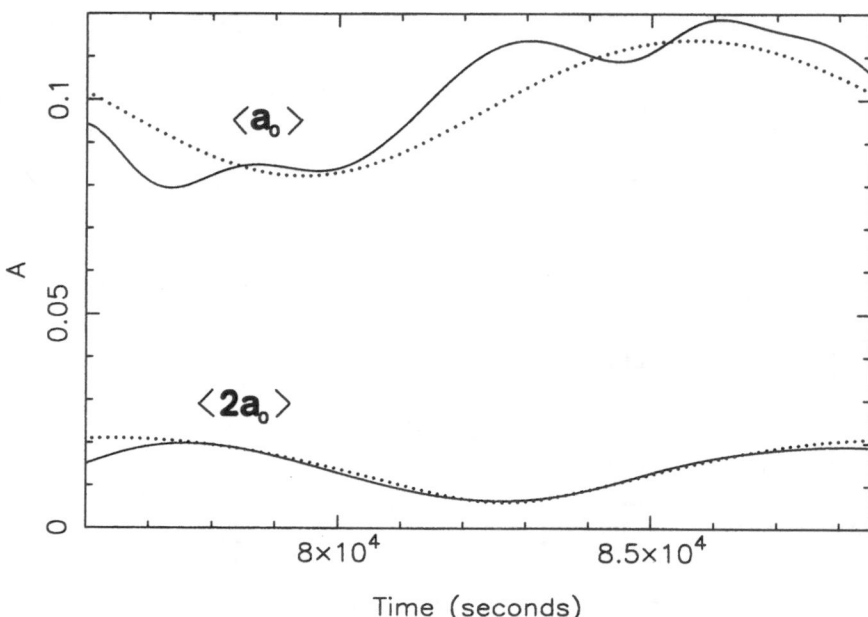

Figure 3. Same as figure 2 but for the oscillation of mean periods $< a_0 >$ and $< 2a_0 >$. The dotted curve is the same as in fig.2 but shifted and scaled for the $< 2a_0 >$ oscillation (see 4.2).

the amplitude variations of the $< 2a_0 >$ oscillation than of the other oscillations of smaller mean periods.

5. Discussion

In order to understand the above results, we must first recall, as mentionned at the end of section 2, that the wavelet analysis does not directly resolve oscillations with close frequencies as separate oscillations but their presence, instead, is detected through the beat phenomenon, and hence the amplitude modulation, they generate. Thus, the amplitude modulated oscillations of G191 16 (fig. 2, 3) are expected to correspond to groups of close frequencies (with the same spacing) in a Fourier spectrum, confirming previous results (Vauclair et $al.$, 1989).

Indeed, the amplitude modulated $< a_0 >$ oscillation gives rise to at least two close frequencies: $\omega_0 = 7.04\ 10^{-3}$ Hz, the main frequency of oscillation during the run considered here and $\omega_1 = \omega_0 + \delta\omega$ with $\delta\omega = 0.0714\ \omega_0$. This small frequency spacing does not exist in the small amplitude regime of oscillations of this star (Auvergne et $al.$, 1990), which excludes its interpretation as a rotational splitting.

Nonlinearities, in generating harmonics and cross frequencies, are responsible for the existence of two other groups of close frequencies, identified as $2\omega_0$, $\omega_0 + \omega_1$ and $3\omega_0$, $2\omega_0 + \omega_1$ which respectively make up the $< a_0/2 >$ and $< a_0/3 >$ oscillations. Similarly, the $< 2a_0 >$

oscillation corresponds to two close frequencies with the same frequency spacing than for the $< a_0 >$ oscillation. One possible interpretation is that these two close frequencies are cross frequencies of ω_0, ω_1 and another frequency $\omega_2 \sim 3\omega_0/2$, also present in this part of the light curve. This would explain the same beat period for all these oscillations and would leave three independent frequencies ω_0, ω_1, ω_2.

Assuming then that ω_0, ω_1, ω_2 are frequencies of pulsation and that the only visible effect of the nonlinearities of the pulsation mechanism is to generate finite pulsation amplitudes while nonlinear corrections to the temperature T and radius r variations are negligible, the only remaining nonlinearities that are able to generate the $< a_0/2 >$, $< a_0/3 >$ and $< 2a_0 >$ oscillations as harmonics and cross frequencies arise from the nonlinear dependence of the luminosity on the temperature and radius, $L = L(T, r)$ (observable effect). Then, it can be shown that, when radius variations are negligible, the above amplitude modulations should have the same phase. This is obviously not the case for the $< 2a_0 >$ oscillation. If, next, are taken into account radius linear variations or nonlinear corrections to the temperature and radius variations, which generally introduce phase shifts due to nonadiabatic effects, we expect none of the above modulations to have the same phase, in contradiction with results of figure 2.

Thus, the pulsations of G191 16 is too strongly nonlinear to be interpreted in a linear framework with nonlinearities acting only as small perturbations. In a strong nonlinear regim, a chaotic pulsation is likely to occur. As an illustration, consider the following simple nonlinear system:

$$\dddot{x} + K_0\ddot{x} + \omega_0^2\dot{x} + R_0 x \left(1 + \beta_0 x\right) = 0 \tag{6a}$$

$$\dddot{y} + K_1\ddot{y} + \omega_1^2\dot{y} + R_1 y \left(1 + \beta_1 y\right) + \delta x y = 0 \tag{6b}$$

For β_0, β_1, $\delta_0 = 0$, the system (6) represents two uncoupled, linear oscillators similar to the oscillator of Baker's one zone model (Baker, 1966). With only $\delta = 0$, the oscillators (6a) and (6b) are nonlinear but still uncoupled and values of the parameters can be chosen such that the equilibrium of (6a), $x = -1/\beta_0$, be vibrationally unstable and $x(t)$ be a period 1 limit cycle (a finite amplitude oscillation with period $\sim 2\pi/\omega_0$) and such that the equilibrium of the second oscillator, $y = 0$, be stable. When the coupling ($\delta \neq 0$) is strong enough, the oscillator (6b) oscillates with both frequencies ω_0 and ω_1.

For the set of parameter values: $\omega_0 = 1$; $K_0 = 0.5$; $R_0 = -0.8$; $\beta_0 = -0.5$, the oscillator (6a) has bifurcated to a period 4 limit cycle. For the second oscillator, parameter values are taken as: $\omega_1 = \omega_0 + 0.0714\,\omega_0$; $K_1 = 0.6$; $R_1 = 0.18$; $\beta_1 = -0.5$ and for the coupling, $\delta = 0.4$. The resulting temporal variation $\ddot{y}(t)$ (which can be assimilated to the temperature of one zone in the Baker's model) mainly remains an amplitude modulated oscillation due to the close frequencies ω_0, ω_1. More drastic consequences of the coupling are observed on the power spectrum of $\ddot{y}(t)$ (Fig. 4) which displays groups of equally spaced, close frequencies with spacing $\omega_0 - \omega_1$. In particular, the frequencies $\omega_0/2$ and $3\omega_0/2$ are splitted into equally spaced frequencies, features that are also observed in power spectra of G191 16. Reconstructed attractors and Poincaré sections for this simple model are very similar to those obtained for G191 16. A wavelet analysis of $\ddot{y}$ with searches for phase correlations has yet to be done. This simplified model, though far from a realistic model of white dwarfs pulsations (!), shows how simple nonlinear mechanisms, which can also be enhanced by the existence of resonances, can be responsible for the puzzling features of the light curves of G191 16 and other white dwarfs.

As shown here, the wavelet analysis, used as a complementary of the Fourier analysis, provides additional informations about pulsating light curves in a conveniently visual way, that must be accounted for in any interpretation.

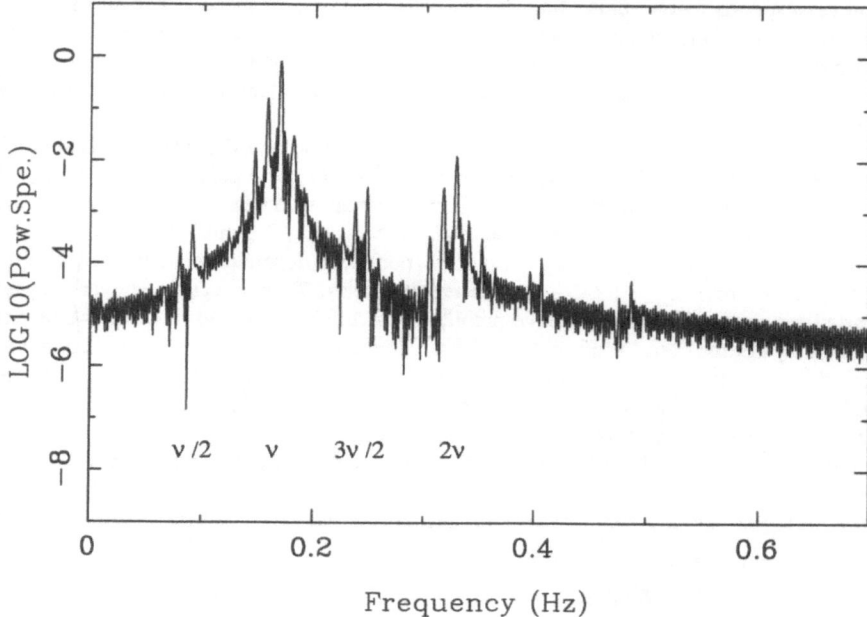

Figure 4. Power spectrum of $\overset{...}{y}(t)$ (eq. 6) calculated over a time interval of about 20 periods. Frequencies in abcissae are $\nu = \omega/2\pi$. The linear frequencies $\nu_0 = \omega_0/2\pi$ and $\nu_1 = \omega_1/2\pi$, their harmonics and subharmonics occur in groups of equally spaced, close frequencies. The spacing is $\nu_0 - \nu_1$.

ACKNOWLEDGMENTS. We gratefully thank Evry Schatzman for helpful discussions.

6. References

Auvergne, M., Chevreton, M., Belmonte, J.A., Vauclair, G., Dolez, N., Goupil, M.J., 1991, these proceedings

Baker, N., 1966, in *Stellar Evolution*, Eds R. F. Stein and A. G. W. Cameron (N.Y.: Plenum Press), p. 333

Goupil, M.J., Auvergne, M., Baglin, A., 1990, *Astron. Astrophys.*, submitted.

Grossmann, A., Kronland-Martinet, R., Morlet, J., 1989, in *Wavelets : Time-Frequency Methods and Phase Space*, ed. J.M. Combes, A. Grossmann, Ph. Tchamitchian, Springer-Verlag, p. 2

Vauclair, G., Goupil, M.J., Baglin, A., Auvergne, M, Chevreton, M., 1989, *Astron. Astrophys.*, **215**, L17

AN ADIABATIC SURVEY FOR ZZ CETI STARS BASED ON A FINITE ELEMENT CODE

P. BRASSARD, G. FONTAINE, C. PELLETIER, and F. WESEMAEL
Département de Physique, Université de Montréal

ABSTRACT. We discuss briefly a new mathematical tool based on finite-element techniques to solve the adiabatic pulsation equations. We then present sample results of an extensive adiabatic survey which has been carried out for ZZ Ceti stars. We discuss the influence of cooling and of the thickness of the hydrogen layer on the profile of the Brunt-Väisälä frequency , on the period spectrum, and on the period changes in ZZ Ceti stars.

1. Introduction

A thorough study of the adiabatic properties of ZZ Ceti stars is required to take advantage of the full potential offered by white dwarf seismology. Such a study necessarily requires an extensive survey of the period structure of evolutionary models of pulsating DA white dwarfs, a task for which existing adiabatic pulsation codes appear inefficient and ill-suited. We have thus been led to develop a new adiabatic code, based on finite-element techniques, which could be used reliably and efficiently to carry out our detailed survey. This code, which forms the basis of the extensive adiabatic analysis reported on here, is now being used in the more complicated study of the non-adiabatic properties of pulsating DA stars.

There are several requirements which must be satisfied by this new pulsation code: we require, first of all, a high accuracy not only for the computation of the periods at high k values, but also to compute reliable values of the period derivative, $\dot{\Pi}$. A high level of accuracy can usually be achieved with the use of high-order numerical methods. A second requirement is a high stability, since convergence problems frequently arise for some modes, with currently available codes. For our complete survey, a robust numerical technique is clearly required to solve those difficult cases. Another requirement is speed. In a survey like ours, there are literally hundreds of models to analyze. Speed is usually achieved through the use of high-order methods or effective algorithms, or both. Finally, because of the magnitude of the task at hand, the code should be able to run with a minimal amount of outside intervention. All these requirements become even more pressing in the context of nonadiabatic studies. After several experiments with various numerical methods, we have found that the finite-element technique was well suited to our needs. We briefly summarize this technique here. Further information can be found in Brassard *et al.* (1991*b*).

193

G. Vauclair and E. Sion (eds.), White Dwarfs, 193–203.

2. The Finite-Element Method

The basic idea behind the finite-element concept is simple: we divide the mesh of integration into small segments, called elements. The method consists in approximating the solution by a linear combination of polynomials forming a complete basis in each element. Linear, quadratic, and cubic elements have all been tried in our code.

The finite-element-method minimizes the error introduced by the approximated solution. We use here the Galerkin minimization scheme, whose main characteristic is to minimize the remainder, defined as the true solution minus the approximated solution, by orthogonalizing it to each basis polynomials. For our specific problem, and in contrast to other minimization schemes, the Galerkin method gives super-convergence, since we can expect a $2n$ order of convergence, instead of the $n + 1$ order associated with regular polynomial interpolations.

3. The Adiabatic Survey of ZZ Ceti Stars

For this survey, we have used evolutionary white dwarf models extracted from the sequences of Tassoul, Fontaine, and Winget (1990). These models, which were specially constructed for pulsation studies of white dwarfs, have a stratified structure consisting of a pure carbon core surrounded by a pure helium envelope, itself surrounded by a pure hydrogen layer. The compositional transition regions (H/He and He/C) are treated under the assumption of diffusive equilibrium.

We have computed all g-modes with $\ell = 1, 2$, and 3 in the period range $40 - 1000s$ ($k_{max} = 95$) with 500 quadratic elements. At this time, we have analyzed over 800 models corresponding to some 80,000 modes, and this without any stability problem. This complete survey has allowed us to study the influence of the stellar mass, of the thickness of the hydrogen and helium layers, of the convection efficiency, and of the cooling on the period structure of ZZ Ceti models. In the following, we present sample results of this survey, with particular emphasis on the effects of cooling and of the hydrogen layer thickness. Complete details on all aspects of this work will be found elsewhere (Brassard et al. 1991a).

4. The Profile of the Brunt-Väisälä frequency

Figure 1 shows the run of the square of the Brunt-Väisälä frequency as a function of depth in two compositionally-stratified DA white dwarf models which differ only by their effective temperature. Our choice of abscissa ($\log q \equiv \log(1 - M(r)/M_\star)$) strongly emphasizes the outer layers; it is only in these outer regions that g-modes have non-negligible amplitudes in white dwarfs. The value $\log q = 0$ corresponds to the center of the star, while large negative values of $\log q$ correspond to the photospheric layers.

Both models are for a 0.6 $M_\odot$ star, and have a similar compositional structure: a pure carbon core surrounded by a mantle containing $10^{-2} M_\star$ of pure helium, itself surrounded by a thin hydrogen envelope containing $10^{-10} M_\star$. The structure in the profiles of N^2 around $\log q \approx -2$ and $\log q \approx -10$ are associated with the changes in chemical composition occuring in the transition zones. A chemically homogeneous model would show a smooth N^2 profile. It is the presence of these quasi-discontinuities associated with composition transition zones that are directly responsible for the filtering capabilities of the models.

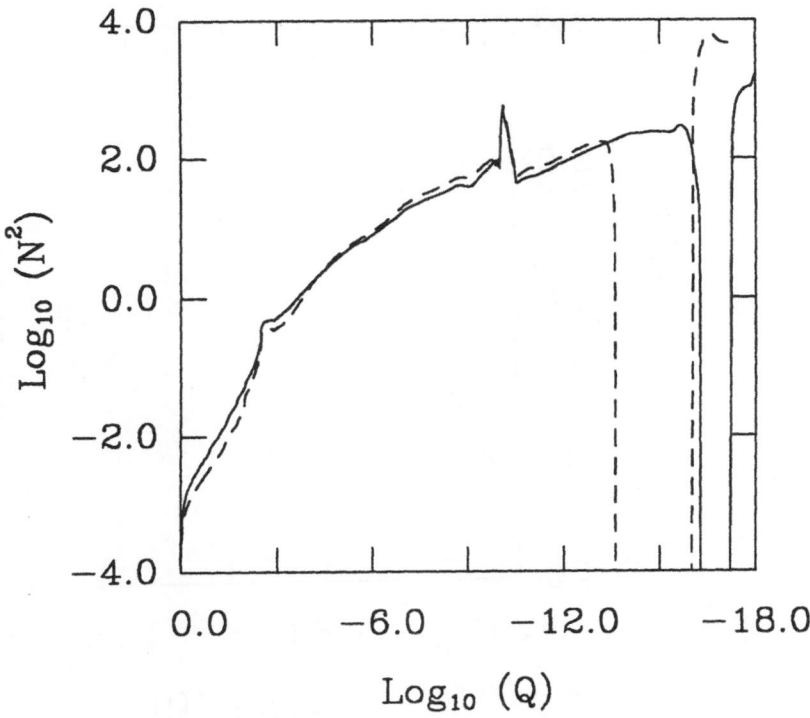

Figure 1. Profile of the Brunt-Väisälä frequency for two ZZ Ceti star models, one at $T_e \approx 14,000K$ (solid line) and one at $T_e \approx 10,000K$ (dashed line).

The figure illustrates the rather mild effects of a change in the effective temperature over the range characteristic of the observed ZZ Ceti instability strip. Except for the outermost layers, in which the cooler model has developed a substantially larger convection zone (regions where $N^2 < 0$) due to the recombination of hydrogen, the N^2 profiles are quite similar. Note that, in the deep degenerate core ($\log q \gtrsim -4$), the Brunt-Väisälä frequency is systematically lower in the cooler model. This small difference reflects the overall increase of degeneracy due to the lower effective temperature of that model. Hence, the period structure of a given model is not expected to change very much as it cools across the instability strip. This should come as no surprise, since the mechanical structure of a white dwarf as cool as a ZZ Ceti is essentially governed by almost complete degeneracy.

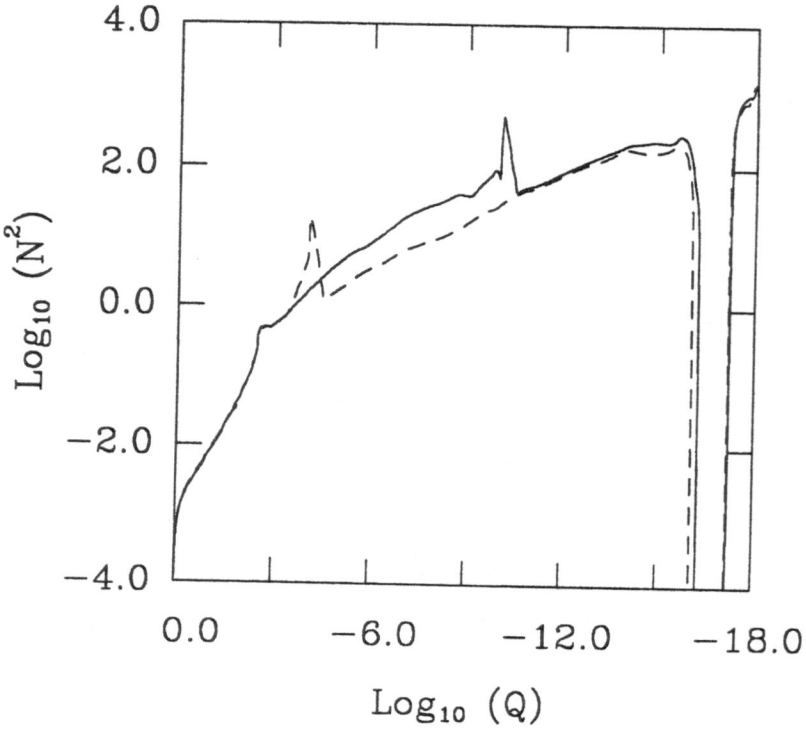

Figure 2. Profile of the Brunt-Väisälä frequency for two ZZ Ceti star models which differ only by the mass of their hydrogen layer. The solid line corresponds to a relatively thin hydrogen layer ($10^{-10}M_\star$), while the dashed line corresponds to a relatively thick layer ($10^{-4}M_\star$). The minor differences in the structure of the hydrogen convection zones are due to the slightly different effective temperatures of the models displayed.

Figure 2 shows the run of the Brunt-Väisälä frequency in two models which have similar effective temperatures (around $T_e \sim 14,000K$), identical helium layers ($M(He) = 10^{-2}M_\star$), but different hydrogen layer masses. One model has a relatively thin hydrogen layer ($10^{-10}M_\star$), while the hydrogen layer in the other model is relatively thick ($10^{-4}M_\star$). Because the spike associated with the H/He transition zone is pushed further down into the star in the thick-layer model, the filtering capacity of the envelope is considerably reduced. This arises because the eigenfunctions already have significantly reduced amplitudes at the depth of the H/He interface in the thick-layer model; the contrast between trapped modes (those confined above the H/He buffer zone) and non-trapped modes is thus considerably reduced. We can then expect that the period structures of models with thick hydrogen layers will not show strong signs of the layered configuration, while the period structures of models with thin hydrogen layers will be dominated by trapping, or resonance, effects.

5. The Period Structure

Figure 3 illustrates the effects of cooling on the period structure of a 0.6 $M_\odot$ DA white dwarf model with $M(He) = 10^{-2}M_*$ and $M(H) = 10^{-10}M_*$. The figure shows the normalized kinetic energy of a mode as a function of the period. Each plotted point corresponds to a radial overtone for g-modes with $l = 2$. The solid curve corresponds to the model with $T_e \sim 14,000K$, while the dashed (dotted) curve corresponds to the same model but further evolved and having $T_e \sim 12,000K$ ($T_e \sim 10,000K$). The profiles of N^2 for the two extreme models were shown in Figure 1.

The primary minima in those curves corresponds to modes with a node at the H/He interface. Such modes have very small amplitudes below the H/He buffer zone; in essence, they resonate with the hydrogen layer thickness or, more accurately, they are efficiently reflected at the composition interface. We can think of those modes as effectively *trapped* in the outer hydrogen layer. Because the kinetic energy of a mode is proportional to the displacement vector integrated over the whole star, modes trapped in the outer hydrogen envelope show a characteristic signature corresponding to minima in kinetic energy (Winget, Van Horn, and Hansen 1981). Note that the H/He transition zone has a finite

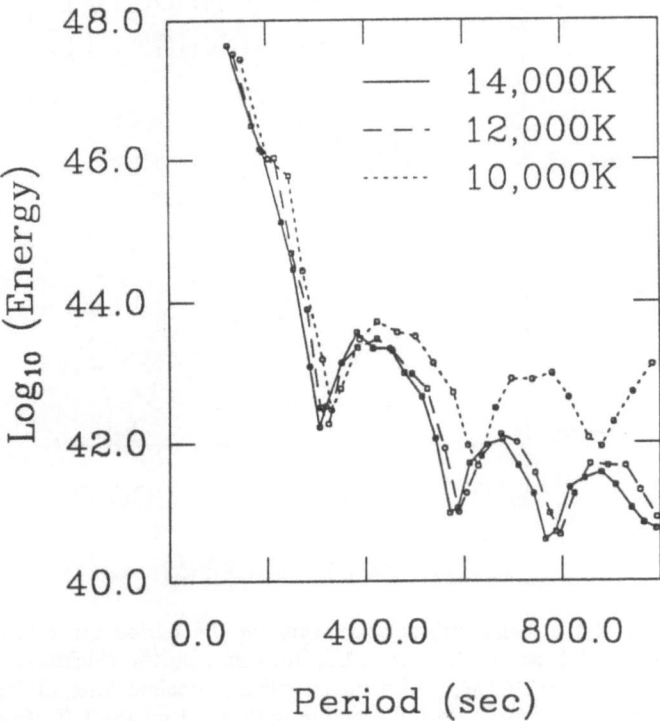

Figure 3. Effect of cooling on the period structure of a $l = 2$ mode. The radial overtones are plotted starting with $k = 1$ on the left of the diagram.

width imposed by diffusion considerations, and is therefore not a true discontinuity. This explains why the kinetic energy minima have finite widths; modes can be *partially trapped* in ZZ Ceti star models. Note also that, in addition to the primary minima shown in Figure 3, there are also secondary features *which are real*, and which correspond to modes with nodes at the He/C interface. Because the He/C transition zone is located much deeper ($\log q \approx -2$) than the H/He transition zone ($\log q \approx -10$), the effect on the kinetic energy is much reduced.

Of course, cooling leads to a small increase in the period of a given mode in a ZZ Ceti star, and hence to positive values of $\dot{\Pi}$. Furthermore, the increased degeneracy associated with cooling pushes the nodes of the eigenfunctions toward the surface, with a resulting increase in the period spacing between trapped modes. Finally, the larger values of the kinetic energy associated with large periods in the cooler model are caused by the presence, in that model, of a very important convection zone, where g-modes cannot propagate.

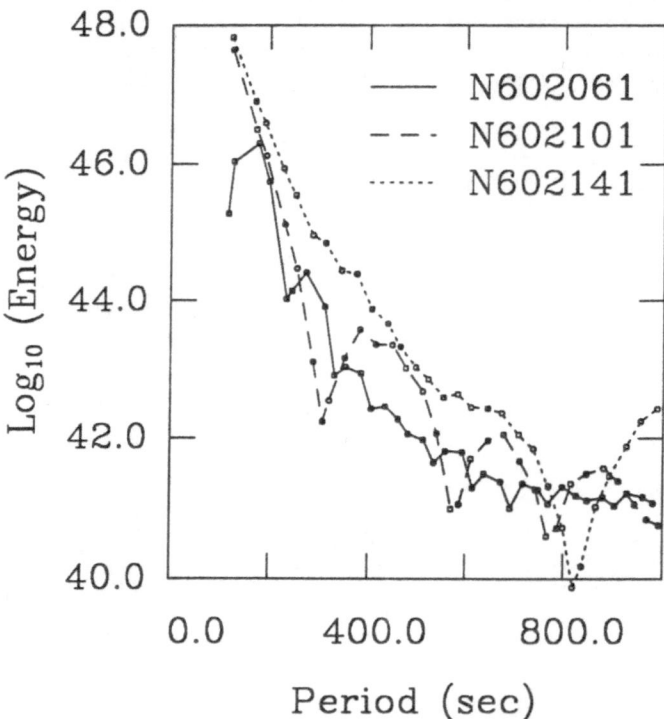

Figure 4. Effect of the compositional structure on the period structure of a $l = 2$ mode. All three models are at $T_e \sim 14,000K$, but differ by the thickness of the outer hydrogen layer ($M(H) = 10^{-6}M_*$: solid line; $10^{-10}M_*$: dashed line; $10^{-14}M_*$: dotted line). The radial overtones are plotted starting with $k = 1$ on the left of the diagram.

Figure 4 shows the period structure for three models with the same effective temperature ($T_e \sim 14,000K$), but which differ by their values of the thickness of the outer hydrogen layer. All models have $M_* = 0.6\ M_\odot$ and $M(He) = 10^{-2}M_*$, but different values of $M(H)$. It is quite apparent that the actual value of $M(H)$ leaves its distinctive imprint on the period structure of a model. Beyond the general decrease of the kinetic energy with increasing period, we can distinguish two trends. Firstly, trapping effects are larger, in the sense that the minima are deeper, as the hydrogen layer gets thinner. This arises, as explained above, because the trapped modes in thin-layer models are confined to a more restricted region above the H/He transition zone. Secondly, for a given mode, the period spacing between trapped modes is smaller for thicker hydrogen layers because there are, in the latter case, many more nodes above the H/He interface. Consequently, more of those nodes can fall near the H/He transition region.

We also remark that the contrast in kinetic energy between the trapped modes and the other modes tends to decrease with increasing order k. This is to be expected, since the spacing between nodes for the high-k overtones tends to become comparable to the H/He transition zone thickness itself. Trapping eventually loses its significance for very high values of k.

6. Trapped Modes

In Figure 5, we show the period distribution of *trapped* modes as a function of the thickness of the outer hydrogen layer ($\log q_H \equiv \log M(H)/M_*$). Modes with values of $l = 1, 2$, and 3 are plotted. To the extent that *only* trapped modes are actually excited in real ZZ Ceti stars, the figure can be considered as a "theoretical Fourier spectrum" (with no information, of course, on the amplitudes). *By matching such distinctive period patterns with actual observations, one can hope to measure the thickness of the outer hydrogen layer in pulsating DA white dwarfs.*

The general trends in this figure can been summarized as follows:

(1) The period spacing between trapped modes increases with decreasing $M(H)$ because of the smaller number of nodes above the H/He interface in a given period range.

(2) Likewise, the number of trapped modes in a given period range increases with increasing l. This is because, for a given period, the value of the radial overtone (the number of radial nodes) increases with l. Hence, more modes have nodes falling at the H/He interface for larger values of l.

Figure 6 shows the values of the periods of models with $l = 2$ for evolving models. This is for an evolutionary sequence of models with $M_* = 0.6\ M_\odot$, $M(He) = 10^{-2}M_*$, and $M(H) = 10^{-12}M_*$. The 22 first overtones (or k–values) are shown. The dotted part of the curves represent the first trapped mode at the H/He interface, while the trapped mode switches from $k = 13$ to $k = 12$ around $T_e \approx 13,000K$.

Clearly, the changes in the periods due to cooling are quite smooth, with the exception of the trapped mode. The overall effect of trapping for a mode is the reduce slightly the slope of the curve. We note also that smaller effects of trapping can be seen for low-k modes. These effects are produced by the trapping at the He/C interface and are more visible at low-k values because the trapping mechanism is more effective for the first trapped modes.

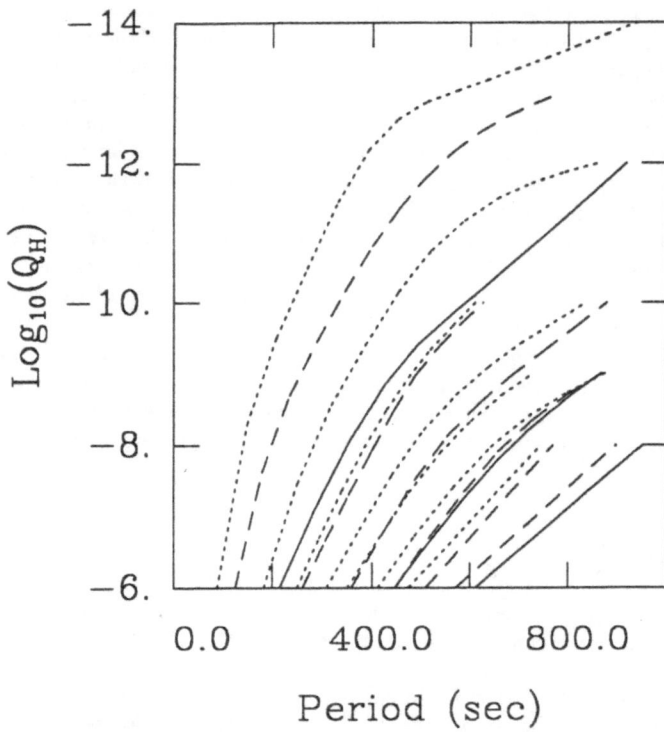

Figure 5. Period distribution of trapped modes as a function of the mass of the outer hydrogen layer for a model with $T_e \sim 10,000K$, $M_\star = 0.6\ M_\odot$, $M(He) = 10^{-2}M_\star$. Solid lines are for $l = 1$, dashed lines are for $l = 2$, and dotted lines for $l = 3$.

7. Period Changes in Cooling ZZ Ceti Stars

Figure 7 shows the variation of $\dot{\Pi}$ with effective temperature for three different overtones of a $l = 2$ mode in a sequence of evolving models of ZZ Ceti stars. Two sequences, characterized by different hydrogen layer masses are shown. The value of $\dot{\Pi}$ thus depends not only on the effective temperature and on the actual mode under consideration (via both l and k), but also on the chemical stratification: thicker hydrogen layers will be characterized by smaller values of $\dot{\Pi}$.

Generally, $\dot{\Pi}$ decreases monotonically with decreasing effective temperature. There are exceptions, however, when modes become trapped during the cooling. This is illustrated by the top continuous curve in Figure 7. The dip in that curve is real, and corresponds to the onset of trapping for the $k = 20$ overtone near $T_e \approx 14,500K$; trapping switches

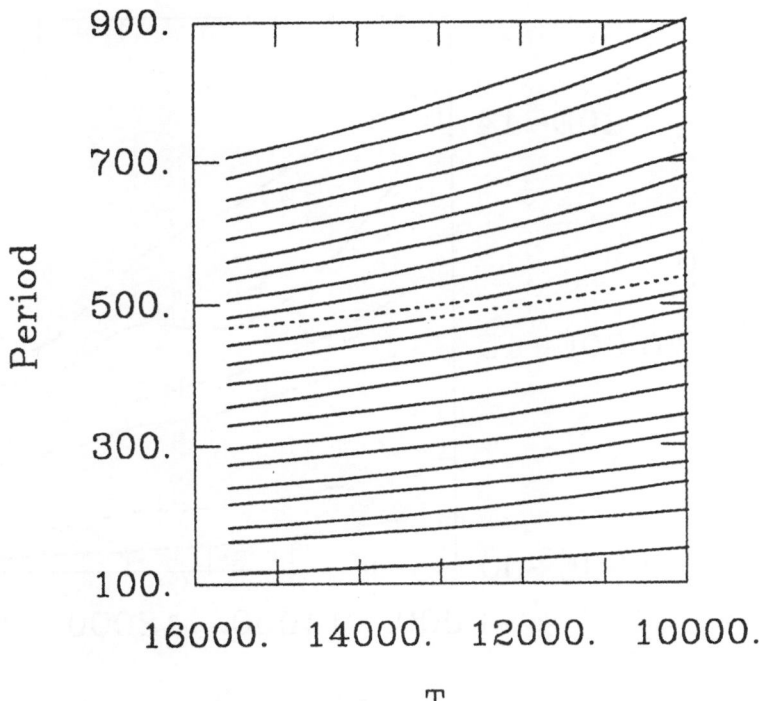

Figure 6. Period of a $l = 2$ mode in evolving ZZ Ceti models. The 22 first overtones (k-values) are shown from the bottom to the top. The dotted part of the curves represent the first trapped mode at the H/He interface.

to the mode with $k = 19$ around $T_e \approx 12,300K$. This diagram underscores the need for detailed calculations of the period changes in cooling ZZ Ceti stars if the true potential of asteroseismological studies of these stars is to be realized. Our extensive survey of the period changes encountered in models of ZZ Ceti has already allowed Fontaine *et al.* (1991) to contrast these predictions with the value of $\dot{\Pi}$ now measured by Kepler *et al.* (1991) in G117-B15A. The current activity in asteroseismological studies of ZZ Ceti stars makes the calculations summarized here particularly timely and relevant.

This work was supported in part by the NSERC Canada, by the fund FCAR (Québec), and by NATO. G. Fontaine acknowledges additional support from a Killam Fellowship.

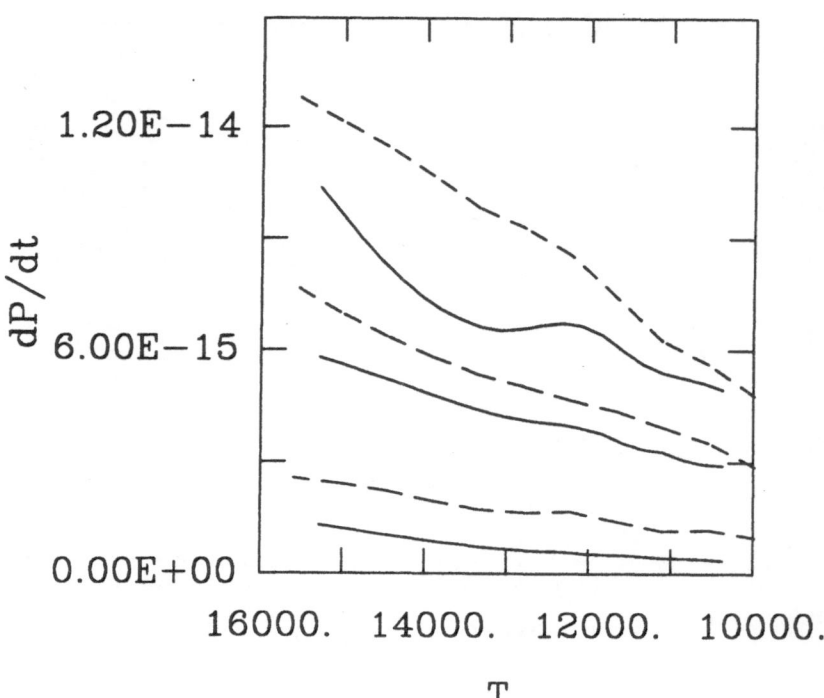

Figure 7. $\ddot{\Pi}$ for evolving models as a function of ·effective temperature. The dashed lines are associated with a sequence of evolutionary models at 0.6 $M_\odot$, with $M(He) = 10^{-2}M_\star$ and $M(H) = 10^{-12}M_\star$. The various curves correspond to a $l = 2$ mode with, from top to bottom, $k = 20$, 10, and 1, respectively. The solid curves are associated with the same mode and overtones, but in a model with a thicker $(M(H) = 10^{-6}M_\star)$ hydrogen layer.

REFERENCE

Brassard, P., Fontaine, G., Wesemael, F., and Tassoul, M. 1991a, *Ap. J.*, in press.
Brassard, P., Pelletier, C. Fontaine, G., and Wesemael, F. 1991b, *Ap. J. Suppl.*, in press.
Fontaine, G., Brassard, P., Wesemael, F., Kepler, S.O., and Wood, M.A. 1991, these Proceedings.
Kepler, S.O., *et al.* 1991, these Proceedings.
Tassoul, M., Fontaine, G., and Winget, D.E. 1990, *Ap. J. Suppl.*, **72**, 335.
Winget, D.E., Van Horn, H.M., and Hansen, C.J. 1981, *Ap. J. (Letters)*, **245**, L33.

DISCUSSION

VAN HORN :

Do you have similar results for the period VS layer mass for He? Are the results similar for H/He/C models and for He/C DB models?

BRASSARD :

– We get mode trapping at the He/C interface but the effects on the kinetic energy are **very** small because this interface is very deep.

– Our survey is aimed only at DA variables. But, because the He layers in the DBs are thick, mode trapping cannot be considered as an effective selection mechanism.

WINGET :

1 –I want to point out that your development of a finite-element code is a profound contribution to the field. It is only with this technique that high radial overtone frequencies can actually be computed for comparison with the observations. Also, it is the method of choice for computing rates of period change. Thank you for developing this important tool.

2 – The "avoided crossings" in your results illustrate the possibility of the other effect of mode trapping on dP/dt, and that is the nonlinear "pulling" of frequencies. This may help to understand the dP/dt in G117 -B15A reported by Kepler.

A STUDY OF PERIOD CHANGE RATES IN POST-AGB STARS
I. PG 1159-035

LETIZIA STANGHELLINI[1,2] AND ARTHUR N.COX[2]
[1]Osservatorio Astronomico di Bologna,
via Zamboni 33, I-40126 Bologna
[2]Los Alamos National Laboratory,
P.O. Box 1663, Los Alamos, NM 87545

ABSTRACT. The observed period decrease of the strongest mode in the post-AGB variable star PG 1159-035 (GW Vir) is investigated with standard evolution models and ones with the surface helium greatly reduced. These latter models were investigated because the presence of even 10% of helium by mass can poison the cyclical CO ionization that produces the κ and γ effects pulsational driving. We use the Lagrangian based pulsation program of Pesnell (1990). As found by others before, we theoretically predict period increases, in spite of a radius decrease, rather than the observed period decrease. If mode trapping to decrease the period of this particular 516 second mode is the explanation, it has to be done with only a slight composition gradient at the few percent level of mass into the model. New evolution models with more mass loss and less surface helium are being investigated to see if they can explain the observed period decrease.

1. Introduction

PG 1159-035 (hereafter PG 1159) was the first DO star to have been observed to oscillate (Mc Graw *et al.* 1979). Its variability has been interpreted as nonradial g^{+}-mode oscillations caused by the rapid opacity and compressibility increases, in the carbon and oxygen partial ionization zones, not very deep below the stellar surface (Starrfield *et al.* 1983, 1984). Given the particular stage of its evolution, between the tip of the asymptotic giant branch (AGB) and the white dwarf cooling line, investigations of the physical quantities and processes of this star via asteroseismology could turn out to be of fundamental importance toward the understanding of the late stages of stellar evolution. These stages further are deeply linked to the galactic enrichment problem and consequently to the evolution of the galaxy as a whole.

One of the intriguing aspects of the data for PG 1159 is that the highest amplitude mode has a period that is observed to decrease with time (Winget *et al.* 1990). In this study we aim to investigate the linear pulsationally nonradial oscillations for two models

205

G. Vauclair and E. Sion (eds.), White Dwarfs, 205–210.
© 1991 *Kluwer Academic Publishers.*

that closely follow the stellar evolution results (Iben 1989, Hollowell 1990) with particular emphasis on the period decrease of the most prominent mode.

2. Models

Both models presented in this analysis are from the Palomar-Green evolutionary sequence calculated by Iben (1989). The original models have a surface helium layer that extends down to $1 - q \simeq 0.05$. The maximum surface helium abundance is $Y = 0.75$. In our models we cut off all except $1 - q \simeq 1 \times 10^{-11}$ of this surface helium; the very thin residual layer has been left to account for the observed helium spectral lines.

In Table 1 we list the physical parameters for the two models used in our study. In column (1) we give the effective temperature of the model. In column (2) is listed the model number, following Iben's nomenclature (see also Stanghellini et al. 1990a). In column (3) the logarithm of the luminosity, in units of the solar luminosity, is given. Column (4) gives the stellar model radius. In columns (5) and (6) the central temperature and density, respectively, are listed, and in column (7) the maximum temperature within the model is given. The evolutionary time elapsed between models 60 and 63 is $\Delta t = 3.426 \times 10^{11} sec$.

3. Pulsation Analysis: Results

We perform our stability analyses with the Lagrangian nonradial, nonadiabatic code of Pesnell (1990). The models are in hydrostatic and thermal equilibrium, and they are tested for instabilities of modes with periods between 350 and 950 seconds. This range includes most of the prominent observed periods. These modes are barely unstable in the models. Some of the resulting modes are listed in Table 2. Column (1) gives the g^+-mode considered, columns (2) and (3) the periods, in seconds, respectively, for models 60 and 63, and column (4) the period change that has occurred during the evolution between models 60 and 63.

The period changes from the most accurately calculated modes, with uncertainties in the periods of less than one percent, is from 5 to 10 seconds. The period increase rate is

$$1.5 \times 10^{-11} \leq d\Pi/dt \leq 2.9 \times 10^{-11} s/s,$$

which is very close to the value observed by the Whole Earth Telescope ($-2.49 \pm 0.06 \times 10^{-11}$ s/s, Winget et al. 1990), but it has opposite sign. A similar theoretical result has been achieved by Kawaler et al. (1985), with a study using only adiabatic periods.

4. Discussion

For more insight, let us discuss the physical causes for a period to change with time. Two factors can cause a period decrease: a decrease of the stellar radius, and an increase of the concentration of the star. From model 60 to model 63, the radius decrease alone produces a period decrease of about 36 seconds for a mode with 500 seconds period. This is because, for fixed pulsation constant, the period goes as the 3/2 power of the radius. Clearly from the results, model deconcentration giving a period increase has the overwhelming influence here. Note that the period changes in Table 2 are smaller for the lower order g-modes,

TABLE 1 Models ($M/M_\odot = 0.6$)

$T_e(K)$	$Model$	$logL/L_\odot$	$R(10^9 cm)$	$T_c(10^7 K)$	$\rho_c(10^6 g/cm^3)$	$T_{max}(10^8 K)$
132,100	60	2.08	1.47	12	2.41	1.7
117,900	63	1.85	1.40	12	2.39	1.7

TABLE 2 Period Analysis

mode	Π (sec) model 60	Π (sec) model 63	$\Delta\Pi$ (sec)
29	774	781	7
28	747	754	7
27	720	728	8
26	693	703	10
25	667	676	9
24	640	648	8
23	614	621	7
22	588	596	8
21	563	571	8
20	538	543	5
19	512	517	5
18	487	492	5
17	462	467	5
16	437	441	4
15	412	414	2

reflecting the fact that these modes sense regions of the models nearer to the surface that are not deconcentrating so rapidly.

To further check this aspect, we compare in figure 1 the density versus (relative) radius plots for our two models. We find that the density gradient of model 60 is steeper than the one for model 63. Model 63 is less concentrated and from that effect alone, the period should increase as found. This density gradient decrease is a direct effect of the cooling of the temperature maximum layer in the interior. If we artificially increase the concentration of model 63, in order to obtain $d\Pi/dt = 0$, we find that both models, which are complete to the very center, have about the same radius. Obviously this result can not correspond to a realistic evolutionary sequence, with model 63 being more evolved.

Kawaler (1990) showed that, for pre-white dwarfs, some modes trapped in the outer shells of the star can exhibit period decreases, since the outer layers and the composition gradient regions of the post-AGB stars are still shrinking at this stage of their evolution. He obtains period decreases similar to those observed.

In order to obtain mode trapping, however, a significant composition gradient is needed. For DOV stars, such a gradient could be the one between helium rich surface layers and the core CO composition. The composition transition zone should be sufficiently deep to involve a large enough fraction of the stellar mass to influence the stellar mechanical properties. But Stanghellini et al. (1990a, 1990b) showed that a thick layer with only 10% helium by mass poisons the CO ionization κ and γ effects for surface effective temperatures higher than 90,000 K. It is not clear whether a very small admixture of helium in the surface pulsating driving layers (10^{-9} of the mass deep) will still allow net pulsation driving while giving a large enough composition gradient a few percent of the mass into the model. In our two models in this paper, the composition gradient from the surface pure helium to the CO core occurs at 10^{-11} of the mass into the star. The modes do not feel that gradient and show the constant period spacing of 26 seconds, as expected from asymptotic theory (Tassoul, 1980).

Recently, Werner et al. (1990) determined $T_e = 140,000$ K from non-LTE spectrum analysis. The temperature error in their determination is estimated to be approximately 10%, and this rules out the possibility for PG 1159 to be as cool as 90,000 K.

We cannot exclude that mode trapping is the cause for the observed period shift. We need to investigate whether a rather low amount of helium in a thick surface layer that includes the pulsating driving layers and extends to perhaps a few percent of the stellar model mass can still allow pulsation driving by the cyclical CO ionization with enough of a composition gradient to affect mode trapping. Modeling of PG 1159 with a new evolutionary sequence that starts with a lower helium surface mass, as even suggested by the large surface carbon and oxygen abundances, is in progress. Even a very small change in the model could effect the periods of the oscillations without having a major effect on the instability strips (Stanghellini et al. 1990a). We plan to further investigate this delicate aspect of the evolution of PG 1159 with the new models.

L. Stanghellini is grateful to the T-6 Group of the Los Alamos National Laboratory for their hospitality during the summer of 1990, when this research was completed.

5. References

Hollowell, D. 1990, *private communication*
Iben, I. Jr. 1989, *private communication*

Kawaler, S. D. 1990,"White Dwarf Evolution: Cradle-to Grave Constraints Via Pulsation", in *Confrontation between Stellar Pulsations and Evolution*, C. Cacciari and G. Clementini (eds.), *in press*

Kawaler, S. D., Hansen, C. J., and Winget, D. E. 1985, "Evolution of the Pulsation Properties of Hot Pre-White Dwarfs Stars", *Ap. J.*, **295**, 547

McGraw, J. T., Starrfield, S., Liebert, J, and Green, R. 1979, "PG1159-035: A New, Hot, Non-DA Pulsating Degenerate", in *IAU Colloquium No 53*, Van Horn and Weidemann (eds.), 377

Pesnell, W. D. 1990,"Nonradial, Nonadiabatic Stellar Pulsations", *Ap. J.*, Nov 2.

Stanghellini, L., Cox, A. N., and Starrfield, S. 1990a, "Post-AGB Non-Radial Instability Strips", *Ap. J.*, *submitted*

Stanghellini, L., Cox, A. N., and Starrfield, S. 1990b, "Nonradial Instability Strips for Post-AGB Stars", in *Confrontation between Stellar Pulsation and Evolution*, C. Cacciari and G. Clementini (eds.), *in press*

Starrfield, S., Cox, A. N., Hodson, S. W., and Pesnell, W. D. 1983, "The Discovery of Nonradial Instability Strips for Hot Evolved Stars", *Ap. J. Lett.*, **268**,L27

Starrfield, S.,Cox, A. N., Kidman, R. B., and Pesnell, W. D. 1984, "Nonradial Instability Strips Based on Carbon and Oxygen Partial Ionization in Hot Evolved Stars", *Ap. J.*, **281**, 800

Tassoul, M. 1980,"Asymptotic Approximations for Stellar Nonradial Pulsations", *Ap. J.Suppl.*, **43**, 469

Winget, D. E., *et al.* 1990, "Asteroseismology of the DOV Star PG1159-035 with the Whole Earth Telescope", *Ap. J.*, *submitted*

Werner, K., Heber, U., and Hunger, K., 1990, " Non-LTE analysis of four PG 1159 Stars ", *Astr. Ap.*, *sumbitted*

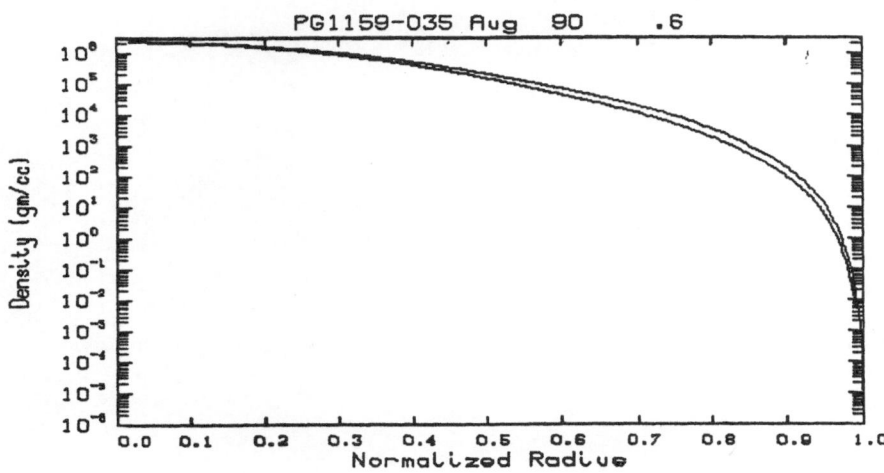

Figure 1. Relative density versus normalized radius for models 60 and 63. The upper curve corresponds to model 63.

DISCUSSION

FONTAINE :

I think that it is very nice that we now see, for the first time, an acknowledgement from the Los Alamos group that considering a layered configuration (i.e. putting a very thin layer of He on top of a C/O envelope) does not change the pulsation properties of their PG1159 models. This is a point which I have argued with Sumner Starrfield over many years. Of course, the direct consequence of this acknowledgement is that the surface (atmospheric) composition of PG1159 stars tell us **nothing** about the composition in the driving regions! This is quite contrary to what has been claimed by the Los Alamos group over many years.

Concerning the blue edge of the ZZ Ceti instability strip, I also find comfort in the fact that the Los Alamos calculations now are apparently able to push the theoretical blue edge towards values more in agreement with the observations. In the Cox et al. paper (1987, Ap. J., 317, 303), it is stated that their maximum possible blue edge was around $T_e \sim 11,500$ K, i.e., well below the observed blue edge. It seems that improvements (of yet unknown origin) have been made in their modelling capabilities of white dwarfs. In this context, it is to be recalled that our group never had any difficulty in reconciling the observed with our theoretical blue edge temperature (see Tassoul, Fontaine, Winget 1990, Ap. J., 72, 335).

NONADIABATIC NONRADIAL PULSATIONS FOR DAV WHITE DWARF STARS

ARTHUR N. COX AND DAVID E. HOLLOWELL
Los Alamos Astrophysics, Los Alamos National Laboratory

ABSTRACT. New complete 0.6 $M_\odot$ models based on evolution studies including element diffusion and gravitational settling for white dwarfs have been studied for their nonadiabatic oscillations. The derived periods and growth rates confirm the earlier result that any thickness of the surface hydrogen can give the observed nonradial pulsations, providing it is at least thick enough to contain the surface convection zone. It is now found that a very unconventional and large ratio of the mixing length to pressure scale height is required to deepen the convection zone so that the pulsation driving mechanisms can operate for the low degree and order g-modes observed at the 13,000K instability strip blue edge. These pulsation results were obtained from a Lagrangian-based eigensolution program.

1. Introduction

The question of the allowed thickness for the surface hydrogen layer in classical DA white dwarfs has been discussed intensively in the last 5 years. Earlier studies by Cox et al. (CSKP, 1987) using both Eulerian and Lagrangian pulsation stability programs found that any thickness of the surface pure hydrogen layer is allowed, providing that the layer is thick enough to contain the surface convection zone. This result was in conflict with a study by Winget (1981), who suggested that the ZZ Ceti white dwarf pulsations constrained the hydrogen layer to be less than 10^{-8} of the mass thick. This paper reinvestigates this disagreement with new models based on recent evolution calculations allowing for element diffusion and gravitational settling that determine the shape of the hydrogen to helium and helium to carbon-oxygen composition gradients in the surface layers.

Evolution studies after the tip of the AGB have produced planetary nebula nuclei and white dwarfs, but it is not easy to justify white dwarf models with thin hydrogen. Generally the expected result is either a hydrogen layer with a mass of near 10^{-4} of the mass or none at all. Our two evolution models, based on work by Iben (private communication) have 6×10^{-6} and 2×10^{-7} of the mass in the surface hydrogen layer after diffusion has operated for the time needed to get to the DA white dwarf state. These thinner than normal hydrogen layers were obtained by artificially applying surface mass loss just after the AGB stage. This wind may actually occur, but it is not readily justified from theory or observation.

Observations, most notably the appearance of non-DA white dwarfs at about 9000K where helium, carbon, and oxygen are dredged-up by a deepening convection zone, do point to thin surface hydrogen layers on all DA white dwarfs. But we

G. Vauclair and E. Sion (eds.), White Dwarfs, 211–218.
© 1991 Kluwer Academic Publishers.

find that the existence of nonradial pulsations does not constitute a thin hydrogen constraint for these stars.

The new calculations reported here were undertaken to improve the results obtained by CSKP. In that paper the composition gradients from hydrogen to helium and from helium to CO were excessively sharp. Detailed profiles were not available until the publication of the Montreal models by Tassoul, Fontaine, and Winget (1990). Actually we have had the capability to make the calculations ourselves with a version of the Iben and MacDonald evolution program, but they have been completed only recently. It appears that the composition gradients we get agree with those for the Montreal models, as best we can tell from the paper. Now the small sharp composition steps between equation of state and opacity tables have been replaced by smooth interpolation between Los Alamos Astrophysical Opacity Library (Huebner *et al.*, 1977) tables for pure hydrogen and pure helium.

It was thought that the difference between the Winget (1981) and the CSKP results might possibly be caused by the sharp composition gradients, because the eigenmodes obtained for these models had strongly varying stabilities typical of sharp composition gradients. This mode trapping produced large fluctuations in growth rates from mode to mode as the eigensolutions sampled the driving differently. However, since the driving is always in the surface hydrogen layer at a mass level always less than 10^{-10} of the mass into the model, the much deeper composition changes would not seem to affect much the pulsation stability results.

2. Models

For this paper five $0.6M_\odot$ models near the pulsation instability strip blue edge on the Hertzsprung-Russell diagram have been calculated using the evolution results for the mass, radius, luminosity, and composition structure. These newly constructed models are complete and self-consistent in all variables, especially for the dP/dr and dT/dr, which are set directly by the momentum and energy equations. Note that the Montreal models do not have this self-consistency or even an accurate dP/dr. Data for the pulsation analyses are: shell masses, radii, and thicknesses, temperatures, densities, pressures, specific heats, gammas, opacities, and the temperature and density derivatives of the opacities for each mass shell.

In Table 1 we list the physical parameters for the five models used in our study. In row (1) we give the effective temperature of the model. In row (2) we give the ratio of the mixing length to pressure scale height. Note that values of 6 and 9 are necessary to produce pulsational instability near the 13,000K instability strip blue edge. Model radii are given in row (3), and in row (4) the model surface luminosities are listed. As found by many others, the internal luminosity is strictly proportional to the internal mass at every interior level. Models 3 and 5 are thin models, but they still are 20 times thicker than considered by Winget (1981) to be the maximum. The bottom of the composition gradient is either 10^{-4} or 10^{-5}, as obtained approximately from the evolution models. As seen in the next column, these "H burn-out" levels range between 5 and over 7 million kelvin. The helium composition deeper decreases to the pure CO at 5% into the mass for all five models. Central conditions are in the next three rows. Their accuracy is not and does not need to be high, because the white dwarf g-modes exist only near the stellar surface. The bottom and top temperatures and mass depths for the surface convection zone are given next. Since the pulsation driving by the convection blocking mechanism

(Pesnell, 1987 and CSKP) occurs right at the bottom of the hydrogen convection zone, one can readily see that the pulsation is produced in the upper 10^{-13} of the stellar mass. Densities at these levels are in the last row.

Table 1. 0.6 $M_\odot$ DA White Dwarf 600 Zone Model Parameters

Model	1	2	3	4	5
T_e (K)	12,500	12,500	12,500	13,000	13,000
ℓ/H_p	6	9	9	9	9
R (10^8 cm)	9.10	9.10	9.10	9.09	9.09
L (10^{31} ergs/s)	1.44	1.44	1.44	1.68	1.68
M_{hyd}/M_* (10^{-6})	6	6	0.2	6	0.2
$(1-q)_{bot}$	10^{-4}	10^{-4}	10^{-5}	10^{-4}	10^{-5}
T_{bot} (10^6)	7.2	7.2	5.2	7.4	5.3
R_{ball}/R_{photo}	0.19	0.19	0.20	0.19	0.20
T_c (10^6 K)	11.5	11.5	11.5	12.2	12.3
ρ_c (10^6 g/cm^3)	2.9	2.9	2.8	2.9	2.8
Convection in hydrogen					
T_{bot} (10^5 K)	0.74	1.02	1.02	0.63	0.63
T_{top} (10^3 K)	11.2	11.2	11.2	11.7	11.7
M_{bot}/M_* (10^{-13})	0.34	1.9	1.9	0.12	0.12
M_{top}/M_*	10^{-17}	10^{-17}	10^{-17}	10^{-17}	10^{-17}
ρ_{bot} (g/cm^3)	1.4×10^{-7}	1.3×10^{-4}	1.3×10^{-4}	1.3×10^{-5}	1.3×10^{-5}

3. Pulsation Results

We make our stability analyses with the Lagrangian nonradial, nonadiabatic code of Pesnell (1990). In this code, and all other current nonradial codes, variations of the convection are assumed to be frozen during the pulsations. This is really a poor approximation for white dwarf models, and further work on this problem is underway. Models are tested in this paper only for instabilities of modes with periods between 500 and 600 seconds for $\ell=1$. This range includes the observed periods for most ZZ Ceti variables. In Table 2 column (1) gives the g-mode considered, and the other columns give the periods and growth rates.

Table 2. Periods (seconds) and Growth Rates (Δ KE/KE per period)

Model	1 12,500K	2 12,500K	3 12,500K	4 13,000K	5 13,000K
g_7			507 6.2×10^{-5}		500 -1.3×10^{-4}
g_8	512 1.9×10^{-6}	513 2.4×10^{-6}	554 5.9×10^{-5}	503 -3.0×10^{-6}	547 -1.3×10^{-4}
g_9	550 1.5×10^{-6}	551 1.8×10^{-6}		550 -4.6×10^{-4}	628 -9.9×10^{-4}

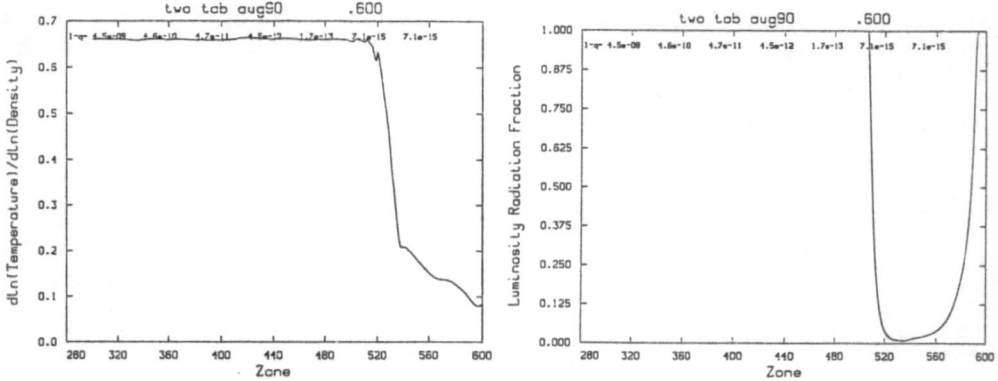

Figure 1. The model 1 $\Gamma_3 - 1$ is plotted versus zone for the outer 300 mass shells.
Figure 2. The fraction of the stellar luminosity carried by radiation is plotted for the outer 300 mass shells of model 1 showing the strong convection zone.

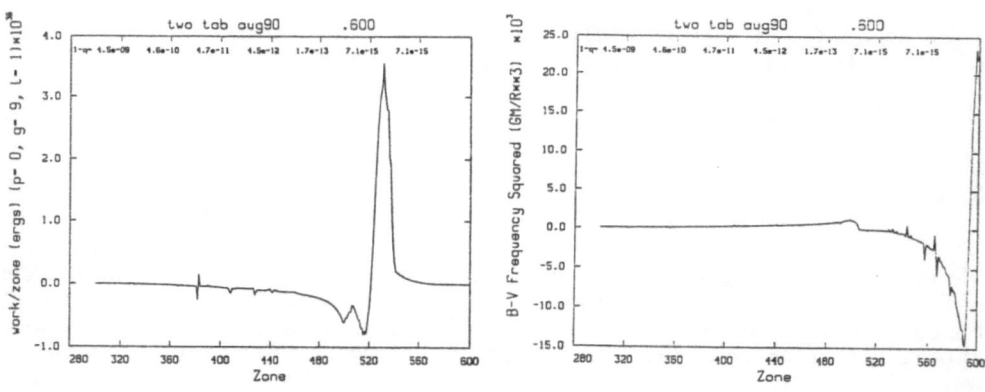

Figure 3. The work per zone versus zone is plotted for the convection blocking, κ, and γ effects pulsation driving and radiative damping layers in model 1 for the g_9 mode with $\ell=1$.
Figure 4. The Brunt-Väisälä frequency for the outermost zones in model 1 showing that the difference $(dln\rho/dr - 1/\Gamma_1 dlnP/dr)$ is precise enough in the pulsating layers to allow accurate periods and growth rates.

Figure 1 for model 1 gives the $\Gamma_3 - 1$ variations in the outer 300 mass shells, showing that the ionization of hydrogen extends down to 10^{-14} of the mass. Figure 2 shows that the convection zone extends down in mass to where the hydrogen ionization is complete. Figure 3 gives the model 1 work plot with positive contributions giving driving and negative contributions damping. Note that there is a bit of convection blocking driving at the convection zone bottom near zone 500, but most of the driving comes from the usual κ and γ hydrogen ionization effects.

4. Points of Possible Misunderstanding

Unlike many recent white dwarf pulsation studies, we have found that the evolution models are not at all suitable for pulsation analyses. The fine zoning needed in the pulsation period and driving determining layers are usually not available in evolution models calculated with other motivations. Thus we have constructed new models using only the total mass, surface luminosity, surface effective temperature, and the internal composition structure from the evolution model. Since we track these models exactly with our reconstructions with different mass zoning, there is no problem of divergence near the new model center.

Our models are constructed by integrating from the surface inward using at each mass shell the pressure (momentum) and luminosity (energy) balance equations. Thus $dlnP/dr$ is known to extremely high precision, and the Brunt-Väisälä frequency, though not used in the Pesnell code, is smoothly varying (except for equation of state table value fluctuations) as seen in Figure 4. Actually at the center of white dwarfs the difference $(dln\rho/dr - 1/\Gamma_1 dlnP/dr)$ is approximately 10^{-4} of either term, but with our precision of one part in 10^7 for the integrations, we have no problem with precision for the pressure and density gradient differences.

A final point of frequent misunderstanding is that the Pesnell code has matrix elements for the eigensolution that somehow are contaminated by the loss of precision discussed above. These elements are not the result of any subtractions of nearly equal numbers, but the model variables that are used in these elements must have adequate precision as for all pulsation analyses.

5. References

Castor, J. I. 1971, "On the Calculation of Linear Nonadiabatic Pulsations of Stellar Models," *Ap. J.*, **166**, 109.

Cox, A. N., Starrfield, S. G., Kidman, R. B., and Pesnell, W. D. 1987, "Pulsations of White Dwarf Stars with Thick Hydrogen or Helium Surface Layers," *Ap. J.*, **317**, 303.

Huebner, W. F., Merts, A. L., Magee, N. H., and Argo, M. F. 1977, "Astrophysical Opacity Library," Los Alamos Report, LA 6760-M.

Pesnell, W. D. 1987, "A New Driving Mechanism for Stellar Pulsations," *Ap. J.*, **314**, 598.

Pesnell, W. D. 1990, "Nonradial, Nonadiabatic Stellar Pulsations," *Ap. J.*, Nov 2.

Tassoul, M., Fontaine, G. and Winget, D. E. 1990, "Evolutionary Models for Pulsational Studies of White Dwarfs", *Ap. J. Suppl.,* **72**, 335.

Winget, D.E. 1981, "Gravity Mode Instabilities in DA White Dwarfs," Ph. D. thesis, Univerity of Rochester.

DISCUSSION

Van HORN :

A couple of years ago at the Dartmouth meeting, Brassard et al. did some calculations that to me seem persuasive that Pesnell's code does not do g-mode calculations reliably. As I recall, the reason is that this Lagrangian calculation effectively cannot represent the Brunt-Väisälä physics accurately, because it computes differences between two large numbers. For this reason I do not trust g-mode results from the Pesnell code, and I do not believe the thick H results derived from those DAV computations.

COX :

The Brassard et al. paper has no bearing on this or the CSKP work. Our new models and those used by CSKP were tightly converged (one part in ten million) with all quantities for all mass shells needed in the pulsation analysis consistent with each other. The Brunt-Väisälä frequency is a property of a model and not a pulsation code. A bad model will surely give a bad eigensolution analysis. The Brassard et al. paper does show that the poorly converged Montreal models, as described by Tassoul, Fontaine, and Winget (1990), (with pure carbon rather than realistic CO cores) are probably not good enough for precision pulsation studies. As these authors discuss, the $dlnP/dr$ term is not as precise as one part in 10^3, and the necessarily interpolated points are unlikely to be consistent with the basic conservation equations. When imprecise models are used, terrible precision problems arise with the Brunt-Väisälä frequency, as discussed in 1979 by Keeley (unpublished) at the Tucson conference and rediscovered by Pesnell. Artificially smoothing the models can make things appear better, but the physics may be lost. The Brassard et al. paper, using only Montreal models, has no bearing on any of the previous or current Los Alamos results, because their paper neither uses the complete and tightly converged Los Alamos self-consistent models nor the Pesnell code. The Pesnell Lagrangian code does not even involve the Eulerian-only concept of a Brunt-Väisälä frequency for its matrix elements. How an Ap. J. paper can criticize models and a pulsation code without even considering either of them is a mystery to many pulsation experts. As explained in the CSKP paper, and by many at this conference, the "suspicious" CSKP fluctuations in period spacings and mode instabilities are due to the steep composition gradients used. These gradients are much smoother and more realistic in this new work that confirms the thick H CSKP stability results.

Incidentally an Eulerian-based eigensolution formulation is not completely physically appropriate, because pulsation variations are not symmetric around Eulerian space points. Further composition variations during pulsations at an Eulerian point are not symmetric in time. That is one reason why the Castor (1971) and Pesnell (1990) pulsation analysis codes are Lagrangian.

Editorial comment :

The paper "Nonadiabatic nonradial pulsations for DAV white dwarf stars", by Cox and Hollowell, was presented at the workshop by L. Stanghellini. Although Dr. Cox was not present at the workshop, we have decided to accept his reponse to Dr Van Horn's remark in view of the seriousness of the issue raised. The readers may like to refer to the original papers under discussion :

1. Brassard, P., Fontaine, G., Wesemael, F., and Kawaler, S.D. 1989, in *Proc. IAU Colloq. No. 114 : White Dwarfs,* ed. G. Wegner (Springer-Verlag : Berlin), p. 263 "Are Pulsations a Useful Probe of the Structure of the Outer Layers of White Dwarfs"

2. Cox, A.N., Starrfield, S.G., Kidman, R.B., and Pesnell, W.D. 1987, *Ap. J.*, **317**, 303.

3. Pesnell, W.D. 1987, in *Stellar Pulsation : A Memorial to John P. Cox*, ed. A.N. Cox, W.M. Sparks, and S.G. Starrfield (Springer-Verlag : Berlin), p. 363.

4. Tassoul, M., Fontaine, G., and Winget, D.E. 1990, *Ap. J. Suppl.*, **72**, 335. "Evolutionary Models for Pulsation Studies of White Dwarfs"

NLTE ANALYSIS OF FOUR PG1159 STARS

K. WERNER*, U. HEBER*, K. HUNGER
Institut für Theoretische Physik und
Sternwarte der Universität Kiel
Olshausenstr. 40, D-2300 Kiel 1
Federal Republic of Germany

ABSTRACT. The results of a quantitative spectral analysis of four PG1159 stars are presented, which is based on a new grid of NLTE model atmospheres and new optical spectra. Accordingly, these stars are amongst the hottest stars known with effective temperatures exceeding 100,000 K. All stars have the same high surface gravity ($\log g = 7$) and show the same "exotic" abundance pattern, which is dominated by carbon (50% by mass) and helium (33%) followed by oxygen (17%), whereas hydrogen and nitrogen are trace elements only. These abundances indicate that the pulsations observed in PG1159-035 and PG1707+427 are driven by cyclic ionisation of carbon and oxygen. However, only two of the four PG 1159 stars analysed are pulsating. The variable star PG1707+427 and the non-variable PG1424+535 have identical atmospheric parameters and abundances (to within observational limits). The same holds for the pair PG1159-035 (variable) and PG1520+525 (non-variable). Hence, the physical parameter which determines whether a PG 1159 star pulsates or not, still remains obscure.

1. Introduction

More than a decade ago a new class of hot hydrogen deficient degenerate stars was discovered: the PG1159 stars named after the prototype PG1159-035 (McGraw et al., 1979). Early analyses showed that these stars are extremely hot with effective temperatures exceeding 80,000 K (Wesemael et al., 1985). Today 17 members of the class are known, out of which seven stars are Central Stars of Planetary Nebulae. The PG1159 stars have attracted the attention of both observers and theoreticians since PG1159-035 and six other members of the class have been discovered to be non-radial g-mode pulsators. It is one of the first great successes of the "Whole Earth Telescope" project (Nather et al., 1990) to have resolved 125 periods in the power spectrum of PG1159-035 (Winget et al., 1990). Applying the techniques of asteroseismology to these observations has led (besides other important results) to a mass determination of unprecedented precision: M = 0.585±0.002M⊙(see Winget, this conference for a detailed description).

A prerequisite for the asteroseismological analysis, however, is the precise knowledge of the photospheric parameters (T_{eff}, $\log g$) and the chemical composition. In particular, knowledge of the element abundances is important to identify the pulsation driving mechanism. Hence, there is a need for quantitative spectroscopic analyses of PG1159

*Visiting Astronomer, German-Spanish Astronomical Center, Calar Alto, operated by the Max-Planck- Institut für Astronomie Heidelberg jointly with the Spanish National Commission for Astronomy

G. Vauclair and E. Sion (eds.), White Dwarfs, 219–227.
© 1991 Kluwer Academic Publishers.

stars. For that purpose we have acquired new spectroscopic observations and developed a new NLTE model atmosphere code capable of treating the "exotic" chemical composition of the PG1159 photospheres. The spectroscopic analyses of four PG1159 stars has already been completed (Werner *et al.*, 1991, WHH 91). Here we shall present our new observations (section 2), outline briefly the problems inherent in their analysis (section 3) and summarize the results of the analyses of four stars in the final sections.

2. Optical Spectra

New optical spectra of PG1159 stars have been obtained in the period 1987 to 1989 at the DSAZ (Calar Alto, Spain) using the 3.5m telescope. A Boller and Chivens (B&C) spectrograph was used in 1987 and 1988 equipped with a RCA CCD detector. In 1989 we used the Cassegrain Twin Spectrograph (CTS), which includes two separate spectroscopic channels behind the common slit aperture. The spectral range between 4000Å and 6800Å was covered by several exposures at a spectral resolution of 1.5Å (CTS) and 2-3Å (B&C), respectively.

In Fig. 1 CTS spectrograms of five representative stars are displayed, comprising the characteristic absorption trough which originates from a dozen of C IV lines and He II 4686Å .

(i) The prototype star, PG1159-035, is shown in the middle of Fig. 1. Its high effective temperature (T_{eff}= 140,000 K and $\log g = 7$, WHH 91) becomes manifest by strong emission cores in C IV 4659Å and He II 4686Å .

(ii) The next spectrum reveals that PG1144+005 is actually a twin of PG1159-035, with the only difference that strong nitrogen emissions are detected in PG1144+005 (the N V 4604Å ,4620Å doublet and the N V 4945Å feature). NLTE calculations (Werner and Heber 1991, in prep.) show that these emissions are of photospheric origin and that the N abundance is much larger than in PG1159-035 (the upper limit for the latter is 0.4%, by mass).

(iii) PG1424+535 (top of Fig. 1) is the "coolest" PG1159 star analysed so far, also having a high surface gravity (T_{eff}= 100,000 K, $\log g = 7$, WHH 91).

(iv) Below that we present the spectrum of the central star of K1-16 (one out of 7 CSPNs in the PG1159 group) which is quite different from the three spectra discussed so far. K1-16 is obviously hotter and more luminous than PG1424+535 and its spectrum appears strikingly similar to that of the central star of NGC 246 (not shown), whose atmospheric parameters were estimated by Husfeld (1987): T_{eff}= 130,000 K, $\log g = 5.7$ and a C/He ratio similar to that of PG1159-035.

(v) The last spectrum in Fig. 1 is from the unique object H1504+65. Surprisingly, no He II lines are noticeable in the whole spectrum (Nousek *et al.* 1986) and in particular, the He II 4686Å emission that usually encompasses the C IV emission is lacking. The question arose if this is really due to absence of helium or simply a consequence of an extremely high temperature causing complete ionisation into He III. Our first exploratory NLTE models for this star indicate that helium is indeed either underabundant or even absent. The remaining absorption wings around 4686Å may be attributed to the highly excited C IV ($n = 6 \rightarrow 8$) and O VI ($n = 9 \rightarrow 12$) transitions. Another broad absorption feature near 4500Å is clearly visible and was already noted by Nousek *et al.* (1986). We find that this is also due to highly excited O VI transitions ($n = 8 \rightarrow 10$ at 4494Å) and it is predicted by our models as soon as $T_{\text{eff}} \approx$ 150,000 K is exceeded. This supports suspicions that H1505+65 is a peculiarly hot, naked C-O core of a star that has lost its entire H and He layers, however, a detailed analysis must be awaited for.

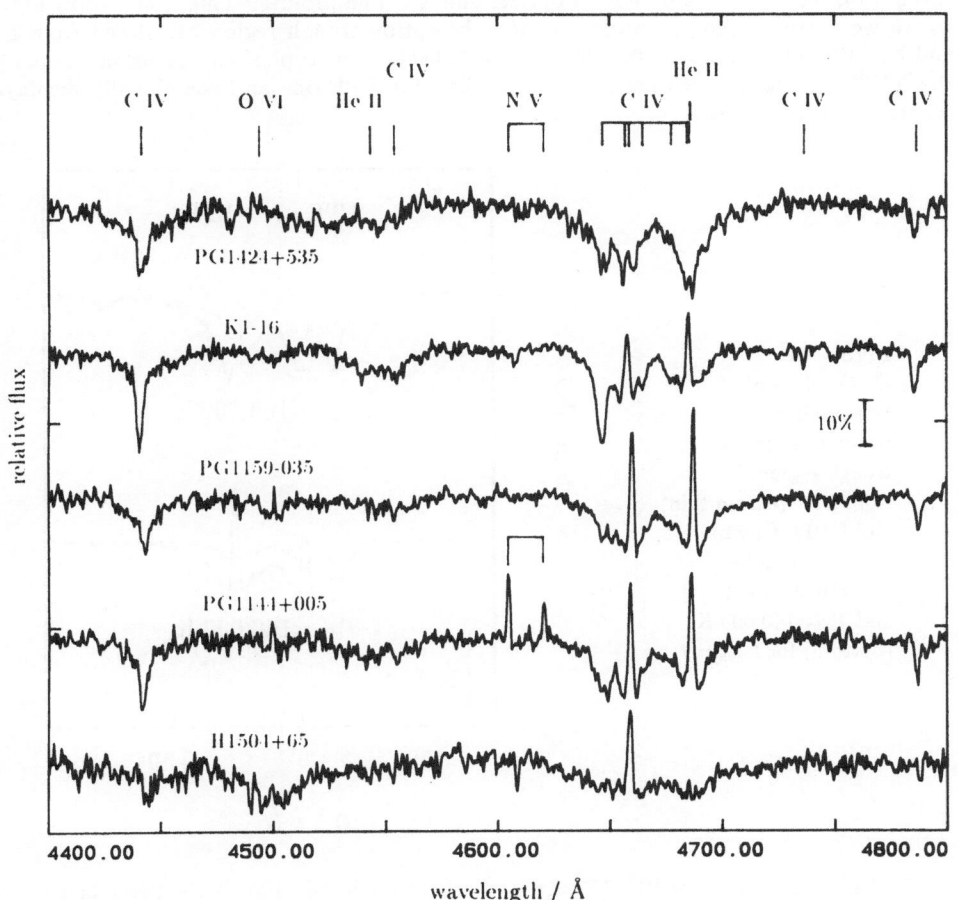

Figure 1: Normalized spectra of five representative members of the PG1159 group, covering the typical absorption trough.

3. The need for extensive NLTE modelling and an adequate line broadening theory

Before we discuss the results of the abundance analysis for the four PG1159 stars PG1159-035, PG1424+535, PG1520+525 und PG1707+427 (WHH 91, see section 4), let us first briefly outline the problems encountered in such an analysis. This will explain why more than a decade elapsed since the discovery of PG1159-035 before the first quantitative abundances could be presented.

Due to the very high effective temperatures and despite of the relatively high gravities ($\log g = 7$), NLTE effects dominate optical and UV line profiles. This is shown in Fig. 2, where we compare the line profiles of the absorption trough region calculated from LTE and NLTE models for T_{eff}=100,000 K and 140,000 K. Two typical effects become apparent: the NLTE profile is considerably stronger than the LTE one and occasionally displays a central emission reversal.

Figure 2:
Comparison of
line profiles
in the absorption
trough region
from LTE (drawn thin)
and NLTE (thick)
model atmospheres at
T_{eff}=100,000 K (top)
and T_{eff}=140,000 K
(bottom) for $\log g = 7$

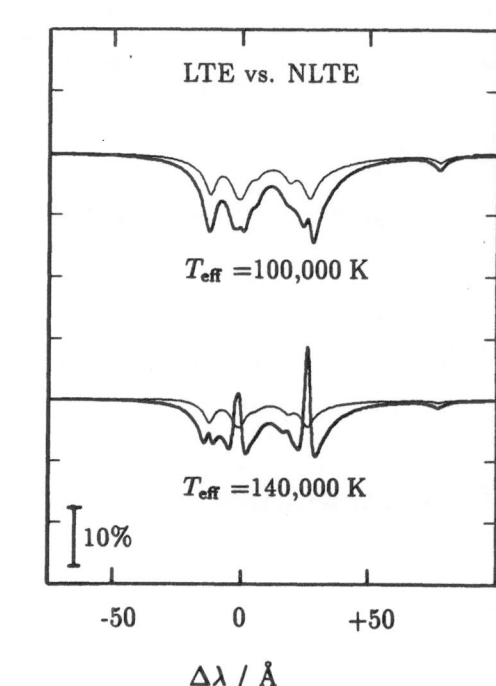

The NLTE modelling of PG1159 atmospheres is rendered difficult for two reasons:

(i) Carbon and oxygen are so abundant that their opacities strongly influence the atmospheric structure and the NLTE model atmosphere code needs to include detailed C and O model atoms (in addition to He).

(ii) The most important spectral lines of carbon and oxygen are due to C IV and O VI and arise from highly excited atomic levels (n=5 or higher, see section 2). In order to obtain reliable occupation numbers for the atomic levels involved, the models have to be extended in line formation calculations to include a very large number of levels and line transitions between them. O VI, for example, has to be represented by 52 NLTE levels and 232 line transitions.

Both items cannot be achieved using the classical NLTE approach, the complete linearisation (CL) technique (Auer and Mihalas, 1969). A new method had to be developed before adequate NLTE model atmospheres could be constructed. This was done by Werner and Husfeld (1985) and Werner (1986, 1988) who used an idea of Scharmer (1981) to tailor the "Accelerated Lambda Iteration" (ALI) method which proved to be very stable and efficient and allowed many more levels and line transitions to be treated than the CL method. The models used for the PG1159 stars included 4 elements (H, He, C and O) in 12 ionisation stages, represented by 100 NLTE levels with 240 line transitions. More than 1500 frequency points were required for a proper treatment of all opacities. New mathematical methods allow even more elaborate model atoms to be treated (Dreizler and Werner, 1990).

Another problem closely related to item (ii) occurs when emergent line profiles have to be calculated from the model atmospheres. This is related to the line broadening theory for C IV and O VI lines. The high lying levels of these ions, namely, are close to degeneracy and a gradual transition from quadratic to linear Stark effect occurs with increasing principal quantum number, i.e. the broadening mechanism ranges between the linear and the quadratic regime. Unfortunately, an exact theory for this intermediate case is not yet available. An approximate treatment was proposed by WHH 91 who determined correction factors to Holtsmark's statistical theory of pressure broadening by comparison with laboratory experiments.

4. Spectral Analysis

For the first spectroscopic analyses we had selected two pulsating members of the class (PG1159-035 and PG1707+427) and two non variables (PG1424+535 and PG1520+525) in order to eventually obtain limits on the instability strip in the HRD.

The analysis is complicated by the fact that besides T_{eff} and $\log g$ three additional atmospheric parameters have to be fitted simultaneously, i.e. the abundance ratios of H, C, and O to He. To determine these five parameters we used twelve optical line profiles: five He II lines, two of which are possibly blended by H; six C IV lines and one O VI line. In Fig. 3 to 6 examples of the fitting procedure are given in the case of PG1159-035. In Fig. 3 the observed absorption trough region is compared to theoretical profiles computed for three different T_{eff}. This line blend is quite temperature sensitive for the T_{eff}'s in question. Four additional lines were used by WHH 91 to constrain T_{eff}. In Fig. 4 the observed C IV doublet (5801Å , 5812Å) is compared to three theoretical profiles calculated for $\log g = 6$, 7 and 8, respectively. As can be seen this doublet is a very good gravity indicator for PG1159-035. In addition four other lines were used to further constrain gravity. The abundance determination is illustrated in Fig. 5 and 6 for oxygen and hydrogen. Only one O VI (5291Å) is observed from which the O abundance can be determined in PG1159-035. Hydrogen is not detactable (see Fig. 6) and only a large upper limit (H/He<1) can be set because of the high temperatures in question. Nevertheless it is likely that the atmospheres are completely devoid of hydrogen.

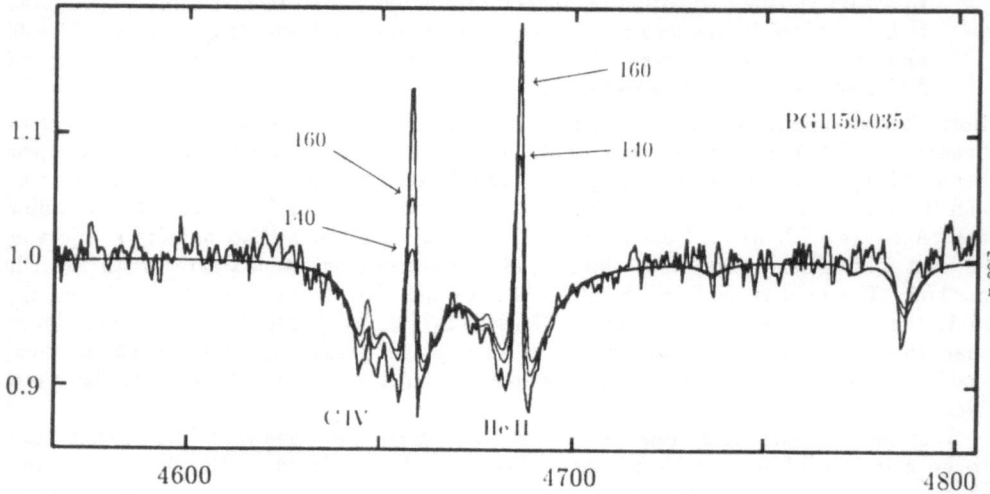

Figure 3: Comparison of the observed spectrum (absorption trough region) of PG1159-035 to theoretical NLTE profiles for $T_{eff} = 120,000$ K, $140,000$ K and $160,000$ K at $\log g = 7$. Note that the observed central emission reversals of C IV and He II can well be reproduced by static NLTE model atmospheres.

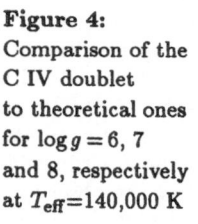

Figure 4:
Comparison of the
C IV doublet
to theoretical ones
for $\log g = 6$, 7
and 8, respectively
at $T_{eff}=140,000$ K

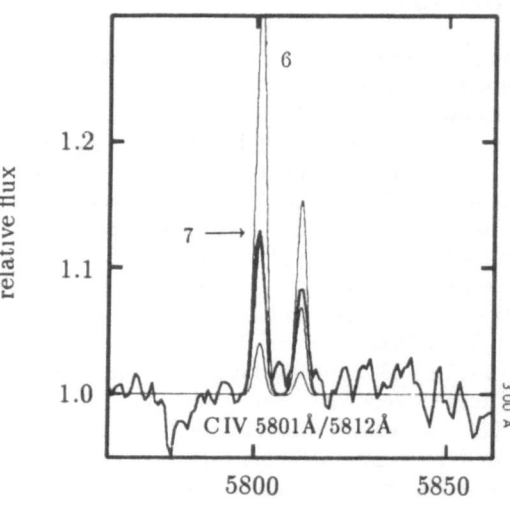

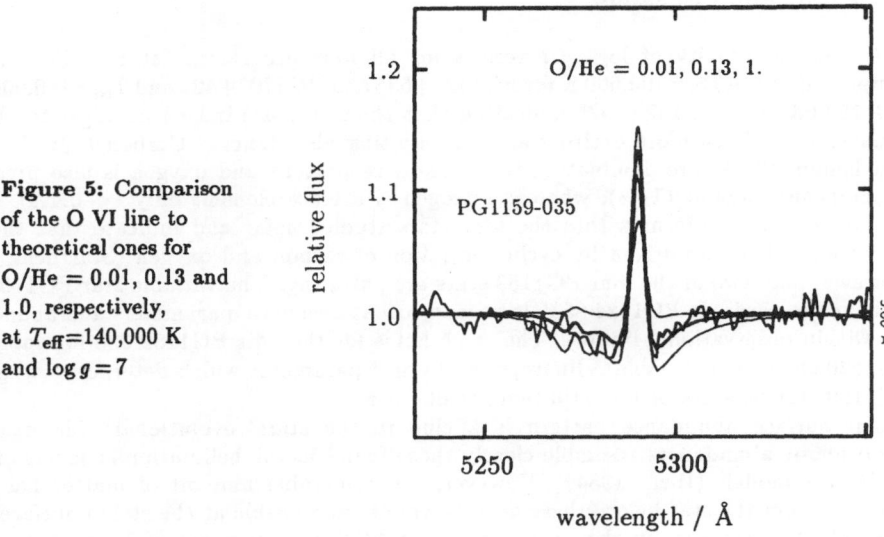

Figure 5: Comparison of the O VI line to theoretical ones for O/He = 0.01, 0.13 and 1.0, respectively, at T_{eff}=140,000 K and $\log g = 7$

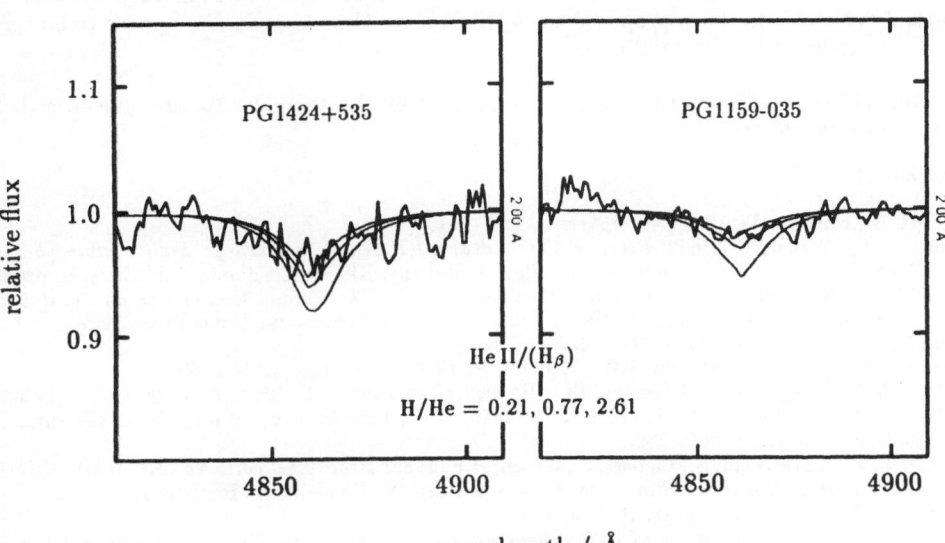

Figure 6: Profile fits to the He II/Hβ regions of PG1424+535 and PG1159-035 by models with an H/He=0.21, 0.77 and 2.61.

5. Results and discussion

A uniform gravity of $\log g = 7$ results for all four programme stars. The effective temperatures are $T_{\mathrm{eff}} = 100,000$ K for PG1424+535 and PG1707+427 and $T_{\mathrm{eff}} = 140,000$ K for PG1159-035 and PG1520+525 indicating that these star are indeed amongst the hottest stars known. Even more exciting are the resulting abundances: Carbon (50% by mass) and helium (33%) are dominating the abundance pattern and oxygen is also present in a significant amount (17%), whereas nitrogen is a trace element only ($< 0.4\%$). These abundances are uniform within the four stars studied sofar and indicate that the non-radial pulsations are driven by cyclic ionisation of carbon and oxygen (Starrfield, 1987). However, only two of the four PG1159 stars are pulsating. The variable star PG1707+427 and the non-variable PG1424+535 have identical atmospheric parameters and abundances (to within observational limits). The same holds for the pair PG1159-035 (variable) and PG1520+525 (non-variable). Hence, the physical parameter which determines whether a PG1159 star pulsates or not, still remains obscure.

The surface abundance pattern is a clue to the stars' evolutionary history. The atmospheric abundances resemble closely those found in the helium buffer layers of post-AGB star models (Iben, 1984). However, a considerable amount of matter has to be removed from the star before these deep layers become visible at the stellar surface. This can only be achieved by the born-again post-AGB star scenario of Iben *et al.* (1983). Calculated surface abundances (Iben, 1984) are in qualitative agreement with observations. An evolutionary link of the PG1159 stars to the Central Stars of Planetary Nebulae of spectral type WC has been proposed by WHH 91 since their abundances appear to be very similar to those of the PG1159 stars.

ACKNOWLEDGEMENTS. This work was supported by the Deutsche Forschungsgemeinschaft under grant Hu 39/28-1.

REFERENCES

Auer, L.H., Mihalas, D. 1969: *Astrophys. J.* **158**, 641.
Dreizler, S., Werner, K. 1990: *Proc. NATO Advanced Research Workshop: Stellar atmospheres: Beyond Classical Models*, eds. L. Crivellari, I. Hubeny, Kluwer Academic Publishers, in press.
Husfeld, D. 1987: *Proceedings of IAU colloquium No. 95: The Second Conference on Faint Blue Stars*, eds. A.G.D. Philip, D.S. Hayes and J. Liebert, Schenectady: Davis Press , 237.
Iben, I., Jr. 1984: *Astrophys. J.* **277**, 333.
Iben, I., Jr., Kaler, J.B., Truran, J.W., Renzini, A. 1983: *Astrophys. J.* **264**, 605.
Nather, R.E., Winget, D.E., Clemens, J.C., Hansen, C.J., Hine, B.P. 1990: *Astrophys. J.* **361**, 309.
Nousek, J.A., Shipman, H.L., Holberg, J.B., Liebert, J., Pravdo, S.H., White, N.E., Giommi, P. 1986: *Astrophys. J.* **309**, 230.
Mc Graw, J.T., Starrfield, S., Liebert, J., Green, R.F. 1989: *Proc. IAU coll. No. 53: White Dwarfs and Variable Degenerate Stars*, eds. H.M. van Horn, V. Weidemann, Rochester, p. 377.
Scharmer, G.B. 1981: *Astrophys. J.* **249**, 720.
Starrfield, S. 1987: *Proceedings of IAU colloquium No. 95: The Second Conference on Faint Blue Stars*, eds. A.G.D. Philip, D.S. Hayes and J. Liebert, Schenectady: Davis Press , p. 309.
Wesemael, F., Green, R., Liebert, J. 1985: *Astrophys. J. Suppl.* **58**, 379.
Werner, K. 1986: *Astron. Astrophys.* **161**, 177.
Werner, K. 1988: *Astron. Astrophys.* **204**, 159.
Werner, K., Husfeld, D. 1985: *Astron. Astrophys.* **148**, 417.
Werner, K., Heber, U., Hunger, K. 1991: *Astron. Astrophys.* , in press, WHH 91.
Winget, D.E., Nather, R.E., Clemens, J.C., Provencal, J., Kleinman, S.J., et al. 1990: *preprint*, .

DISCUSSION

D'ANTONA :

The chemical composition of PG1159-035 poses serious problems to the late phase of stellar evolution f...., as you must find a reasonable way of stripping away the helium buffer and expose **even** some oxygen. But it poses problems also to the pulsations as this star may be badly out of equilibrium if this hoppened not too long ago. This may make a significant difference between two stars travelling across the same (log g -log T_{eff}) region, one may pulsate and the other not.

HEBER :

At present, only the born-again post-AGB scenario of Iben (1984, Ap. J. **277**, 333) can explain how the hydrogen-rich envelope **and** the helium buffer layer is removed. In this scenario the star retraves its post-AGB evolution for a second time. In the (log g, log T_{eff}) diagram, our programme stars lie close to the position on Iben's track were carbon to become visible at the surface.

BARSTOW :

Although the temperature is quite well constrained by your analysis, given the difficulty of finding computing time to produce model grids covering a range of abundances, how well do you think the abundances are constrained by these results.

HEBER :

Errors of the abundances are difficult to investigate systematically. We estimate that a factor of up to three in the individual abundances is entirely possible.

LIEBERT :

Did you attempt to get an actual NV abundance from our high dispersion IUE detection of NV 1240 Å (albeit with low signal-to-noise ratio)?

HEBER :

Yes. We used the equivalent widths. The results are consistent with the upper limit from the optical spectra. However, the NV resonance lines proved to be quite insensitive to nitrogen abundance since it is on the flat part of the curve of growth. No improvenient over the upper limit from the optical could be obtained.

FONTAINE :

COMMENT : I would like to reiterate the point I made following Letizia Stanghellini's talk. The atmospheric abundances at PG1159 stars do **not** constrain the chemical composition in the driving regions. We have no way of knowing that the composition is uniform from the surface down into the envelope.

A SEARCH FOR TRACE AMOUNTS OF HYDROGEN IN DB STARS

HARRY L. SHIPMAN, PETER THEJLL, and SUNITA BHATIA
Physics and Astronomy Department
University of Delaware
Newark, Delaware 19711

JAMES LIEBERT
Steward Observatory
University of Arizona
Tucson, Arizona 85721

ABSTRACT. A search for trace amounts of hydrogen in DB stars revealed no hydrogen in a sample of four objects. Upper limits are considerably lower than the quantities of hydrogen found in the DBA stars, stars with He-dominated photospheres and trace quantities of H. This search was designed with some care to serve as an experiment to decide whether the H in the DBA stars was produced from accretion or from mixing. These results suggest that the H comes from accretion.

1. INTRODUCTION

The existence of the DBA stars, stars with He-dominated photospheres but with trace quantities of H, has been one of those conundrums which have confounded those who have attempted to interpret and explain the puzzling, diverse chemical compositions of white dwarf stars (see Sion 1986 for a review of the complex phenomenology). Shipman, Liebert, and Green (1987) determined that 20 % of a sample of DB stars from the Palomar-Green survey contained trace quantities of hydrogen. Hydrogen abundances are typically of order 10^{-4} in these objects, which otherwise generally resemble the DB stars. The source of this hydrogen could provide a clue to the chemical evolution of white dwarf stars, one of the central puzzles in our field since the pioneering work by Schatzman in 1958.

Shipman, Liebert, and Green discussed two scenarios which could in principle explain the existence of H in the DBA stars. An obvious source for the H is interstellar accretion, since most of the material in the interstellar medium is in the form of hydrogen. The amount required is not implausibly large. Another possibility, particularly interesting in view of the single-channel evolutionary scenario discussed by Fontaine and Wesemael (1988) among others, is that the H in the DBA stars could come

G. Vauclair and E. Sion (eds.), White Dwarfs, 229–234.

from convective dredge-up. This H would then be an indicator of mixing between the white dwarf photosphere and subphotosphere. Shipman, Liebert, and Green discussed this possibility also, and could not favor one or the other based on optical data or physical plausibility alone.

The objective of the present study is to decide between the two possibilities by investigating an additional sample of DB white dwarfs to seek interstellar hydrogen. Because the question was formulated before the sample was selected, we are performing an "astrophysical experiment" in the sense that we are selecting another sample of objects and subjecting it to a test. It is one of a few astrophysical investigations where the term "experiment," rather than "observation," is called for.

2. EXPERIMENTAL DESIGN

The two proposed origins for the H in the DBA stars -- convective mixing and interstellar accretion -- suggest an experimental test. While detailed theories of convective mixing and accretion would produce their own predictions, on relatively general grounds one would expect that something like the following would emerge from any such theories.

2.1 Consequences of Interstellar Accretion

If accretion is the source of the H in the DBA stars, then the DBA stars would constitute a subset of the total sample of DB stars. Stars which had undergone an appropriate accretion episode relatively recently would be visible as DBA stars. Stars which had not undergone such an episode would be garden-variety DB stars, with no trace of interstellar H. The PG survey suggests that in this scenario the fraction of DB stars which are DBA's is approximately 20%.

The reason for some stars being DB's and some being DBA's in this context depends, of course, on a precise theory. If it is time scales that make the DBA stars different from the DB's, one possiblity is that the DBA's are stars in which there was insufficient time to be mixedthe H downwards or to expel it from the surface of the star by a process yet to be investigated in any quantitative detail. It is also possible that it is the character of the accretion episode which distinguishes DB's, DBA's, and possibly DA's. For example, those stars which accreted heavily could turn into DA's. Those which had accreted modest amounts of material could be DBA's. And those which had never passed through a giant molecular cloud would remain DB's.

However, whatever the details are, the character of interstellar accretion suggests the following as a reasonable working hypothesis: If accretion is responsible for the DBA phenomenon, then any sample of DB stars will consist of 20 % DBA stars.

2.2 Consequences of Convective Mixing

If convective mixing is the origin of the H in the DBA stars, then the observations suggest an interesting possibility. The observed abundances of H in the DBA stars are just barely above the threshold of detectability in the optical. H abundances substantially less than 10^{-4} would be difficult to detect using the current generation of telescopes and spectrographs. It is quite possible that the DBA stars detected in the Palomar-Green survey are just the "tip of the iceberg," in that most DB stars contain trace amounts of hydrogen but lie below the threshold of optical detectability.

While the expectations of any mixing origin for the DBA stars must be dependent on the precise nature of the theory, one reasonable possibility is that the DBA stars appear when a white dwarf with a thin layer of H on top of a He layer becomes convectively unstable and the H mixes with a He subphotosphere. General considerations suggest that the resulting H abundance could take on a wide range of values. The H abundance would presumably depend on the thickness of the convective layer and the thickness of the H layer with which it was mixed. There seems to be no a priori reason why the H abundance would be restricted to 10^{-4}. One would expect to find a substantial population of DB stars with H abundances somewhat, but not too much, less than this. If stars with greater H abundances existed in profusion, they would have been readily discovered through ground-based spectroscopy.

In the absence of a detailed theory, the foregoing qualitative discussion suggests the following as an alternative working hypothesis:

If convective mixing is responsible for the existence of the DBA stars, then any DB star will contain trace amounts of hydrogen if you look hard enough.

2.3 Hypothesis Testing

Examination of a new sample of DB stars with a new, more sensitive technique can determine which of the above working hypotheses is consistent with the data. The technique we used was to obtain small aperture spectra of a set of DB stars using the International Ultraviolet Explorer (IUE) satellite. Our expectation, based on broadening theories and models that were available at the time, that H/He ratios of order 10^{-8} could be reached with reasonable exposures. An improvement in sensitivity of four orders of magnitude should be sufficient to test the two hypotheses in reasonable detail.

How many stars do we need to examine? The most statistically vexing case is one where all the sample population contains hydrogen. The obvious conclusion to draw is that mixing is the cause, but what are the chances that this conclusion would be erroneous? The accretion hypothesis, as formulated above, suggests that something like 20 % of the DB stars contain hydrogen. If a sample of N stars were observed and all contained H, the probability that this sample of N was drawn from a population where the frequency of H pollution was really 0.2 would be $(0.2)^N$. For N=6, this probability is $10^{-4.2}$. This sample size is clearly sufficient; in fact N=5 would be reasonable. We prepared our telescope request for N=6, allowing for contingencies which would only permit us to observe 5 stars.

The other situation is a bit cleaner. If some, or most, of the stars in the observed sample contain no hydrogen, then accretion is suggested as the source of H in the DBA stars according to the arguments above. The convection hypothesis (as formulated here) could only remain viable if the hydrogen-free stars in the sample were freaks. Under this alternative, observations are needed of a sample that is big enough so that freaks can be ruled out. 3-4 stars should suffice in this instance.

3. RESULTS

After the first set of observations was made in the summer of 1987, it became clear that some objects had little or no hydrogen in them. It also became clear that acquisition of visually faint targets in the small aperture had become more difficult than was the case in our previous experience, presumably because of the loss of one additional gyro on the satellite. Because it became clear at this time that it was likely that the accretion hypothesis was correct, the earlier observing strategy was modified and we only obtained data on four stars. Target selection was largely based on the availability of targets during the shift and an anticipation that by observing stars with higher temperatures we could obtain better data and thus better limits on the H/He abundance.

Analysis of the data was based on the new temperature scale for DB stars in this region determined by Thejll, Vennes, and Shipman (1990).

The latest broadening theory for He II lines was used in the data analysis. For some of the hotter stars, the possible presence of a He II component in the core of Lyman alpha was more of a limiting factor than the quality of the spectroscopy. Because of the larger than expected intensity of He II Lyman alpha, our expected sensitivity was not in fact reached and we could only limit H contents to 10^{-7}.

4. CONCLUSIONS

We examined four stars and found none with detectable hydrogen. This result is not consistent with the mixing hypothesis as illustrated above. It is consistent with the idea that the H in the DBA stars comes from interstellar accretion. The finding of Ca II H and K lines in some DBA stars, resulting in their being classified as DBAZ stars (Kenyon, Shipman, Sion, and Aannestad 1988, Kenyon, Sion, and Aannestad 1988) also suggested accretion as the origin of the H in the DBA stars. However, that result depends implicitly if weakly on a model for interstellar accretion which assumes that Ca and H are accreted in solar proportions. The present result makes no such assumptions, though it makes a presumption that a mixing explanation of the H in the DBA stars would predict that a fair number of stars should have H/He ratios somewhat below the values observed in DBA stars.

The limits given above are the most sensitive yet derived for the H content of white dwarf stars with He dominated photospheres. We expect that these limits should present useful tests for models in which DB stars just below the "DB gap" from 30,000-45,000 K are created by convective mixing. They should also present interesting challenges to models of interstellar accretion by white dwarf stars.

This research has been supported by the NSF and by NASA.

TABLE 1. Stars, Numbers of Observations,
and Upper Limits to their Hydrogen Abundance

Star	$T(eff)/10^3$ K	Number of Spectra	H/He Upper Limit
PG0112+104	27.5	3	$< 3 \times 10^{-5}$
G 270-124	23	1	$< 10^{-7}$
GD 358	25.2	4	$< 10^{-7}$
GD 325	14	2	$< 3 \times 10^{-6}$

REFERENCES

Fontaine, G., and Wesemael, F. (1988) in A.G.D. Philip, D.S. Hayes, and J. Liebert (eds.), Proceedings of IAU Colloquium 95: The Second Conference on Faint Blue Stars, L. Davis Press, Schenectady, N.Y., p. 319.

Kenyon, S.J., Sion, E.M., and Aannestad, P.A. 1988, Astrophysical Journal (Letters) 330, L55.

Kenyon, S.J, Shipman, H.L., Sion, E.M., and Aannestad, P.A. 1988, Astrophysical Journal (Letters) 328, L65.

Shipman, H.L., Liebert, J., and Green, R.F. (1987) Astrophysical Journal 315, 239.

Sion, E.M. (1986). Publ. Astron. Soc. Pacific 98, 821.

Thejll, P.A., Vennes, S., and Shipman, H.L. (1990) Astrophysical Journal, submitted for publication.

DISCUSSION

FONTAINE :

I would like to argue in favor of the accretion scenario for the presence of traces of hydrogen in DBA stars. The alternative case of dilution of small quantities of hydrogen left over from previous evolutionary phases has been studied in detail by C. Pelletier (unpublished Ph.D. thesis, 1986). His results indicate very clearly that the expected H/He ratios after dilution are **much** smaller than the current visibility limits (H/He $< 10^{-5}$). Thus, it seem that, indeed, the observed traces of H in cool DB white dwarfs do represent the "tip of the iceberg". I fully expect that DB white dwarfs contain variable quantities of hydrogen caused by accretion. We are currently witnessing only those ($\sim 20\%$) which are the most heavily polluted.

VAN HORN : You can use your limits on the H/He ratio to place limits on the accretion rates, what result does that give?

SHIPMAN :

Something like 10^{-22} $M_{\odot}$ per year, if you assume that any H which was accreted during the $\sim 10^7$ year cooling history of one of these objects stayed on its surface. However, there are so many assumptions which enter that estimate that I question whether it's meaningful.

ABUNDANCES OF TRACE HEAVY ELEMENTS IN HOT DA WHITE DWARFS

S. VENNES, P. THEJLL, AND H.L. SHIPMAN
Department of Physics and Astronomy, University of Delaware
Newark, Delaware, 19716 , U.S.A.

ABSTRACT. We have examined the *IUE* high-dispersion spectra of 25 DA white dwarfs in order to establish the nature of the relationship between the EUV/soft X-ray opacities discovered by observations from the *EXOSAT* and *Einstein* satellites and the heavy element traces often apparent in the *IUE* ultraviolet spectra of these stars. The study of the ionization balance of silicon and carbon atoms as well as the wavelength shift of the lines relative to the photosphere give important clues to the physical nature of the line absorption. Some spectra suggest the existence of a relatively close and relatively dense shell of ionized material of uncertain composition surrounding at least 6 and maybe as many as 13 of the white dwarfs. The heavy elements may be photospheric only in the case of Feige 24. The inferred EUV/soft X-ray opacity may be provided by heavy elements in some cases and by helium beneath a thin layer of hydrogen in some other cases.

1. Introduction

The acquisition, over the past 10 years, of high-dispersion spectra of the brightest of the DA white dwarfs by the *IUE* satellite has resulted in the build-up of a complex and yet to be explained picture of the elemental pattern in these stars. The early observation of Feige 24 by Dupree and Raymond (1982) had already revealed the dualistic nature of the absorption, some of the heavy element lines apparently originating from the photosphere of that star and some others being labeled "circumstellar". The interstellar absorption lines can be easily recognized among the others from simple physical arguments concerning the population levels. The detection of highly ionized species of carbon, nitrogen and silicon in Feige 24 challenges our knowledge of the processes at play within the atmosphere of the white dwarfs. The so-far pure hydrogen atmospheres of the DA white dwarfs were not only supposedly contaminated with helium seen through the EUV/soft X-ray window (Kahn *et al.* 1984) but the clear assignment of C, N, and Si ions to the photosphere of the DA Feige 24 by Dupree and Raymond (1982) has raised the question of the origin of these elements. The first mechanism invoked to explain the presence of elements heavier than hydrogen, and that can effectively counteract the effect of gravitational settling is support by selective radiation pressure. This mechanism was found less efficient than expected by Vennes *et al.* (1988) in the case of helium and Chayer *et al.* (1987) found that the predictions of the radiative support theory do not reproduce the observed C abundances in both Feige 24 (T_{eff} = 55 000 K) and the hot DO white dwarf PG 1034+001 ($T_{eff} \approx$ 80 000 K; Poulin *et al.* 1989). The diffusion problem in white dwarfs may not be solved until the effect of mass loss is taken into account. The presence of He may still be understood in the context of a thin hydrogen layer on top of the He envelope (Vennes *et al.* 1988). This explanation put forward to explain the bulk of the X-Ray observations (see Paerels and Heise 1989 for a large sample) remains a viable possibility.

G. Vauclair and E. Sion (eds.), White Dwarfs, 235–247.
© 1991 *Kluwer Academic Publishers.*

The study of the other DA white dwarfs of interest reveals even more complexity relative to a simple picture in which the observed elements are at diffusive equilibrium in a stable photosphere. While the cooler DA white dwarf Wolf 1346 ($T_{eff} = 20\,700$ K) may show *photospheric* traces of silicon according to Bruhweiler and Kondo (1983), the origin of the C IV, N V, and Si IV lines in the *IUE* spectrum of G191-B2B ($T_{eff} = 62\,250$ K) is still not clear. Apparently altered by a typographical error, the measurement of the systemic velocity in the companion of G191-B2B (Trimble and Greenstein 1972) led Bruhweiler and Kondo (1981) to suggest an origin outside the photosphere for these lines. A reanalysis of the photographic plates by Reid and Wegner (1988) resulted in a new systemic velocity such that the C,N and Si lines are apparently redshifted by a velocity corresponding roughly to the Einstein gravitational redshift expected for G191-B2B. Moreover the detection of an Hα emission core at a redshift similar to the C,N and Si lines suggested, if the emission core is photospheric in origin, that these elements are also sitting in the photosphere.

The DA star CD-38°10980 ($T_{eff} = 25\,000$ K) is also a difficult case. The detection of Si II and Si III lines in the *IUE* spectrum of that star by Holberg *et al.* (1985) suggested in first instance a phenomenon similar to that observed in Wolf 1346 but the very precise measurements of Balmer line shift by Wegner (1978), and Koester (1987) altogether rules out the possibility of the silicon being photospheric. Thus the origin of the heavy element features remains to be determined. The DA GD71, studied by Bruhweiler (1984), possibly shows the resonance lines of the ions C IV, N V, and Si IV. The comparison of the radial velocity from the Balmer lines (Trimble and Greenstein 1972) and the measurement of the heavy element line shifts of GD 71 by Bruhweiler (1984) suggested an expanding halo as the site of these absorption features. Finally the DA star GD 394 exhibits the Si III λ1298.9Å sextuplet and the Si IV λ1396.7Å doublet shortward shifted, possibly in an expanding shell associated with the star (Bruhweiler and Kondo 1983).

The early compilation of the observations of heavy elements in the hot DA white dwarfs by Bruhweiler and Kondo (1983) led them to suggest that the blueshifted (relative to the stellar reference frame) lines in some of the DA stars originate from a *halo* around the star and in some case, where a residual velocity is inferred, from an expanding shell, hypothetically related to a *stellar wind*. Apart from this, negative detections were also reported by Bruhweiler and Kondo (1983) in the case of HZ43, Sirius B, 40 Eri B. The reason for this apparent dichotomy has yet to be provided.

We present an analysis of all the *IUE* high dispersion spectra available in the *IUE* archives. Considerable improvement of the analysis of some stars is achieved by using "co-added" images to enhance the signal-to-noise ratio. The spectra have been searched for Si II, Si III, Si IV, C II, C III, C IV, N I, N V, and O I spectral lines. The wavelength shift and the equivalent width of the significant features have been obtained in order to investigate the physical nature of the absorbing medium. Our primary goal is to determine the origin of the absorption features, i.e. whether it is photospheric or not. This has some impacts on the radiative support theory and in general on the problem of the diffusion of the elements in white dwarf envelopes. We also examine the consequences of the new results on the mass loss scenario and on the existence of a Strömgren sphere related to the DA white dwarfs. In §2, §3, and §4 we present the observational material, the synthetic spectra and the analysis respectively. In § 5 we examine the implications of the properties of the heavy element lines on the physics of the photosphere and on the nature of the interaction between the white dwarf and its environment.

2. The Sample of High-Dispersion Spectra for the DA White Dwarfs

We have collected 37 high-dispersion images for 25 DA white dwarfs and one image for the DAO white dwarf PG 1210+533. These images are generally characterized by a low signal to noise (< 10) but according to the current calibration, the wavelength scale (Thompson, Turnrose, and Bohlin 1982) is precise to 3 km s^{-1}. Previous studies have taken advantage of this performance to overcome the limitation in signal-to-noise ratio. All images have their wavelength scale corrected for the earth/spacecraft motion and for the time/temperature spectral distortion (Thompson, Turnrose, and Bohlin 1982). In the present stage of this work no account has been taken for eventual absolute flux errors due to a bad background subtraction. However this can be done relatively simply by a comparison with the corresponding low dispersion images. After an initial inspection, the individual stars GD 419, EG 15, EG 257, and G231-040 were not studied any further due to the very low signal-to-noise ratio of their high-dispersion *IUE* images.

3. The Synthetic Spectra for the Heavy Element Lines

In some *IUE* high-dispersion spectra of the DA stars the signal-to-noise ratio allows measurement of the equivalent width of a line down to 10 mÅ. This extremely low value has necessitated the calculation of new tables of line equivalent widths to complete the tables in Henry, Shipman, and Wesemael (1985). These last tables only extend to an abundance (C, N, O, Si, Mg) of 1×10^{-7}. Model equivalent widths are calculated at abundances of log $n(X)/n(H)$ = -5,-6,-7,-8,-9 (T_{eff} = 18 to 65 x 10^3 K) for the C II λ1335.3Å triplet, C III λ1175.7Å sextuplet, C IV λ1549.1Å doublet, Si II λ1263.3Å triplet, Si III λ1298.9Å sextuplet, and Si IV λ1396.7Å doublet. We make use of the N V λ1240.1Å doublet calculations of Henry, Shipman, and Wesemael (1985). The N V doublet is constraining only at very high temperature (T > 60,000 K) and is thus not a strong indicator on nitrogen abundances for most of the DA stars. Figure 1 presents a sample result of these calculations.

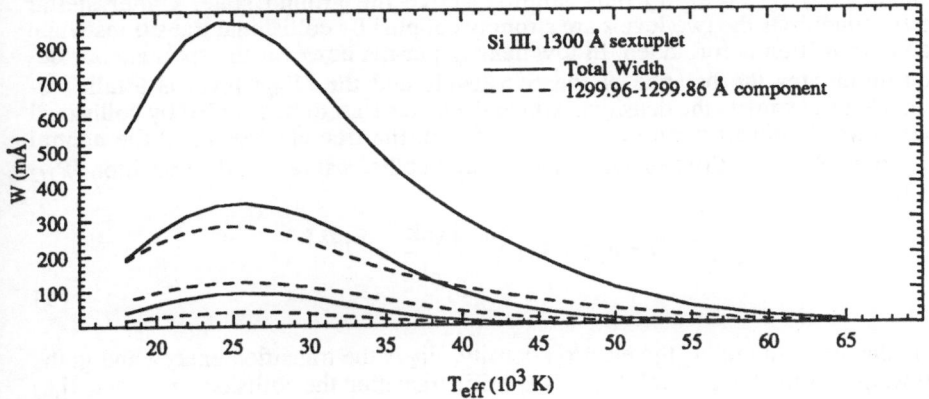

Figure 1. Equivalent width of the Si III λ 1298.9 Å line as a function of the stellar effective temperature. The full lines are, from top to bottom, the total equivalent widths of the sextuplet at log $n(Si)/n(H)$ = -6,-7,-8. The dash lines are at the same abundances but for the 1299.96-1299.86 component.

4. Measurement and Interpretation of the Spectral Lines

To establish the physical nature of a particular line absorption in the high dispersion spectra of the DA white dwarfs, two important characteristics are extracted: the equivalent width and the line wavelength centroid of that feature. The first characteristic is directly related to the column density in the line of sight or in the relevant cases to the photospheric abundance and the second characteristic can be simply translated to a velocity relative to the photosphere:

$$\frac{v_{rel}+v_g+v_{star}}{c} = \frac{(\lambda-\lambda_{lab})}{\lambda_{lab}},$$
(1)

where v_{rel}, v_g, and v_{star} are the velocity of the absorbing region relative to the photosphere, the gravitational redshift, and the heliocentric velocity of the star respectively. The equivalent width and the centroid of the feature are obtained by fitting a Gaussian with the central depth, the width and the central wavelength kept as free parameters. The limit of detectability (W_{min}) in a spectrum is a function of the signal-to-noise ratio (σ) and the spectral resolution (Δ):

$$W_{min} = \frac{\Delta}{\sigma-1}.$$
(2)

The absorption lines of the neutral or singly ionized elements are often of interstellar origin, i.e. characteristic of a low density medium. The simple physical properties of these atomic levels can be used to constrain the density in the absorbing region. The distinction between lines originating in a low density medium and those originating from a higher density medium is easily exemplified by the Si II triplet $\lambda 1260.40$, 1264.73, 1265.02 Å. The first member originates from the ground state level $^2P_{1/2}$ while the two others originate from an excited level $^2P_{3/2}$ (0.036 ev) only slightly above the ground state. Under stellar atmospheric conditions the two levels are strongly coupled by collisional transitions while the radiative transition is forbidden. In low density plasma however the spontaneous de-excitation dominates the balance of the two levels and the $^2P_{3/2}$ level is totally de-populated. We can estimate the density at which the levels start to be coupled by collisional transitions. Two dominant perturbers are considered: the free electrons and the neutral hydrogen atoms. Spitzer (1968) estimates the electron collisional rate of the transition $^2P_{1/2}$ - $^2P_{3/2}$ to be:

$$n_e\gamma_{jk}(s^{-1}) = n_e \, 8.63 \times 10^{-6} \frac{\Omega(j,k)}{g_j T^{1/2}} e^{-E_{jk}/kT},$$
(3)

where T is the temperature, n_e the electron density, E_{jk} is the transition energy and g_j the statistical weight of the lower level. For this specific transition the collision strength $\Omega(j,k)$ is equal to 7.7. The dominating de-excitation process under low-density conditions is spontaneous de-excitation, for which $A_{kj} = 2.1 \times 10^{-4}$ s^{-1}. The ratio of the upward to the downward rate is thus:

$$R = 0.16 \frac{n_e}{T^{1/2}} \cdot \tag{4}$$

The upper level starts to be significantly populated for a ratio around $R \approx 0.1$. This means, for $T = 10^4$ K, an electron density higher than 63 cm^{-3}. This value of n_e, around 100 cm^{-3}, traces a boundary between what will be defined as the interstellar medium and what will be called the immediate stellar environment (circumstellar region: a halo or a dense Strömgren sphere). This limiting value strongly constrains the size of a Strömgren sphere to be less than 0.01 pc or 10^{16} cm (Dupree and Raymond 1983). The H collisional rate is given by Spitzer (1968):

$$n_H \gamma_{jk} (s^{-1}) = n_H 1.46 \times 10^{-9} 10^{-5040 E_{jk}(ev)/T}, \tag{5}$$

where n_H is neutral hydrogen density. For T around 10^3 K the ratio R of the collisional excitation to the spontaneous de-excitation is:

$$R = 7 \times 10^{-6} n_H. \tag{6}$$

From equations (5) and (6) it is obvious that in ISM physical conditions the Si II level $^2P_{3/2}$ is totally depopulated and that the most probable mechanism to populate this level is the collision with electrons ($n_e > 10^2$ cm^{-3}). The C II $\lambda 1335.3$Å triplet offers a similar but less constraining conclusion ($n_e > 4$ cm^{-3}). For those lines which are clearly not originating from the ISM, the equivalent widths are converted in terms of photospheric abundances n(X)/n(H) by interpolating in the relevant tables, even if the line is not clearly associated with the photosphere. This will allow an analysis of the ionization balance of the atomic elements. Table 1a presents the limits on the abundances for those stars which do not show photospheric or circumstellar C, N or Si lines. The temperature and gravity are obtained, when available, from Holberg, Wesemael, and Basile (1986), Wesemael, Green and Liebert (1985) for the DAO star PG 1210+533 and Kidder, Holberg, and Wesemael (1989) for Sirius B. Otherwise, we estimate them from the Lyman α line profile in the small aperture low-dispersion *IUE* spectrum, using the grid of theoretical line profiles presented by Vennes, Shipman, and Petre (1990).

Table 1a
Limits on the photospheric abundances

name	T_{eff}	log g	n(C)/n(H)[a]	n(N)/n(H)[a]	n(Si)/n(H)[a]
GD 50	39,300	8.95	< -6.4	...	< -6.5
40 Eri B[b]	16,325	7.65	< -9.0	...	< -9.0
Sirius B	25,000	8.60	< -8.8	...	< -8.8
EG 70	25,800	8.05	< -7.6	...	< -7.9
GD 153	42,375	8.23	< -7.9	< -5.1	< -8.8
HZ 43	57,500	8.50	< -8.5	< -6.6	< -8.0
EG 118	32,300	8.40	< -7.5	...	< -6.3
Lanning 18	29,700	7.95	< -7.2	...	< -7.9
GD 391	24,800	8.20	< -7.1	...	< -7.7

[a] Logarithmic values.   [b] The limits on the abundance are taken at 18,000 K.

In Tables 1b and 1c are presented, divided in two groups, the abundances obtained assuming a photospheric origin to all features, even if the comparison of the Balmer line shifts with the heavy element line shifts reveals that it is not appropriate. In group I are the non-ambiguous abundances obtained from the equivalent width of the heavy element lines. In group II are collected those abundances which are based on the uncertain detection of a line feature. Clearly, in many cases among the group II, the addition of one or two more high-dispersion spectra might help settle the question of the heavy element abundances in these stars. In the next section we present the interpretation of these detections and non-detections.

Table 1b
Abundances under the photospheric hypothesis (group I)

name	T_{eff}	log g	n(C)/n(H)[a]	n(N)/n(H)[a]	n(Si)/n(H)[a]	v_{Balmer}	v_{CNSi}
Feige 24[b]	55,000	7.23	-6.4	-5.3	-6.3	...	orbital
G191 B2B	62,250	7.55	(IV) -5.6 (III) < -6.0	-5.5	-6.5	+ 18	+18
EG 102	20,900	8.00	-5.8	...	-7.1	...	- 13
CD-38°10980	24,500	8.10	≤ -8.2[c]	...	(II) -7.1 (III) -7.9 (IV) < -8.6	+ 40	+ 30
Wolf 1346	20,680	8.10	< -7.9	...	(II) -7.0 (III) < -7.9	+ 36	+ 20
GD 394	36,125	8.10	< -7.4	...	(II) < -7.0 (III) -5.8 (IV) -5.8	+94	+24

[a] Logarithmic values, in parenthesis the ionization level used in the determination.
[b] Abundances from Chayer et al. (1987).
[c] The "≤" symbol refers to a possible but not firm detection of the element in the *IUE* spectrum.

Table 1c
Abundances under the photospheric hypothesis (group II)

name	T_{eff}	log g	n(C)/n(H)[a]	n(N)/n(H)[a]	n(Si)/n(H)[a]	v_{Balmer}	v_{CNSi}
GD 2	46,100	8.20	≤ -6.4[b]	< -4.5	< -7.3	+ 76	-5
GD 659	36,850	8.17	≤ -6.8	≤ -3	≤ -8.5	- 37	+31
GD 71	33,300	7.78	(II) < -7.4 (IV) ≤ -6.4	...	< -8.8	+ 63	+37
EG 46	25,500	8.05	< -7.2	...	(II) < -6.9 (III) ≤ -6.6 (IV) ≤ -5.9	...	+47
GD 140	21,375	8.45	≤ -7.9	...	≤ -8.0	...	+56
PG 1210+533	50,000	8.00	≤ -6.0	< -4.3	< -6.8	...	-4
GD 246	53,600	7.64	≤ -6.3	< -5.5	≤ -7.5	...	-24

[a] Logarithmic values, in parenthesis the ionization level used in the determination.
[b] The "≤" symbol refers to a possible but not firm detection of the element in the *IUE* spectrum.

5. The Nature of the Heavy Element Absorption in the DA White Dwarfs

A case by case study of the heavy element lines and the EUV/soft X-ray opacities discovered in the DA white dwarfs instructs us on the nature of this absorption. In the following we group together the stars that offer similar properties, allowing us to build a phenomenological classification (Table 2).

Table 2
A comparative study of the EUV/soft X-ray and *IUE* /high-dispersion data

name	T_{eff}	log M_H	n(C)/n(H)[a]	n(N)/n(H)[a]	n(Si)/n(H)[a]
GD 2	46,100	-13.6	$\leq$ -6.4[b]	< -4.5	< -7.3
GD 659	36,850	-13.5	$\leq$ -6.8	$\leq$ -3:	$\leq$ -8.5
GD 153	42,375	-13.7	< -7.9	< -5.1	< -8.8
GD 246	53,600	-14.1	$\leq$ -6.3	< -5.5	$\leq$ -7.5
Feige 24[c]	57,500	complex	-6.4	-5.3	-6.3
G191 B2B	62,250	-15.1	-5.6	-5.5	-6.5
GD 394	36,125	-13.8	< -7.4	...	-5.8
GD 71	33,300	> -13.3	$\leq$ -6.4	...	< -8.8
Sirius B[d]	25,000	pure H	< -8.8	...	< -8.8
EG 70	25,800	> -13.5	< -7.6	...	< -7.9
CD-38°10980	24,500	> -13.5	$\leq$ -8.2	...	-7.1
GD 391	24,800	> -13.5	< -7.1	...	< -7.7
EG 46	25,500	< -14.2	< -7.2	...	$\leq$ -5.9
HZ 43	57,500	pure H	< -8.5	< -6.6	< -8.0

[a] Logarithmic values.

[b] The "$\leq$" sign refers to a possible but not firm detection of the element in the spectrum.

[c] Vennes *et al.* (1989) have suggested that the *EXOSAT* spectrum can be explained by the presence in the photosphere of a variety of trace elements from carbon to calcium.

[d] Paerels *et al.* (1988) have explained the *EXOSAT* spectrum in the context of pure hydrogen atmosphere.

1) Feige 24 presents two unique properties which allow a detailed interpretation of its heavy element lines. First, the relatively high orbital velocity of this double system (DA+dM1-2) permits the *IUE* spectrograph to follow the wavelength displacement of the photospheric features due to orbital motion, and at appropriate moment in the orbital motion the photospheric and circumstellar line components clearly separate and reveal the bimodal origin of the heavy element lines in Feige 24 (Figure 2). The fact, shown in Figure 3, that the width of the "circumstellar" carbon line does not vary with time implies that the circumstellar bubble is much larger than the binary orbit (>> 10 R_0 or 10^{12} cm). Also the fact that the "photospheric" lines of carbon, nitrogen and silicon seem to follow the orbital motion of Feige 24 imposes that these lines are truly close to the white dwarf, i.e. photospheric in origin. Simultaneous *IUE* observations of the spectral lines in the white dwarf and optical spectroscopy of the companion may provide good orbital parameters (Thorstensen *et al.* 1978) and firmly establish the photospheric nature of these lines in Feige 24.

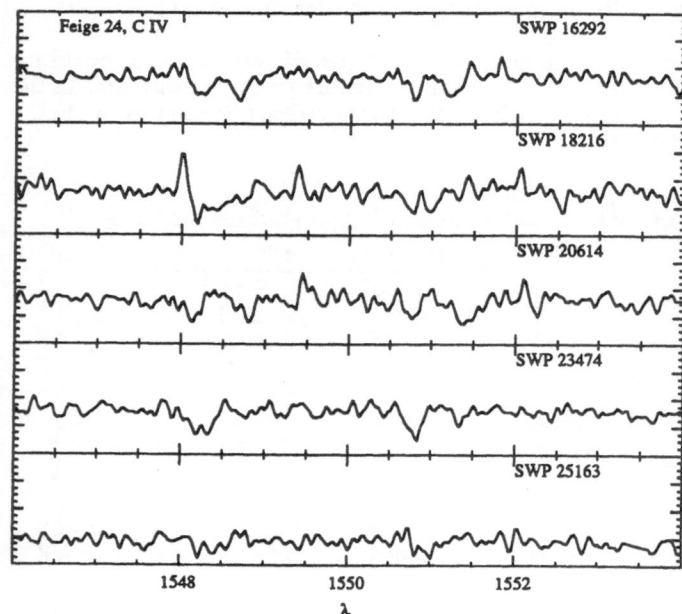

Figure 2. The C IV doublet at 5 different epochs. Note the component at steady velocity (circumstellar) and the orbiting component (most probably photospheric).

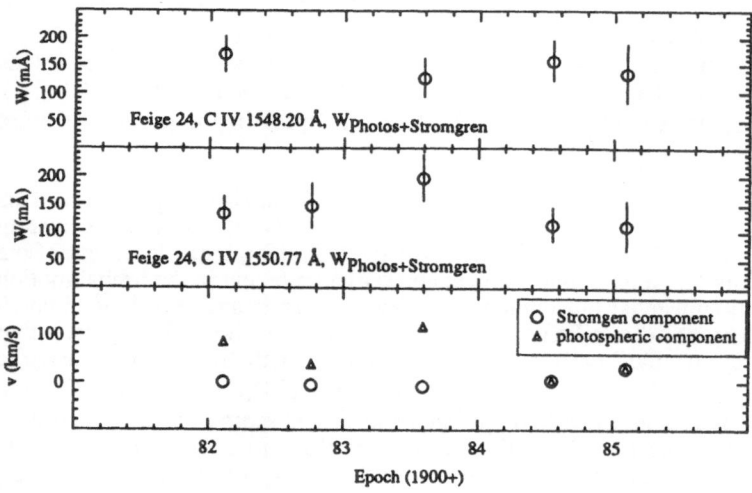

Figure 3. The total width and velocity of the C IV doublet at 5 different epochs.

2) The group of hot DA stars GD 2, GD 659, GD 153, and GD 246 is the most interesting one as it shows very low traces of heavy elements, if any, but on the other hand *a significant opacity source is inferred from the EUV/soft X-ray spectral range* (Vennes, Fontaine, and Wesemael 1989; Vennes, Shipman, and Petre 1990). The star GD 153 is one good example of the EUV/soft X-ray properties of these stars (Figure 4). This dilemma can be resolved either by postulating that other elements than carbon and silicon can still be present, though not detectable by *IUE*, or by postulating as suggested by Vennes *et al.* (1988) that *the atmosphere of these stars is layered with a thin hydrogen atmosphere on top of the helium envelope*. We cannot at the moment establish with certitude this possibility, but it remains the only suitable model to explain the properties of the EUV/soft X-ray emission from these four stars (Vennes, Fontaine and Wesemael 1991).

3) In addition to Feige 24 (discussed above), the stars G191 B2B and GD 394 exhibit strong ultraviolet heavy element lines. These lines are possibly of photospheric origin in the case of G191 B2B, but are clearly not in the case of GD 394. In the first case, except for a slight discrepancy between the abundance determinations from the ions of C III and C IV, the wavelength shifts of the Balmer and C,N, and Si lines seem to favor a photospheric origin. This possibility is supported by the fact that new calculations, presented here, of NLTE line profiles (9 H I levels, 36 line transitions) in DA white dwarfs suggest the onset of an emission core in the hydrogen Balmer line as detected by Reid and Wegner (1988) (Figure 5). In the case of GD 394, the discrepancy between the abundance determination from the Si II ion and the ions of Si III and Si IV, tells that the ionization balance is far from the Saha-Boltzman prescription, imposing thus a non-photospheric origin. On the other hand, no matter where the C and Si atoms in both G191 B2B and GD 394 are sitting-in they provide large EUV/soft X-ray opacity, and if they are not photospheric they suggest the existence of an ionized region surrounding the white dwarfs. This ionized "bubble" is likely to affect the interpretation of the EUV/soft X-ray emission from these stars.

4) At the opposite of the two previous cases, the DA stars GD 71, Sirius B, EG 70, CD-38°10980, and GD 391 show neither a EUV/soft X-ray source of opacity or any significant traces of heavy elements. The atmospheres of these stars are, to a very large degree, made of pure hydrogen.

5) The star EG 46 shows a very large EUV/soft X-ray opacity, but a very low heavy element abundance. The discovery by Holberg, Kidder, and Wesemael (1991) of optical He I lines in this star suggests a phenomenon similar to that of GD 323 (Liebert *et al.*1984) and is likely to be the explanation for the non-detection of this star by *EXOSAT*.

6) Although some investigations (Jordan *et al.* 1987; Koester 1989) have suggested that helium may be the source of opacity in the EUV/soft X-ray spectrum of that star, Paerels *et al.* (1986) have clearly demonstrated that the featureless *EXOSAT* spectrum of HZ 43 excludes any sources of opacity other than hydrogen. The fact that we do not see any heavy elements in the *IUE* high-dispersion spectrum of that star supports the conclusion that the atmosphere of HZ 43 is dominated by hydrogen to a very high degree of purity. To be interpreted correctly the complete spectral coverage from *IUE* and *EXOSAT* the effective temperature needs to be lowered to around 50,000K and the population ratio n(H I)/n(He I) in the ISM cannot be kept at 10 but must be lowered to a value < 1 (Heise *et al.* 1988; Paerels and Vennes 1991).

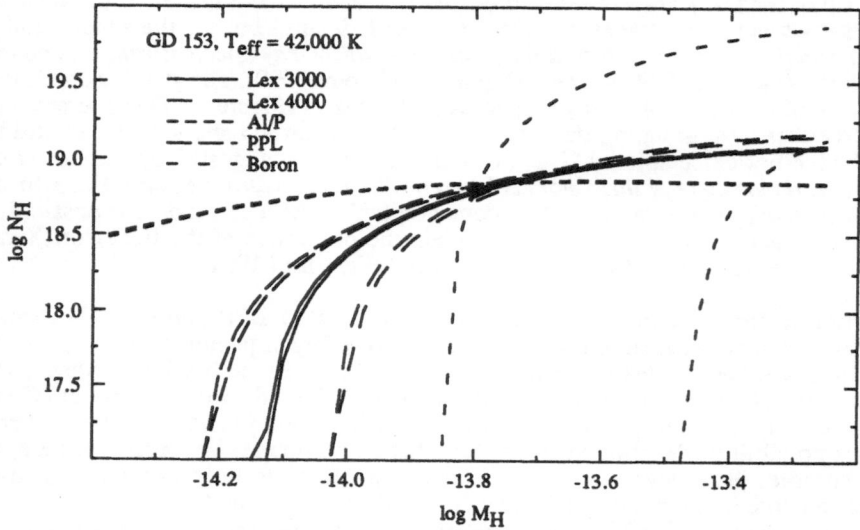

Figure 4. Analysis of the *EXOSAT* count rate of the star GD 153. The solution in the $(\log M_H, \log N_H)$ plane is (-13.7, 18.8).

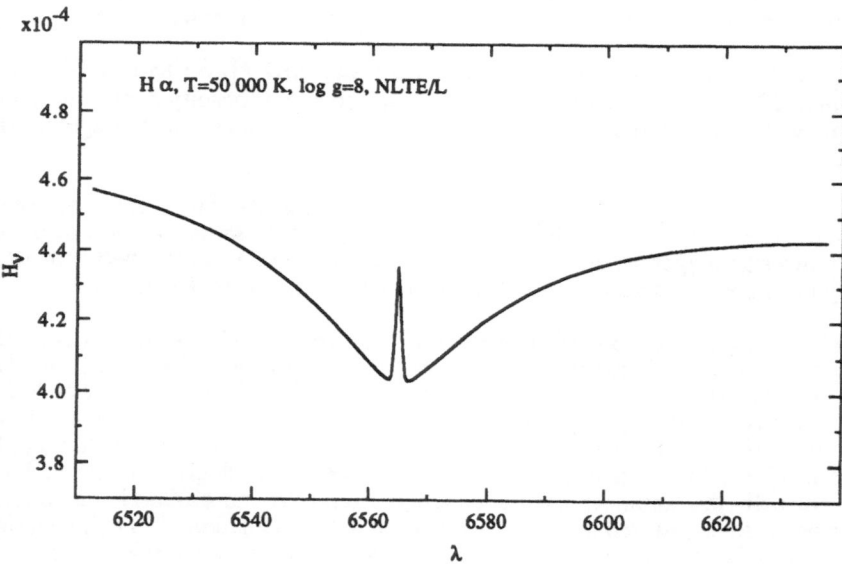

Figure 5. NLTE Hα line profile at T_{eff} = 50,000 K and log g =8 appropriate for some of the hot DA white dwarfs.

We must finally mention that the case of the DAO PG1210+533 supports the existence of a stratified atmosphere in the DA/DAO stars. In a general effort to interpret quantitatively the spectral properties of white dwarfs, MacDonald and Vennes (1990) have considered the spectral properties of the DA/DAO/DO white dwarfs in the context of stratified atmospheres. The firm upper limit on the carbon abundance in PG 1210+533 corresponds to 1/600 th of the cosmic ratio while the helium abundance is between 1/5 th and 1/10 th of the solar ratio. This discrepancy indicates that PG 1210+533 is not currently accreting and that the presence of helium in the photosphere of that DAO is more likely related to the presence of a thin layer of hydrogen.

What is the nature of the heavy element absorption in the DA white dwarfs and what is the connection to the EUV/soft X-ray opacity ? There are several facts to keep in mind: the detection of the Si II $\lambda 1263.3$Å triplet in EG 102, CD-38°10980 and Wolf 1346 implies n_e > 100 cm^{-2} for the absorbing region. We know that in, at least, CD-38°10980, Wolf 1346 and GD 394 the lines cannot originate from the photosphere because the ionization balance is far from that expected under photospheric conditions. The conclusion is that *at least some of the hot DA white dwarfs are surrounded by a dense (n_e > 100 cm^{-3}) and compact (r < 0.01 pc) shell of ionized material.* There is no direct evidence, as claimed in the past, for a wind in the velocity measurements because after taking into account the gravitational redshift from the data in Tables 1b and 1c the velocities relative to the photosphere are in some cases positive and in others negative and are more likely to reflect relative motions rather than concentric motions. Except maybe for G191 B2B for which there are strong reasons to believe that the heavy elements are photospheric, only Feige 24 presents a clearly photospheric origin to some of its lines. Many DA white dwarfs present traces of an ionized shell surrounding it, among them Feige 24.

A second striking feature is that many stars (between 40 and 70 % of our sample) at all temperatures and gravities do not show any traces of heavy element absorptions in their UV spectra. So the existence of a circumstellar ionized material is not a general property of the hot DA white dwarfs.

To conclude we note that there is a strong indication that the interpretation of the EUV/soft X-ray emission from the hot DA white dwarfs should be carried out in the lights of two complementary models. In some DA stars (Feige 24, GD 394, and G191 B2B) the EUV/soft X-ray opacity is probably provided by a host of trace heavy elements, while in some other DA stars (GD 2, GD153 and GD 246) the opacity can be provided by helium in a stratified configuration.

We acknowledge support from the grant NSF AST 87-20530 and NASA NAG 5-972. We wish to thank Jim MacDonald for his comments.

6. References

Bruhweiler, F.C. 1984, in *Future of Ultraviolet Astronomy Based on Six Years of IUE Research*, ed. J.M. Mead, Y. Kondo, and R. Chapman (NASA CP-2349), p.269.

Bruhweiler, F.C., and Kondo, Y. 1981, *Ap.J. (Letters)*, **248**, L123.

Bruhweiler, F.C., and Kondo, Y. 1983, *Ap.J.*, **269**, 657.

Chayer, P., Fontaine, G., Wesemael, F., and Michaud, G. 1987, in *IAU Colloquium 95, The Second Conference on Faint Blue Stars*, A.G.D. Philip, D.S. Hayes and J. Liebert eds. (Schenectady: L.Davis Press), p.653.

Dupree, A.K., and Raymond, J.C. 1982, *Ap.J.(Letters)*, **263**, L63.
Dupree, A.K., and Raymond, J.C. 1983, *Ap.J.(Letters)*, **275**, L71.
Heise, J., Paerels, F.B.S., Bleeker, J.A.M., and Brinkman, A.C. 1988, *Ap.J.*, **334**, 958.
Henry, R.B.C., Shipman, H.L., and Wesemael, F. 1985, *Ap.J.Suppl.*, **57**, 145.
Holberg, J.B., Kidder, K., and Wesemael, F. 1991, these proceedings.
Holberg, J.B., Wesemael, F., and Basile, J. 1986, *Ap.J.*, **306**, 629.
Holberg, J.B., Wesemael, F., Wegner, G., and Bruhweiler, F.C. 1985, *Ap.J.*, **293**, 294.
Jordan, S., Koester, D., Wulf-Mathies, C., and Brunner, H. 1987, *Astron.Astrophys.*, **185**, 253.
Kahn, S.M., Wesemael, F., Liebert, J., Raymond, J.C., Steiner, J.E., and Shipman, H.L. 1984, *Ap.J.*, **278**, 255.
Kidder, K., Holberg, J.B., and Wesemael, F. 1989, in *IAU Colloquium 114, White Dwarfs*, ed. G.Wegner (Berlin: Springer), p.350.
Koester, D. 1987, *Ap.J.*, **322**, 852.
Koester, D. 1989, in *IAU Colloquium 114, White Dwarfs*, ed. G.Wegner (Berlin: Springer), p.206.
Liebert, J., Wesemael, F., Sion, E.M., and Wegner, G. 1984, *Ap.J.*, **277**, 692.
MacDonald, J., and Vennes, S. 1990, *Ap.J.*, submitted.
Paerels, F.B.S., Bleeker, J.A.M., Brinkman, A.C., Gronenschild, E.H.B.M., and Heise, J. 1986, *Ap.J.*, **308**, 190.
Paerels, F.B.S., Bleeker, J.A.M., Brinkman, A.C., and Heise, J. 1988, *Ap.J.*, **329**, 849.
Paerels, F.B.S., and Heise, J. 1989, *Ap.J.*, **339**, 1000.
Paerels, F.B.S., and Vennes, S. 1991, in preparation.
Poulin, E., Wesemael, F., Holberg, J.B., and Fontaine, G. 1989, in *IAU Colloquium 114, White Dwarfs*, ed. G.Wegner (Berlin: Springer), p.144.
Reid, N., and Wegner, G. 1988, *Ap.J.*, **335**, 953.
Spitzer, L. 1968, *Diffuse Matter in Space* (New York: Wiley)
Thompson, R.W., Turnrose, B.E., Bohlin, R.C. 1982, *Astron.Astrophys.*, **107**, 11.
Thorstensen, J.R., Charles, P.A., Margon, B., and Bowyer, S. 1978, *Ap.J.*, **223**, 260.
Trimble, V.T., and Greenstein, J.L. 1972, *Ap.J.*, **177**, 441.
Vennes, S., Chayer, P., Fontaine, G., and Wesemael, F. 1989, *Ap.J.(Letters)*, **336**, L25.
Vennes, S., Fontaine, G., and Wesemael, F. 1989, in *IAU Colloquium 114, White Dwarfs*, ed. G.Wegner (Berlin: Springer), p.368.
Vennes, S., Fontaine, G., and Wesemael, F. 1991, in preparation.
Vennes, S., Pelletier, C., Fontaine, G., Wesemael, F. 1988, *Ap.J.*, **331**, 876.
Vennes, S., Shipman, H.L., and Petre, R. 1990, *Ap.J.*, in press.
Wegner, G. 1978, *M.N.R.A.S.*, **182**, 111.
Wesemael, F., Green, R.F., and Liebert, J. 1985, *Ap.J.Suppl.*, **58**, 379.

DISCUSSION

HOLBERG :

I was unclear about your use for a gaussian to fit the IUE high dispersion metal features. Isn't it true that for the most part these lines are unresolved and that the equivalent width at the line centroïd are the only recoverable quantities?

VENNES :

This is true, the lines are unresolved, and particularily the faint ones. We are using only the integrated width of the Gaussian and its centroïd. The use of a Gaussian was motivated by its capability to provide easily an estimate of the uncertainty of the fitting parameters. For the very strong, saturated, resolved lines, a lorentzian function is more appropriate.

HEBER :

Concerning the Hα emission core in G191-B2B : Stefan Dreizler and Klaus Werner in our Institute have included Stark-broadening in the statistical equilibrium calculations of their NLTE code and confirm the Hα emission being a NLTE effect.

NEW RESULTS ON RADIATIVE FORCES ON IRON IN HOT WHITE DWARFS

CHAYER, P., FONTAINE, G., and WESEMAEL, F.
Département de Physique
Université de Montréal
C.P. 6128
Montréal, Québec, Canada, H3C 3J7.

ABSTRACT. Detailed computations of radiative forces on Fe are performed in H-rich and He-rich envelope models of hot white dwarfs, and equilibrium abundance profiles of Fe are obtained from those calculations. In the absence of other competing mechanisms, detectable amounts of Fe would be visible at the photosphere of a 0.6 $M_\odot$ white dwarf when $T_e \geq 50\,000$ K for H-rich models, and when $T_e \geq 35\,000$ K for He-rich models. In general, the abundance of Fe supported by the radiative forces increases monotonically with increasing effective temperature. Equilibrium abundances of Fe are larger in He-rich models than in H-rich ones at the same effective temperature.

1. Introduction

Feibelman and Bruhweiler (1990) have examined the entire *IUE* archival data base of high-dispersion SWP spectra of hot white dwarfs in search of Fe VII absorption lines. They have found four very hot white dwarfs showing such absorption lines. These stars are the pulsating hot white dwarf PG 1159-035, the very hot DO white dwarfs PG 1034+001 and KPD 005+5106, and the central star of K1-16. Until recently, only absorption lines of C, N, O, and Si have been observed in hot hydrogen-rich (DA) and helium-rich (non-DA) white dwarfs. Vennes, Chayer, Fontaine, and Wesemael (1989) have also considered the possibility that many different metals (supported by radiative forces) could be present in low abundances in the atmosphere of the hot DA white dwarf Feige 24. These abundances would be too small for individual line detection, but their combined effects reproduce quite well the observed EUV spectrum. The recent identification of iron found in the photospheres of a few very hot white dwarfs gives further impetus for studying the process of radiative levitation in white dwarfs.

As is well known, gravitational settling is very efficient in white dwarfs, and heavy elements diffuse very rapidly toward the base of the envelope (Fontaine, Michaud 1979; Vauclair, Vauclair, and Greenstein 1979). However, for hot white dwarfs ($T_e \geq 20\,000$ K), gravitational settling must compete with selective radiative forces that can be very important, due to the relatively large radiative flux in these hot stars. Therefore, many heavy

249

G. Vauclair and E. Sion (eds.), White Dwarfs, 249–255.

elements can be pushed toward the surface and significant quantities of these elements can be supported against settling when equilibrium between radiative and gravitational forces occurs. Chayer, Fontaine, and Wesemael (1989) have computed equilibrium abundances for C, N, O, and Si in a set of DA and non-DA envelope models, and showed that these elements should be observed in the photosphere of hot white dwarfs under some conditions. In this paper, we extend the work of Chayer, Fontaine, and Wesemael (1989) to iron, and verify the validity of the radiative support hypothesis to explain the appearance of this element in the photosphere of the stars observed by Feibelman and Bruhweiler (1990). Before discussing the results, a brief discussion of the method of computation is presented.

2. Calculation of radiative forces

The radiation flux that flows toward the surface carries momentum that can be transferred to trace elements. The magnitude of the momentum transfer depends on the extinction coefficient of these elements and the nature of the flux. This leads to the expression,

$$g_{rad,\nu}(Fe)d\nu = \frac{4\pi}{c}\frac{1}{X(Fe)}\kappa_\nu(Fe)H_\nu d\nu, \tag{1}$$

for the radiative acceleration that is transferred to iron in the frequency interval $d\nu$. $X(Fe)$ and $\kappa_\nu(Fe)$ are, respectively, the mass fraction of iron, and the contribution of this element to the monochromatic opacity; H_ν is the Eddington flux and c the speed of light. H_ν can be written as

$$H_\nu \approx -\frac{1}{3}\left(\frac{1}{\kappa_\nu\rho}\frac{\partial B_\nu}{\partial T}\right)\frac{\partial T}{\partial r}, \tag{2}$$

when we use the diffusion approximation. It applies when τ_ν, the monochromatic optical depth, is larger than unity. In the latter expression, κ_ν is the total monochromatic opacity, ρ the mass density, T the temperature, and B_ν the Planck distribution. The derivative of B_ν with respect to T can easily be found and when τ_R, the Rosseland mean optical depth, is greater than unity, we can approximate the radiative gradient by

$$\frac{\partial T}{\partial r} = -\frac{3}{16}\rho\kappa_R\frac{R^2}{r^2}\frac{T_e^4}{T^3}, \tag{3}$$

where κ_R is the Rosseland mean opacity, R the radius of the star, r the distance from the center of the star, and T_e the effective temperature. Therefore, the flux becomes with these approximations (τ_ν and $\tau_R \gg 1$)

$$H_\nu = cst\frac{\kappa_R}{\kappa_\nu}\frac{R^2}{r^2}\frac{T_e^4}{T}\frac{u^4e^u}{(e^u-1)^2}, \tag{4}$$

where $u \equiv h\nu/kT$, and cst contains all the numerical constants.

On the other hand, we explicitly write the monochromatic opacity of iron as

$$\kappa_\nu(Fe) = \frac{\pi e^2}{mc}\sum_{i,j}\frac{N_{ij}}{\rho}\sum_k f_{i,j-k}\phi_\nu, \tag{5}$$

where the summation is performed on ionization state i and energy levels j and k. The population of each level of energy N_{ij}, is obtained with the Saha and Boltzman-Maxwell equations appropriate for local thermodynamic equilibrium. These energy levels can be found in the compilation of Corliss and Sugar (1982). We take about 150 energy levels by ionization state. The oscillator strengths $f_{i,j-k}$ (transition from the lower level j to the upper level k in ionization state i) are taken from Fuhr *et al.* (1981) and, when they are not available, we use estimated oscillator strengths like the ones developed in Michaud, Charland, Vauclair, and Vauclair (1976). In our calculations, all transitions permitted by the selection rules are included, but we keep only the transitions originating from the ten first energy levels. The line profile ϕ_ν is a Voigt profile which combines the natural width and pressure broadening using equation (3.1) of Michaud, Vauclair, and Vauclair (1983). Finally, equation (1) is integrated numerically over all frequencies for each line of each state of ionization by means of the expressions given by equations (4) and (5). Because several states of ionization are simultaneously present, the total radiative acceleration is computed with the prescription of Montmerle and Michaud (1976) that takes account the different diffusion velocity of each type of ion.

We have calculated radiative accelerations on Fe in envelope models of DA and non-DA white dwarfs obtained from the code of Fontaine and Van Horn (1976) which uses the equation of state of Fontaine, Graboske and Van Horn (1977). The convective zone is described by the ML3 version of the mixing length theory following the work of Fontaine, Tassoul and Wesemael (1984) on pulsating white dwarfs. H-rich and He-rich envelope models have been computed in the range of 20 000 K $\leq T_e \leq$ 100 000 K using steps of 5000 K in effective temperature. The Rosseland opacities used in the envelope code are those of Cox and Stewart (1970; Iben I and Iben V mixtures). The parameters of the envelope are taken from a 0.6 $M_\odot$ evolutionary track where the gravity increases (as it should) with decreasing effective temperature. In the range of temperatures that we are concerned with, the convective zone is not present at the surface of DA models. But in non-DA models, convection appears at the surface around $T_e \approx$ 65 000 K, and the base of the convection zone sinks deeper and deeper under the photosphere with decreasing effective temperature. In these cases, we compute the radiative acceleration directly under the convective zone.

3. Results

Figure 1 shows the maximum radiative acceleration profiles for DA and non-DA models of different effective temperatures with $M/M_\odot = 0.6$. The number corresponding to each curve stands for the effective temperature (in units of 10^3 K) of each model, and the horizontal dashed line indicates the maximum surface gravity reached by our cooling 0.6 $M_\odot$ star. The first shell at the top of the envelope (on the right-hand side of the figure) is at $\tau_R = 1$, except in non-DA models where a convective zone appears at the surface. The differences in the shape of each radiative acceleration profile depend on the variations of the physical conditions with depth. These changes in the physical conditions modify the Eddington flux and the extinction coefficient of the dominant and trace elements and, consequently, the radiative acceleration. We note that in both sets of models the radiative acceleration increases monotonically with effective temperature. This is also the case for C, N, and O but not Si in hot white dwarfs according to Chayer, Fontaine, and Wesemael (1989).

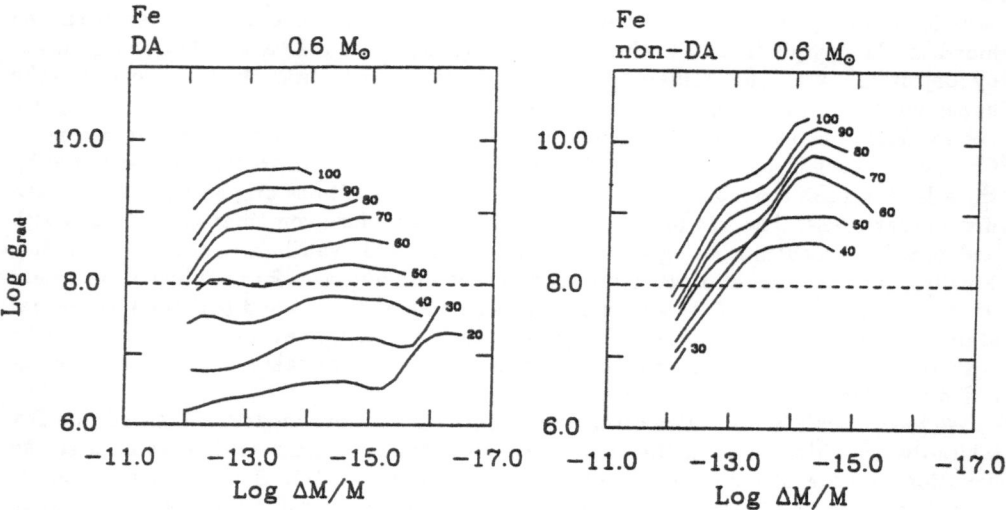

Fig. 1.- Radiative accelerations on Fe for an abundance reduced by 8 orders of magnitude with respect to solar as a function of fractional mass in envelopes of DA and non-DA white dwarfs of 0.6 $M_\odot$. Each curve corresponds to an envelope model of different effective temperature. The horizontal dashed line indicates the surface gravity of our cooling white dwarf model at minimum radius.

In Fig. 1, the radiative acceleration was computed with an abundance that is 8 orders of magnitude smaller than the solar abundance. In this manner, we want to make sure that the lines are completely unsaturated to obtain the *maximum* possible radiative acceleration (see Morvan, Vauclair, and Vauclair 1986). By doing that, we can compare directly the radiative acceleration g_{rad} and the gravitational acceleration g, and if, for example, $g_{rad} < g$, then Fe can no longer be supported. In fact, we must actually compare g_{rad} with g_E, the effective local gravity which takes into account the electric field. From the diffusion equation (see equation (1) of Vennes *et al.* 1988), we arrive at the expression

$$g_E = g\left(1 - \frac{m_1 \, (Z_2 + 1)}{m_2 \, (Z_1 + 1)}\right), \tag{6}$$

for the effective gravity, where g is the gravitational acceleration, m_1, Z_1, and m_2, Z_2 are the mass and the mean charge of the dominant element and the trace element, respectively. When $g_{rad} < g_E$, no equilibrium is longer possible between radiative and gravitational acceleration, and Fe inevitably diffuses to the bottom of the envelope. Therefore, radiative levitation alone cannot give rise to observable abundances of iron in the atmosphere of DA white dwarfs with $T_e < 50\,000$ K, and non-DA white dwarfs with $T_e < 35\,000$ K. However, for the hotter models where $g_{rad} > g_E$ in the envelope, a finite abundance of iron can be supported by the radiative acceleration.

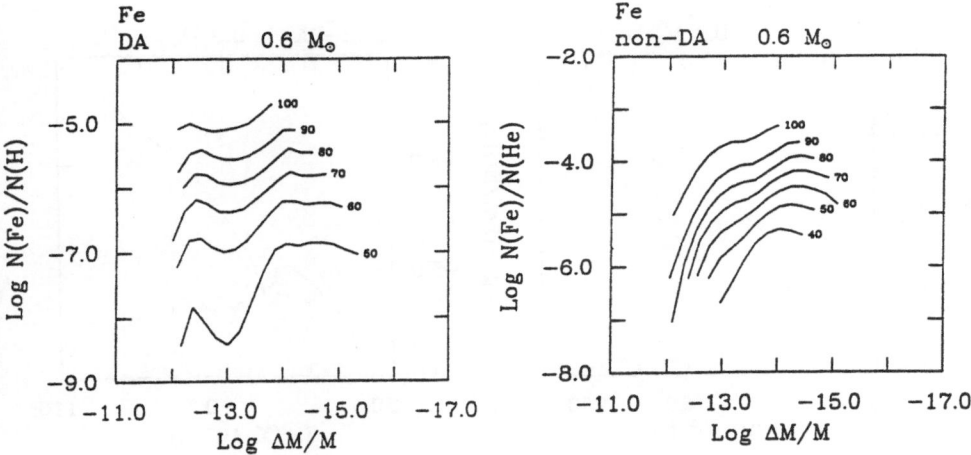

Fig. 2.- Equilibrium abundance profiles for Fe at different effective temperatures (in units of 10^3 K) in envelopes of DA and non-DA white dwarfs of 0.6 $M_\odot$.

Figure 2 illustrates such equilibrium abundance profiles for Fe with respect to H and He in DA and non-DA models, respectively. We computed these equilibrium abundance profiles by iterating the radiative acceleration as a function of the iron abundance until $g_{rad} = g_E$ for each layer in the envelope. When we compare the equilibrium abundances of iron relative to the main atmospheric constituent in DA and non-DA models of the same effective temperature, we note that they are larger in non-DA models. For instance, the relative equilibrium abundance of Fe at $\tau_R = 1$ in a non-DA model is greater than in a DA star by a factor of about 7 when $T_e = 80\ 000$ K. The greater mass density in non-DA models is responsible, in part, for this difference. Indeed, a greater mass density increases the width of the lines which can absorb a greater number of photons now available in the wings of the lines. This increases the radiative acceleration and a larger abundance of Fe will be radiatively supported.

Finally, Figure 3 illustrates the evolution of the equilibrium abundance of Fe at the photosphere of our 0.6 $M_\odot$ DA and non-DA white dwarf models. This, of course, assumes that radiative levitation is the only competing mechanism against settling. In both sets of models, the surface abundance of Fe generally decreases monotonically with decreasing effective temperature. The exception is for non-DA models in a narrow range of effective temperature around $T_e \approx 65\ 000$ K where convection appears at the surface. Also, as we mentioned just above, the relative surface abundance of Fe is greater in non-DA models.

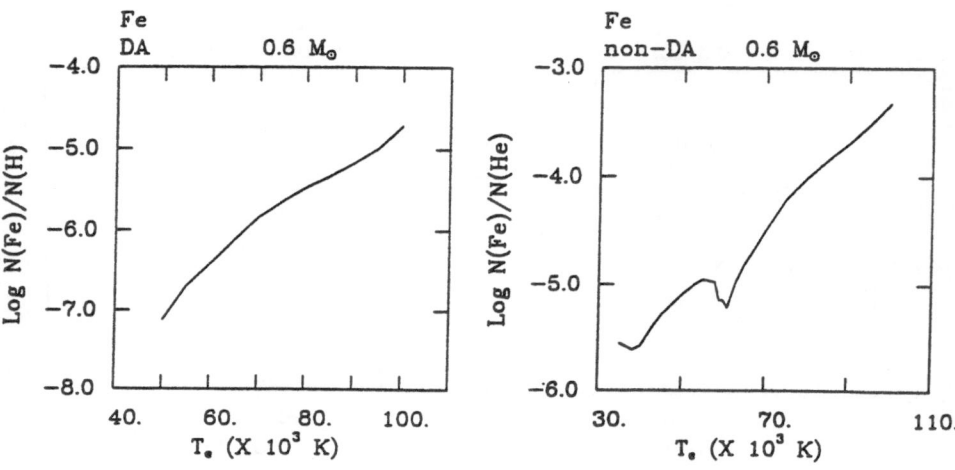

Fig. 3.- Equilibrium abundance of Fe at the photosphere of DA and non-DA white dwarfs of 0.6 $M_\odot$ as a function of effective temperature.

4. Conclusion

From our calculations, we conclude that it is quite plausible that radiative levitation can be a key mechanism to explain the observations of Fe VII absorption lines reported by Feibelman and Bruhweiler (1990) in a few very hot white dwarfs. Indeed, in such objects, radiative acceleration on Fe can counterbalance the effect of gravity, and some quantity of Fe can be radiatively supported in the atmosphere of these stars. Unfortunately, we cannot yet use our calculations in a detailed comparison of the predictions of the radiative support theory with the observations because detailed abundance analyses for iron have not been carried out yet. Qualitatively, however, the radiative support theory seems to be a most natural way to explain the occurrence of Fe in hot white dwarfs. It remains to be seen how the influence of weak winds (Chayer, Fontaine, and Wesemael 1989) can alter the abundances supported by radiative forces.

This work has been supported in part by NSREC Canada, by the fund FCAR (Québec), and by a Killam Fellowship to one of us (G.F.).

References

Chayer, P., Fontaine, G., and Wesemael, F. 1989, in *IAU Colloquium No 114: White Dwarfs*, ed. G. Wegner (New York: Springer), p. 253.

Corliss, C., and Sugar, J. 1982, *J. Phys. Chem. Ref. Data*, **11**, 135.

Cox, A.N., and Stewart, J.N. 1970, *Ap. J. Suppl.*, **19**, 261.

Feibelman, W.A., and Bruhweiler, F.C. 1990, *Ap. J.*, **357**, 548.

Fontaine, G., Graboske, H.C., Jr., and Van Horn, H.M. 1977, *Ap. J. Suppl.*, **35**, 293.

Fontaine, G., and Michaud, G. 1979, *Ap. J.*, **231**, 826.

Fontaine, G., Tassoul, M., and Wesemael, F. 1984, in *Proc. 25 th Liège Astrophysical Colloquium, Theoretical Problems in Stellar Stability and Oscillations*, ed. A. Noels and M. Gabriel (Liège: Université de Liège), p. 328.

Fontaine, G., and Van Horn, H.M. 1976, *Ap. J. Suppl.*, **31**, 467.

Fuhr, J.R., Martin, G.A., Wiese, W.L., and Younger, S.M. 1981, *J. Phys. Chem. Ref. Data*, **10**, 305.

Michaud, G., Charland, Y., Vauclair, S., and Vauclair, G. 1976, *Ap. J.*, **210**, 447.

Michaud, G., Vauclair, G., and Vauclair, S. 1983, *Ap. J.*, **267**, 256.

Montmerle, T., and Michaud, G. 1976, *Ap. J. Suppl.*, **31**, 489.

Morvan, E., Vauclair, G., and Vauclair, S. 1986, *Astr. Ap.*, **163**, 145.

Vauclair, G., Vauclair, S., and Greenstein, J.L. 1979, *Astr. Ap.*, **80**, 79.

Vennes, S., Pelletier, C., Fontaine, G., and Wesemael, F. 1988, *Ap. J.*, **331**, 876.

Vennes, S., Chayer, P., Fontaine, G., and Wesemael, F. 1989, *Ap. J. (Letters)*, **336**, L25.

THE EFFECTIVE TEMPERATURE OF THE DBV's, AND THE SENSITIVITY OF DB MODEL ATMOSPHERES TO INPUT PHYSICS.

Peter Thejll[1], Stephane Vennes and Harry L. Shipman
Department of Physics and Astronomy, University of Delaware,
Newark, DE 19716, USA

ABSTRACT. A new grid of DB models is applied to the problem of the DBV temperatures and the DB gap. It is found that the DBV instability strip lies lower than thought before. This has consequences for the calibration of mixing-length theories and the reality of the DB gap. The DBV GD358 is discussed in detail.

1 Introduction

The effective temperatures of the majority of the pulsating DB stars (DBV's) were determined by Liebert et al. (1986). This determination proved that the overall understanding of the pulsation phenomenon was good in that the predicted range of temperatures was close to the measured one. The exact range of temperatures predicted for DBV's is a sensitive function of the convective efficiency as well as other input parameters that enter the basic physical theories. The effective temperatures found by Liebert et al. were determined by fitting the UV flux in the IUE range. At the time a discrepancy was found between the effective temperature, found from fitting of the UV flux, and that found from fitting optical line profiles, as well as those UV temperatures found by using different sets of model grids. Koester et al. (1985) had previously found a lower effective temperature for GD358.

The convective efficiency is parametrized, in most current models of convection in white dwarfs, by mixing-length parameters, and at one point Fontaine et al. (1984) calibrated the mixing length parameter by finding the best fit between observations of pulsations and their theoretically predicted temperature range, based on mixing length theories. They found that the so called ML3 theory had a better fit than ML2 or ML1.

The existence of a gap in the observed temperatures of helium rich white dwarfs, the so called 'DB gap' that stretched from 45000K to about 30000K, is the focus of much attention. Any successful theory of evolution for white dwarfs must explain why there are no $\log(g)=8.0$ stars observed in this range. It has been suggested by D'Antona (1988) that the gap is merely a statistical fluke since few stars are

---

[1]present address: NORDITA, Blegdamsvej 17, DK-2100 Copenhagen Ø, Denmark

G. Vauclair and E. Sion (eds.), White Dwarfs, 257–265.
© 1991 Kluwer Academic Publishers.

expected in the temperature range anyw~y, because of the short cooling times at these high temperatures. By accurately ⅃nding the temperature of the hottest DB belcw the gap one can say something about the size and hence the reality of the gaₚ.

We have set out to calculate a new grid of DB models, in the effective temperature range of the DBV's, in order to better understand the sensitivity of model atmospheres to the uncertainties that exist in the input physics, and to see if the DBV range of temperatures is well known at the moment, and finally in order to see how hot the hottest DB below the gap really is. The purpose of this short paper is to discuss the models and to describe how they influence the questions mentioned above. The reader interested in the full technical details of our investigation is referred to Thejll, Vennes and Shipman (1990).

2 The models - and their sensitivity to input physics.

We calculated a grid of models between 19000K and 30000K, with log(g) at 7.5, 8.0 and 8.5, and at various (small) H abundances, as well as with changes in the input physics, such as the mixing length theory, the treatment of the equation of state, the influence of convergence errors in the model calculation stage and the importance of choosing different continuous opacity sources (bound-free and free-free) as well as bound-bound opacities. The models are all chemically homogenous, line-blanketed, plane parallel, LTE models in hydrostatic equilibrium, with radiative and convective energy transport. We used our version of Kurucz' (1970) code ATLAS.

We chose to measure the sensitivity of the models to changes in the inputs by measuring the change in effective temperature that would be found if a standard spectrum was fitted to the calculated IUE flux using least-χ^2 fitting after normalization at the V-magnitude.

We found that the influence of slightly different input physics was quite small. In the worst case the errors can add up, if the largest possible error has been made in each choice, to an uncertainty in the fitted effective temperature of ±500K.

3 Comparing models

We compared our grid of DB models to others, published and unpublished. The grid presented by Wegner and Nelan (1987) has good agreement at 20000K/log(g)=8.0, the differences being so small that the effective temperature deduced using our models or Wegner and Nelan's would not be different by more than 500K, in agreement with the error estimate above.

The agreement between models given to us by Detlev Koester (1989, unpublished) and Francois Wesemael (1990, unpublished) of the 25000K/log(g)=8.0 blanketed model was extremely good, the T_{eff} difference being less than 200K.

However, when we compared our models to the cooler models of Wesemael (1981) - from now on denoted W81 in this paper - we found that there was a large disagreement between out 25000K and 30000K models and Wesemael's blanketed models. In figure 1 we show the difference in the calculated fluxes and in figure 2 we compare the actual model structures for our and Detlev Koester's 25000/8 model. We see that our, Francois Wesemael's and Detlev Koester's recent models agree very well, but that the W81 model does not agree with either. The flux in the new models is higher than in the W81 model.

We were unable to track down the difference in the models, as is Francois Wesemael, but we all agree that there is something wrong with the 25000K/log(g)=8, LTE, blanketed model of W81. There is furthermore evidence that at least the 30000K/8, LTE, blanketed model of W81 also is in error. While we have no proof that our and Detlev Koester's and the new 1990 Francois Wesemael model are correct we can show by integrating the emitted flux of the tabulated W81 25000K/log(g)=8 model that it is wrong. The emitted flux does not add up to $\frac{\sigma T_{eff}^4}{4\pi}$ as it must in a converged model.

From comparisons that others have made (see Barstow 1990, for an example) between their models and the hotter ones of W81 we know that the hotter of the W81 models have been found to be fine, to the extent that comparisons have been made.

At least one piece of important work has been based on the cool W81 models and we must examine what new results emerge if our new model grid is used to interpret observational data.

4 The DB and DBV temperature scale

In Liebert et al. (1986) the available IUE data for the DB's and the then newly discovered DBV's were analysed and temperatures were derived by fitting the IUE flux to model fluxes using the W81 models and other model grids as well. Effective temperatures derived from fits to the optical spectra were available and the different temperatures were compared and discussed. In general the IUE temperatures based on W81 models were higher than those derived from other model grids as well as from optical data. We can now understand this; if the flux is too low in the W81 models, compared to other models, then a hotter model must be chosen from the W81 grid to fit the observed IUE continuum slope.

In Liebert et al. (1986) the emphasis was placed on the temperature scale based on the W81 models. The non pulsating stars just above and below the range of the DBV's set the limits on this instability strip. Based on L86 it appeared to lie

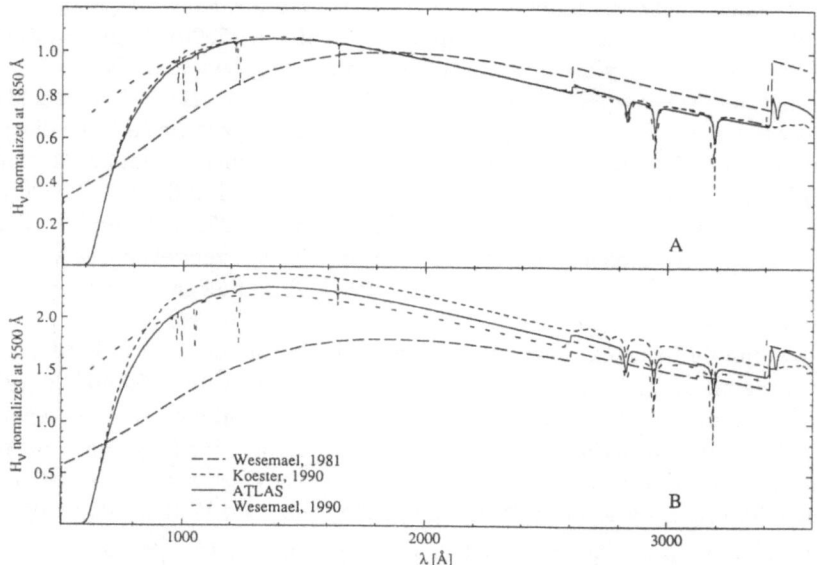

Figure 1: UV fluxes in $erg\ cm^{-2}\ s^{-1}\ Hz^{-1}\ sr^{-1}$ from various models at 25,000K, log(g)=8.0. a) All spectra normalized at λ=1850 Å. b) All spectra normalized at λ=5500 Å. The Wesemael (1981) model gives the lowest flux, while the other models agree. The discrepancies shortwards of 1000 Å is due to the inclusion of the strong HeI resonance lines in some models.

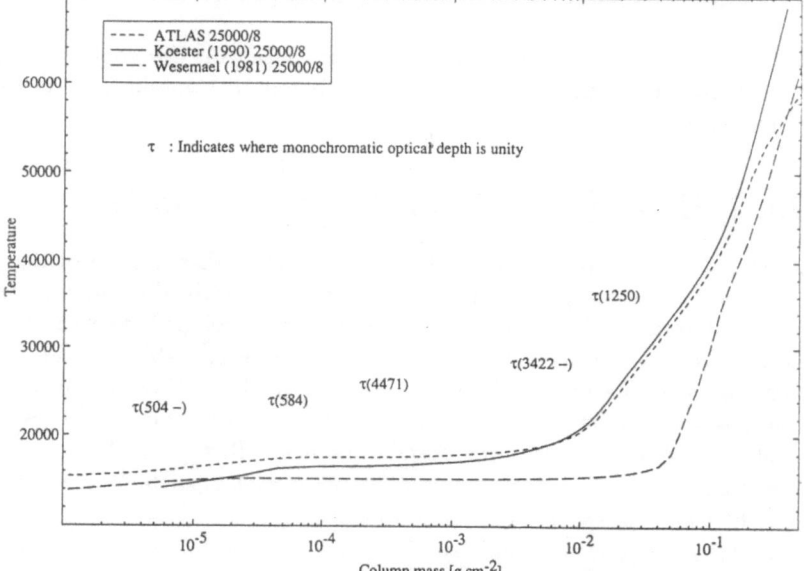

Figure 2: Comparison of the structures of the models from figure 1. The Wesemael (1981) model is noticeably different from the ATLAS model and the Detlev Koester (1990) model. T_{eff}=25,000K, log(g)=8.0.

between 24,000 and 29,000; based on our new models it lies between 24000K and 21500K - both ranges with uncertainties of at least ± 1000K at either end. What is the consequence of this lowering of the DBV temperature range?

The pulsations in the DBV's take place at the bottom of the surface convection zone under conditions where the thermal time scale is that of possible oscillations. Since the thermal structure of the envelope is governed by such things as the convective efficiency we get a coupling between theories of convection and theories of pulsations. The oscillations are well understood but the theory of convection is less so, thus a way of calibrating convection theories exists. A common theory of convection is the mixing length theory, where the convective efficiency - and other assumptions - are parametrized (see Fontaine, Villeneuve and Wilson (1981) for a description of mixing length theories). By matching the predicted thermal time scales against the observed window of pulsations, Fontaine et al. (1984) sought to determine which of the mixing length theories ML1, ML2 and ML3 were favored by observations. They found that the convectively efficient ML3 theory matched best.

Now that we have lowered the temperature range of the DBV's we find no such match and ML2 seems to give a better fit. This is one of the major results of our redetermination of the DB and DBV temperatures. Figure 3 shows the old and new DBV temperatures. It is interesting to note that the two BPM stars in the new gap have been observed by Kepler (1990, private communication) to be stable. The size of the error bars on the temperatures are so large that it is not evident if these stars fall in, above or below the DBV strip. They appear to be at the lower end of the strip. The star forming the present lower limit (1654+160) has data of such low quality that we have not attempted to give an error estimate on T_{eff}. BPM 17088 and 70524 may be better candidates for definition of the lower end of the instability strip since they both have data that has allowed the calculation of meaningful error bars on T_{eff}. GD190, a stable DB, with the new temperature 23,000±1,500K is also of interest in this context as it may represent the upper end of the instability strip.

The other result is that the hottest DB below the so-called DB gap - in which no DB stars are observed - is now at 28,000K, and not at 30,000K as was the case before. Widening the gap a little has thus added weight to the viewpoint that the gap is really there - it is not based on a statistical fluctuation.

5   GD 358

The first DBV found was GD358 (1645+32, EG239). Koester et al. (1985) found that they could fit the spectrum of GD358 with models having T_{eff}=24000± 1000K, log(g)=8±0.3, The temperature is completely in agreement with our IUE temperature re-determination presented in this paper. In Liebert et al. (1986) Koester models were used by the authors of that paper to fit the IUE flux of GD358, as well as with the W81 models. They found T_{eff}=28000K±1000K based on the W81 mod-

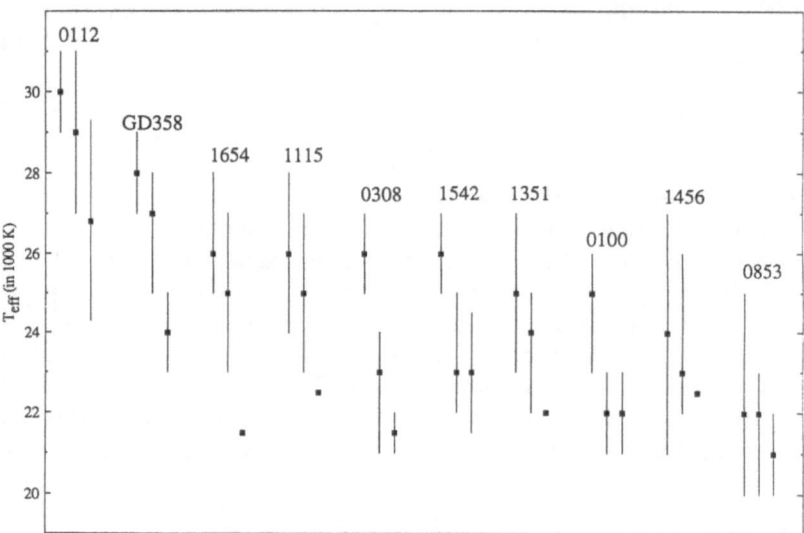

Figure 3: Effective temperatures of some DB and all the DBV's using different models. In each group of three points the leftmost is the temperature due to the Liebert *et al.*(1986) work, using the Wesemael (1981) models, the second bar is from the same work, using Koester models, and the last bar is from this work using our new grid of DB models. '0112'= PG0112+104, 'GD358'= 1645+325, '1654'= PG1654+160, '1115'= PG1115+158, '0308'= BPM17088, '1542'= GD190= 1542+182, '1351'= PG1351+489, '0100'= BPM70524, '1456'= PG1456+103, '0853'= LB8827= 0853+163.

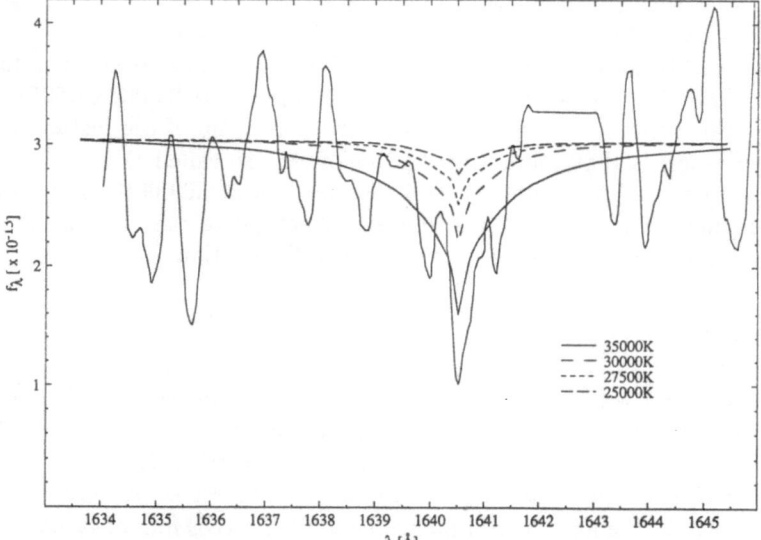

Figure 4: The 1640 A feature in GD358 reported by Sion *et al.*(1988). Superimposed are line profiles from our model grid. The feature does not have the shape predicted by our photospheric models.

els and $T_{eff}=27,000K^{+1000}_{-2000}$ based on the Koester models. We have focused special interest on this object in view of the new grid of models. Sion et al. (1988) reported that there were signs of a HeII feature at 1640 Å and that CII could be seen at 1335 Å in the high resolution IUE images SWP31432 and SWP33681. With the best fit to the IUE continuum flux level that we get, the models near the derived T_{eff} of 25000K do not show any sign of a feature with as strong a core as the one reported by Sion et al. at 1640 Å. In figure 4 we show the reported feature as reproduced in Sion et al. (1988) with model profiles superimposed. There is no reasonable fit between the models and the observations, so we conclude that the feature is not real or not of photospheric origin. We speculate that a sharp-cored feature might be formed in material close to the star but not in the actual photosphere.

Our new temperature for GD358 rules out the suggestion by Sion et al. that T_{eff} could be as high as to place GD358 in the DB gap.

We wish to thank all the people who provided us with support during this work, and especially those people who gave us models for comparison; Detlev Koester, Francois Wesemael, Sylvie Beaulieu at Johns Hopkins and Franziska Eber in München. We gratefully acknowledge support from NSF (grant AST-8720530), NASA (grant NAG 5-348) and the Danish Research Academy (grant 880070) as well as NORDITA for providing support for the trip to Toulouse. And many thanks to Jim Liebert for the wine!

6 References

Barstow, M.A., (1990) 'LTE model atmospheres for hot, helium-rich degenerate stars', M.N.R.A.S., 243, 182-191.

D'Antona, F., (1988) 'Evolution of White Dwarfs: Starting from planetary nebulae' in IAU Colloqium 114: 'White Dwarfs', ed. G. Wegner, Berlin: Springer Verlag, p.44.

Fontaine, G., Tassoul, M., and Wesemael, F. (1984) 'On the calibration of the mixing-length theory in pulsating white dwarfs' in Proceedings of the 25th Liege International Astrophysical Colloqium: 'Theoretical problems in stellar Stability and Oscillations', Liege: Universite de Liege, p. 328.

Fontaine, G., Villeneuve, B., and Wilson, J., (1981) 'On the acoustic flux of Sirius A', Astrophysical Journal, 243, p.550.

Koester, D., Vauclair, G., Dolez, N., Oke, J.B., Greenstein, J.L., and Weidemann, V. (1985) 'Atmospheric parameters of the variable DB white dwarf GD358', Astronomy and Astrophysics, 149, 423-428.

Kurucz, R.L., (1970) Smithsonian Ap. Obs. Report No 309.

Liebert, J., Wesemael, F., Hansen, C.J., Fontaine, G., Shipman, H.L., Sion, E.M., Winget, D.E., and Green, R.F. (1986) 'Temperatures for hot and pulsating DB white dwarfs obtained with the IUE observatory',Astrophysical Journal, 309, p.241.

Sion, E.M., Liebert, J., Vauclair, G., and Wegner G., (1988) 'The detection of ionized helium and carbon in the pulsating DB degenerate GD358' in IAU Colloqium 114:'White Dwarfs', ed. G. Wegner, Berlin:Springer Verlag, p.354.

Thejll, P., Vennes, S., and Shipman, H.L., (1990, in press) 'A critical analysis of the UV temperature scale of the Helium-dominated DB and DBV White Dwarfs', Astrophysical Journal.

Wegner, G., and Nelan, E.P.,(1987) 'Ultraviolet and visual spectroscopy of DB white dwarfs', 319, p.916.

Wesemael, F., (1981) 'Atmospheres for hot, high-gravity stars II. Pure Helium models', Astrophysical Journal Supp. S., 45, p.177.

DISCUSSION

KEPLER :

I have observed bothBPM stars in the instability strip and they are not variables, so there are non-variables in your "new" instability strip.

D'ANTONA :

Your results which shift the DB variables to lower T_{eff} actually are in support of the new Canuto and Mazzitelli turbulent convection theory. At T_{eff} smaller than 25000 K, CM theory begins to be more efficient than the classical mixing length, or, at least, the behaviour of convective gradient is such that we get larger convective masses than in the classical MLT.

FONTAINE :

I am somewhat surprised that the sensitivity of your results on the assumed convective efficiency is so small. Is this because convection starts relatively deep into your model atmospheres?

THEJLL :

The convection affects the temperature structure from mass-loading $\sim 10^{-1}$ g.cm and down into the photosphere, at 25000 K/log(g) = 8. This is below $\tau_v = 1$ at all v.

STANGHELLINI :

Your results lower the red and blue edges of the DB instability strip. This is in agreement with our non-adiabatic results, that allow for colder pulsators (at the red edge of the strip) than previously derived from the observations.

THE MODIFIED HYDROSTATIC EQUILIBRIUM EQUATIONS FOR STRATIFIED HIGH GRAVITY STELLAR ATMOSPHERES

K. UNGLAUB, I. BUES
Remeis - Sternwarte Bamberg
Sternwartstr. 7
D-8600 Bamberg
Federal Republic of Germany

ABSTRACT. A model atmosphere code for hot high gravity stars is in progress which takes into account the chemical stratification of hydrogen, helium and carbon. The normal hydrostatic equation is splitted into one equation for each species. These equations are coupled by the electric field and the ionization terms. The local electric field is obtained from the charge neutrality condition, the ionization terms are computed according to the method by Wildt(1936, 1937), solved analytically. Integration of the hydrostatic equilibrium equations from an inner layer outwards yields the run of partial pressures for each species with geometrical depth.

1. Introduction

Stratified model atmospheres for white dwarfs made up of hydrogen and helium have already been computed, e.g. by Jordan and Koester (1986). They obtained an abundance profile of hydrogen over helium as a function of gas pressure and the total hydrogen mass on top of a helium rich layer. The hydrostatic equation has been integrated from outer regions inwards with a grey or scaled temperature stratification of a flux constant model. The iteration scheme for flux constancy was a Feautrier - type method.

Our aim now is to extend calculations to a model atmosphere of hot high gravity stars of various elements, where chemical stratification of elements directly is taken into account in the hydrostatic equation.

2. Methods

The normal hydrostatic equation may be splitted into one equation for each species k of particles:

$$\frac{dp_k}{dr} = p_k * (\frac{-A_k \, m_p \, g}{k \, T} + \frac{Z_k \, e \, E}{k \, T} + \phi_k + \frac{F_k}{k \, T}) \tag{1}$$

Here p_k is the partial pressure of particle species k, A_k the atomic weight, Z_k the charge. E is the local electric field, Θ_k is the ionization term of particle species k. F_k is the sum of all forces besides gravitational and electrical.

In a mixture made up of hydrogen and helium six different kinds of particles are present: H, H^+, He, He^+, He^{++} and free electrons. Therefore a system of six equations is required for the computation of partial pressures, for a composition of helium and carbon 11 equations would

G. Vauclair and E. Sion (eds.), White Dwarfs, 267–274.
© 1991 Kluwer Academic Publishers.

be necessary. By integration of this system of equations the partial pressure of each kind can be obtained as a function of geometrical depth. Forces other than gravitational and electrical can also be taken into account, e.g. selective radiative forces which act only on one species. An example would be the radiative force acting on C^{3+} due to the absorption in the strong resonance line at 1548 Å.

The hydrostatic equations for each species are coupled by the electric field and the ionization terms. The electric field can be computed from the charge neutrality condition, which is e.g. for a mixture of hydrogen and helium:

$$\frac{dp_e}{dr} = \frac{dp_{H^+}}{dr} + \frac{dp_{He^+}}{dr} + 2\frac{dp_{He^{++}}}{dr} \tag{2}$$

To obtain the local electric field the hydrostatic equilibrium equations (1) are inserted into equation (2). The resulting equation is solved for the electric field.

The ionization term for each species multiplied by its partial pressure can be understood as a pressure gradient due to ionization effects:

$$\phi_k \, p_k = (\frac{dp_k}{dr})_{IONIZATION} \tag{3}$$

The hydrostatic equilibrium of the whole mixture remains unchanged by ionization effects, these terms only guarantee the ionization equilibrium:

$$\sum_k p_k \, \phi_k = 0 \tag{4}$$

The equilibrium of a stellar atmosphere made up of a mixture of several gases in chemical equilibrium has already been studied by Wildt (1936, 1937). Wildt's method has been applied to a mixture of partially ionized hydrogen and helium by Montmerle and Michaud (1976). They show explicitly the linear system of equations from which the ionization terms can be computed (their Appendix B, equation B 17). For our purposes of a composition of carbon and helium this system which consits of 11 equations had to be solved. We used Kramer's rule and obtained an analytic solution.

Now the hydrostatic equilibrium equations can be integrated. To obtain self - consistent model atmospheres the transfer equation has to be incorporated into the code or a temperature iteration procedure has to be applied. Previous estimates about the temperature structure in the atmosphere can be obtained from our flux - constant, chemically homogenous model atmospheres of various compositions. For hot high - gravity stars such models have been computed with Lucy's iteration scheme for abundance ratios of C/He = 10.0, pure helium composition, He/H = 0.03 and He/H = 0.1. In the deepest layers the gas pressures are about $\log P_G = 8.0$ (SI) and the temperatures are in the range from about 100000 K to 150000 K.

For our stratified atmospheres the hydrostatic equilibrium equations are integrated from an inner layer outwards, where a plane - parallel stratification is assumed. The total atmosphere is subdivided into about 10000 intervals, the thickness Δx of one interval is 1 m for a white dwarf atmosphere. The integration variable is the geometrical height x from the bottom layer.

If the partial pressures of all kinds of particles in an integration interval i are known, the local electric field and the ionization terms can be computed. Then the simplest method to obtain the partial pressure of species k in the interval i+1 is to use an integration method which is linear in the step - width Δx (see equation (5)). It could be improved by using a taylor series, for example. Taking into account terms up to fourth order would permit to use a greater step width

$$p_k(l+1) = p_k(l) + \alpha(l) \, p_k(l) \, \Delta x \tag{5}$$

$$\alpha(l) = -\frac{A_k \, m_p \, g}{k \, T(l)} + \frac{e \, E(l)}{k \, T(l)} + \phi_k(l) + \frac{F_k(l)}{k \, T(l)}$$

of about 50 m. But this poses some problems, e.g. to obtain the local electric field an equation of fourth order has to be solved then.

2.1. INPUT PARAMETERS

1) The total stellar mass. The results in the next section have been obtained with $M_* = 0.5 \, M_\odot$
2) The total mass of the atmosphere, here $M_{atm} = 10^{-12} \, M_*$.
3) The gravity of the star, here $\log g = 8$.
With a plane - parallel stratification and an atmospheric mass negligible small compared to the total stellar mass the radius of the star and the gas pressure at the bottom of the atmosphere can be obtained from this three parameters.
4) The temperature structure of the atmosphere. A temperature of 120000 K has been assumed in the bottom layer, as estimated from the chemically homogenous model atmospheres. Then the temperature is assumed to decrease linearly with geometrical height. So at a height from the bottom interval of x = 5.5 km the temperature is about 50000 K, at x = 8 km the temperature is about 20000 K. The computation of self - consistent model atmospheres is a future project, the intention here is to test the code available up to now and to show more qualitative results.
5) The chemical composition at the bottom of the atmosphere.
From the gas - pressure, temperature and chemical composition in the bottom interval the partial pressures of all species are computed and the integration of the hydrostatic equilibrium equations starts.

2.2. ABSORPTION COEFFICIENTS

The absorption coefficients of H, He, He^+, C, C^+, C^{2+}, C^{3+}, C^{4+} and free electron scattering are explicitly taken into account and the Rosseland mean value is computed. From the absorption coefficients per heavy particle in each integration interval the monochromatic and Rosseland - mean optical depths are obtained by integration from the outermost layer inwards.

3. Results

In Fig. 1a the run of partial pressures with geometric height from the bottom interval outwards is shown for a mixture made up of hydrogen and helium. The helium to hydrogen ratio in the bottom interval is 100 here. The temperature decreases from 120000 K in the bottom interval to about 7000 K in the outermost region.
The partial pressure of ionized hydrogen H^+ first increases outwards, because here the electric force exceeds the gravitational force acting on a proton.
The partial pressure of neutral hydrogen H first decreases because of the gravitational force, but in the outer layers at a temperature of about 15000 K the ionization term of H dominates the gravitational term and the partial pressure increases to a point where hydrogen becomes neutral.
The partial pressures of He^+ and He^{++} in the bottom layer are in the same order of magnitude. In general, the partial pressures of helium ions decrease faster than those of

hydrogen, because the gravitational force acting on helium is four times higher than the gravitational force acting on a proton. At a temperature of about 50000 K the singly ionized state of helium becomes dominant over the doubly ionized state.

a)  b)

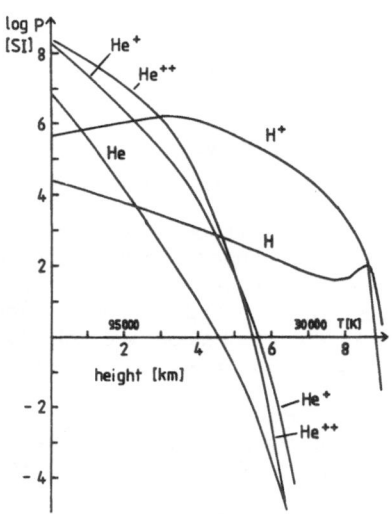

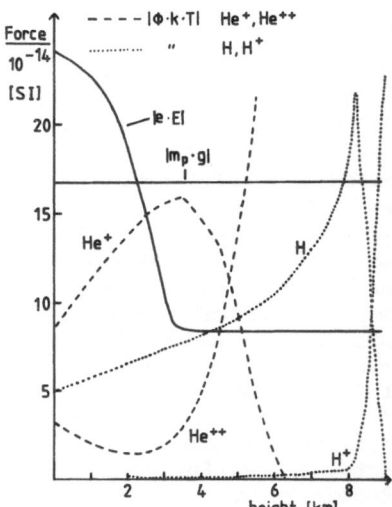

Figure 1. a) Run of partial pressures with height starting at the bottom layer for a mixture of hydrogen and helium. b) Comparison of gravitational and electric forces acting on a proton with the "ionization force" acting on H, H^+, He^+, He^{++}.

In Fig. 1b the various forces appearing in the hydrostatic equilibrium equations are compared. The gravitational force acting on a proton is constant throughout the atmosphere for a given gravity. The electric force is largest in the inner regions, where helium and hydrogen are highly ionized. In the outer regions with hydrogen as the dominant element, the electric force has half the value of the gravitational one (neglecting the electron mass). Note that the gravitational and the electric force acting on a proton have an opposite sign.

The ionization term of H^+ is negligibly small, except in the outermost regions, where it rapidly increases. If an element is preferably in one ionization stage (e.g. hydrogen in the interior), the ionization term of the dominant ionization stage becomes very small. But in the outer layers where almost all hydrogen is neutral, the ionization term of H^+ becomes very large. With the same sign as the gravitational term it causes the partial pressure of H^+ to vanish. For helium the partial pressures of He^+ and He^{++} are in the same order of magnitude in the hotter regions. Therefore none of the corresponding ionization terms is negligible small. At a height of $\approx$ 6 km, however, helium is mainly in the singly ionized state and therefore the ionization term of He^+

a)

b)

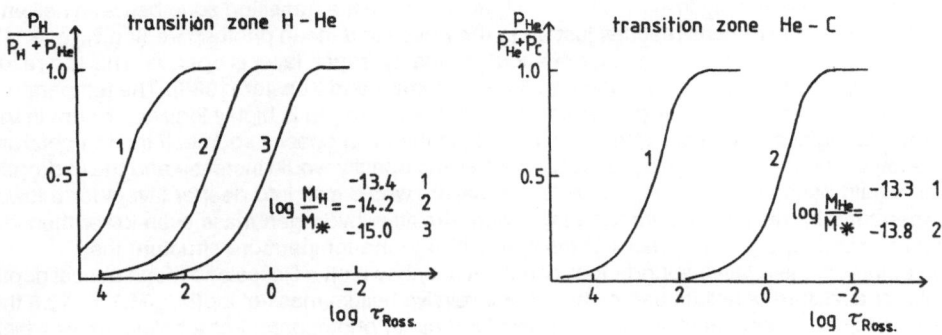

Figure 2. Relative abundances on a Rosseland mean optical depth scale for a) mixtures of hydrogen and helium and b) mixtures of helium and carbon. Parameter is the relative hydrogen or helium mass in the atmosphere, respectively. The total stellar mass is $M_* = 0.5\,M_\odot$.

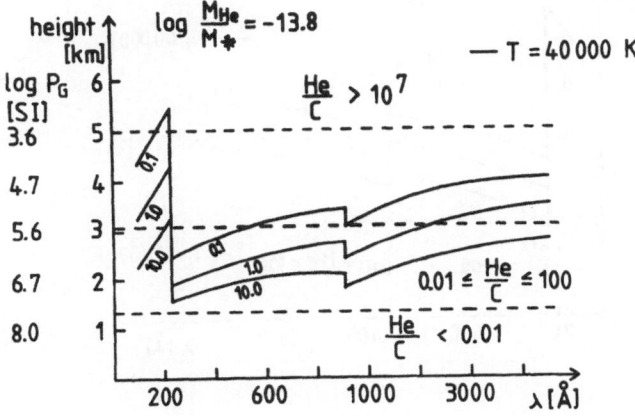

Figure 3. Lines of constant monochromatic optical depth for $\tau = 0.1$, 1.0 and 10.0, respectively, for a mixture of helium and carbon. *Ordinate:* geometric height from the bottom layer; also indicated are the corresponding gas pressures. *Abscissa:* wavelength. The relative helium mass in the atmosphere is $\log(M_{He}/M_*) = -13.8$, where $M_* = 0.5\,M_\odot$. With dotted lines the regions are indicated where the abundance ratios helium to carbon are He/C $\leq$ 0.01, 0.01 $\leq$ He/C $\leq$ 100 and He/C $\geq 10^7$, respectively. The temperature is 120000 K in the bottom layer, the height is indicated where the temperature is $T = 40000$ K.

goes to zero.

In Figure 2a the relative abundances of hydrogen are shown on a Rosseland mean optical depth scale. Parameter is the ratio of hydrogen mass to total stellar mass $\log(M_H/M_*)$. It can be seen that for a relative hydrogen mass $\log(M_H/M_*) = -15.0$ the transition zone between helium - and hydrogen rich regions reaches just up to the Rosseland mean photosphere ($\log \tau_{Ross} \approx -1$), whereas for a relative hydrogen mass of about -14 the hydrogen layer is opaque. This is a result which is already known from previous works, e.g. Jordan and Koester (1986). The temperature and gas pressure in the hydrogen layer are such that hydrogen is highly ionized, except in the outermost regions. Therefore electron scattering is the main opacity source. If the temperature were lower and hydrogen were preferably neutral, the opacity would increase and the hydrogen layer would become more opaque. Moreover helium would sink into deeper layers for a lower temperature. In this case a hydrogen layer with a relative hydrogen mass even lower than -15 could become opaque. The result is more sensitive to the temperature structure then.

In Figure 2b the relative abundances of helium are shown on a Rosseland mean optical depth scale for a mixture of helium and carbon. For a relative helium mass of $\log(M_{He}/M_*) = -13.8$ the transition zone clearly reaches into the Rosseland mean photosphere. For a helium mass which is 5 times higher the helium layer is opaque. Helium is mainly singly ionized througout the atmosphere.

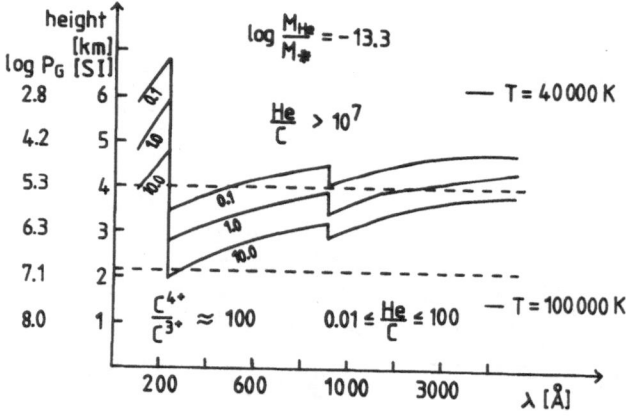

Figure 4. The same as figure 3 for a relative helium mass $\log(M_{He}/M_*) = -13.3$. The heights are indicated where the temperature is 100000 K and 40000 K, respectively. The ratio of partial pressures of C^{4+} and C^{3+} is about 100 in the deepest layers.

In figures 3 and 4 lines of constant monochromatic optical depths are plotted for $\tau = 0.1$, 1.0 and 10.0, respectively. The strong absorbtion edges of He^+ at 228 Å and 912 Å can clearly be seen. A transition zone between carbon rich and helium rich composition is defined by an abundance ratio helium to carbon $0.01 \leq He/C \leq 100$. For a relative helium mass of -13.8 (figure 3) the helium rich layer is transparent for a wavelength region between the absorption edge of He^{++} at 228 Å and about 1500 Å. Here the monochromatic optical depth of the upper boundary of the transition zone is clearly smaller than 1.0.

For a relative helium mass of -13.3 (figure 4) the helium rich layer is opaque at all wavelengths. Apart from a wavelength region between 228 Å and about 500 Å the helium to carbon ratio is even smaller than 10^7 in the outer atmospheric regions, where the monochromatic optical depths are $\tau \le 0.1$. Here carbon is only a trace element in the line - forming region.

The run of gas pressure with geometrical height is indicated at the ordinates of figures 3 and 4. The temperature is 120000 K in the bottom layer. In figure 4 are also indicated the heights where the temperature is 40000 K and 100000 K. Carbon is preferably in the fourfold ionized state. The ratio of the partial pressures of C^{4+} and C^{3+} is about 100 in the inner regions. This value does not change significantly up to the outer regions. A future project is to include radiative forces into the hydrostatic equilibrium equations and to investigate whether they can prevent carbon from gravitational settling.

4. Discussion

The results obtained up to now could as well be obtained with the method used by Jordan and Koester (1986). The aim is to improve the computation of model atmospheres and envelopes, so that changes of element abundance with depth can be taken into account directly in the hydrostatic equation. So detailed depth - dependent abundances of elements could be obtained. With this method forces other than gravitational and electrical can be included quite easily. Detailed radiative accelerations on e.g. carbon in model envelopes and atmospheres of hot subdwarfs have been computed e.g. by Bergeron et al. (1988), where a chemically homogenous composition of the atmosphere has been assumed. Our intention is to do similar calculations for hot white dwarfs and subdwarfs, where inhomogenous element distributions are taken into account.

*Acknowledgements.*Thanks go to the Regionales Rechenzentrum Erlangen, where the computations were performed on a CYBER 995 computer.

References

Bergeron, P., Wesemael, F., Michaud, G. and Fontaine, G. (1988) 'Studies of hot B Subdwarfs VI. Detailed calculations of radiative forces on metals in the envelopes and atmospheres of hot hydrogen - rich Subdwarfs', Astrophys. J. **332**, 964

Jordan, S., Koester, D. (1986) 'Model atmospheres and synthetic spectra for White Dwarfs with chemically stratified atmospheres', Astron. Astropys. Suppl. Ser. **65**, 367

Montmerle, T., Michaud, G. (1976) 'Diffusion in stars: ionization and abundance effects', Astrophys. J. Suppl. Ser. **31**, 489

Wildt, R. (1936) 'Equilibrium of stellar atmospheres under a temperature gradient', Astrophys. J. **83**, 202

Wildt, R. (1937) 'Note on stellar ionization and electric fields', Monthly Notes Roy. Astron. Soc. **97**, 225

DISCUSSION

VAN HORN :

I have two questions : (1) Can you do fully self-consistent calculations that solve the radiative transfer equations -and thus get the temperature structure -together with your hydrostatic equilibrium equations? (2) Do you also plan to do models with constant mass-flow rates, thus determining the composition of the outflowing winds?

UNGLAUB :

1) This is a future project. Up to now an approximate temperature structure is taken from chemically homogenous model atmospheres. But this does not take into account feed-back effects between temperature structure and chemical composition.

2) First static model atmospheres shall be computed. With self-consistent models of various abundance ratios mass flow can be considered in a next step. The method of using one hydrostatic equation for each species of particles should be appropriate to determine the composition of the out flowing winds.

THE DBAQ G35-26

Peter Thejll[1], James MacDonald and Harry L. Shipman
Department of Physics and Astronomy, University of Delaware,
Newark, DE 19716, USA

ABSTRACT. The DQ G35-26 is spectroscopically analyzed and the results discussed in the context of dredge-up of carbon and the accretion of hydrogen. The best fit is T_{eff}=13,000K with firm limits between 11,000K and 14,000K, log(g) between 9.0 and 9.5. The hydrogen abundance is close to 1% by number and the carbon abundance close to 3%. All other elements have less than solar abundance.

1   Introduction

The star G35-26 (Gr 469, WD0203+207) has for some time been known in the literature as an interesting object due to its peculiar spectrum: It shows hydrogen lines with a strong Balmer decrement along with C I lines and a suggestion of a HeI line at 4471 Å. Before the C I lines had been identified by Liebert (1983) the features where thought to be Zeeman components of the Balmer lines. Better spectroscopic data unveiled the true nature of the features.

The strong Balmer decrement is another interesting feature: The β and γ lines are clearly visible but the δ line is nearly absent - how can such a decrement be explained?

G35-26 is - apart from the hydrogen - a classical example of 'dredge-up' of carbon in cool DB's and for some time the spectroscopic data needed to carry out an abundance analysis has been available in the literature. We report here the results of such an analysis and comment upon two interesting features of the results: The best fitted gravity is apparently greater than what is usual for DB's and the hydrogen abundance is quite large. The reader interested in the full technical details of this investigation is refered to Thejll *et al.* (1990).

2   The analysis

Based on the spectra published by Hintzen and Jensen (1979) and Liebert (1983), as well as the multichannel data published by Greenstein in 1978 we have found the best fit between the, admittedly not very good, data (the object is faint, V=17.5,

---

[1]present address: NORDITA, Blegdamsvej 17, DK-2100 Copenhagen Ø, Denmark

G. Vauclair and E. Sion (eds.), White Dwarfs, 275–283.
© 1991 Kluwer Academic Publishers.

Table 1: The observed elemental abundances in G35-26 compared to the solar abundances. Abundances relative to the total particle density have been calculated from the abundances given by Allen (1973).

	$n(x)/n(tot)$	$\frac{[n(x)/n(tot)]_{G35-26}}{[n(x)/n(tot)]_{\odot}}$
H	0.005-0.01	$\sim 10^{-2}$
C	0.03	$\sim 10^{2}$
He	0.97	~ 12
N	$< 10^{-4}$	< 1
O	$< 10^{-3}$	< 1
Ca	$< 10^{-6}$	< 1

hence the low quality of the data) and homogeneous, convective, LTE models in hydrostatic equilibrium, blanketed by H and C I and He I lines.

The models were calculated using Kurucz' program ATLAS (1970) where we included all important sources of opacity; bound-bound transitions for H, C I, C II, He I as well as several lines of O I, N I, Ca I and Ca II - UV resonance liens as well as optical. O, N and Ca lines are not seen in the spectra and therefore serve merely to set upper limits on the abundances of those elements. We include broadening of the H lines by Van der Waals interactions due to neutral Helium as well as neutral carbon. Continuous sources of opacity from Rayleigh scattering in H, He and C I as well as bound-free and free-free opacity in C^- has been added to those already in ATLAS.

Since the observational data does not have homogenous quality it has not been possible for us to arrive at a fit with quality-of-fit information. We can merely report where the best fit is - T_{eff} is between 11,000K and 14,000K, log(g) is between 9.0 and 9.5, n(C)/n(tot)=3% and n(H)/n(tot)=1% with the rest of the composition being made up of helium. O is less abundant than 1 in 10^3, N is less abundant than 1 in 10^4 and Ca is less abundant than 1 in 10^6 - this is less than what is seen in the Sun. See table 1.

3 Other data

The gravity above is contested by the existing trigonometric parallax of 0.0060" ± 0.0054" which has been measured by C.C. Dahn and collaborators (1990, private communication). The gravity that follows from assuming this parallax does not agree with the high value we have found. log(g)=9.0 corresponds to a 2 σ deviation of the parallax from its measured value while log(g)=9.4 corresponds to 3 σ. As explained above we cannot state how significant our gravity determination is.

The existing proper motion of 0.34"/year measured by Greenstein (1978) lends indirect support to a high gravity. At constant apparent magnitude the gravity of a star must increase as it is brought towards the observer to make the radiating area smaller. The measured proper motion, combined with the distance derived from the parallax gives the rather high tangential velocity of 263 km/s. This is not typical for DB's. Sion *et al.* (1988) quote the value $\approx 57 \pm 37$ km/s as being the mean for other DB's and $\approx 87 \pm 100$ km/s for DQ's. DQ's are thus known as a group with larger than average space velocities, but G35-26's velocity is still a lot above the average for DQ's. If the gravity lies in the range from log(g)=9.0 to 9.5 we get the more reasonable value of 72 to 95 km/s. Alternatively one might assume that G35-26 is a member of the galactic halo population and is on its way through the galactic plane; the angle of the proper motion with respect to ecliptic North is roughly 220°, counted positive East of North, and while the actual angle between the velocity vector of G35-26 and the galactic plane cannot be deduced from the position angle of the proper motion without knowing the radial velocity, it does show that G35-26 certainly does not move perpendicularly to the plane - the angle is no more than 30° below the galactic plane.

So we have several pieces of independent contradictory information. A more accurate parallax is not easy to come by due to the field of stars that G35-26 is in - the field is not rich enough to allow the use of a CCD detector at the Naval Observatory and less accurate photographic plates must be used. Better spectra should be easy to come by with the investment of one night of observation on a large (4m or more) telescope. A proper motion would be very useful to have but the star is very faint, making it hard to get high resolution spectra of a line core, and the photosphere is dominated by helium - Doppler shifts are very hard to distinguish from Stark shifts in a helium photosphere.

4   The elemental abundances

The 3% of neutral carbon found in the analysis places G35-26 on the theoretically predicted relation between T_{eff} and carbon concentration. The dredge-up process is by now the accepted explanation of the phenomenon of DB stars with C - DQ's. As the object cools, the bottom of the surface convection zone moves down and starts to mix the end of the diffusion–distributed carbon tail into the photospheric layers of helium. The depth to which the convection zone reaches is sensitive to the effective temperature and as the temperature drops below 16,000K the zone moves down and a sharp increase in the observed C is predicted. G35-26 is close to the peak of the $n(C) - T_{eff}$ relation and by plotting it along with the other DQ's we see that, given the scatter, G35-26 does not contradict the other results. See figure 4 of Pelletier *et al.* (1986).

The 1% of hydrogen is another matter: where does this much hydrogen come from? Since we have not found any traces of O, N or Ca, as we would expect if the carbon was due to accreted grains of interstellar matter, we feel it is justified to say that the hydrogen has been accreted, with probably other materials in solar abundances - but only the H lines are strong enough to be seen. This explanation is consistent with the available data.

Is it possible to accrete that much hydrogen? Since the model photosphere is convective up to very small optical depths we can estimate the mass of hydrogen present in at least the convection zone, under the assumption that the material is homogenous and that no hydrogen is found below the convection zone. Roughly 10^{-9} $M_\odot$ of the star is in the convection zone and the number abundance of hydrogen is 1 in 100, giving a mass of hydrogen in the convection zone of about 2.5×10^{-12} $M_\odot$. If we allow all of the cooling time on the WD sequence for accretion - roughly 10^9 years - then the accretion rate would have to be on the order of $10^{-21} M_\odot/yr$ which is not an unreasonable rate, especially in view of the fact that there is virtually no UV flux below 911 Å with which to ionize the nearby ISM and thus increase the accretion rate - the purely geometric accretion rate expected is on the order of $10^{-21} M_\odot/\text{year}$. Of course, while the accretion theory is possible it does not answer the old question of why *all* DB's do not show this much H, or indeed much more. Perhaps various mechanisms, which depend on a stars magnetic field, the exact history of passage through the ISM etc, must be invoked to explain the differences seen between H abundances in DB's.

5   The carbon abundance and convective dredge-up.

The derived carbon abundance is near $n(C)/n(\text{tot}) = 0.03$. Models for the pollution of helium rich white dwarf atmospheres by convective dredge-up of carbon have been studied by Vauclair and Fontaine (1979), Fontaine *et al.* (1984) and Pelletier *et al.* (1986). Pelletier *et al.* find that the observed relationship between carbon abundance and effective temperature is that expected from the evolution of a 0.6 $M_\odot$ CO white dwarf with a $2 \times 10^{-4} M_\odot$ helium envelope. Although the carbon concentration for G35-26 fits in well with the observed $n(C)/n(He)$ - T_{eff} relation, the surface gravity of G35-26 may well be significantly higher than that of a 0.6 $M_\odot$ white dwarf and the results of Pelletier *et al.* (1986) cannot then be directly applied to G35-26. Instead, we have calculated stratified envelope models of mixtures of helium and carbon. The composition profile is determined from the balance between forces due to gravity, partial pressure gradients and induced electric fields. In addition, mixing in convection zones is treated as a diffusion process. Further details of the technique used for constructing the models can be found in MacDonald and Vennes (1990). In general, for 0.6 $M_\odot$, our results agree with the results of Pelletier *et al.* (1986).

The top of the convection zone in both the stratified envelope models and the homogeneous atmosphere models is at a Rosseland optical depth of ≈ 0.4. Hence we can make the assumption that the abundance ratios found from the best fitting homogeneous atmosphere model for G35-26 are representative of the abundance ratios at a Rosseland optical depth of unity in the stratified envelope. For the composition parameters of G35-26 and mass $1.2M_\odot$, we find a helium layer mass of 5×10^{-5} $M_\odot$. In addition, stratified envelope calculations, in which the diffusive equilibrium equations are solved, for mixtures of hydrogen, helium and carbon, give a hydrogen mass of $2 - 5 \times 10^{-12}$ $M_\odot$ and a helium mass of 3×10^{-5} $M_\odot$. Much of the hydrogen is mixed in the surface convection zone of mass $\approx 10^{-9}$ $M_\odot$.

The mass of helium found here for G35-26 can be compared to the minimum helium layer mass necessary to sustain nuclear burning by the triple alpha reaction on a CO core. For 1.2 $M_\odot$ Kawai, Saio and Nomoto (1988) find, for pure helium envelopes, a minimum helium layer mass of about $10^{-4}M_\odot$. Near the end of the Asymptotic Giant Branch phase of the pre-white dwarf evolution the helium envelope may contain up to 20% carbon due to convective dredge up (see e.g. Iben 1984) and the molecular weight will be higher than for pure helium. This results in a higher temperature in the burning shell for given helium mass and hence the total helium mass when nuclear reactions cease will be less than for the pure helium case. Thus there is no obvious inconsistency between the helium mass found from carbon dredge-up models for G35-26 and the helium layer mass that may be expected from stellar evolution models.

6 Is the high gravity true?

Before we go on to discuss the significance of high mass white dwarfs we should look at the confidence we have in the high gravity finding. Could we be wrong so that G35-26 really has a normal gravity of log(g)=8.0?

If we underestimate the Van der Waal's broadening of the H lines then we might well be forced to choose a high gravity model to make the Balmer decrement fit the observations. We have however included V. d. W's broadening due to all the important sources : neutral H broadening H, neutral He broadening He and neutral C broadening He. Our treatment of the Van der Waal's broadening is simple - it is based on the impact approximation (See Gray (1976) p. 231) - we also treat the influence of neutral C perturbers by scaling the amount of neutral helium. It turns out that one C atom is as efficient at perturbing H atoms as 1.7 He atoms. To check our work an independent approach should be tried.

G227-5 is another DQ, quite similar to G35-26 in the Balmer decrement - $H\gamma$ is the last clearly visible Balmer line. It has been analysed by Wegner and Koester (1985), but they held log(g) fixed at 8. Further analysis of high quality data by Greenstein is in preparation.

7 The evolution of massive WD's

In this meeting we have seen several other mentions of massive WD's - GD50 (Saffer, Bergeron *et al.*), WDHS1 (Schönberner) and in the literature EG56 (Koester, Schulz and Weidemann (1979), as well as Madej and Grabowski, (1990)) and recently 1E2259+586 (Paczynski, 1990). How do we expect white dwarfs with masses of $\log(g)=9.0$ and upwards to form?

A few scenarios exist to explain the creation of massive white dwarfs. 'Normal' WD's might merge with each other Iben and Webbink (1988) discuss qualitatively the evolution of close binary white dwarfs and it is clear that there is a rich field of possibilities for the outcome of WD mergers, but at the present there is not enough information to draw quantitative conclusions.

It is also possible for a single star to evolve into a massive white dwarf. Nomoto (1984), and others, have discussed the evolution of single stars into massive WD's. Nomoto's calculations show that it should be possible for main sequence stars with masses in the range from 8 to 10 $M_\odot$ to evolve into O+Ne+Mg white dwarfs with masses of 1.2 $M_\odot$ - these white dwarfs would have $\log(g)=9.0$ and above. The ratio of the number of stars on the main sequence with masses between 8 and 10 $M_\odot$ and 1 and 8 $M_\odot$ is roughly 1 to 300. About 1300 white dwarfs are known and we know of 5 WD with high mass - close to 1 in 300.

It is interesting to see that of the $\log(g)=9$ or above WD's that are known or suspected all but G35-26 have been DA's. Most WD's of any mass are DA's and it would be interesting to increase the number of known massive white dwarfs for the purpose of getting better statistics. One might look for massive white dwarfs in young clusters, but it may also be possible to find massive WD's in the existing body of data because it happens that people fitting a WD spectrum assume that $\log(g)$ is 8 rather than performing a fit in the $T_{\text{eff}} - log(g)$ plane. The sort of sophisticated spectral fitting techniques displayed at this meeting must become the standard in the future. If substantially more massive white dwarfs are found in the existing data it seems that evolution of single massive white dwarfs may not be able to produce sufficient numbers.

The tell-tale sign of a massive white dwarf is the sharp Balmer decrement, this is displayed in G35-26 and in GD50. Spectroscopic data of good quality is needed to detect and satisfactorily model this - we do not think it can be done with photometric data, but getting good data, and reviewing the existing data is worthwhile because massive white dwarfs provide yet another tool for understanding stellar evolution - both for single and binary systems - and G35-26, and the other DQ's, help us understand chemical evolution of cooling white dwarfs. Massive white dwarfs also give us a unique opportunity to study the very end of the Chandrasekhar mass-radius relation.

The authors wish to thank the following sources for support: NASA (grant NAG 5-972), NSF (grant AST-8720530), the Danish Research Academy (grant 880070),

281

and NORDITA.

8 References

Allen, C.W., (1973) 'Astrophysical Quantities', 3 ed, Athlone Press:London.

Fontaine, G.,Villeneuve, B., Wesemael, F., and Wegner, G., (1984), Astrophysical Journal Letters, 277, L61.

Gray, D. F., (1976) 'The observation and analysis of stellar photospheres', John Wiley & Sons, Inc. London.

Greenstein, J.L., (1978) Publ. A.S.P., 90, p.303.

Hintzen, P., and Jensen, E., (1979), Publ. A.S.P., 91, p.492.

Iben, I., Jr, (1984), in IAU Symposium 104, Observational Tests of Stellar Evolution Theory, ed. A. Maeder and A. Renzini (Dordrecht: Reidel), P.3.

Iben, I., Jr, and Webbink, R.F., (1988) 'On the formation of close binary white dwarfs' in IAU Colloquium 114: 'White Dwarfs', ed G. Wegner, Berlin: Springer Verlag, p. 477.

Kawai, Y., Saio, H., and Nomoto. K., (1988), Astrophysical Journal, 328, p.207.

Kurucz, R.L., (1970) Smithsonian Astrophysical Observatory Report Number 309.

Koester, D., Schulz, H., and Weidemann, V., (1979) 'Atmospheric parameters and mass distribution of DA white dwarfs', Astronomy and Astrophysics, 76, 262.

Liebert, J., (1983) Publ.A.S.P., 95, 878.

MacDonald, J., and Vennes, S., (1990), submitted to Astrophysical Journal.

Madej, J., and Grabowski, B., (1990) 'Significance of the pressure shift of Balmer lines in white dwarfs with extremely high gravity', Astronomy and Astrophysics, 229, p.467.

Nomoto, K., (1984) 'Evolution of 8-10 $M_\odot$ stars toward electron capture supernovae. I Formation of electron-degenerate O+Ne+Mg cores' Astrophysical Journal, 277, p.791.

Paczynski, B., (1990), 'X-Ray Pulsar 1E2259+586: A merged white dwarf with a 7 second rotation period?', preprint.

Pelletier, C., Fontaine. G., Wesemael, F., Michaud, G., and Wegner, G., (1986), Astrophysical Journal, 307, P.242.

Sion, E.M., Fritz, M.L., McMullin, J.P., and Lallo, M.D., (1988), 'Kinematical tests of white dwarf formation channels and evolution' Astronomical Journal, 96, p.251.

Thejll, P, MacDonald J, Shipman H.L., (1990) 'An atmospheric analysis of the carbon-rich white dwarf G35-26', Astrophysical Journal, 361:197-206.

Vauclair, G., and Fontaine, G., (1979), Astrophysical Journal, 230, p.563.

Wegner, G., and Koester, D., (1985) 'Atmospheric analysis of the carbon white dwarf G227-5', Astrophysical Journal, 288, 746-750.

DISCUSSION

BARSTOW :

COMMENT on IE 2259+586 :

Paczynski proposed that this is a recent WD merger with $\log g > 9$. I will be computing model atmospheres to investigate whether the observed X-ray flux is consistent with this idea.

LP 790-29: PRELIMINARY MODEL ATMOSPHERES FOR THIS STRONGLY POLARIZED CARBON WHITE DWARF

I. BUES
Remeis-Sternwarte Bamberg
Sternwartstr. 7
D-8600 Bamberg
Federal Republic of Germany

ABSTRACT. Helium-rich model atmospheres with various amounts of carbon have been computed in the range of 8000 $\geq T_{eff} \geq$ 6000 K, log g = 8. Magnetic field effects have been taken into account for the absorption of molecular carbon. Semiempirical parameters for the Zeeman-shifted band heads of the Swan transition have been applied to field strengths of $5 \times 10.^7$ to 2×10^8 Gauß. With the modified Unno scheme, the proper treatment of circular and linear polarization, together with radiative transfer, yields synthetic spectra with polarization, which can be compared to observations of LP 790-29.

INTRODUCTION

Since the discovery of the first strongly polarized white dwarf EG 129= Grw +70 8247 by Kemp et al.(1970) 27 white dwarfs with magnetic fields larger than 5 Megagauß have been observed. For some of the stars with field strengths $\leq$ 8 Megagauß the element composition in the atmospheres could be determined as hydrogen-rich, and even for Grw +70 8247 an identification of the strange 4135 feature as due to Zeeman-shifted components of the Balmer lines is explained by the energy spectrum modeling by Wickramasinghe and Ferrario (1988) and Jordan (1988).

For white dwarfs with observed parallaxes in the solar neighbourhood, 5 objects within 25 pc show polarization, 4 of them are cooleer than 10 000 K. That is why we expect a high percentage of old white dwarfs with magnetic fields, mainly in that range of temperature, where normal white dwarfs have carbon features in their spectra due to hydrogen decrease and carbon enhancement in a helium-rich atmosphere.

The hottest magnetic white dwarf, Feige 7, has hydrogen and helium lines in its spectrum with a shifted and splitted structure, thus showing an intermediate abundance ratio of hydrogen over helium if compared to other hot white dwarfs (Liebert et al. 1977). Most of the polarized objects show a quasicontinuous spectrum with some special features, likely for an overlapping of weak components during rotation and inclination of a dipole field in oblique rotator structure.

G 99-37 is the only star with shifted and splitted hydrogen and carbon lines as well as slightly shifted strong bands of CH and C_2 in its spectrum and a period of 4.117h (Bues, Pragal, 1988). This white dwarf can be used as a target to get semiempirical parameters for the shift of band heads of the Swan and Deslandres-d'Azambuja bands due to strong magnetic fields as is shown by Pragal et al. (1986) and used for a new analysis of G 99-37 by Pragal (1990). Here we would like to demonstrate the use of these susceptibility parameters for some model atmospheres of higher magnetic field strengths and apply them to LP 790-29, the polarized object with the strongest features at all.

G. Vauclair and E. Sion (eds.), White Dwarfs, 285–293.
© 1991 Kluwer Academic Publishers.

LP 790-29

LP 790-29=LHS 2293 has been observed with spectrometry and spectropoplarimetry by Liebert et al.(1978), spectrometry by Wickramasinghe and Bessell (1979) and polarimetry by West (1989). It exhibits the strongest deep features with equivalentwidths of more than 250 A in the spectrum of a white dwarf. In addition, broad band circular polarization of more than 10 % in the region of the band features suggests magnetic origin of the shifts. Thought to be unique in appearance by Liebert et al. and suggested to be carbon-rich in abundance, this object should be a tool for a test of abundance changes with strong magnetic fields.

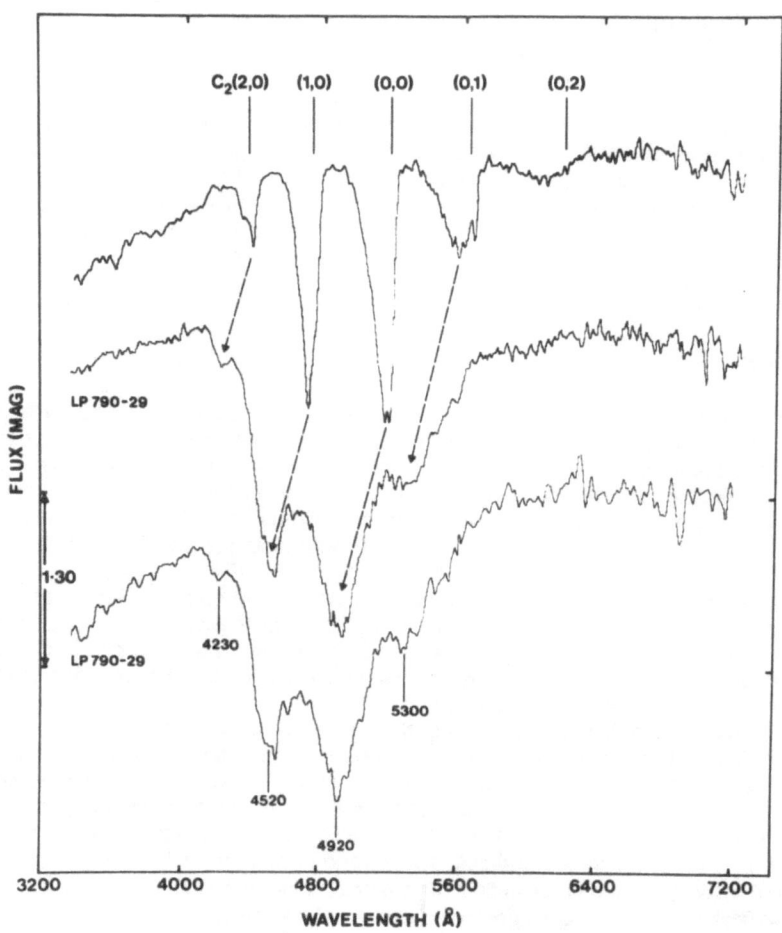

Fig.1: Spectrum of the the polarized WD LP 790-29 compared to BPM 27606, the WD with the strongest Swan-band features of C_2 close to laboratory wavelengths. The polarized star shows the indicated shift of 240 A due to the quadratic Zeeman-effect. (from Wickramasinghe and Bessell,1979)

A direct comparison of the Swan bandhead positions with those of BPM 27606 is shown in Fig.1. Although the shape of the non polarized band transitions is steeper, the identification of the bands as due to C_2 is convincingly demonstrated. Pressure effects by magnetic pressure broaden the shifted features additionally and depress the continuum. Our own CCD spectra, taken with the ESO 1.52m telescope at La Silla, show two more features in the blue region of the spectrum, but no variability of the bands.

Ruiz and Mazda (1988) and Hintzen et al. (1989) observed with ESO439-162 and LP 77-57, respectively, two other white dwarfs with similar features. The ESO object is slightly hotter than LP 729-29, LP77-57 with $T_{eff} = 5800$ K cooler . Both objects have not been observed with polarimetry up to now.. Yet the shift of the features and the derived magnetic field is of equal order of magnitude as LP 729-29. More weak objects of this type are likely to exist in the extended solar neighbourhood and could be important for proper motion investigations, even for binaries.

MODEL ATMOSPHERES

Model atmospheres for white dwarfs with observed strong polarization meet several difficulties in addition to normal LTE-atmospheres suitable for white dwarfs in the lower temperature range.Although hydrogen or helium are the main constituents of the atmosphere, the appearance of the observed spectra is due to shifted components of metal lines and carbon bands. Even if present, the components of the Balmer lines do not affect the temperature stratification. Strong absorption features are formed in the deeper layers of the atmospheres, just next to the main portion of the emergent flux.

That is why various abundances of carbon have to be taken into account for the computations of the absorption coefficients in the depths scale from the first steps of iteration of the flux constant model atmospheres. Our first attempts started without magnetic shifts and were applied to white dwarfs like G 47-18 or BPM 27606, the object in the upper line of Fig.1. The effective temperatures are above 8000 K and slightly higher than those of the magnetic WDs with carbon, which can be determined within 500 degrees from the flux distribution.

The band structure of the blue degraded Swan band has been included according to a modified Golden(1967) smeared band model, where the most important contributions correspond to rotation line numbers 10 to 30. Transitions with $\Delta v = 0, ^+1$ have been taken into account for Swan and Deslandres-d'Azambuja bands of C_2 up to $v' = 6$ with the f-values from Danylewych et al.(1974).

Flux constant LTE-model atmospheres have been computed for 8 000 K $\geq T_{eff} \geq$ 6 000 K, log g =7, 8 and 8.5, respectively, with ratios of $10^4 \geq$ He/C $\geq$ 300, 100 $\geq$ C/H $\geq$ 1.

These temperature and pressure stratifications have been used as initial models for the calculation of intensities and polarizations in a modified Unno scheme. The latter calculations started in intermediate depths of the Rosseland mean scale and solved the equations for intensity, linear and circular polarization simultaneously by a numerical procedure with more stepths than the Martin and Wickramasinghe (1979) solutions. The results are intensity and polarization for one point on the stellar surface.

For the integration of the flux, the surface is divided into 72 areas, which give nearly equal contributions to the flux depending on inclination angle and line of sight.

Our synthetic spectra have been calculated with a field structure of a centred dipole model in mind. The effective field strength, however, can be obtained similarly for a decentred dipole, just the choice and addition of the segments has to be changed.

Special emphasis is taken in evaluating the proper shift and polarization in the absorption coefficients along the band regime of the Zeeman-shifted C_2 bands, most important for the overall absorption in the cool strong-band stars, where the spectrum in the visible is dominated by carbon. First attempts by Rupprecht (1983) for the calculation of the $\Delta v=0$ transition of the Swan band in a quasihydrogenic manner showed a splitting of the branches due to the linear Zeeman effect up to field strengths of 100 Megagauß, but not at all shifts like the ones observed in Fig.1.

The main computational problem is the fact that molecular parameters with magnetic interaction do not exist for symmetric molecules like C_2. A special method for the treatment of molecules affected by magnetic fields has been developed by Flygare and Benson (1971) in terms of the susceptibility. Just one parameter is necessary to be known for the molecule, the effect of larger field strength then enters linearly up to 500 Megagauß. For C_2 this parameter was not known from laboratory measurements or theoretical computations.So we decided to try a semiempirical method for the determination of this parameter and use our observations of the low field white dwarf G 99-37. In his PhD Thesis M.Pragal (1990) worked out a scheme to derive constants from 14 spectra taken with the ESO 1.52m telescope during 1983 to 1988, which could be used for higher field strengths in our synthetic spectra.

An earlier analysis of G 99-37 by Angel et al.(1974) by use of the polarization profile of the G-band yielded an effective field strength of 3.6 Megagauß. Rupprecht(1983) had just one spectrum and obtained 8 Megagauß for the G-band with first order approximation. Spectra of better resolution as well as polarization and photometric measurements (Bues,Pragal,1989) showed variability and a period of 4.117h. In addition, Balmer and CI line components are clearly present in some of the spectra. Thus, for all the spectra, an effective magnetic field strength could be derived, dependent on phase. It varied between 8 and 25 Megagauß.
From the shift of the single band heads for the $\Delta v=0$ and +-1 transitions the susceptibility parameter for C_2 has been taken as a constant, true also for higher field strengths (limit 500 Megagauß). The absorption coefficients in the wavelength range of the Swan transitions have been calculated with the corresponding g-factors for some polar and effective field strengths for the model atmospheres mentioned above.

RESULTS

Figs.2-5 show synthetic spectra for the range of T_{eff} and abundances most likely to account for LP 790-29. For Fig.2 a) and b) the ratio of He/C is of the order most of the nonmagnetic white dwarfs with carbon features can be explained by. As C/O is just slightly shifted to carbon and temperatures in the outer layers of the atmospheres are rather high, the molecular abundances are small and the absolute value of the depression due to the Swan band is weak. Although the synthetic spectra for high fields show the shift of single band heads, the corresponding observed spectra would look like quasicontinuous fluxes. For a higher log g, a more severe broadening affects each transition separately. For the largest field strength, 50 Megagauß, a broad depression around 5070 A instead of a $\Delta v= 0$ transition will not change the V- colour compared to a pure continuous model atmosphere.
Whereas these fluxes demonstrate mainly the change of the π- components, Figs. 3 and 4 show overlapping effects of σ- and π- components for various inclination angles. For large depressions of the flux it is necessary to increase the carbon abundance with respect to helium, hydrogen and oxygen.The flux of a pure continuous model atmosphere comparable to Fig.2 is indicated by c, just to give an impression of the overall depression in the visible region, the main portion of the energy distribution is shifted to the UV-region around 3000 A.

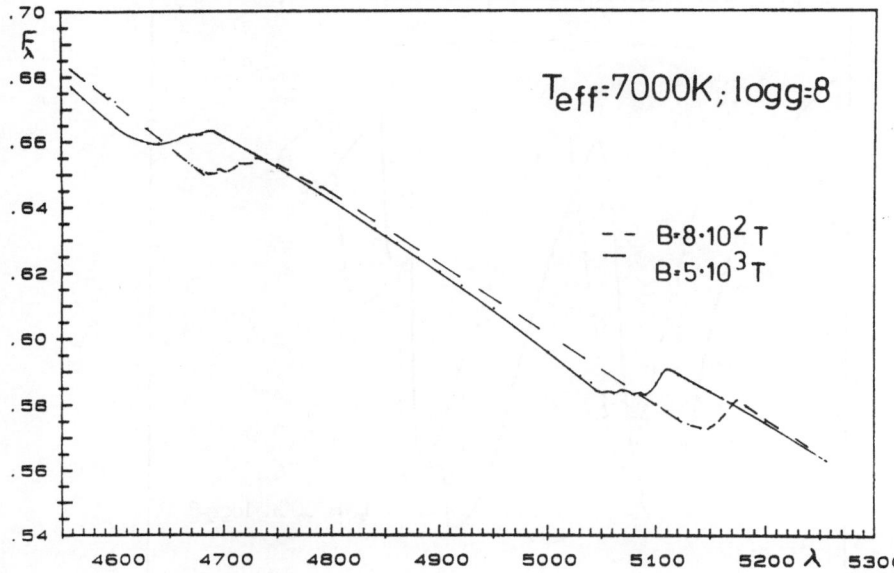

Fig. 2: a) Synthetic spectrum for a model atmosphere (T_{eff} = 7000 K, log g = 8) with a ratio of He/C = 3000, C/H = 10, C/0 = 3, B_{pol} = 10^3 Tesla and B_{pol} = 2 x10^3 Tesla. Shifts of the band heads are clearly seen.

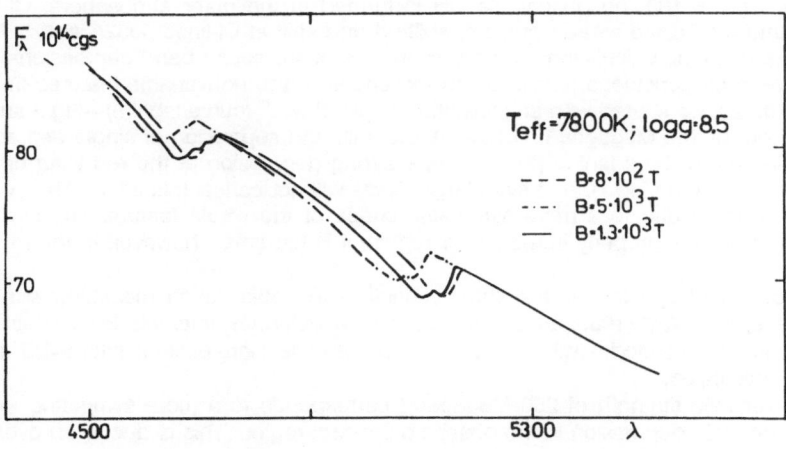

b) Same as a) for T_{eff} = 7800 K, log g = 8.5

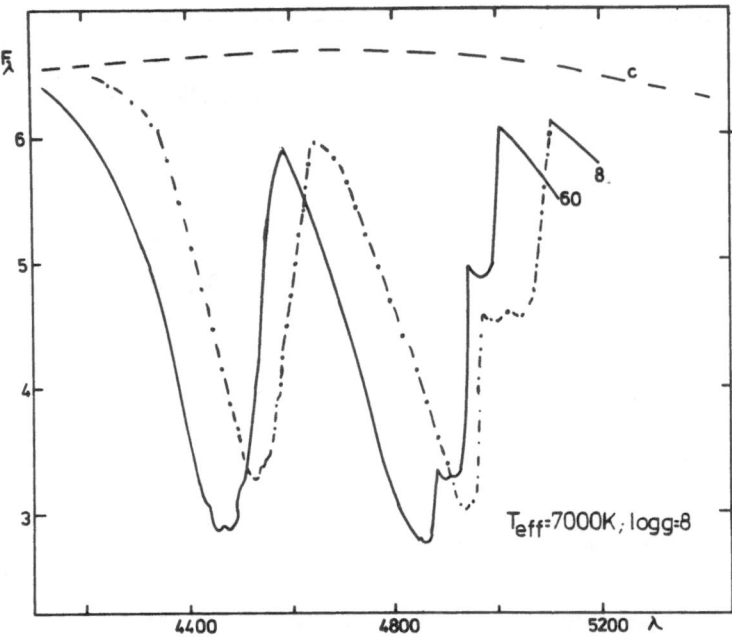

Fig. 3: Synthetic spectrum for a model atmosphere (T_{eff} = 7000 K, log g = 8) He/C = 300; C/H = 100, C/O = 10 , ----B_{pol} = 10^4 Tesla, i = 0° ---B_{pol} = 2.5 x 10^4 Tesla, i = 60°, c = pure continuum, F_λ is in units of 10^{14} cgs.

With a ratio of C/H ≥ 100, no observable CH is formed in the outer atmosphere. CI-line transitions around 4223A are weakly present, splitted and shifted CI-lines 5052A and 5380A have been included in the calculations, yet are overcome by molecular band depressions.

In order to investigate conditions applicable to the shape of the nonvariable features in the spectra of LP 790-29, we started with an inclination angle of i=0° (curve label 8)---Fig.3 and a polar field strength of 100 Megagauß. For ∆v=0, the shift and separation of single degraded band heads with severe broadening prevents **one** strong depression at the red wing of the feature, as observed(comp.Fig.1), and even larger fields will not cancel this effect. The ∆v=-1 transition can be fitted due to a more symmetric shape of the whole feature. The circular polarization also behaves properly in the ∆v=-1 region; it is too small, however, in the region around 5000A.

Although the observed spectra do not show variability, the polarization measurements by Liebert et al.(1978) and West (1989) are different in some wavelengths intervals. So we decided to apply our central dipole model with several inclination angles. Here examples for i=30° and 60° have been calculated.

For 60°, the polar field strength of 250 Megagauß corresponds to a more symmetric ∆v=0 transition with the main depression in the observed spectral region. This is due to an overlap of the σ⁺-transitions 1,1 and 2,2 with the Ŧ-transitions of 0,0 and 1,1, respectively.

The change in the degraded ∆v=-1 separated band heads is strongest for an angle of 60° with increasing broadening and decreasing polarization. In addition, the σ⁻-components become visible in the blue wing of the degraded band. If compared to the overall polarization,

291

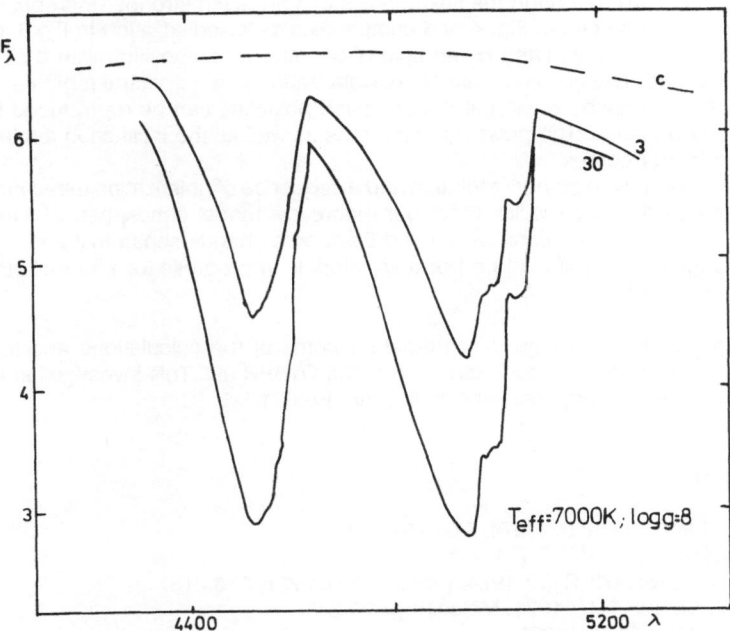

Fig. 4: Synthetic spectrum for a model atmosphere (T_{eff} = 7000 K, log g = 8)
lower curve: same parameters as in Fig. 3, yet i = 30°
upper curve: He/C = 1000, C/H = 33, C/0 = 100, i = 30°

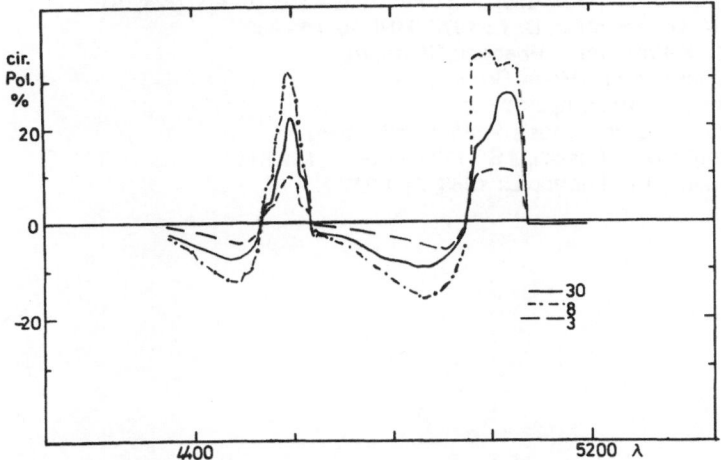

Fig. 5: Circular polarization for the fluxes of Fig 4 and --- of Fig. 3. The order of magnitude and the relative intensity for LP 790-29 is obtained for T_{eff} = 7000 K, B_{pol}
= 2.5 x 10^4 Tesla, i = 60°.

the values are smaller, although the flux depressions are even stronger than observed.
For conditions represented in Fig. 4 and compared circular polarization in Fig.5, the shape of the observed fluxes is best fitted by an agle of 30° and a composition with a slightly smaller carbon abundance. The absolute value of of polarization in the spectral region of 4400-5000A and especially the change of sign at the observed positions can be reproduced by the same intermediate abundance. The polar field strengths as well as the inclination angle seem to be accurate within 15 degrees.
Further observations with higher resolution and a sequence of polarization measurments should improve these preliminary results. From our theoretical model atmospheres, a more detailed treatment of all the C_2 transitions (Swan and Deslandres-bands separately and more wavelengths steps) should be possible. Work is in progress for a model atmosphere of 7500K and log g=8.5.

Acknowledgements: Thanks go to M.Pragal for some of the calculations and to Regionales Rechenzentrum Erlangen for computer time at the CYBER 995.This investigation is supported by the Deutsche Forschungsgemeinschaft, grant Bu 321/5-3.

REFERENCES

Angel,J.R.P., Landstreet,J.D.: 1974, ApJ **191**,457
Bues,I.,Pragal,M.: 1989, IAU Coll.**114**,329
Danylewych,L.L., Nicholls,R.W.: 1974, Proc.R.Soc.Lond.**A.339**,213
Flygare,W.H., Benson,R.L.: 1971, Molec.Phys.**20**,225
Golden,S.A.: 1967,JQSRT,**7**,225
Hintzen,P., Oswalt,T.D., Liebert,J., Sion,E.M.: 1989, ApJ **346**,454
Jordan,S.: 1988, PhD Thesis Kiel
Kemp,J.C., Swedlund,J.B., Landstreet,J.D., Angel,J.R.P.: 1970, ApJ **162**,L67
Liebert,J., Angel,J.R.P., Stockman,H.S., Spinrad,H.: 1977, ApJ **214**,457
Liebert,J., Angel,J.R.P., Stockman,H.S., Beaver,E.A.: 1978, ApJ **225**,181
Martin, B., Wickramasinghe. D. T.: 1979, MNRAS **189**,883
Pragal,M.: 1990,PhD Thesis Erlangen-Nürnberg
Pragal,M., Bues,I.: 1986, Meme.Soc.It. **58**,97
Ruiz,M.,T., Maza,J.:1988, ApJ **335**,L15
Rupprecht,G.: 1983,PhD Thesis Erlangen-Nürnberg
Wickramasinghe,D.T., Bessell,M.S.: 1979, MNRAS **188**,841
Wickramasinghe,D.T., Ferrario,L.: 1988, ApJ **327**,222

DISCUSSION

WEIDEMANN :

Are there polarization measurements of ESO 439-162?

BUES :

The object is far too weak to be observed with the PISCO equipment of the 2.2 m telescope at ESO.

SOME EFFECTS OF THE UV RADIATION FROM WHITE DWARFS ON THE ACCRETION OF INTERSTELLAR HYDROGEN

C. P. VERDON, R. L. MCCRORY and R. EPSTEIN
Laboratory for Laser Energetics
University of Rochester
Rochester, NY 14627-0011

H. M. VAN HORN, M. P. SAVEDOFF
Department of Physics and Astronomy and C. E. Kenneth Mees Observatory
University of Rochester
Rochester, NY 14627-0011

Abstract. We have carried out preliminary hydrodynamic calculations of the accretion of interstellar hydrogen by white dwarfs, and we are beginning to study the effects of the ionizing UV radiation from the star upon the flow. Even for velocities as low as ~ 2.82 km s^{-1}, with a stellar temperature $T_{\text{eff}} = 10,000$ K, the parameter values we have selected for the trial calculations discussed here, the HII region around the star is highly distorted. For this case, we find that the ionizing radiation *does* reduce the accretion rate of hydrogen, by about a factor of three. Further work is in progress and will be reported elsewhere.

1. Introduction

The element abundances observed in white dwarfs are generally thought to result from a balance between the rate at which matter is accreted and the rate at which the heavier elements sink out of the base of the surface convection zone. Recent calculations (*cf.* Dupuis 1990, and references therein) show that the observed metal abundances can be explained if a white dwarf accretes interstellar matter with solar proportions at a rate $\sim 2.3 \times 10^{-15}$ $M_{\odot}$ yr^{-1}, typical of accretion from cold interstellar clouds. If accretion were to occur at such rates, however, one would expect the amount of accreted H to be so large that the spectrum would be completely dominated by it. For the He-atmosphere white dwarfs this is completely contrary to the observations. Instead, most white dwarfs must accrete *hydrogen* at a rate $< 3 \times 10^{-20} M_{\odot}$ yr^{-1}. Thus, if accretion is the correct explanation for the observed metal abundances, *something* must inhibit the accretion of H onto these stars.

This inhibiting mechanism may be associated with the ionizing radiation from the hot white dwarfs (*cf.* Mestel 1954; Vauclair, Vauclair, and Greenstein 1979; Alcock and Ilarioniv 1980). In support of this hypothesis, Sion *et al.* (1986), have shown observationally that

G. Vauclair and E. Sion (eds.), White Dwarfs, 295–304.

the screening mechanism is operative for white dwarfs with $T_{\text{eff}} \gtrsim 11,000$ K, suggestively near the approximate temperature necessary for significant ionization of atomic H. More recently, Sion *et al.* (1990) have concluded that the screening mechanism must operate, with variable efficiency, even down to T_{eff} as low as ~ 5800 K.

The issue of precisely what prevents accretion of H onto a white dwarf is still unresolved today, perhaps in part because of the difficulty of performing satisfactory, two-dimensional accretion calculations. We have recently begun a series of calculations to address this problem, and in the present paper we describe our initial results. For our preliminary studies, we have investigated accretion of pure H gas, neglecting the effects due to He or heavier elements which are present in the real interstellar medium. While we are able to show that the UV ionizing radiation from the white dwarf *does* reduce the rate of accretion, the stellar temperature we have selected for these initial calculations is sufficiently low that the effect we find here is not yet sufficient to explain the the orders-of-magnitude reduction needed. We anticipate that calculations using larger stellar temperatures will have larger effects, but this remains to be investigated. We are currently engaged in additional calculations to explore this issue.

2. Sketch of Accretion Physics

We consider a star moving with speed v through a uniform gas of pure hydrogen, with mass density $\rho_\infty \equiv H n_\infty$, where n_∞ is the corresponding number density, and $H = 1.66044 \times 10^{-24}$ g is the unit of atomic mass. To a first approximation, mass elements with $\frac{1}{2}v^2 > GM/r$ are essentialy unaffected by the gravity of the star, while those with $\frac{1}{2}v^2 < GM/r$ are almost all captured. Thus, the quantity

$$R_A \equiv 2GM/v^2 \tag{1}$$

defines an effective "accretion radius" for a gravitating body (Hunt 1971; Alcock and Illarionov 1980). For a stationary star, the accretion flow is limited by the sound speed c_∞, and the corresponding length scale is the so-called Bondi radius[1] (*cf.* Hunt 1971)

$$R_B \equiv GM/c_\infty^2. \tag{2}$$

More physically, the flow is deflected by the gravitation of the star and forms a high-density accretion column in its wake. Those parts of the flow with relative velocities less than the escape velocity in this column fall back onto the star and are accreted (Hoyle and Lyttleton 1939; Bondi and Hoyle 1944). For cases in which the flow is subsonic or only mildly supersonic, Bondi (1952) has suggested the approximate interpolation formula

$$\dot{M} \approx \frac{2\pi (GM)^2 \rho_\infty}{(c_\infty^2 + v^2)^{3/2}}. \tag{3}$$

The typical parameter values determined in the recent analysis by Dupuis (1990) correspond to a 0.6 $M_\odot$ white dwarf moving at 20 $km \ s^{-1}$ through a dense ($n_\infty = 10 \ cm^{-3}$), cool ($T = 100$ K, corresponding to $c_\infty = 1.18 \ km \ s^{-1}$) interstellar cloud. These yield $R_A = 3.98 \times 10^{13}$ cm, $R_B = 5.74 \times 10^{15}$ cm, and $\dot{M} = 1.30 \times 10^{-15} \ M_\odot \ yr^{-1}$.

Radiation from the white dwarf has a substantial influence on the flow, however (*cf.* Vauclair, Vauclair, and Greenstein 1979; Alcock and Illarionov 1980). This can be seen as

---

[1] Note that Alcock and Illarioniv (1980) instead define $R_B = 2GM/c_\infty^2$.

follows. For a stationary star, the UV radiation ionizes the gas out to the Strömgren radius R_S. This radius is determined by the condition that the total number of ionizing photons Q emitted by the star per second exactly balance the total number of recombinations per second in the entire HII region. This gives (cf. Osterbrock 1974, p. 21)

$$Q = \frac{4\pi}{3} R_S^3 n_{H^0}^2 \alpha_B,$$ (4)

where n_{H^0} is the number density of *neutral* hydrogen, $\alpha_B(T) \approx 2.6 \times 10^{-13}$ cm^3 s^{-1} is the net recombination coefficient to all levels except the ground state, at a typical HII region temperature $T = 10^4$ K (Osterbrock 1974, p. 17), and Q is the total number of Lyman continuum photons radiated by the star. If the white dwarf can be assumed to radiate like a black body, this is given by

$$Q = 4\pi R_*^2 \int_{\nu_0}^{\infty} \frac{\pi F_\nu}{h\nu} d\nu = 4\pi R_*^2 \int_{\nu_0}^{\infty} \frac{\pi B_\nu}{h\nu} d\nu.$$ (5)

Here $h\nu_0$ is the ionization potential, and B_ν is the Planck function.

For a 30,000 K white dwarf, the radius of the Strömgren sphere is $R_S \sim 0.2$ pc $n_H^{-2/3}$. This is much larger than the accretion radius for the typical parameter values quoted above. Thus, the presence of the ionized HII region surrounding the white dwarf cannot be neglected in computing the accretion flow. For accretion from a cold interstellar cloud with $T \sim 100$ K, this effect is likely to be very large indeed. The pressure inside the HII region is increased by a factor ~ 200 over that in the neutral gas; the temperature is 100 times larger, and twice as many free particles contribute to the pressure (photo-ejected electrons as well as ions).

The radius of the Strömgren sphere is *not* the appropriate characteristic dimension of the HII region in the flow around an accreting white dwarf, however. The reason is that the matter "blows by" the star faster than it can be ionized. The probability per second of photoionization of a hydrogen atom at a distance r from the white dwarf is given by

$$\frac{1}{t_{ion}} \equiv \frac{Q}{4\pi r^2} a_{\nu_0},$$ (6)

where $a_{\nu_0} = 6.30 \times 10^{-18}$ cm^2 is the photoionization cross-section from the ground state of hydrogen (Osterbrock 1974, p. 14), and for this dimensional argument we have neglected the attenuation of the photon flux due to absorption of the ionizing radiation. For comparison, the timescale for an atom to move a distance r is r/v. Equating these two timescales gives the characteristic dimension of the ionized region in a wind (cf. Alcock and Illarioniv 1980):

$$R_I = \frac{Q a_{\nu_0}}{4\pi v}.$$ (7)

For $T_{eff} \lesssim 7000$ K, R_I is smaller than the radius of the white dwarf itself; for these cases, the ionizing radiation is clearly too weak to have any effect upon the accretion rate, and accretion from dense clouds proceeds at the Bondi rate (eq. [3]). Conversely, for very hot white dwarfs, with $R_I > R_S$, the full Strömgren sphere does form, but it is distorted by the flow.

3. Method of Calculation

The numerical calculations we have performed treat the accretion flow using the time-dependent, mixed Euler-Lagrange code ORCHID (McCrory and Verdon 1988), which was previously developed to study laser-driven pellet implosions for inertial confinement fusion applications. We have modified the code by including (1) the gravitational attraction of the white dwarf, (2) a sink in the central mass zone representing the accreting star, and (3) the effects of ionizing radiation. The central, sink boundary condition for the numerical calculation has proven to be critically important for evaluating the accretion rate; the boundary condition we have used consumes in each timestep either the total mass in the innermost zone surrounding the star or the maximum fraction of that mass allowed by the analytical rate (eq. [3]), whichever is less. The calculations are intrinsically two-dimensional, and we use cylindrical coordinates (r, z), with 122×122 zones, more than used in any previous calculation for similar problems. The grid for our computations was chosen to extend a distance of $\sim 5\ R_A$ in the radial direction and to range from about $-2\ R_A$ to about $+3\ R_A$ in z. Negative (positive) values of z correspond to the upstream (downstream) side of the star. At the upstream boundary, the gas temperature was set to 100 K, the number density to 10 cm^{-3}, and the velocity to that corresponding to the desired Mach number. The pressure inside the grid is set equal to the pressure at the upstream boundary of the grid, and there is a step change in density and temperature at the point where the gas flows onto the grid. As the calculation evolves in time, the higher density material flows onto the grid at the velocity of interest.

4. Accretion Without UV Radiation and Comparison With Previous Work

As a test of our code, we first performed a calculation of the flow at Mach number $M = 2.4$ *without* including the effects of the ionizing radiation. This value of M is much smaller than the "typical" value expected for a white dwarf, but it is identical to the largest value treated in calculations done previously by Hunt (1971), thus allowing us to compare with these results. Hunt used spherical (r, θ) coordinates, with 36 grid steps in $\log r$ and 12 in θ, and he solved the time-dependent hydrodynamic equations to second-order accuracy. His results showed a prominent bowshock, with the standoff point located on the axis at a distance 0.11 R_B upstream from the star. The magnitude of the velocity increases as the bowshock is approached from the upstream side, reaching $M > 4$ just before the shock. This is just the result of the gravitational acceleration due to the star. The flow rapidly decelerates through the shock, dropping to values $M \sim 1.4$ just behind it. The stagnation point on the downstream side of the flow lies at a distance of 0.19 R_B behind the star.

The values of the parameters employed in our comparable calculations are: $v_\infty = 2.82\ km\,s^{-1}$; $T = 100$ K, giving $c_\infty = 1.18\ km\,s^{-1}$ and Mach number $M = 2.4$; and $n_\infty = 10\ cm^{-3}$. For a white dwarf with mass $M_* = 0.6 M_\odot$, and corresponding radius $R_* = 0.012 R_\odot$, this gives $R_A = 1.99 \times 10^{15}$ cm and $R_B = 5.74 \times 10^{15}$ cm. The characteristic flow timescale for this problem is $t_{hyd} = R_A/v = 223$ years, and our calculation reaches a steady state in about this time. Our results agree quite well with Hunt's, as can be seen by comparing our Fig. 1 with his Fig. 6. For example, we also find that the flow accelerates to $M > 4$ just before the shock and decelerates rapidly to $M \sim 0.5$ behind it. The standoff point is located at $z \approx -0.32\ R_A = -0.11\ R_B$ for the parameters of this calculation, identical with Hunt's value of $-0.11 R_B$, where the negative value of z indicates the upstream side of the

results. Hunt's calculations also give good agreement with Bondi's accretion rate formula. In our calculations, the accretion rate was initially $\sim 10^{-16} M_\odot$ yr^{-1} and asymptotically approached the value $\sim 4 \times 10^{-13} M_\odot$ yr^{-1} after ~ 300 years, in good asymptotic agreement with the Bondi (1952) result from eq. (3): $\dot{M} = 3.67 \times 10^{-13}$ $M_\odot$ yr^{-1}.

5. Accretion With UV Radiation

We have also carried out an accretion calculation for exactly the same parameters as employed in our comparison with Hunt's computation, but now including the effects of the ionizing UV radiation. For this computation, we have assumed the white dwarf to radiate like a black body of temperature $T_{\text{eff}} = 10,000$ K, giving the rate of emission of ionizing photons $Q = 2.20 \times 10^{37}$ s^{-1}. For the parameters of our calculation, the characteristic radius R_I given by eq. (7) is $R_I = 3.91 \times 10^{13}$ cm ≈ 0.02 R_A; for comparison, the radius of the Strömgren sphere is $R_S = 5.86 \times 10^{15}$ cm $= 2.95$ R_A. This hydrodynamic study is much more time-consuming than the calculation without UV radiation, because it takes many time-steps to ionize the region around the star. It is necessary to have good spatial resolution near the star, because the UV flux increases as $1/r^2$ with decreasing r. Thus the ratio of the timescale for photoionization to that for hydrodynamic flow is (cf. eqs. [6], [7])

$$t_{ion} = \frac{r^2}{R_I v} \equiv \frac{r}{R_I} \times t_{hyd}(r). \tag{8}$$

The timescale for changes in the ionization structure near the star consequently is much shorter than the hydrodynamic timescale, requiring much finer time-resolution in the calculations.

Another difficulty, as expected from the discussion in §2, is that only the matter near the star becomes ionized, and the size of this region is $R_I << R_S$ (cf. eq. 7). The HII region is not even approximately spherical but is strongly deformed by the flow, gradually fanning out behind the star. The ionization also is significantly higher downstream from the star, as the photons remain in the gas longer and have more time to ionize it.

The results of these hydrodynamics-plus-radiation calculations are illustrated by the velocity-vector plot shown in Fig. 2. This figure shows the velocity contours at an elapsed time $t = 554.4$ years for the $M = 2.4$ hydrodynamical calculation *including* the effects of the UV radiation from the white dwarf. This is very close to the elapsed time for Fig. 1, and the two figures can be compared directly. The UV radiation preheats the gas, forming a higher temperature region around the star and substantially reducing the strength of the bowshock. Note that the accretion column has now completely disappeared, and the flow is actually accelerated *away* from the star on the downstream side by the increased pressure of the hot ionized gas.

The integrated mass-accretion rates for both cases are shown as functions of time in Fig. 3. The two horizontal lines shown in this figure correspond to two different estimates for the accretion rate. The lower curve is Bondi's interpolation formula, eq. (3), while the upper one is the approximation $\dot{M} = \pi R_A^2 \rho_\infty v$. As the starting transients die out, on the hydrodynamic timescale ~ 300 years, the calculation that neglects the effects of the ionizing radiation (the curve labeled "Hydro only") asymptotically approaches the Bondi value, 3.67×10^{-13} $M_\odot$ yr^{-1}, given by eq. (3). The calculation that includes the UV effects (the curve labeled "Hydro + Radiation") clearly approaches an asymptotic value about $1/3$ as large.

star. The stagnation point in our calculations lies at $z \approx +0.60$ $R_A = 0.20$ R_B, as compared with the value of 0.19 R_B from Hunt's calculations.

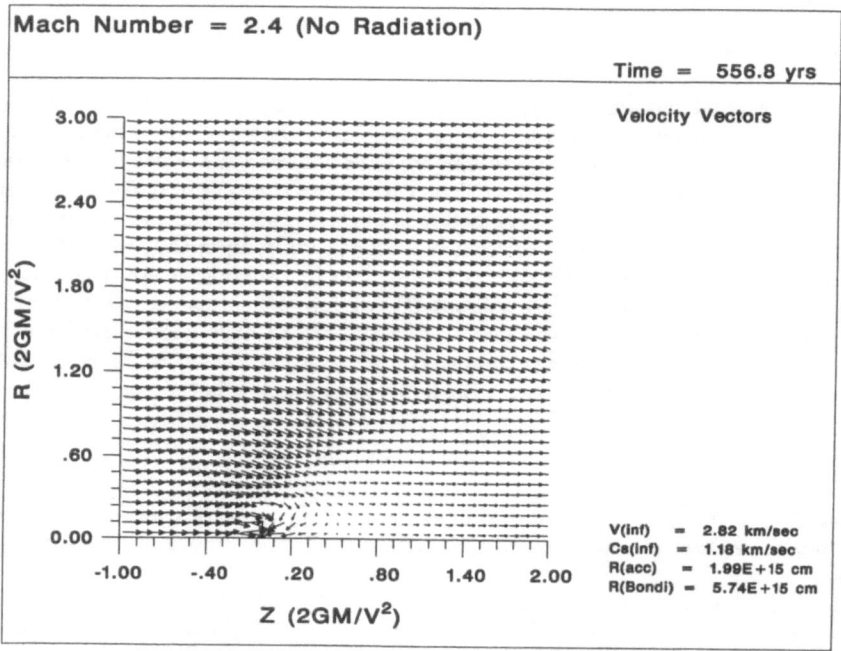

Fig. 1 Velocity contours at $t = 556.8$ years for the $M = 2.4$ hydrodynamical calculation *without* including the effects of the UV radiation from the white dwarf. The length of the arrow indicating each velocity vector and the size of the arrowhead are proportional to the magnitude of the velocity. Both r and z are expressed in units of R_A, and the white dwarf is located at $(r, z) = (0, 0)$. The position of the bowshock is apparent as an abrupt decrease in the magnitude of the velocity. The standoff point is located at $z \approx -0.32$ $R_A = -0.11$ R_B for the parameters of this calculation, where R_A and R_B are respectively the accretion radius (*cf.* eq. [1]) and the Bondi radius (*cf.* eq. [2]). This result is identical with Hunt's value of -0.11 R_B. The formation of the accretion column on the downstream side also is apparent in this figure.

The gravitational deflection of the flow is apparent in Fig. 1 as the flow passes the star and, as expected from earlier studies, there is a counter- streaming "accretion column" of matter falling onto the star from the downstream side. Hunt (1971) has shown that as $M \to \infty$, the accretion flow becomes increasingly strongly peaked behind the star, forming the Bondi-Hoyle accretion column. While there is not much evidence for the accretion column in Hunt's $M = 2.4$ plots, the beginnings of this flow are clearly apparent in our

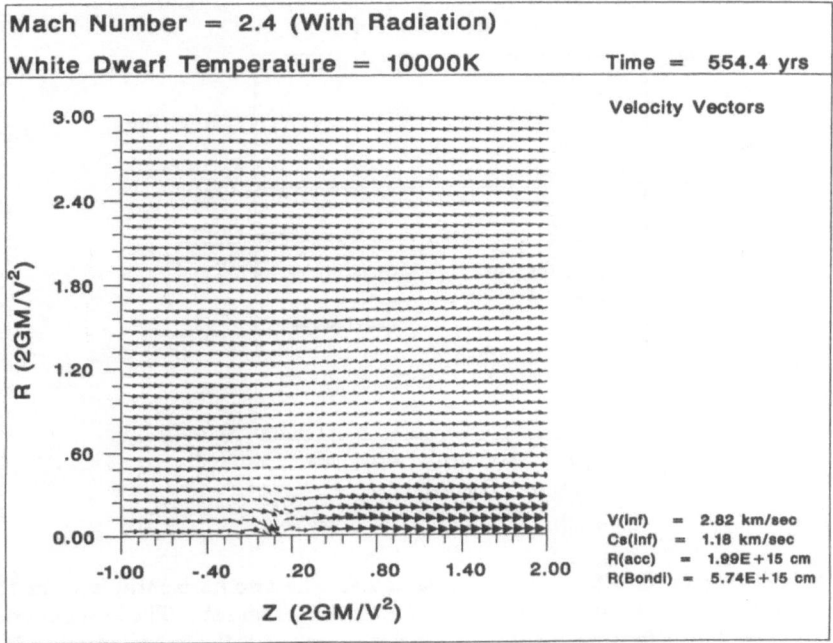

Fig. 2 Velocity contours at $t = 554.4$ years for the $M = 2.4$ hydrodynamical calculation *including* the effects of the UV radiation from the white dwarf. The scales and notations are the same as those in Fig. 1. In this case, the UV radiation preheats the gas, substantially reducing the strength of the bowshock. Note that the accretion column has now completely disappeared, and the flow is actually accelerated *away* from the star on the downstream side by the increased pressure of the hot ionized gas.

6. Summary and Conclusions

We have carried out preliminary hydrodynamic calculations of the accretion of interstellar hydrogen by white dwarfs, and we are beginning to study the effects of the ionizing UV radiation from the star upon the flow. In the calculations reported here, we have taken the white dwarf temperature to be $T_{eff} = 10,000$ K, near the lower limit of stellar temperatures for which we may expect any significant effects due to ionization of the accreting gas. This temperature is low enough that the characteristic dimension R_I of the ionized region is small compared to the accretion radius R_S, and the net influence of the UV radiation is relatively small. Even for this marginal case, however, we find that the ionizing radiation *does* reduce the accretion rate of hydrogen, to $\sim 1/3$ of the Bondi rate given by eq. (3). In addition, even at velocities as low as ~ 2.82 km s^{-1}, which we have selected for these

302

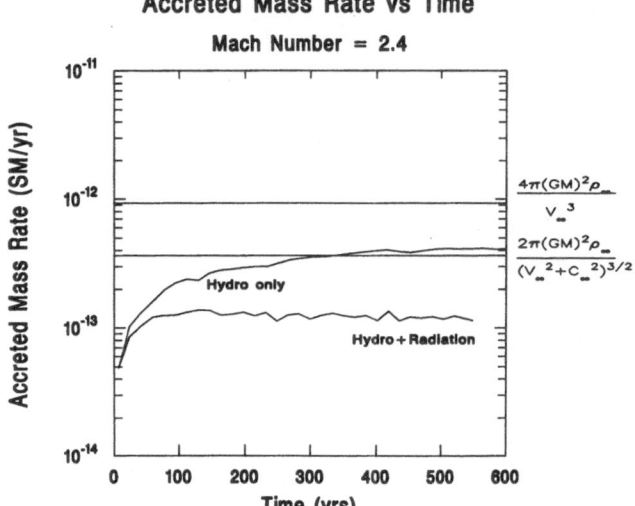

Fig. 3 Mass accretion rate *vs.* time for $M = 2.4$. The two horizontal lines in this figure correspond to two different estimates for the accretion rate. The lower curve is Bondi's interpolation formula, eq. (3), while the upper one is the approximation $\dot{M} = \pi R_A^2 \rho_\infty v$. As the starting transients die out, the calculation that neglects the effects of the ionizing radiation (the curve labeled "Hydro only") asymptotically approaches the Bondi value, 3.67×10^{-13} $M_\odot$ yr^{-1}, given by eq. (3). The calculation that includes the UV effects (the curve labeled "Hydro + Radiation") clearly approaches an asymptotic value about 1/3 the Bondi value.

preliminary calculations, the HII region around the star is highly distorted, forming a narrow cylinder around the white dwarf that fans out downstream of the star. These results clearly demonstrate the need for further investigations of this effect. Additional calculations are currently in progress and will be reported elsewhere.

Acknowledgements

This work has been supported in part by the National Science Foundation under grant AST 88-20322 through the University of Rochester. It has also been supported in part by the U.S. Department of Energy Division of Inertial Fusion under agreement No. DE-FC03-85DP40200 and by the Laser Fusion Feasibility Project at the Laboratory for Laser

Energetics which has the following sponsors: Empire State Electric Energy Research Corporation, New York State Energy Research and Development Authority, Ontario Hydro, and the University of Rochester.

References

Alcock, C., and Illarionov, A. 1980, ApJ, 235, 541.

Bondi, H. 1952, MNRAS, 112, 195.

Bondi, H., and Hoyle, F. 1944, MNRAS, 104, 273.

Dupuis, J. 1990, Ph.D. thesis, Université de Montréal.

Hoyle, F., and Lyttleton, R. A. 1939, Proc. Cambridge Phil. Soc., 35, 405.

Hunt, R. 1971, MNRAS, 154, 141.

McCrory, R. L., and Verdon, C. P. 1988, in *Inertial Confinement Fusion*, ed. A. Caruso and E. Sindoni (Societá Italiana di Fisica: Milan, Italy), p. 83.

Mestel, L. 1954, MNRAS, 114, 437.

Osterbrock, D. E. 1974, *Astrophysics of Gaseous Nebulae* (W. H. Freeman & Co.: San Francisco).

Sion, E. M., Hammond, G., Wagner, R. M., Starrfield, S. G., and Liebert, J. 1990, ApJ, 362, 691.

Sion, E. M., Shipman, H. L., Wagner, R. M., Liebert, J., and Starrfield, S. G. 1986, ApJ, 308, L67.

Vauclair, G., Vauclair, S., and Greenstein, J. L. 1979, AA, 80, 79.

DISCUSSION

SHIPMAN :

I think this is a very interesting problem, because a number of authors have pointed out the coincidence of the onset of chemical peculiarities in DB stars with the turn-off of a significant UV radiation field (around $T_{eff} = 10^4 K$). My question is a simple one : was your stellar radiation field that of a black body or model atmosphere, and do you have any plans to include realistic model atmosphere ?

VAN HORN :

We have used black body models for these calculations. We have no plans at present to use more realistic model atmospheres until we have established, if we can, whether the effects of UV radiation for higher stellar temperatures can inhibit accretion enough to be really interesting.

THEJLL :

Do you perform radiative transfer calculations in the ISM ?

VAN HORN :

Yes and no. The ORCHID code does solve the combined equations for radiative transfer and ionization equilibrium, using a multi-group photon energy average to obtain the photoionization cross-section. We choose the spatial zones small enough so that the radiation density is reasonably constant throughout each one. We ignore the "hardening" of the spectrum in the outer parts of the ionization zone, due to differential absorption of photons of different energies, however, and we use the "on-the-spot" approximation for recombination photons instead of dealing with the diffuse recombination radiation emitted throughout the ionized region.

CONVECTION IN WHITE DWARFS:
APPLICATION OF CM THEORY TO HELIUM ENVELOPE WDS

ITALO MAZZITELLI
Istituto di Astrofisica Spaziale CNR
C.P. 67
I-00044 Frascati
Italy

FRANCESCA D'ANTONA
Osservatorio Astronomico di Roma
I-00040 Monte Porzio
Italy

ABSTRACT. We summarize the main physical properties of the new model for stellar turbolent convection by Canuto and Mazzitelli, which constitutes an improvement of description with respect to the classical Mixing Length Theory (MLT), and has been successfully applied to the solar model.

We apply this theory to the study of convection in the envelope of a typical White Dwarf (WD) of $0.55M_\odot$, and compare the results with those obtained by employing the MLT and leaving all the other physical and numerical inputs inaltered. The results are conforting in the description of the location of the blue edge of the DB instability strip, if this is at $T_{\rm eff} \sim 25000K$, but a blue edge at $T_{\rm eff} \sim 29000K$ would be in strong contradiction with the new theory.

1. Introduction

Study of convection in WDs has always been regarded as one of the capital points in order to derive both structural and cooling properties of these stars, their possible spectral evolution and their pulsational properties.

At temperatures low enough that the degeneracy boundary penetrates through the bottom of the convective layer, the core temperature –and consequently the cooling properties– depend directly on the adiabatic gradient through the convective regions (D'Antona and Mazzitelli 1989, 1990, Tassoul *et al.* 1990).

In turn, the $T_{\rm eff}$ at which degeneracy reaches the convective bottom depends on the surface layers opacity, and therefore on the dominant envelope composition (hydrogen or helium), which may have been the result of possible transition between spectral types due to convective mixing. The achievement of mixing depends, mainly, on the mass extension of convection as a function of $T_{\rm eff}$.

A further tool to understand the WD structure and evolution are the instability strips, mainly the DB variable strip at $T_{\rm eff} \sim 25000K$ and the pulsating DA strip (ZZ Ceti) at $T_{\rm eff} \sim 12000K$. The driving of pulsations is identified with the κ–mechanism operating in

G. Vauclair and E. Sion (eds.), White Dwarfs, 305–316.

the partial ionization zones, mainly at the base of the convective envelope, and the pulsations which are excited have periods close to the thermal timescale (t_{th}) of the envelope.
Therefore, *the exact location in mass* of the base of the convective region is critical in terms
of potential driving of pulsation modes (e.g. Tassoul *et al.*1990).

Study of convection in WDs has known a period of careful exploration in the 1970s,
starting with the pioneering work of K. H. Böhm (1969) and H. M. Van Horn (1968).
Attention of researchers has been longly focused on the role of the equation of state in the
partial ionization region (the adiabatic gradient is determinant as most of the convective
region is adiabatic), and key explorations in this respect have been done by D'Antona and
Mazzitelli (1975, 1979) and Fontaine and Van Horn (1976).

An important aspect of these studies is the definition of convective boundaries for He−
and H− envelopes, to be compared with the observations of the instability strips. Fontaine
and coworkers have widely explored the results obtainable by modifications of the classical
Mixing Length Theory (MLT) (Böhm Vitense 1958): for a recent updated set of results see
Tassoul *et al.*1990.

It is clear that considerable improvement in this field can be achieved only if we improve
the theory of turbolent convection with respect to the MLT. The versions of MLT used up
today mainly consist in slight modifications of the input parameters of the classical MLT
so that more or less efficient convection is provided, but the *functional* dependence of
convection on the physical parameters remains the same as in the MLT.

In the following we shortly describe a new theory of turbolence in stars (Canuto and
Mazzitelli, 1990, hereinafter referred to as CM theory) which represents a sensible improvement with respect to the MLT, and show its first application to the study of convection in
the He-envelope WDs.

2. A new theory of turbolent convection and its accomplishments

A new theory of stellar turbolent convection has been developed and successfully applied
to the solar model by Canuto and Mazzitelli (1990). We summarize here the main improvements that this theory represents with respect to the MLT:

  i) it accounts for the whole distribution of turbolent eddies, including the growth region
     and the Kolmogoroff region, instead of being a 'one–eddy' representation as the MLT;
 ii) it contains no free parameters.

The first improvement comes from the straightforward integration of the Eddy Damped
Quasi Normal Markovian approximation of the Navier–Stokes equations (Orszag, 1977,
Lesieur, 1987). The whole spectrum of eddies and relative characteristic times are taken
into account, from the largest ones excited directly by the growth mechanism, to the smallest
ones, where molecular viscosity cuts off the spectrum, and not only direct transmission of
kinetic energy from larger to smaller eddies is allowed, but also backscatter of kinetic energy
to large eddies.

The Prandtl number (σ) was set to 10^{-3}. In stellar interiors actually values much
smaller than this are expected, but trial computations have shown that the total energy
flux was not sensitive any more to σ, if chosen below this value.

The second improvement comes from the fact that, since the beginning of the MLT,
it was clear that, in a correct treatment of the vertical eddies, the obvious choice of the
characteristic scale length for the stellar turbolence (the mixing length l) was the geometric
thickness z of the convective layers above any given point in the convective envelope (Böhm
and Stuckl, 1967), and it was only the inability of the MLT to fit the solar T_{eff} with this
choice that forced the −less consistent− adoption of a given fraction (actually a multiple)
of the pressure scale heigth H_p. Since the new theory nicely and immediately fits the
solar radius with mixing length equal to the geometric depth ($l = z$), the free parameter
($\alpha = l/H_p$), with all its uncertainties associated, is no longer required.

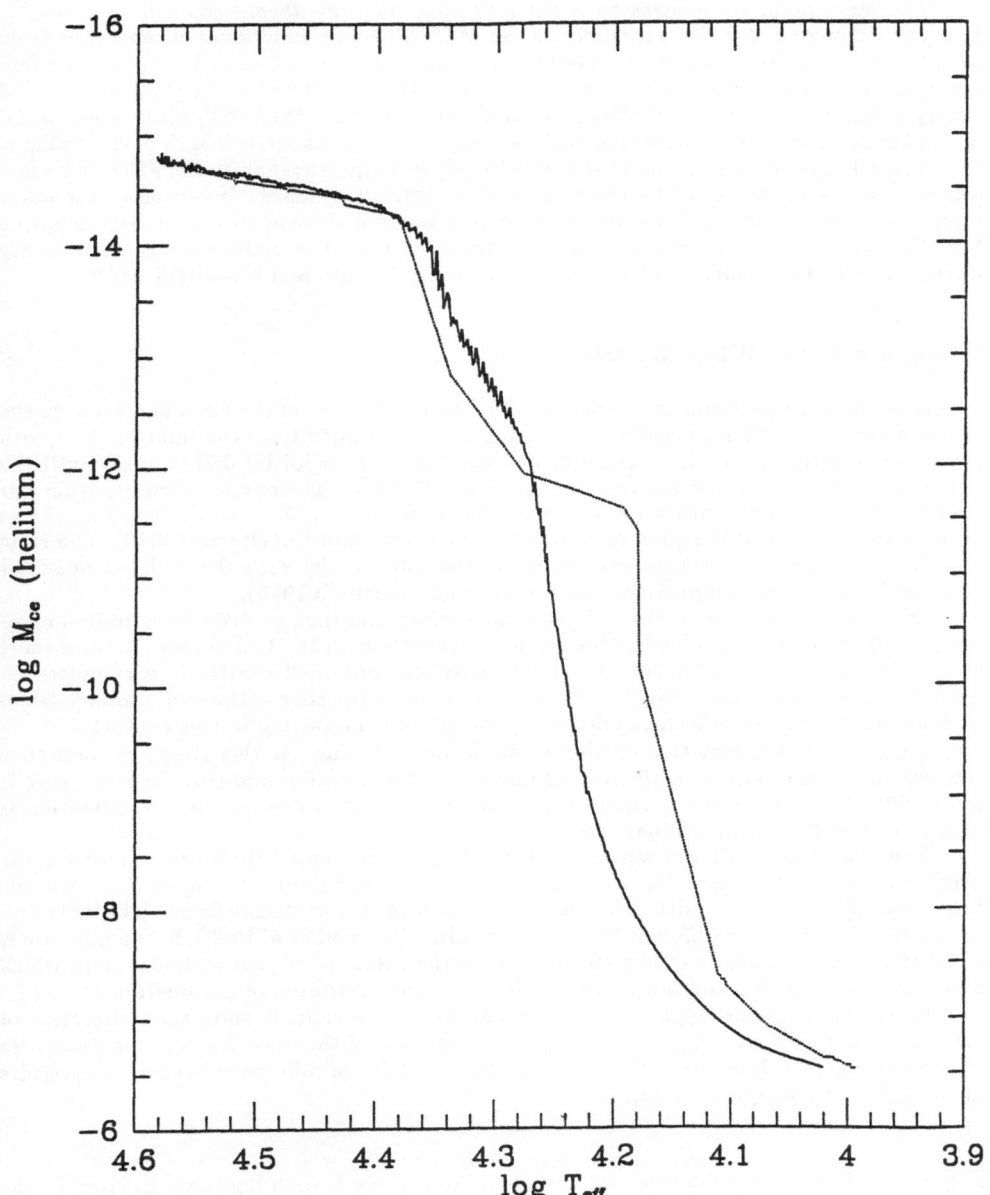

Figure 1: Convective mass fraction M_{ce} as a function of $\log T_{eff}$ for the sequences of $0.55M_\odot$ computed with MLT (continuous line) or CM (dotted) convection treatment.

The main point to be stressed is the following: *the new theory can not be mimicked by careful choices of the free parameter in the MLT*. The very functional dependences from the physical quantities (gravity, temperature, density, opacity and so on) of the energy flux transported by convection, are in fact different in the two theories: in the average, CM theory accounts for a larger efficiency in energy transfer than the MLT, *up to a factor 10*, but in the regions close to the stellar surface, where the characteristic length of the eddies is only a small fraction of H_p, *convection is definitely less efficient than in the MLT*. This fact produces a steep gradient of temperature and a consequent density inversion in the outer layers of the solar model. This density inversion was not present in the models adopting the MLT, but is probably required for a better fit of the solar oscillations spectrum. For further details, the reader is referred to the paper by Canuto and Mazzitelli, 1990.

3. Application to White Dwarfs

We report now the preliminary results obtained by application of the CM convection to the evolution of WDs. We have evolved a starting model obtained from the full evolution with mass loss, starting from the horizontal branch phase, of an initial $0.7 M_\odot$ star, until the WD stage was reached and the remnant mass was $0.55 M_\odot$. The envelope composition was assumed to be helium, with a metal content in mass fraction $Z = 10^{-3}$. The model was evolved through the WD region assuming for the convection the classical MLT. The ratio $\alpha = l/H_p = 1.4$ was chosen, as this value fits the solar model with the updated opacities adopted for the solar composition (see Canuto and Mazzitelli 1990).

Starting models at several T_{eff} 's were memorized and then evolved for a limited number (~ 30) of models by changing the input of convection to the CM theory. After a short relaxation (few models were necessary) the structures obtained constitute a complete sequence of models corresponding to the CM convection (of course the evolutionary times may be not entirely significant, and in any case are not interesting in this context).

We decided to adopt this somewhat artificial procedure, at this stage, as numerical convergence of the models with the inclusion of CM convection *and* turbolent pressure is more difficult than for the normal convection theory. Of course further investigation is needed to refine these preliminary results.

The run of convective mass as function of T_{eff} is presented in figure 1 both for the MLT (continuous line) and for the CM convection (dotted line). It can be seen that the behaviour of M_{ce} is very much dependent on the choice of convection theory. This of course occurs also by assuming different treatments of MLT (Tassoul *et al.*1990), but in our case it is the functional behaviour of flux efficiency with the extent of the convective regions which determines the results, and not a more or less arbitrary variation of parameters.

It is instructive, although it may seem somewhat didascalic, to show the comparison of the run of temperature, ∇_{rad}, ∇_{conv}, ∇_{ad} as a function of the mass fraction for structures at several T_{eff}, to show the main causes of the similarity of difference between the results of the different convection treatment.

Figure 2, 3, 4, 5, 6: the top part shows the run of $\log T$ with log mass fraction in the envelope of the WD of $0.55 M_\odot$ having external envelope composition $Y = 0.999$, $Z = 10^{-3}$ with the two different convection treatments ('conv=0' indicates the MLT results with $\alpha = 1.4$), 'conv=3' the results including CM convection with no free parameters). The lower parts of the figure present the run of gradients through the structure in the two different cases. The continuous line is the adiabatic gradient ∇_{ad}, the dashed line represents the radiative gradient ∇_{rad} and the dotted line represents the superadiabatic (convective) gradient ∇_{conv}. The values of $\log L/L_\odot$ and $\log T_{\text{eff}}$ of the models are indicated.

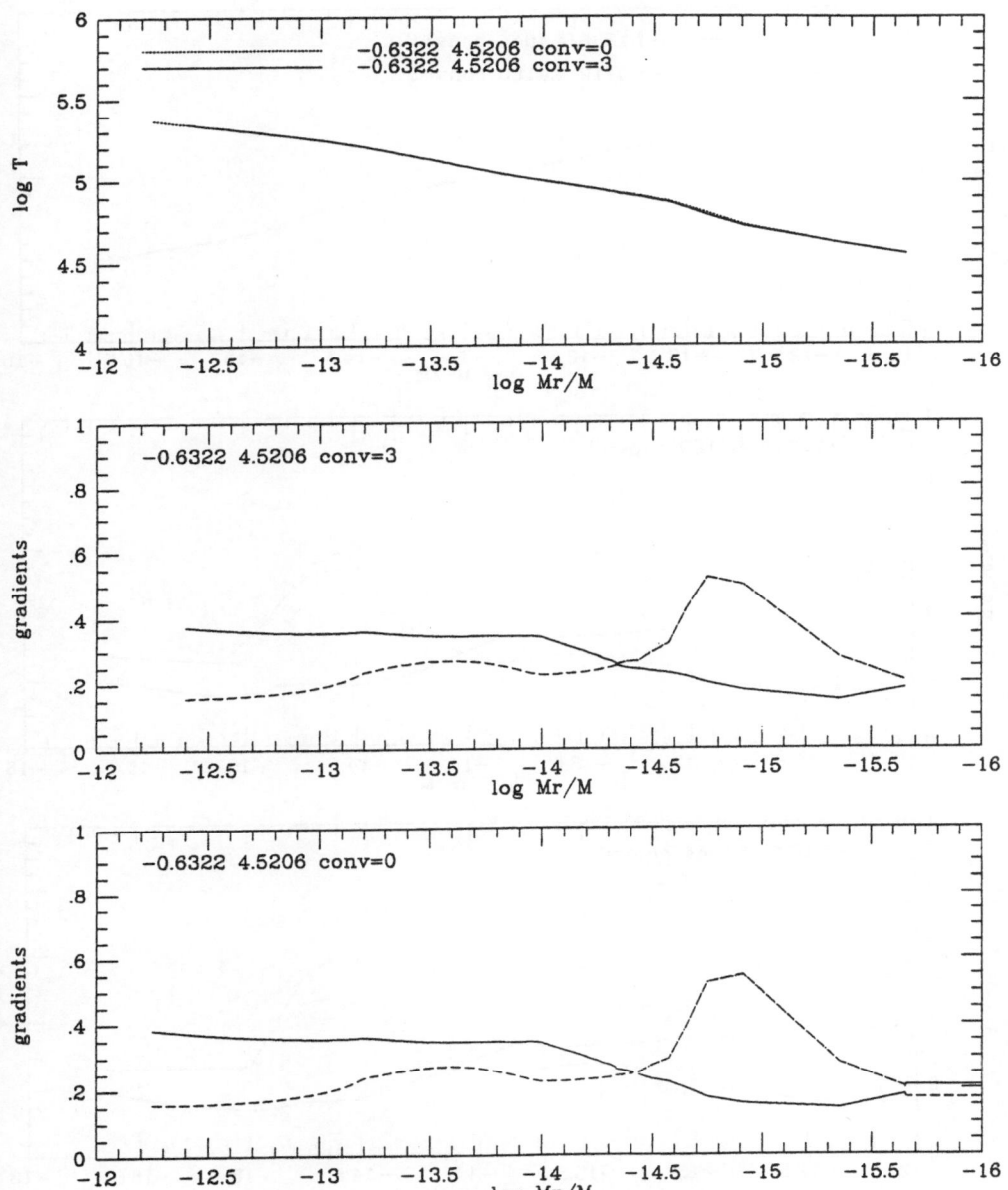

Figure 2

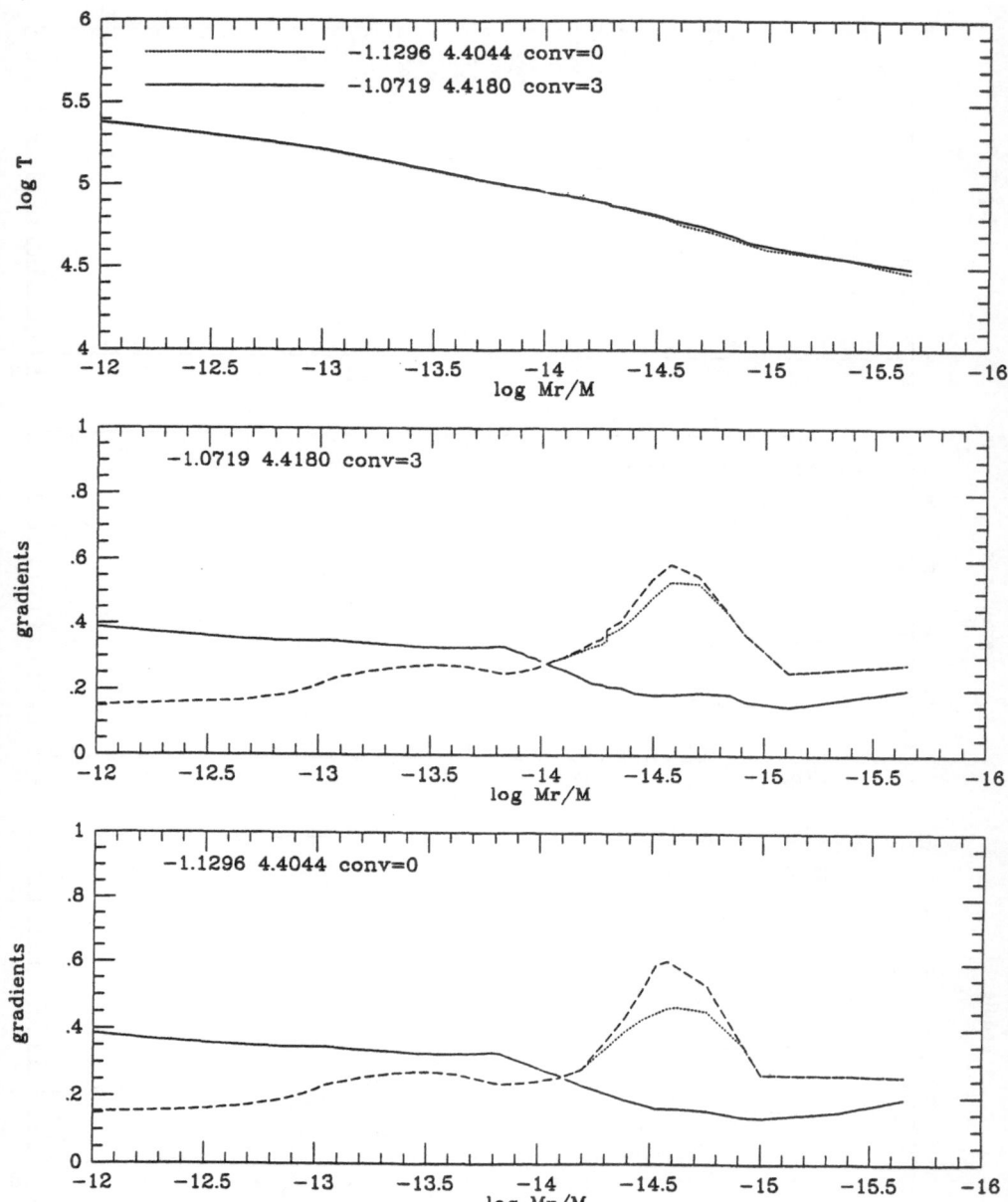

Figure 3

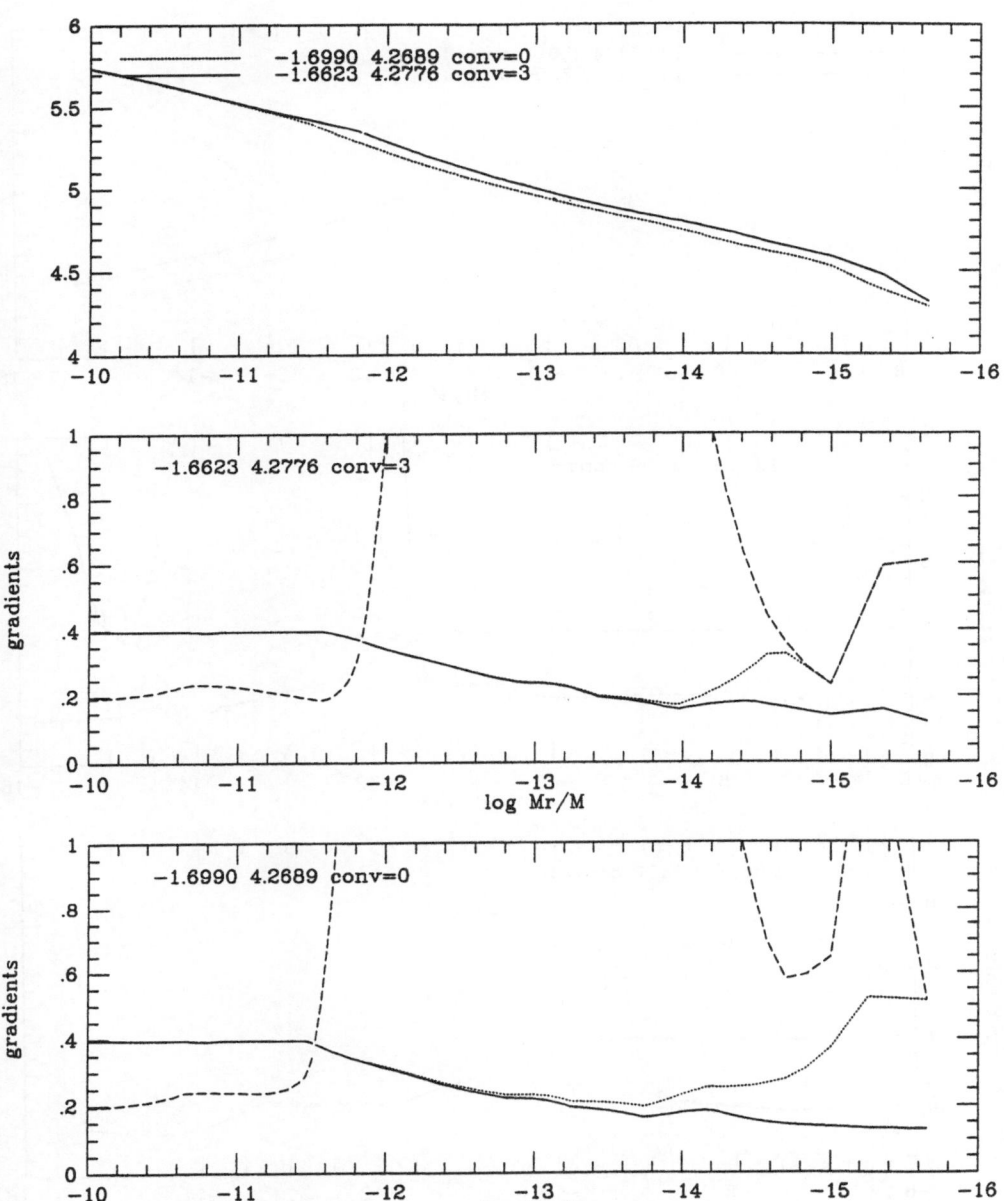

Figure 4

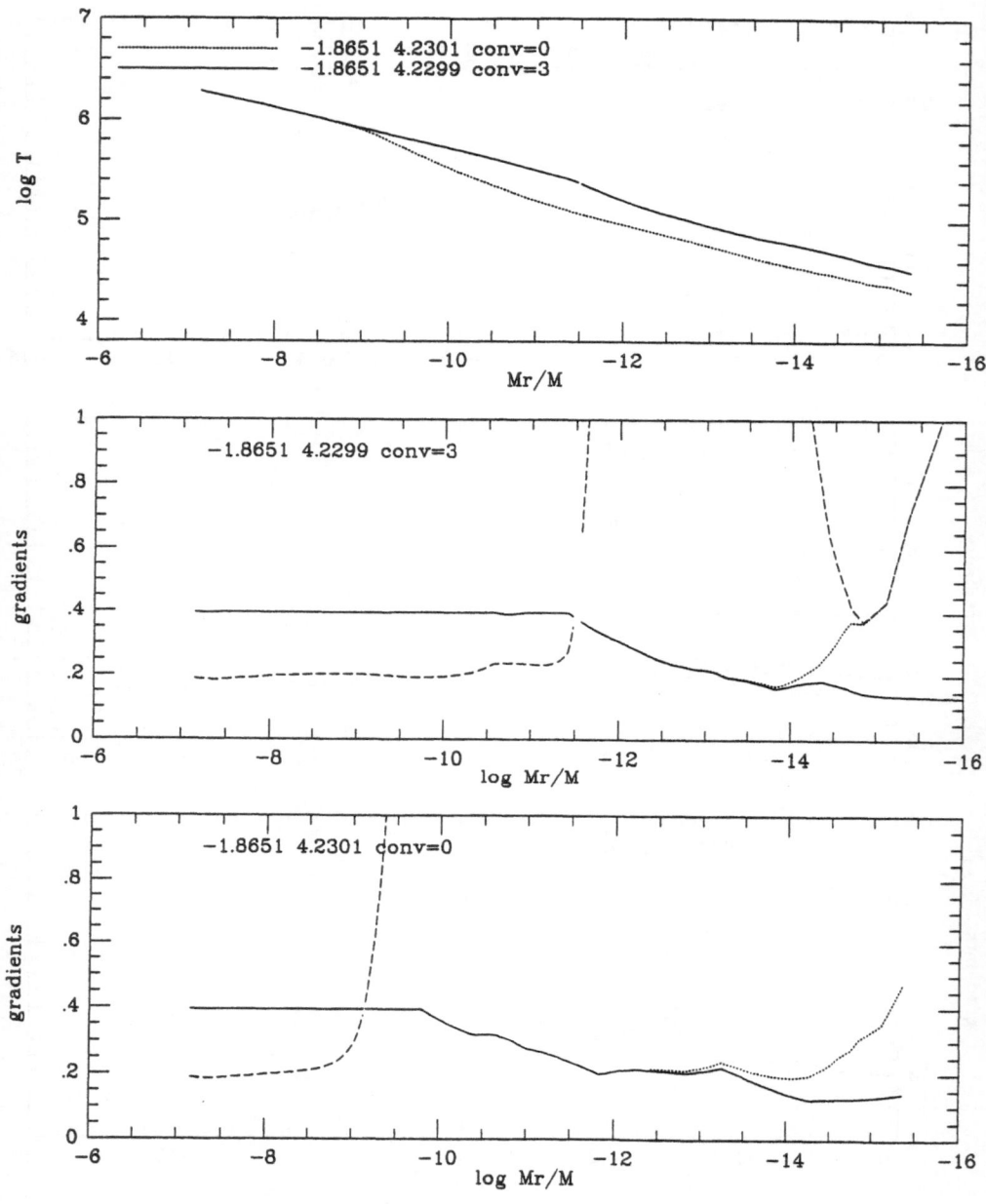

Figure 5

313

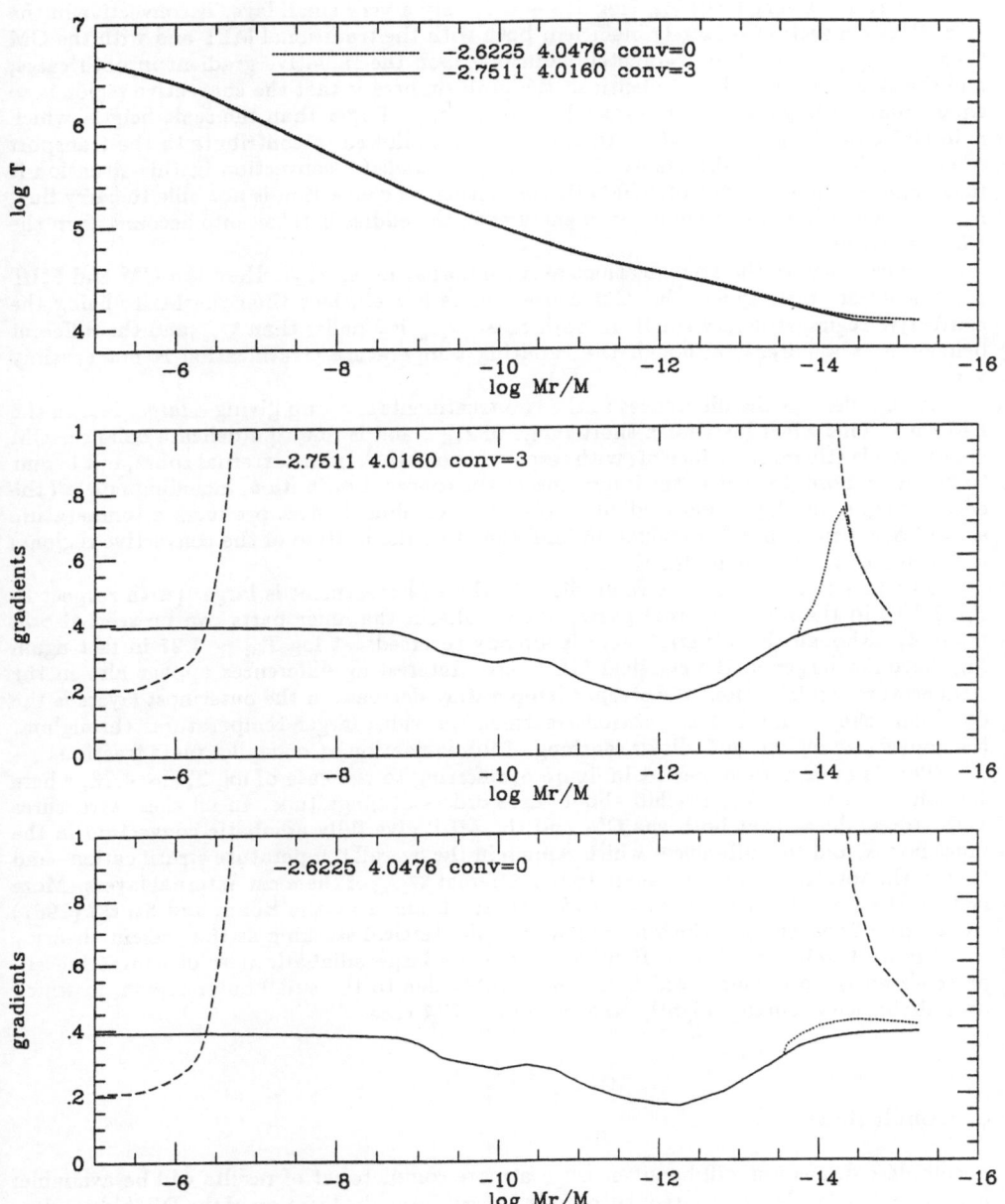

Figure 6

Figure 2: at very large T_{eff} ($\log T_{eff} \sim 4.5$) only a very small layer is convective in the star. Here convection is largely inefficient both with the traditional MLT and with the CM theory: ∇_{conv} is therefore completely adjusted upon the radiative gradient in both cases, and the structures are almost identical: the problem here is that the convective region is so small that the large, energetic eddies have dimensions larger than the scale heigth, which is in CM case the geometrical depth z, and are not allowed to contribute to the transport of energy. This result also shows that talking of *adiabatic* convection in this situation is meaningless: there can not be adiabatic convection, as convection is not able to carry flux, not even when the full distribution of energies in the eddies is taken into account as in the CM treatment.

Figure 3 shows the stratifications at a somewhat lower T_{eff} . Here the CM and MLT treatments are distinguishable: CM convection is less efficient than the MLT, being the convective region still very small. In both cases ∇_{conv} is smaller than ∇_{rad} and the different behaviour is apparent, although the resulting temperature stratification is not sensibly varied.

At smaller T_{eff} the differences in the two treatments show up giving a *larger* M_{ce} in the CM case. This effect lasts for a short range of T_{eff} , and is due to a delicate balance: CM convection is still more inefficient, with respect to the MLT, in the external zones, but begins to be much more efficient in the inner ones as the energy distribution, including now *all* the eddies, begins to play a non negligible role. This combined effect produces a temperature stratification such that full ionization (and therefore the bottom of the convective regions) is reached at a larger mass fraction.

The fact that the convective gradient in the CM treatment is larger (with respect to the MLT) in the more external parts, but smaller in the inner parts can be seen also in figure 4, although the situation here is already reversed: at $\log T_{eff} \sim 4.27$ in fact again M_{ce} becomes larger in the classical MLT case. Interesting differences appear also in the temperature stratification: the steeper temperature decrease in the outermost layers is the dominant effect, and CM convection treatment provides larger temperatures throughout the convective region, and allows reaching of full ionization at a smaller mass fraction.

This is even more apparent in figure 5, referring to the case of $\log T_{eff} \sim 4.23$, where the difference in the M_{ce} reaches about three orders of magnitude. In all these structures WDs are so dense that both the CM and the MLT give fully adiabatic convection in the inner layers, and the differences which remain in the overall temperature stratification –and thus in the total M_{ce} – are due again to the different ∇_{conv} of the most external layers. More or less, these results are comparable with those obtained by the Böhm and Stuckl (1967) treatment of convection, which also allows for the vertical stacking as the present theory.

Figure 6 refers to $T_{eff} \sim 10000K$, where the large adiabatic part of convection supersedes all the differences which can be possibly due to the small outer region, in which overadiabaticity remains slightly stronger in the CM case.

4. Conclusions

A complete discussion will be given when a more complete set of results will be available: here we only consider the consequences for what concerns the location of the DBV blue edge. There has been a controversia between different determinations of the T_{eff} of the hottest DBV (GD 358) as the T_{eff} derived from optical data is $\sim 25000K$ (Oke *et al.*1984) while the T_{eff} from IUE was $\sim 29000K$ (Liebert *et al.*1986). The UV result has generally been regarded as suspect, as the IUE flux calibration at large T_{eff} is not accurate enough to allow precise determinations (Koester *et al.*1983), and Thejll's recent results (this conference) seem to go towards a revision of the T_{eff} scale in general agreement on the lower T_{eff} value.

Interestingly enough, the temperature scale is crucial to establish whether the presented new convection theory, which has already passed the test of the fit of the solar model, although *no free parameters* such as the ratio l/H_p were left to play around, also provides good results for what concerns WD structures. In fact, the new treatment of convection *does not allow* the blue edge of the DB instability strip to be as hot as $29000K$, as convective transport in the envelopes at that T_{eff} is very inefficient, and the convective region extension is mainly determined by the values of the radiative gradient, so that it is the same both for CM and MLT case. The convective mass at so large T_{eff} is so small that t_{th} at its bottom is certainly much shorter than the several hundred seconds of the pulsation period of DBs.

At $T_{eff} \sim 25000K$ or smaller, we are just in the region in which our scanty results predict M_{ce} larger than in the MLT case, that is exactly what is needed to go towards the direction of a good fit between t_{th} and the DBV periods. Therefore, only with the 'low temperature' scale for DB variables these first results can be regarded as satisfactory enough as a test of the ability of this new convection treatment to provide an useful tool for the investigation of convection in WDs also.

References

Böhm, K. H., Stuckl, E. 1967. *Zs. f. Ap.* **66**:487

Böhm, K. H. 1969. In *Low–Luminosity Stars* Gordon & and Breach Science Publishers, p. 393

Böhm–Vitense, E. 1958. *Zs. f. Ap.* **46**:108

Canuto, V. M., Mazzitelli, I. 1990. *Ap. J.*, in press

D'Antona, F., Mazzitelli, I. 1975. *Astron. Astrophys.* **44**: 253

D'Antona, F., Mazzitelli, I. 1979. *Astron. Astrophys.* **74**: 161

D'Antona, F., Mazzitelli, I. 1989. *Ap. J.* **347**: 934

D'Antona, F., Mazzitelli, 1990. *Ann. Rev. Astron. Astrophys.* **28**: 139

Fontaine, G., Van Horn, H. M. 1976. *Ap. J. Suppl. Series* **31**: 467

Koester, D., Weidemann, V., Vauclair, G. 1983. *Astron. Astrophys.* 123: L11

Lesieur, M. 1987. *Turbulence in Fluids*, M. Nijhoff Publ., Dordrecht

Liebert, J., Wesemael, F., Hansen, C. J., Fontaine, G. 1986. *Ap. J.* **309**: 241

Oke, J. B., Weidemann, V., Koester, D. 1984. *Ap. J.* **281**: 276

Orszag, S. A. 1977. *Les Houches Summer School of Theoretical Physics*, ed. by R. Balian and J. L. Peube, Gordon and Breach, New York

Tassoul, M., Fontaine, G., Winget, D. E. 1990. *Ap. J. Suppl.* **72**: 335

Van Horn, H. M. 1968. *Ap. J.* **151**: 227

DISCUSSION

VAN HORN :

Could you show again your graph of convection zone mass vs T_{eff}? I'd like to be sure I understand at what range of T_{eff} the effects are largest.

D'ANTONA :

The effects are largest in the region of fast deepening of M_{ce} when decreasing T_{eff} : as we expect, the extent of the partial ionization zone there is very much dependent on the outer boundary conditions, and in this case, as well as with different choices of α in the MLT, the "outer boundary" is in practice the region of *superadiabatic* convection, close to the surface, where the actual values of ∇_{conv} determine the temperature profile. But there are sensible effects also in the region around $T_{eff} \sim 25000K$, where the larger efficiency of CM convection in the inner layers seem to overcome the lower efficiency in the outermost ones, and provide as a final result a larger M_{ce}.

STANGHELLINI :

The density inversion that follows from the CM convection theory could be important for period changes with stellar evolution. Could this be important even for PG1159-035? (In our models a very thin surface convective layer is present).

D'ANTONA :

It should be investigated. Our present hot WD models, either with hydrogen on top or not, have no Carbon enhancement in the outer layers, so they are purely radiative and convection is not present. I take the occasion to recall that Steve Kawaler has examined the adiabatic pulsational properties of some of our models at $T_{e}ff \sim 100000K$ and found modes trapped in the outer *hydrogen* layer, having periods and (negative) period derivatives of the order of those observed in PG 1159-035. These models are certainly not apt to describe PG 1159-035, first because this star in H-deprived and carbon rich (see the results presented here by Uli Heber), second because there is no known driving for such modes, but they hint that envelope trapping is probably the way to achieve negative period derivatives.

ABUNDANCES IN COOL DZA AND DAZ WHITE DWARFS: NEW RESULTS USING LABORATORY DAMPING CONSTANTS

GORDON L. HAMMOND
Department of Mathematics
University of South Florida, Tampa, FL 33620

EDWARD M. SION
Observatoire Midi-Pyrenees
CNRS, URA285, Toulouse, France
and
Department of Astronomy and Astrophysics
Villanova University, Villanova, PA 19085

SCOTT J. KENYON
Harvard-Smithsonian Center for Astrophysics
60 Garden Street, Cambridge, MA 02138

PER A. AANNESTAD
Department of Physics
Arizona State University, Tempe, AZ 85207

ABSTRACT. We have determined H/He ratios, effective temperatures, and metal abundances for two cool white dwarfs, G77-50 and G74-7, that exhibit CaII lines and hydrogen Balmer lines. Using model atmosphere techniques and our recently published MMT spectra, we find helium to be the dominant atmospheric element in both stars, and the [Ca/H] ratios to be extremely metal deficient relative to the solar value. The observed Balmer decrements for these stars are compatible with the derived H/He ratios when a line broadening treatment properly combining resonant and non-resonant neutral interactions is applied. The derived abundances are consistent with predictions of accretion from the local interstellar medium.

1. Introduction

Helium-rich degenerates cooler than 10,000K, whose spectra exhibit metal absorption lines, seem to be of two types: a majority which show no detectable hydrogen, classified DZ, and a handful of objects showing the presence of both metal lines and hydrogen lines, classified DZA or DAZ depending on the strengths of the Balmer lines relative to CaII. All of these objects are widely thought to have accreted their photospheric metals from the interstellar medium (cf. the review by Koester and Chanmugam 1990; Greenstein and Liebert 1990 and references therein, hereafter denoted GL). In the group showing metal lines but no hydrogen, it would appear that interstellar accretion must occur with the selective screening out of hydrogen, as grains and gas accrete (cf. Wesemael and Truran 1982). However, the operation of this "propeller" mechanism in a slowly rotating, weakly magnetic white dwarf appears to be

317

G. Vauclair and E. Sion (eds.), White Dwarfs, 317–331.
© 1991 *Kluwer Academic Publishers.*

inefficient and erratic when the theory is compared to a comprehensive observational data set (Sion, Hammond, Wagner, Starrfield, and Liebert 1990b). Bergeron, Wesemael, Fontaine, and Liebert (1990a) have recently presented a diagnostic for H/He ratios in a large sample of cool DA stars, some of which show calcium. Since the DBAZ stars (Sion, Aannestad, and Kenyon 1988) and the DZA (and possibly the DAZ) stars almost certainly accreted without the operation of a propeller, they may bridge a gap between ordinary cool DA stars, some of which may owe their hydrogen to accretion, and the DZ stars with accreted metals but no hydrogen. Therefore it is critical to determine accurate H/He abundance ratios (or upper limits) for the DZA and DAZ stars in order to (1) compare them to hotter DBAZ stars and cooler DA and DZ stars, and (2) test existing accretion hypotheses and explore new approaches. These considerations were the motivations for analyzing first the DZA and DAZ stars in the MMT sample of Sion, Kenyon, and Aannestad (1990, hereafter denoted SKA).

However, these cool DZA and DAZ stars present a severe challenge in abundance determinations by means of synthetic spectra: they combine all of the collisional line broadening processes that can operate in warm plasmas. Furthermore, the processes that are best understood, Stark and resonance broadening, make only minor contributions to the profiles of most of the observed lines. Therefore, the reliability of abundance determinations depends on our knowledge of the neutral collision broadening processes.

2. The Neutral Collision Broadening Problem

Earlier analyses of cool white dwarfs employed damping parameters for neutral non-resonant collision broadening, occasionally (imprecisely) called van der Waals broadening, that were derived from the Unsöld (1955) relationship and then multiplied by enhancement factors. Some of these factors have been derived from solar analyses, some from laboratory experiments, and some from calculations based on more realistic interaction potentials than the van der Waals potential used by Unsöld, but few of these factors have been listed in the literature.

We have attacked the line broadening problem in these cool white dwarfs by using, as far as possible, laboratory measured damping constants. We have evaluated the experimental results for broadening by neutral hydrogen and helium, and selected results from experiments that were conducted under cool white dwarf atmospheric conditions, that had the best state diagnostics, that had reliable corrections for other broadening processes, and that had checks for self-absorption effects. Although few experiments passed all these tests, the absorber/perturber pairs for which we found usable data (for transitions detected in cool DZA spectra) are listed in Table 1. Also listed are some references to theoretical works, using realistic model potentials, that confirm some of the laboratory results, and references to the Stark and resonance broadening theories used in this work.

The enhancement factors shown in column 4 of Table 1 are scale factors that multiply the classical van der Waals damping constants to yield agreement with the laboratory values. We find that all the enhancement factors derived from both helium and hydrogen broadening experiments cluster within a limited range from 1.4 to 2.2, and we note that the factors obtained from the theoretical work of Monteiro et al.(1986, 1988), and from the empirical fitting by Ayres (1977) to the solar CaI resonance line and the similar fitting to the MgI 3835 Å triplet in the DZA star Ross 640 (Hammond 1990), also fall within this range. The reliability of the enhancement factors for the neutral broadening of the Balmer lines is less than that for the metal lines in Table 1. Except for the experiment by Eckart & Baldwin (1978), these experiments suffered from inadequate diagnostics or from temperatures too low for reliable extrapolation to white dwarf atmospheric conditions. In spite of these difficulties, we expect the use of all these experimental damping constants to yield more reliable abundances than those from earlier analyses.

TABLE 1.   ENHANCEMENT FACTORS & TEMPERATURE EXPONENTS
FOR COLLISION BROADENING IN COOL WHITE DWARFS

INTERACTION	REFERENCE	Expt/ Theor.	ENHANC. FACTOR	TEMP. EXP.	NOTES
CaII/Stark	Jones, W.,et al. (1972)	E			
"	Dimitrijevic,Konjevic(81)	T		-.11	
CaII/Helium	Hammond, G. (1975)	E	1.37(H)		
			1.87(K)		
"	Monteiro,T., et al.(1986)	T	1.4 (H)	.37	
			1.9 (K)	.39	
CaII/At.Hyd	Monteiro,T., et al.(1988)	T	2.1 (H)	.29	
			2.2 (K)	.31	
CaI/Helium	Driver & Snider (1976)	E	1.8	.43	(1)
"	Bowman & Lewis (1978)	E	2.2		(1)
CaI/At.Hyd.	Ayres (1977) (solar)	e	2.2		
MgI/Stark	Griem, H. (1964)	T		-.19	
MgI/He,/H	Hammond, G. (1990) (R640)	e	2,2,1		(2)
MgI/Helium	Deridder,van Rensbergen,	T		.32	
MgI/At.Hyd.	" (1976)	T		.28	
Balmer/Strk	Kepple & Griem (1969)	T			
Balmer/Res.	Lortet & Roueff (1969)	T			
Hα/At. Hyd.	Eckart & Baldwin (1979)	E	1.77	.3	(3)
Hα/Helium	Kielkopf, J. (1975)	E	1.6	.3	(4)
"	Sidell & Ch'en (1977)	E	1.9		(5)
Hβ/Helium	"	E	1.8		(6)

NOTES:
(1) Temp. exp. derived from these expts. (765 K to 6000 K).
(2) Enhancement factors for the J=5/2, 3/2, 1/2 components.
(3) Exptl. T=10,000 K; more reliable than the 3 expts. below.
(4) Exptl. T=800 K; more reliable than the 2 expts. below.
(5) Exptl. T=6300 K; uncertain diagnostics.
(6)   "   " ;   "   "   , only one measurement.

In cool DA or DAZ atmospheres, several authors have considered the contribution of resonance broadening to the Balmer line profiles. However, there is some confusion in the white dwarf literature regarding the correct treatment of the combined resonant and non-resonant effects produced by the collision of a ground state hydrogen atom perturber with an excited state hydrogen absorber. This problem has been treated by Lewis (1967) and applied to the Sun by Lortet & Roueff (1969), and we apply their results here to these cool white dwarf atmospheres. Their principal finding is that the broadening of the lower ($n=2$) level is due largely to the resonance interaction while the higher levels are perturbed mainly by non-resonant interactions, and the combined effect is far less than that obtained by simply adding the damping constants for the two interactions. The non-resonant broadening by hydrogen collisions has been neglected by other authors. We also point out that resonance broadening is nearly constant along the Balmer series; to a close approximation, it is proportional to the ratio of the oscillator strength and the frequency of Lyman-α. Therefore, the observed increase in the Balmer decrement with decreasing H/He ratio (Greenstein 1986, hereafter denoted GR) is due not to a decrease in resonance broadening along the Balmer series, as suggested by Wehrse (1977), but rather to increasing non-resonant broadening by helium. We explore this point below in our analysis of the DAZ star G74-7.

A point related to the neutral line broadening problem is the question of the importance of quenching in modifying the line profiles in these high density atmospheres. Quenching is defined as the destruction (Wehrse 1977) or the partial depopulation (Hummer & Mihalas 1988, hereafter denoted HM) of excited states due to close perturber collisions. Bergeron et al.(1990a) have applied the HM formalism to both hot and cool hydrogen rich white dwarfs, and they claim that its use provides a diagnostic for the H/He abundance ratio in the cool DA stars. Although this formalism is appropriate in the hot DA stars in which the Coulomb interaction dominates, the hard sphere approximation used for neutral interactions is much too severe. There is ample laboratory spectroscopic evidence that high-lying Rydberg states persist at neutral perturber densities such that several perturbers are within the excited state orbit. The review by Allard & Kielkopf (1982) discusses those experiments and the supporting theories. We find no justification for including hard-sphere quenching in our line profile calculations for cool DZA and DAZ stars (see Section 5. below).

3. The Model Atmosphere Code

The code used in this work is an extensively updated version of the one described by Hammond (1974). The basic structure of the code was taken from Mihalas (1967); some opacity subroutines came from Kurucz (1970). The ODF metal line blanketing treatment is similar to Carbon & Gingerich (1969), and Wehrse (1972); blanketing by CaII H&K and by the first five Balmer lines is included explicitly. Lyman-α resonance broadening (Sando & Wormhoudt 1973) and collision induced absorption of H_2 by H, H_2 and He (Borysow & Frommhold 1989) have recently been added to the usual H, H^-, He^-, Rayleigh scattering, and Mg, Ca, Si, C, Fe photoionization continuous opacities. Molecular hydrogen is the only diatomic specie included in the equation of state.

The depth structure is calculated at 50 points evenly spaced in log τ over 5 decades and at 70 wavelengths (912 Å to 4.5 μm) for each depth point. Convective energy transport is treated with conventional mixing length theory (ML1) as detailed in Mihalas (1978). The Avrett-Krook temperature correction procedure is used with a simplification suggested by Swihart (1965). The mass of hydrogen, $(M_H/M\odot)$, in the atmosphere is calculated by integration from $\log(\tau) = -4$ to $+1$. The present code, which runs on the Univ. So. Fla. VAX 3400, has been tested by comparing the runs with depth of temperature, gas pressure, electron pressure, continuous opacity, and radiative flux with those from both hydrogen-rich and helium-rich, log g = 8, ATLAS-5 models (Kurucz 1970) run on the Univ. Md. UNIVAC 1108 (Hammond 1974). The differences found in all these variables from the two codes average 0.1 percent.

Broad band, Johnson UBV colors are calculated using the filter transmission functions and zero points given by Ažusienis and Straižys (1969). The Cousins R-I color is calculated with a zero point derived by convolving the Bessell (1990) R and I filter transmissions with the photoelectric scanner fluxes for the cool DZ van Maanen 2 (Greenstein 1983). The bolometric correction is calculated using the suggestion of Weidemann & Bues (1967) to determine the integration constant from black-body fluxes. Strömgren uvby colors are calculated from the filter functions and zero points given by Matsushima (1969). The monochromatic MCSP colors originally defined by Greenstein (1976) and the additional DBSP colors defined in Greenstein & Liebert (GL) are also calculated. The UBV and uvby color calculations are blocked by the CaII H&K lines and by the lowest five Balmer lines.

4. Observations and Abundance Analyses

The observed spectra analyzed here are part of the large body of data (SKA) taken to test directly the interstellar accretion hypothesis for the source of surface metals in DZ stars. The test requires not only abundance determinations but also radial velocity measurements to derive complete space motions and thus relate abundances to the structure in the local interstellar medium (the LISM). For a preliminary study of this test, see Aannestad & Sion (1985). It is now recognized that pressure shifts due to non-resonant collisions between metal atoms and helium and hydrogen atoms are important corrections for radial velocity measurements in these cool, dense atmospheres, and we derive important diagnostic information from these shifts.

We describe below in some detail the results for G77-50, a very cool DZA that has no previously published analysis, and our results for G74-7, a cool DAZ for which a recent paper (Bergeron et al. 1990a) obtained quite different results by applying the HM hard sphere quenching and by neglecting non-resonant broadening of the Balmer lines by hydrogen collisions. We also briefly describe the results for Ross 640 and L745-46A, two other DZA stars in the MMT sample that have previously been analyzed by others (with results similar to ours), and then compare the H/He abundance ratios and metal abundances of these four DZA and DAZ stars with the hot DBZ and DBAZ stars (Kenyon et al. 1988; Sion, Aannestad and Kenyon 1988).

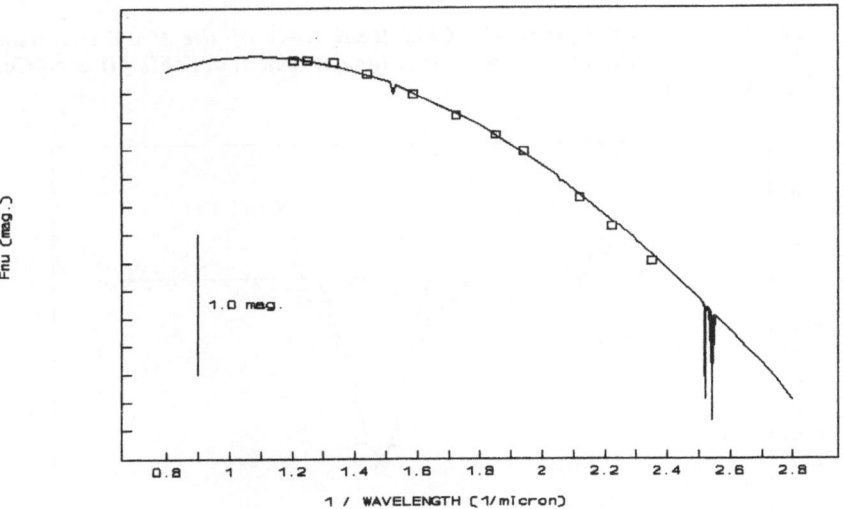

Figure 1. The model fluxes for $T_{eff} = 5200$ K have been normalized to the DBSP fluxes at 5405 Å.

4.1 G77-50 ABUNDANCE ANALYSIS

Spectra for this very cool DZ (WD0322-019, Gr 566, LHS 1547) were first reported by Hintzen & Strittmatter (1974), who classified it as DG. Greenstein (1986, GR) resolved an EW = 2.1 Å

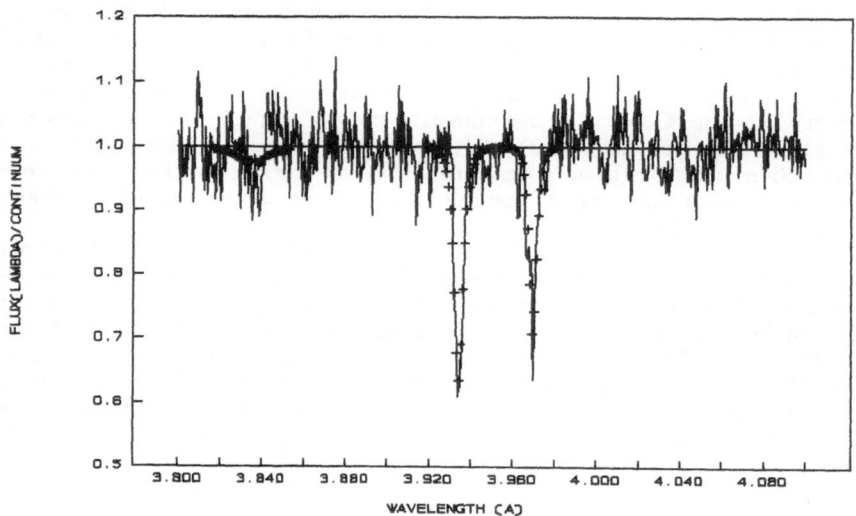

Figure 2. The 5200 K model fit to the CaII H&K lines of the MMT spectrum of G77-50. The metal abundances relative to solar derived from this fit are [Ca/H] = -2.8, [Mg/Ca] < -0.3.

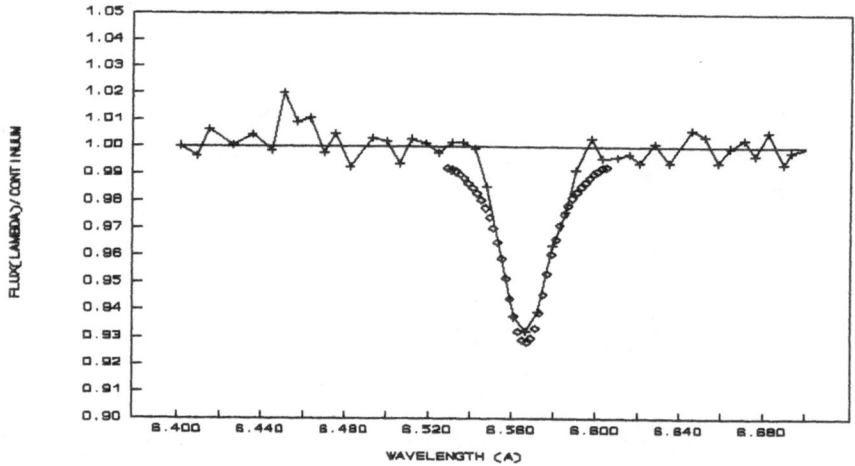

Figure 3. Model fit to the observed (GR) profile of Hα in G77-50.

Hα line and classified it DA10+ based on DBSP colors. McCook & Sion (1987) list $M_v = 15.1$ based on MCSP colors (Greenstein 1984). The blue spectrum at high resolution showing sharp, weak CaII H&K lines is plotted in SKA. The model that provides the best fit to all the observational data, when the surface gravity is fixed at log g=8.0, has the following parameters and error estimates:

T_{eff} = 5200($\pm$100)K
$\log(N_H/N_{He}) \equiv$ (H/He) = -1.6 ($\pm$ 0.1)
(Ca/He) = -10.1 ($\pm$ 0.3), [Ca/H] = -2.8 ($\pm$ 0.4)
(Mg/He) < -9.3, [Mg/Ca] < -0.3
Bolometric Correction (BC) = -0.14

The parentheses denote the logarithm of number ratios, and the square brackets denote the logarithmic ratio relative to the solar value.

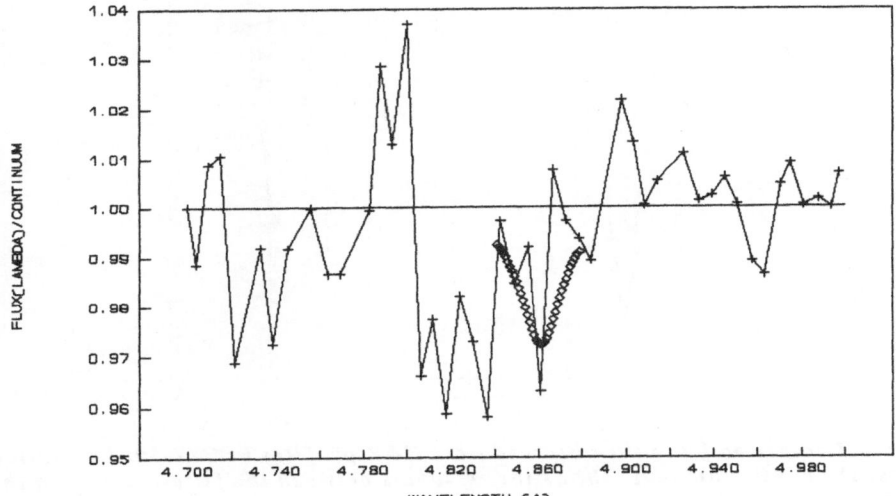

Figure 4. *The Hβ region of G77-50, and the model Hβ profile.*

The effective temperature is derived from the fit of the calculated continuous flux to the DBSP fluxes listed in GL. The fit is shown in Figure 1. The calcium abundance is obtained by fits to the CaII H&K lines. The best fit is shown in Figure 2; the model profiles have been convolved with a Gaussian instrumental profile with full width at half-power (FWHP) = 1.0 Å. The upper limit for the magnesium abundance is estimated from the weak calculated MgI multiplet at 3835 Å

The H/He abundance ratio is determined from fits to the observed Hα profile shown in Fig. 8 of GR. An instrumental profile of FWHP = 9.0 Å was convolved with the calculated profiles of both Hα and Hβ. The fits to the GR observations are shown in Figures 3 and 4. The Balmer decrement, EW(Hβ)/EW(Hα), for this model is 0.2, a value consistent with the GR observations if the noise level in Fig. 4 is considered.

An independent measure of the H/He abundance ratio is obtained from the differential pressure shift of the CaII H&K doublet. The shift parameters for both helium and hydrogen collisions have been included in the line absorption coefficient calculations in the model code.

324

Measurements of the velocity difference of the two lines have been made both for a large grid of models with variable T_{eff} and variable H/He number ratios {5000K < T_{eff} < 12000K and -6 < (H/He) < +1} and also for most of the DZ and DZA stars in the DZ MMT Atlas (SKA). The physics underlying this scheme was presented by Hammond (1989); the scheme is now

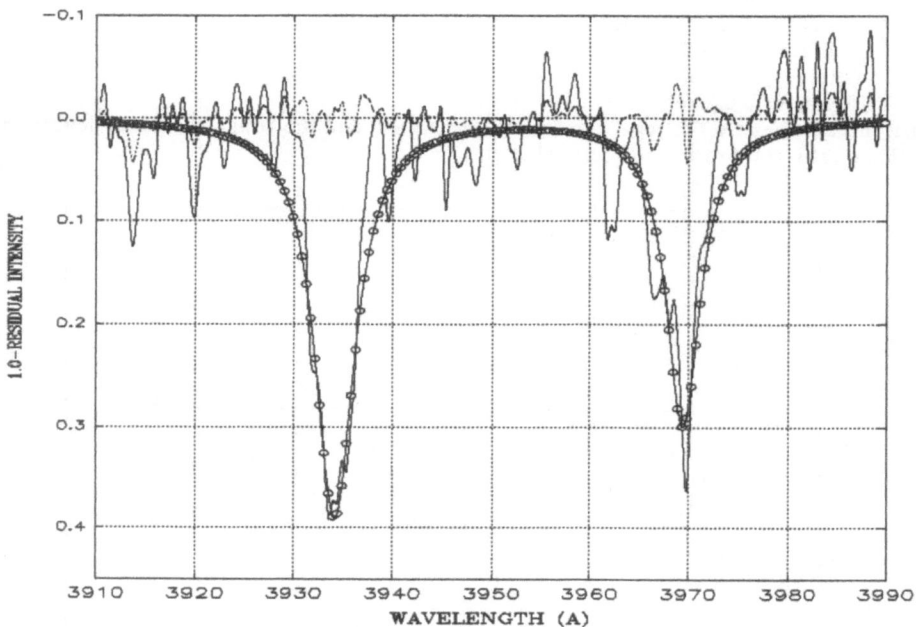

Figure 5. *Lorentzian fit to the observed CaII H&K profiles in G77-50. The dotted line near the continuum level shows the residuals between the fitted curve and the MMT data.*

enhanced with the inclusion of the pressure shifts due to hydrogen collisions (Monteiro et al. 1988). The central wavelengths of the two components, in both the synthetic spectra and also the observed spectra, are now determined by nonlinear least squares fits to the sum of two Lorentzian profiles of arbitrary width, depth, and central wavelength using the Levenberg-Marquardt method (Press et al. 1989). From the shape of the shift-(H/He)-T_{eff} surface in the T_{eff} = 5200K region we find that the differential shift measurement error of ± 7 km/s corresponds to an error of ± 0.3 dex in (H/He). The Lorentzian fit to the G77-50 spectrum is shown in Figure 5 and yields a differential shift of 34 km/s, and a similar fit to the synthetic H&K profile of the best model for G77-50 yields a differential pressure shift of 28 km/s. The good agreement between these shift values supports the value of (H/He) = -1.6 determined from the Balmer line fits.

4.2 G74-7 ABUNDANCE ANALYSIS

The presence of calcium in G74-7 (WD0208+396, EG168, LHS 151) was suggested by Eggen & Greenstein (1965), and confirmed by Lacombe, Liebert, Wesemael, and Fontaine (1983, hereafter LLWF). The MMT spectrum shown in SKA at 1 Å resolution yields EW for the

CaII(K) and Balmer lines similar to those reported in LLWF. Liebert, Dahn, and Monet (1988, hereafter LDM) derive $M_v = 13.46(\pm.13)$ from the trigonometric parallax, and they obtain BC = -0.40 from the hydrogen-rich models by Shipman (1979).

We find $T_{eff} = 7300(\pm 100)$K by fitting the observed UBV (LDM), uvby (Wegner 1983), and MCSP colors. The MCSP colors show little sensitivity to the H/He ratio or to the metallicity [M/H], for (H/He) < +1. An extensive grid of models was required to obtain simultaneous fits to the CaII and the Balmer lines, especially for fitting the profile of the $H\varepsilon + CaII(H)$ blend. The fits of the synthetic profiles to the MMT data are shown in Figure 6. The chemical composition that yields those fits, and the other parameters for the model with log g = 8.0, are listed below:

T_{eff} = 7300 ($\pm$ 100)K
(Ca/He) = -10.6 ($\pm$ 0.3), [Ca/H] = -3.0 ($\pm$ 0.5)
(Mg/He) < -9.5, [Mg/Ca] < 0.0
Bolometric Correction (BC) = -0.26

The upper limit for (Mg/He) quoted here provides a MgI 3835 Å triplet with EW = 0.2 Å, which significantly perturbs the profile of the H9 line.

Using the experimental damping constants and the broadening theories described in Section 2, we have calculated Balmer profiles for a wide range of H/He ratios, and some of these profiles are shown in Figure 7. The lower set of profiles are the best fits to the MMT data and the predicted $H\alpha$ and $H\beta$ profiles for (H/He) = -1.4, and the upper sets come from H/He ratios ±1 dex from the best fit composition. Log g = 8.0 for all three sets. The profiles for (H/He) = -0.4 are clearly too narrow, and the profiles for (H/He) = -2.4 exhibit an extremely steep Balmer decrement. The gas pressures in the most helium-rich model are about 0.8 dex higher than those in the best fit model, and only the relatively small damping of $H\alpha$ by helium allows $H\alpha$ to survive in the (H/He) = -2.4 atmosphere. The damping due to

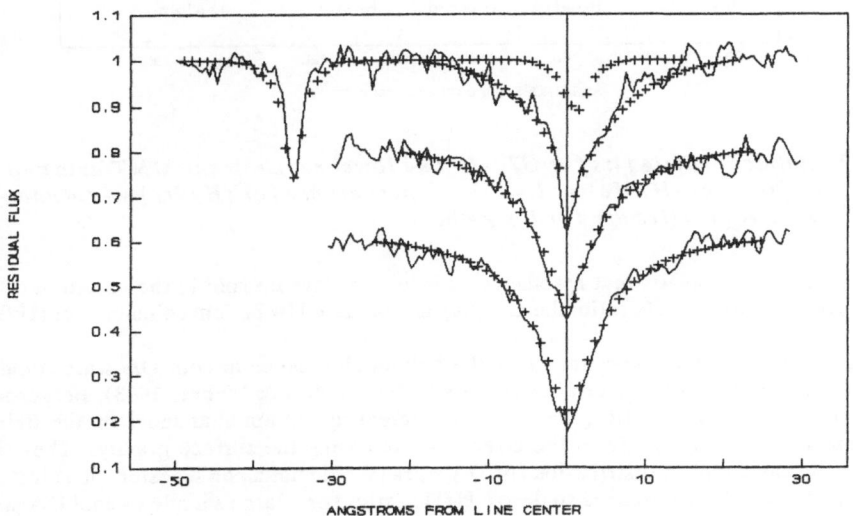

Figure 6. *CaII and Balmer line fits to the MMT spectrum of G74-7. From bottom to top, the profiles are $H\gamma$, $H\delta$, and the $H\varepsilon + CaII(H)$ blend. The CaII H&K lines are also shown on the top line.*

326

combined resonant and non-resonant hydrogen collisions relative to the damping due to helium collisions (for any Balmer line) varies in the following way for these three compositions:

(H/He)	$(\Gamma^*N)_H/(\Gamma^*N)_{He}$
-0.4	10.
-1.4	0.06
-2.4	0.004

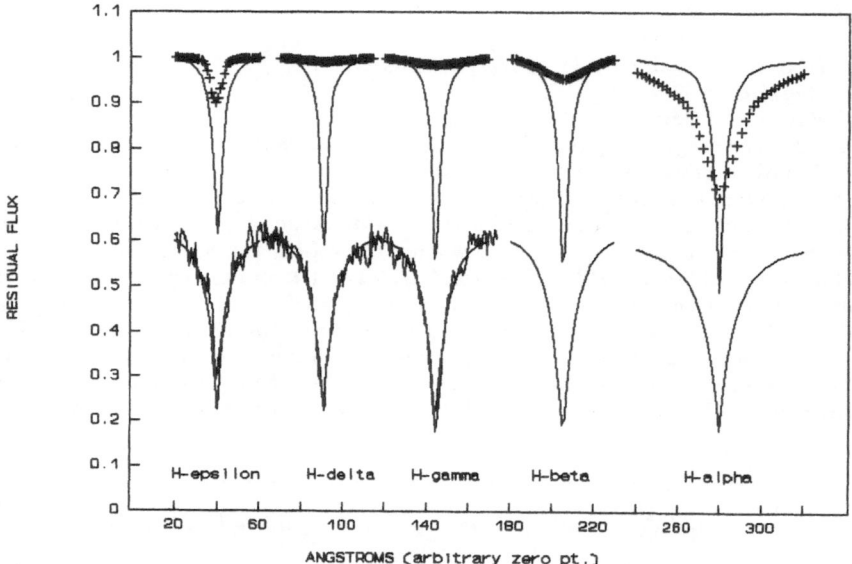

Figure 7. Balmer profile fits for G74-7. The lower set show the MMT data and the synthetic profiles for (H/He) = -1.4. The upper sets are for (H/He) = -0.4 and -2.4 (+ symbols). All profiles are for log g = 8.

It is clear from these results that resonance broadening plays no role in the variation of the Balmer decrement with H/He ratio: the damping is dominated by helium collisions for (H/He) < -1.

In spite of earlier suggestions that the helium abundance in cool DA stars could be deduced from high Balmer line profiles (Wehrse 1977, Liebert & Wehrse 1983), Bergeron et al. (1990a) claim that, in practice, the effects of increasing helium abundance on the Balmer profiles are indistinguishable from the effects of increasing the surface gravity. They then argue that the narrow mass distribution for DA stars (Weidemann and Koester 1984) justifies fixing log g at 8.0 and then proceed to derive H/He ratios for a large sample of cool DA stars. We find this claim is true for the Hγ, Hδ, and Hε profiles in G74-7. The sharp profiles for (H/He) = -0.4 in Fig. 7 widen to become satisfactory fits to the MMT profiles when the surface gravity is increased to log g = 8.2. However, when we calculate the mass for G74-7 from T_{eff}, M_{bol}, and gravity, we obtain M/M⊙ = 0.57 for log g = 8.0 and M/M⊙ = 0.91 for log g = 8.2. This latter mass is statistically unlikely, but we have to allow its possibility in this case, because there is no independent diagnostic for surface gravity in cool white dwarfs, and because we are deprived of the measurement of the H/He ratio from the differential pressure shift of the CaII H&K lines due to the blend of Hε and CaII(H).

4.3 ROSS 640 AND L745-46A ABUNDANCE ANALYSES

These two cool DZA stars have been analyzed by model atmosphere techniques by numerous authors in the past two decades. The most recent studies, using IUE spectra, are by Zeidler, Weidemann and Koester (1986, hereafter denoted ZWK), but the damping constants they used are not fully described. Our results for these stars have been presented by Aannestad, Hammond, Sion and Kenyon (1990), and by Sion, Hammond, Aannestad and Kenyon (1990a), but full details will be included in the analysis of all the MMT spectra (Hammond et al. 1991). We include our results for these two stars here to improve the statistics for DZA and DAZ stars in our attempt to understand the source of hydrogen and metals in the atmospheres of these helium-rich objects.

For Ross 640, using the damping constants described in Table 1, we obtain excellent fits to the CaII H&K lines and the MgI 3835 Å triplet in the MMT (SKA) spectrum, to the Hα and Hβ profiles published by Liebert (1977), and to the MCSP colors (GR) with a model with T_{eff}=8300 ($\pm$200)K, BC= -0.39, (H/He)= -3.4($\pm$0.2), (Ca/He)= -8.7($\pm$0.3), and (Mg/He)=-7.5($\pm$0.3). These abundances differ (present results - other results) by 0.1 dex, 0.4 dex, and -0.3 dex for H, Ca, and Mg, respectively, from those obtained by ZWK, differences well within the errors of both analyses.

For L745-46A, we obtain excellent fits to the CaII H&K lines in the MMT spectrum, to the EW=3.5 Å Hα quoted in GL, and to the MCSP colors (GR) with a model with T_{eff}=7500($\pm$200)K, BC= -0.26, (H/He)= -3.2 ($\pm$0.4), (Ca/He)= -11.1($\pm$0.3), and (Mg/He) $\leq$ -10.0 based on the non-detection of an EW=0.1 Å MgI 3835 Å triplet. These results differ by 2.7 dex in (H/He) from those of Koester (1987) who reported an Hα EW=0.1 Å; they differ by 0.7 dex, and >-1.0 dex for Ca and Mg, respectively, from those of ZWK who used the MgI and MgII resonance lines in the IUE spectrum. Thus the ZWK result may be the more reliable Mg abundance for L745-46A.

4.4 ABUNDANCE SUMMARY

We list in Table 2 our abundance results for the four DZA and DAZ stars discussed in this section. We also include the values for the mass of hydrogen in the atmosphere and the values for [Mg/Ca]. Both quantities play a role in the discussion below of the origins of these three elements in the helium-rich outer layers of these cool degenerate stars.

TABLE 2.   COOL DZA and DAZ ABUNDANCE SUMMARY

STAR	T_{eff}(K)	(H/He)	($M_H/M_{\odot}$)	(Ca/He)	[Ca/H]	[Mg/Ca]
G77-50	5200	-1.6	-14.8	-10.1	-2.8	<-0.3
G74-7	7300	-1.4	-15.6	-10.6	-3.0	<-0.5
L745-46A	7500	-3.2	-16.3	-11.1	-2.3	< 0.0
Ross 640	8300	-3.4	-16.8	-8.7	+0.3	+0.1

5. Discussion and Implications

Our sample of cool DZ stars that show some spectroscopic evidence of hydrogen is admittedly small, but the temptation is strong to look for correlations or trends with T_{eff}, hydrogen and metal abundance ratios, and space motions relative to the LISM. To enhance the sample, we include results for an additional DZA star, K789-37, recently analyzed by Koester, Wegner, and Kilkenny (1990). We also include abundance results for the three hot DB stars that show helium and both hydrogen and calcium lines, the DBAZ stars reported by Kenyon, Shipman, Sion, and Aannestad (1988) and by Sion, Aannestad, and Kenyon (1988); these DBAZ stars have been analyzed with Shipman's model code.

We first look for a relationship between the H/He abundance ratio and T_{eff}. If these helium-rich DZA and DBAZ stars acquired their hydrogen contamination by accretion during passage through clouds in the LISM, we would expect more hydrogen in the older, cooler objects. The line in Figure 8 terminating near G77-50 is an 'eye-ball' fit to the DZA and DBAZ data that agrees with this hypothesis, but this line clearly gives great weight to the G77-50 datum. The triangle symbols indicate upper limits for (H/He) due to the non-detection of hydrogen in DZ and DBZ stars. Discovery of a few DZA stars in the Θ_{eff} = 0.7 to 1.0 range would certainly improve the testing of this hypothesis. The mass of hydrogen in the atmospheres of the stars in Table 2 increases by 2 dex for the T_{eff} range 8300 to 5200K, a trend that reinforces the accretion hypothesis. These hydrogen masses are consistent with accretion rates mid-way between the Eddington rate and the Bondi rate (see, e.g., Vauclair, Vauclair, and Greenstein 1979), and are thus consistent with episodic cloud encounters.

G74-7 has generally been thought to have a hydrogen-dominated atmosphere (see, e.g., Shipman 1979; LLWF; GR) based on its shallow Balmer decrement. Bergeron et al.(1990a)

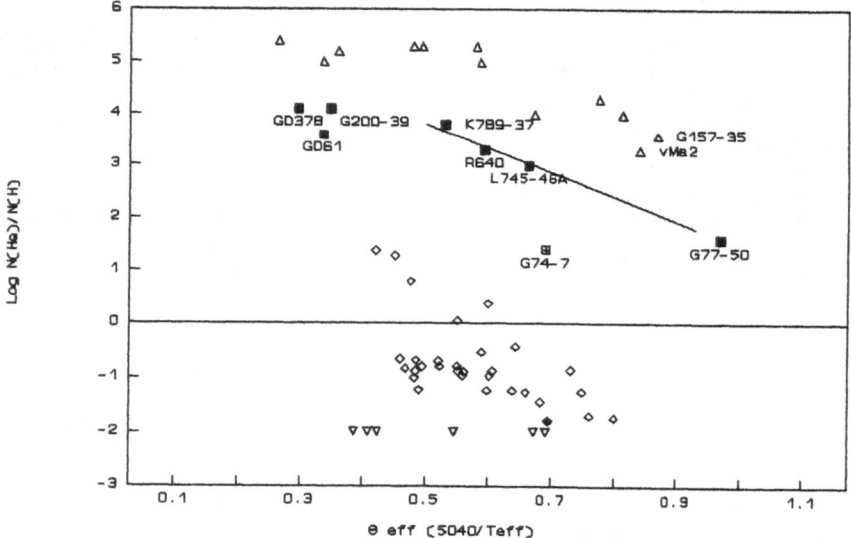

Figure 8. The H/He ratio for warm and cool white dwarfs vs. Θ_{eff}. The ordinate is -(H/He), see Table 2. The squares identify the objects with both hydrogen and calcium lines.

derive (H/He) = +1.9 for G74-7 by applying the HM quenching formalism to the high Balmer lines, ignoring the influence of metal lines on the Balmer profiles, omitting the non-resonant broadening by hydrogen collisions, and assuming log g = 8.0. In addition to our objection to the application of the HM quenching scheme to cool white dwarfs, we now find that Bergeron et al. (1990b) have had to reduce the quenching efficiency nearly to zero for the high Balmer lines of G74-7 and other cool DA stars if pure hydrogen compositions are assumed. It would seem, if the Bergeron et al. treatment of collision broadening and quenching were correct, that not only are the effects of varying gravity and varying (H/He) inseparable, but also the effects of varying quenching efficiency and varying (H/He) would be indistinguishable, certainly within the noise levels at the weak high Balmer lines. The Bergeron et al. (1990a) results are shown in Fig. 8 with diamond symbols, the filled diamond indicates G74-7, and the inverted triangles are for objects with helium abundance less than the scheme's detection limit of 1%. Taking the Bergeron et al. (1990a) results at face value, Figure 8. may be revealing two sequences of degenerates having similar slope in the (He/H)-Θ_{eff} plane. However, the discrepancy between the two results for G74-7 is large, due partially to differing treatments of collision broadening, but it may diminish considerably when the additional ambiguity due to hard-sphere quenching is removed.

The Ca/H abundance ratios relative to the solar value, [Ca/H], are very low for G77-50, G74-7, and L745-46A, but the Mg/Ca ratios (or upper limits) are nearly solar for all four stars in Table 2. The low calcium abundance for these three weak CaII stars correlates well with their space motion and position in the low density inter-cloud region (see Aannestad and Sion 1985), but the fact that they show some calcium would indicate accretion from either in situ, or recent passage through, small cloudlets, or the accretion of comets (Alcock, Fristrom, and Siegelman 1986; Shipman and Greenstein 1983) or asteroidal material (dust?) depleted of volatiles (Sion et al. 1990b). The derived Mg/Ca ratios do not contradict the most recent predictions of gravitational diffusion (Paquette et al. 1986) if the accretion episodes were recent, as the location of Ross 640 would indicate. However, we would expect magnesium to be spectroscopically visible in some of the DZ stars in the inter-cloud region if the accreted material had the solar Mg/Ca abundance ratio. Although the Mg 3835 Å feature is visible in 4 of the 14 cool DZ stars, it is unblended only in Ross 640. Good metal abundance ratios, using the MgI and MgII resonance lines, may require Hubble Space Telescope observations.

6. Acknowledgements

This work was supported by NSF Grant AST 88-02689 to Villanova University, and by grants of computer time from the College of Natural Sciences, University of South Florida. GLH is grateful to the conference chairman, G. Vauclair, for NATO travel support. EMS is grateful to the Observatoire Midi-Pyrenees, CNRS (URA285) for its hospitality and financial support.

7. References

Aannestad, P., and Sion, E. 1985, AJ, **90**, 1832
Aannestad, P., Hammond,G., Sion,E., and Kenyon,S. 1990, BAAS, **21**, 1193
Alcock, C., Fristrom, C., and Siegelman, R. 1986, ApJ, **320, 462**
Allard, N., and Kielkopf, J. 1982, Rev. Mod. Phys., **54**, 1103
Ayres, T. 1977, ApJ, **213**, 296
Ažusienis, A., and Straižys, V. 1969, Sov. Astr., **13**, 316
Bergeron, P., Wesemael,F., Fontaine,G., Liebert,J. 1990a, ApJ, **351**, L21
Bergeron, P., Wesemael,F., Fontaine,G. 1990b, ApJ, in press
Bessell, M. 1990, PASP, **102**, 1181
Borysow, A., and Frommhold, L. 1989, ApJ, **341**, 549

330

Bowman, N., and Lewis, E. 1978, J. Phys. B, **11**, 1703
Carbon, D., and Gingerich, O. 1969, in **Proceedings, Third Harvard-Smithsonian Conference on Stellar Atmospheres**, ed. O. Gingerich, (Cambridge: MIT Press)
Deridder, G., and van Rensbergen, W. 1976, A&AS, **23**, 147
Dimitrijević, M., and Konjević, N. 1981, in **Spectral Line Shapes**, ed. B. Wende, (Berlin: de Gruyter)
Driver, R., and Snider, J. 1976, ApJ, **208**, 518
Eckart, M., and Baldwin, K. 1979, J. Phys. B, **12**, L319
Eggen, O., and Greenstein, J. 1965, ApJ, **142**, 925
Greenstein, J. 1976, AJ, **81**, 323
Greenstein, J. 1983, MNRAS, **203**, 1213
Greenstein, J. 1984, ApJ, **276**, 602
Greenstein, J. 1986, ApJ, **304**, 334 (GR)
Greenstein, J., and Liebert, J. 1990, ApJ, in press (GL)
Griem, H. 1964, **Plasma Spectroscopy**, (New York: McGraw-Hill)
Hammond, G. 1974, Ph.D. dissertation, Univ. Maryland (College Park)
Hammond, G. 1975, ApJ, **196**, 291
Hammond, G. 1989, in IAU Coll. #114, **White Dwarfs**, ed. G. Wegner, (Berlin: Springer-Verlag)
Hammond, G. 1990, in preparation
Hammond, G., Sion, E., Aannestad, P., Kenyon, S. 1991, in preparation
Hintzen, P., and Strittmatter, P. 1974, ApJ, **193**, L111
Hummer, D., and Mihalas, D. 1988, ApJ, **331**, 794 (HM)
Jones, W., Sanchez, A., Greig, J., Griem, H. 1972, Phys. Rev. A, **5**, 2318
Kenyon, S., Shipman, H., Sion, E., and Aannestad, P. 1988, ApJ, **328**, L65
Kepple, P., and Griem, H. 1969, Phys. Rev., **173**, 317
Kielkopf, J. 1975, J. Chem. Phys., **62**, 3784
Koester, D. 1987, IAU Coll. #95, **Second Conference on Faint Blue Stars**, eds. A. Philip, D. Hayes, J. Liebert (Schenectady: L. Davis Press)
Koester, D., Wegner, G., and Kilkenny, D. 1990, ApJ, **350**, 329
Koester, D., and Chanmugam, G. 1990, **Reports on Progress in Physics**, in press
Kurucz, R. 1970, SAO Special Report No. 309
Lacombe, P., Liebert, J., Wesemael, F., and Fontaine, G. 1983, ApJ, **272**, 660 (LLWF)
Lewis, E. 1967, Proc. Phys. Soc., **92**, 817
Liebert, J. 1977, A&A, **60**, 101
Liebert, J., and Wehrse, R. 1983, A&A, **122**, 297
Liebert, J., Dahn, C., and Monet, D. 1988, ApJ, **332**, 891 (LDM)
Lortet, M., and Roueff, E. 1969, A&A, **3**, 462
Matsushima, S. 1969, ApJ, **158**, 1137
McCook, G., and Sion, E. 1987, ApJS, **65**, 603
Mihalas, D. 1967, in **Methods in Computational Physics, v. 7**, ed. B. Alder (New York: Academic Press)
Mihalas, D. 1978, **Stellar Atmospheres** (San Francisco: W.H. Freeman & Co)
Monteiro, T., Cooper,I., Dickinson,A., Lewis,E. 1986, J.Phys.B, **19**, 4087
Monteiro, T., Danby,G., Cooper,I., Dickinson,A., Lewis,E. 1988, J. Phys. B, **21**, 4165
Paquette, C., Pelletier,C., Fontaine,G., Michaud,G. 1986, ApJS, **61**, 177 Press, W., Flannery, B., Teukolsky, S., Vetterling, W. 1989, **Numerical Recipes (FORTRAN Version)**, (New York: Cambridge Univ. Press)
Sando, K., and Wormhoudt, J. 1973, Phys. Rev. A, **7**, 1889
Shipman, H. 1979, ApJ, **228**, 240

Shipman, H., and Greenstein, J. 1983, ApJ, **266**, 761
Sidell, N., and Ch'en, S. Y. 1977, JQSRT, **17**, 117
Sion, E., Aannestad, P., and Kenyon, S. 1988, ApJ, **330**, L55
Sion, E., Kenyon, S., and Aannestad, P. 1990, ApJS, **72**, 707 (SKA)
Sion, E., Hammond, G., Aannestad, P., Kenyon, S. 1990a, IAU Coll. #145, Druzba, Bulgaria
Sion, E., Hammond, G., Wagner,R., Starrfield,S., and Liebert, J. 1990b, ApJ, **362**, 691
Swihart, T. 1965, ApJ, **141**, 821
Unsöld, A, 1955, **Physik der Sternatmosphären**, (Berlin: Springer-Verlag)
Vauclair, G., Vauclair, S., and Greenstein, J. 1979, A&A, **80**, 79
Wegner, G. 1983, AJ, **88**, 109
Wehrse, R. 1972, A&A, **19**, 453
Wehrse, R. 1977, Mem. Soc. Astr. Italiana, **48**, 13
Wesemael, F., and Truran, J. 1982, ApJ, **260**, 807
Weidemann, V., and Bues, I. 1967, Zs. f. Ap., **67**, 415
Weidemann, V., and Koester, D. 1984, A&A, **132**, 195
Zeidler-K.T., E.-M., Weidemann, V., Koester,D. 1986, A&A, **155**, 356 (ZWK)

EVIDENCE FOR FRACTIONATED ACCRETION OF METALS ON COOL WHITE DWARFS

J. Dupuis
Department of Physics and Astronomy, Dartmouth College
Wilder Laboratory, Hanover, NH 03755 USA
and
G. Fontaine, and F. Wesemael
Département de Physique, Université de Montréal
C.P. 6128, Succ. A, Montréal, Québec, Canada, H3C 3J7

ABSTRACT. We compare the metal-to-metal number abundances ratios predicted by our two-phase accretion model with the abundances determinations currently available for cool helium-rich white dwarfs. Although most of the stars in our sample have metal abundances consistent with accretion of metals in solar proportions, a handful of them appear to have accreted material with abundances anomalies. We have found a way to explain this fact which avoids invoking a variable composition of the interstellar medium or a filtering mechanism acting only on a star-to-star basis. *All* the observed metal-to-metal abundances ratios become consistent with the predictions of the two-phase accretion model if a unique composition of the accreted material is assumed in which calcium is overabundant by a factor of ~ 3 and silicon is underabundant by a factor of ~ 7. This strongly suggests that a mild fractionation process is altering the relative abundances of heavy elements in the accreted material by the time it reaches the surface of cool white dwarfs.

1. Introduction

With the help of a powerful diffusion code based on finite-elements techniques (Pelletier, Fontaine and Wesemael 1989), it is possible to fully develop the accretion-diffusion model giving an explanation for the traces of metals detected in a minority of cool white dwarfs. Using a simple two-phase model of the interstellar medium as in Wesemael (1979), we have investigated the time evolution of the distributions of various metals inside envelopes of cool white dwarfs. The full details of these calculations will be described in Dupuis *et al.* (1990). Cool white dwarfs are old enough to have experienced several encounters with interstellar clouds during which they may accrete at rates sufficiently large to leave transient detectable traces of accreted material which, of course, contains metals. The metal-line phenomenon in cool white dwarfs is expected to be observed during and shortly after the crossing of a cloud. A majority of cool white dwarfs do not show metal lines because they accrete at a

333

G. Vauclair and E. Sion (eds.), White Dwarfs, 333–341.
© 1991 *Kluwer Academic Publishers.*

much reduced rate from the hot and tenuous component of the ISM. In that case, accretion sustains only small amounts of metals in their atmospheres, too small in fact to be detected. The two-phase accretion model takes representative values of the cloud encounter time, the average time between cloud encounters, the accretion rate during a cloud collision, and the accretion rate between cloud encounters which are calculated using standard parameters for the ISM. We were then able, using these values, to simulate accretion episodes onto cool white dwarfs. These calculations lead to prediction of metals abundances in cool white dwarfs and it is the goal of this work to compare these predictions with the observations. The emphasis, here, will be put on the metal-to-metal abundances ratios, which as will be explained in the next section, can be used to constrain the composition of the material reaching the surface of cool white dwarfs.

2. Comparison with the Observations

As a simplifying approximation, we have chosen the composition of the accreted material in our simulations to be equal to the solar composition. We expect that any deviations from strict solar composition should be revealed in the relative abundances patterns obtained from observations. In order to determine if such deviations occur during accretion of interstellar matter due to some kind of fractionation process, we have compared the metals abundances ratios predicted by the two-phase accretion model to the abundances ratios currently available in the literature for cool helium-rich white dwarfs. In table 1 we have compiled the metal-to-metal number abundances ratios for cool helium-rich white dwarfs showing more than one metal in their spectra. The first column in that table lists the usual name of the star. The next six columns give the logarithms of the metal-to-metal abundances ratios for Mg/Ca, Si/Ca, Fe/Ca, Si/Mg, Si/Fe, and Fe/Mg. The next column gives the minimal deviation expected from solar composition in the accreted material. Finally, in the last column we give the references to the papers in which the abundance analysis were carried on.

On the Figures 1 and 2, we have superposed the range of predicted relative metal abundances by our simulations of accretion episodes to the observed abundances compiled in Table 1. The small filled circles correspond, as usual, to abundances determinations while the short horizontal lines are upper limits on relative abundances (the upward-looking arrow in the central panel of Figure 2 is the lower limit on the Si/Fe ratio in L745-46A). The typical uncertainty assumed on the relative abundances is chosen to be approximately equal to a factor of five; this is shown by the vertical bar in the upper left corner of Figures 1 and 2. The predicted range of relative abundances shown by the dashed area is the result of time-dependent calculations which will be fully described in Dupuis *et al.* (1990). These curves were obtained by assuming that the metals are accreted in solar proportions and that the ratio between the high and low accretion rates is equal to 10^5. If the ratio of the accretion rates is large, the elements have more time to segregate after a cloud encounter and large abundance differences will develop, and the more so if the difference in diffusion timescales is large. Also, the values of the theoretical limits depend on the assumed composition of the accreted material. Therefore, any observational point falling within the expected range in Figures 1 and 2 is consistent with accretion in solar proportions. As explained above, relative abundance anomalies are built up by diffusion after an encounter with an interstellar cloud. Hence, depending on the properties of the cloud and the time elapsed since the encounter, the model predicts that 1) the abundances should show no correlation with effective temperature and 2) the abundances should be scattered within the allowed domain if accretion in solar proportions has occurred. The observations are certainly consistent on the first count. In addition, most of the observational points indeed fall within the expected ranges when due account of the uncertainties is taken.

TABLE 1. Compilation of metal-to-metal number abundances ratios in cool helium-rich white dwarfs

Name	Mg/Ca	Si/Ca	Fe/Ca	Si/Mg	Si/Fe	Fe/Mg	Proportions	References
GD 40	0.2	≤0.7	0.2	≤0.5	≤0.5	0.0	Ca ↑~2[a]	2,3,4
PG1225-079	0.7	≤-0.2	0.8	≤-0.9	≤-1.0	0.1	Si ↓~5	5,7
L119-34	1.6	0.1	1.2	-1.5	-1.1	-0.4	Si ↓~8	8,9
Ross 640	1.9	1.8	0.8	-0.1	1.0	-1.1	solar	5,8,10
GD 401	≤0.0	≤-0.5	≤-0.2	...	...	...	Ca ↑~3	11,12
							Si ↓~4	
G105-B2B	≤2.3	≤2.0	...	...	...	...	solar	8
GD 95	2.4	...	1.1	...	...	-1.3	solar	5
L745-46A	1.4	0.9	≤-0.6	-0.5	≥1.5	≤-2.0	Ca ↑~8	8,12
G165-7	1.6	...	1.8	...	...	0.2	solar	13
G163-28	≤2.7	...	...	...	...	...	solar	8
G139-13	1.3	...	1.2	...	...	-0.1	solar	1,6
G157-35	1.8	...	1.3	...	...	-0.5	solar	5,8
VMa 2	1.8	...	1.8	...	...	0.0	solar	5,8
Sun	1.1	1.2	1.3	0.1	-0.1	0.2		14

References.-(1) Dupuis et al. 1990. (2) Shipman, Greenstein, and Boksenberg 1977. (3) Shipman 1984. (4) Shipman and Greenstein 1983. (5) Liebert, Wehrse, and Green 1987. (6) Sion, Kenyon, and Aannestad 1990. (7) Koester, Wegner, and Kilkenny 1990. (8) Zeidler-K.T., Weidemann and Koester 1986. (9) Weidemann and Koester 1989. (10) Hammond 1989. (11) Cottrell and Greenstein 1980. (12) Koester 1987. (13) Wehrse and Liebert 1980. (14) Allen 1973. (a) The upward or downward looking arrows in the Table 1 mean that the corresponding metal is either overabundant or underabundant in the accreted material by the indicated factor.

There are few notable exceptions (i.e. values falling outside the theoretical limits by more than a factor 5) however, and these must reflect non-solar proportions in the accreted material. Taking the data at their face value, we have estimated the minimal changes from solar proportions required to explain the observed relative abundances in the exceptional cases. For example, GD 40 shows Si/Mg, Si/Fe, and Fe/Mg ratios which are consistent with solar proportions in the accreted material (Fig. 2). By contrast, it also shows Mg/Ca, Si/Ca, and Fe/Ca ratios which are all smaller than the expected lower limits (Fig. 1) and which are explained by a relative overabundance of calcium in the accreted material. Based

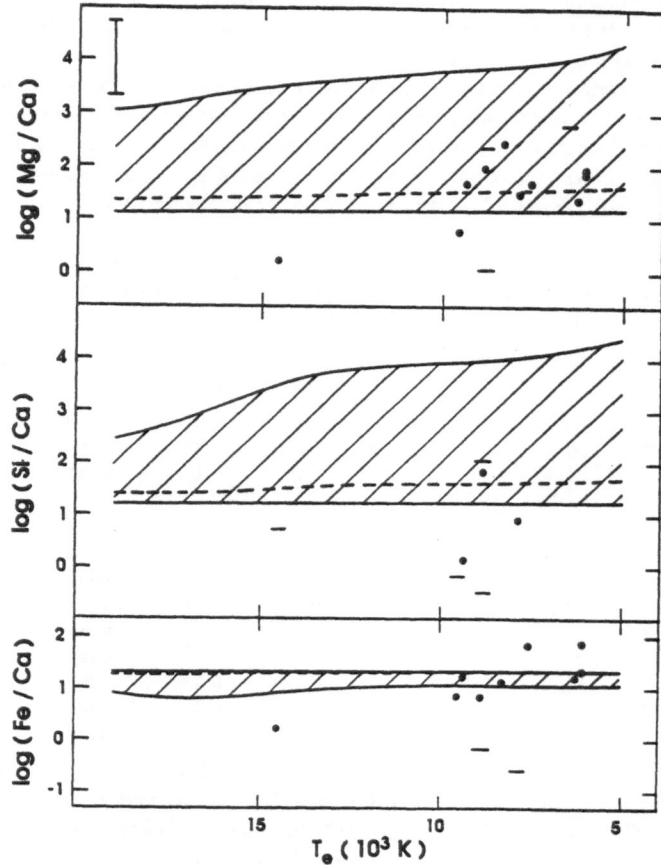

Figure 1. Metal-to-metal abundance ratios in cool He-rich white dwarfs as funtions of effective temperature. Small filled circles give actual values while short horizontal line segments give upper limits. The error bar is representative of the observational uncertainties. The regions defined by the two continuous curves joined together by diagonal lines correspond to the range of expected abundances on the basis of accretion in solar proportions and assuming a contrast of 10^5 between the high and low accretion rates. One of the limits of a given domain (the continuous horizontal line) corresponds to the solar value of the abundance. The dashed line inside a domain gives the steady state abundance. The top, middle, and bottom panels give, respectively, the Mg/Ca, Si/Ca, and Fe/Ca abundance ratios.

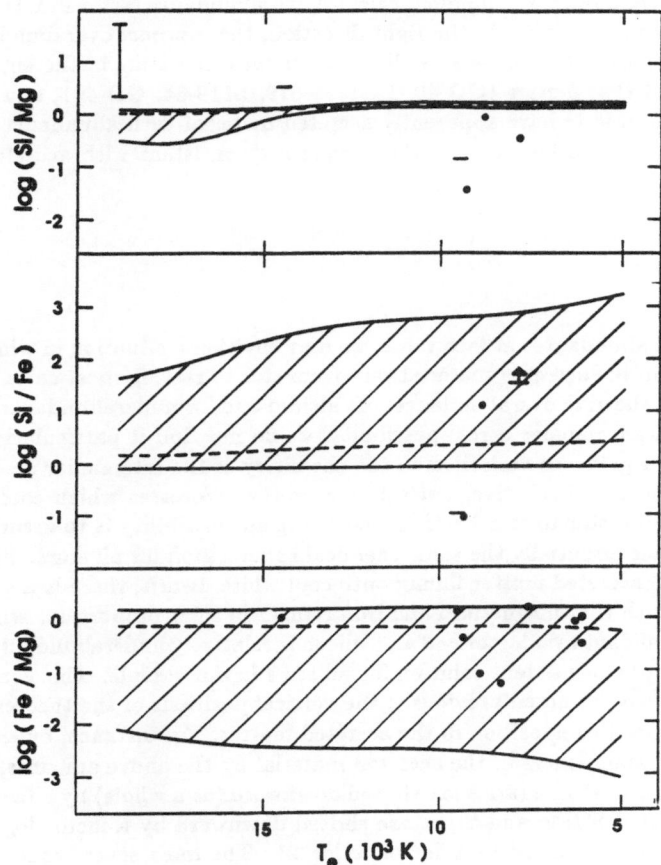

Figure 2. Same as Fig 2, but, this time, the top, middle, and bottom panels give, respectively, the Si/Mg, Si/Fe, and Fe/Mg abundance ratios.

on the observed Mg/Ca ratio, the nominal calcium overabundance factor in GD 40 is about 10. By pushing the uncertainties in the right direction, the *minimal* overabundance factor is ~ 2. We have reported this value as well as others for other stars in the eighth column of Table 1. We find that 5 stars (GD 40, PG1225-079, L119-34, GD 401, and L745-46A) in our sample of 13 objects have apparently accreted material with abundance anomalies. The observations for the other stars in the sample are consistent with accretion in solar proportions.

3. Discussion

Taking the inferred abundances at face value, we may envision a situation in which the proportions of metals in the accreted material vary from star to star, in most cases being solar but, in other cases, showing overabundances of calcium and/or underabundances of silicon. While this explanation remains a real possibility, we do not find it particularly appealing because it suggests significant variations in the chemistry of the interstellar medium in the solar neighbourhood or, alternatively, selective accretion processes which could filter the accreted material on a star-to-star basis. A more elegant possibility is to assume that the accreted material has essentially the same chemical composition for all stars. Indeed, if we assume that, in the accreted matter falling onto cool white dwarfs, there is a general fractionation process such that magnesium and iron remain in solar proportions while calcium is relatively overabundant by a factor ~ 3 and silicon is relatively underabundant by a factor ~ 7, we find a marginal consistent solution for all stars in our sample. This can be seen in Figures 1 and 2, where, as noted previously, the vertical positions of the theoretical ranges depend on the assumed composition in the accreted matter. For instance, by changing the calcium and silicon abundances in the accreted material by the above amounts, the ranges of expected Mg/Ca and Fe/Ca ratios are shifted downward (as a whole) by a factor of log 3, the ranges of expected Si/Mg and Si/Fe are shifted downward by a factor log 7, and the explained Si/Ca ratios is shifted by a factor of log 21. The most severe constraints come from the Si/Mg and Fe/Ca values, which reflect mostly the proportions in the accreted material because the diffusion timescales of the two respective elements are quite similar. By adopting an overabundance factor of ~ 3 for calcium and an underabundance factor of ~ 7 for silicon, *all* of the observations can be marginally reconciled (the Fe/Ca ratio remains somewhat low in L745-46A) with the idea of a unique composition for the accreted material.

It remains to be seen whether or not a fractionation process could lead to a general enrichment of calcium as well as a general depletion of silicon in the material reaching the surfaces of cool white dwarfs. This makes resurface the idea of selective accretion already suggested by the seemingly reduced accretion rates for hydrogen compared to metals accretion rates. The hydrodynamical calculations of accretion on white dwarfs including the effect of ionizing radiation presented by H. Van Horn at this meeting support the idea of selective accretion. We feel that to reach a better comprehension of the atmospheric compositions of cool white dwarfs, it is now necessary to further investigate the physical and chemical properties of metals in the interstellar medium.

Partial financial support for this work has been provided by NSERC Canada and the fund FCAR (Québec). J.D. acknowledges support from a NSERC postdoctoral fellowship, an AAS international grant and the Margaret Cullinan Wray Charitable Lead Annuity Trust.

REFERENCES

Allen, C.W. (1973) Astrophysical Quantities, Athlone, London.

Cottrell, P.L. and Greenstein, J.L. (1983) 'Calcium and magnesium(?) in the helium rich white dwarf GD401', Astrophys. J. 242, 195-198.

Dupuis, J. et al. (1990) 'A study of metal abundances patterns in cool white dwarfs. II. Simulations of accretion episodes', in preparation.

Dupuis, J. et al. (1990) 'A study of metal abundances patterns in cool white dwarfs. III. Comparison of the predictions of the two-phase accretion model with the observations', in preparation.

Hammond, G.L. (1989) 'Pressure shifts of metal lines in cool white dwarfs', in G. Wegner (ed.), White Dwarfs, Springer-Verlag, Berlin, pp. 346-349.

Koester, D. (1987) 'Evolution of surface abundances in cool white dwarfs', in A.G.D. Philip, D.S. Hayes, and J.Liebert (eds.), The Second Conference on Faint Blue Stars, L. Davis Press, Schenectady, pp. 329-340.

Koester, D., Wegner, G., and Kilkenny ,D. (1990) 'Model atmosphere analysis of the DZ white dwarf K789-37', Astrophys. J. 350, 329-333.

Liebert, J., Wehrse, R., and Green, R.F. (1987) 'White dwarfs with metallic line spectra', Astron. Astrophys. 175, 173-178.

Pelletier, C., Fontaine, G., and Wesemael, F. (1989) 'Finite element analysis of diffusion processes in white dwarfs', in G. Wegner (ed.), White Dwarfs, Springer-Verlag, Berlin, pp. 249-252.

Shipman, H.L. (1984) 'Carbon, silicon, and oxygen in the white dwarf GD 40', in J.M. Mead, R.D. Chapman, and Y. Kondo (eds.), The Future of Ultraviolet Astronomy Based on Six Years of IUE Research, NASA Conference Publication 2349, pp. 281-284.

Shipman, H.L., and Greenstein, J.L. (1983) 'Iron and magnesium in the white dwarf GD 40: a test of diffusion theory', Astrophys. J. 266, 761-768.

Shipman, H.L., Greenstein, J.L., and Boksenberg, A. (1977) 'Calcium in the helium white dwarf GD 40', Astron. J. 82, 480-486.

Sion, E.M., Kenyon, S.J., and Annestad, P.A. (1990) 'An atlas of optical spectra of DZ white dwarfs and related objects', Astrophys. J. Suppl., in press.

Wehrse, R., and Liebert ,J. (1980) 'A spectrum analysis for the unusual metallic line white dwarf G 165-7', Astron. Astrophys. 86, 139-148.

Weidemann, V., and Koester, K. (1989) 'The presence of carbon in DZ star atmospheres', Astron. Astrophys. 210, 311-312.

Wesemael, F. (1979) 'Accretion from interstellar clouds and white dwarf spectral evolution', Astron. Astrophys. 72, 104-110.

Zeidler-K.T., E.M., Weidemann, V., and Koester, D. (1986) 'Metal abundances in helium-rich white dwarf atmospheres', Astron. Astrophys. 155, 356-370.

DISCUSSION

WEIDEMANN :

You did not display the **relative** metal abundances as a function of time since the last accretion event in comparison with the observed values. I remember from our investigation that it is not possible to reconcile these data with the predictions.

DUPUIS :

On the contrary, I have shown the comparison between the observed **relative** abundances (Si/Ca, Fe/Ca, Mg/Ca, Si/Mg, Si/Fe and Fe/Mg) and the prediction at our models. The observed abundances fall in on very near the bands of expected abundance ratios. The correlation between different abundances ratios depends on both the accretion rate and the time since the encounter. We have decided to put constraints on the accretion via the composition of accreted material first.

A NEW LOOK AT OLD FRIENDS: 40 ERI B AND GD323

DETLEV KOESTER
Department of Physics and Astronomy
Louisiana State University, Baton Rouge, USA

1. Introduction

In this paper I will present some ideas on apparently unrelated topics, which, I hope, can be justified by the spirit of a workshop. What they have in common is that they are related to what I believe are hot topics: the structure of the outer layers in white dwarfs, the mass distribution, and the He abundance in cool DA.

2. 40 Eri B

As everybody at this meeting knows, 40 Eri B was the first white dwarf identified to be peculiar. It is also the second brightest and by far the easiest to observe, because it is much farther away from its bright companion as is Sirius B. Supposedly it also has a very well determined astrometric mass of 0.43 ± 0.02 $M_{\odot}$ (Heintz 1974) and parallax (Gliese and Jahreiß1987). This object should be well suited to serve as a test case for the interpretation of white dwarf spectra and evolution.

Unfortunately this testing has not been straightforward and the results so far are not satisfactory. The last studies of 40 Eri B that I am aware of (Wegner 1979, 1980) found a significant discrepancy between the measured gravitational redshift and the value expected from the astrometric mass.

Our study (Koester and Weidemann 1990) was also motivated mainly by a new determination of the gravitational redshift. Fig. 1 shows the spectra around Hα for all three stars in the system, obtained within an hour with exactly the same instrument configuration (CASPEC spectrograph at the ESO 3.6 m telescope on La Silla, resolution $\approx$ 0.25 Å). It is obvious that the relative velocity shifts of A, B, and C can be measured with high accuracy from such spectra. Because Hα is in emission in component C, one might be concerned about any intrinsic velocities due to e.g. a stellar wind. We have therefore in addition to Hα measured four CaI absorption lines ($\lambda\lambda$ 6439.073, 6462.566, 6493.780, 6572.781), with essentially the same result.

Because we need only relative radial velocities no corrections for solar and terrestrial motions were applied. The velocities have, however, been corrected for small gravitational shifts of A and B, and for the orbital motion B - C, determined from the orbital elements of Heintz (1974). The final value obtained using Hα in component C is 26.51±1.5 km s^{-1} (see Table 1).

The other piece of new information since 1980 is the parallax value of 0.2084±0.0023 in the 3rd edition of the *Catalog of Nearby Stars* (Gliese and Jahreiß1987), which is very close to the value used by Heintz (1974) in his final analysis (0.2073), but has a smaller error.

The last parameter we will need is the effective temperature, from which the radius can be derived using theoretical model atmospheres and the parallax. We have made a new determination from the published Strömgren colors and a theoretical model grid, with the result T_{eff} = 17000±200 K. This agrees well with other data in the literature: 16940 K (Koester et al. 1979), 17000 (Greenstein

343

G. Vauclair and E. Sion (eds.), White Dwarfs, 343–351.
© 1991 *Kluwer Academic Publishers.*

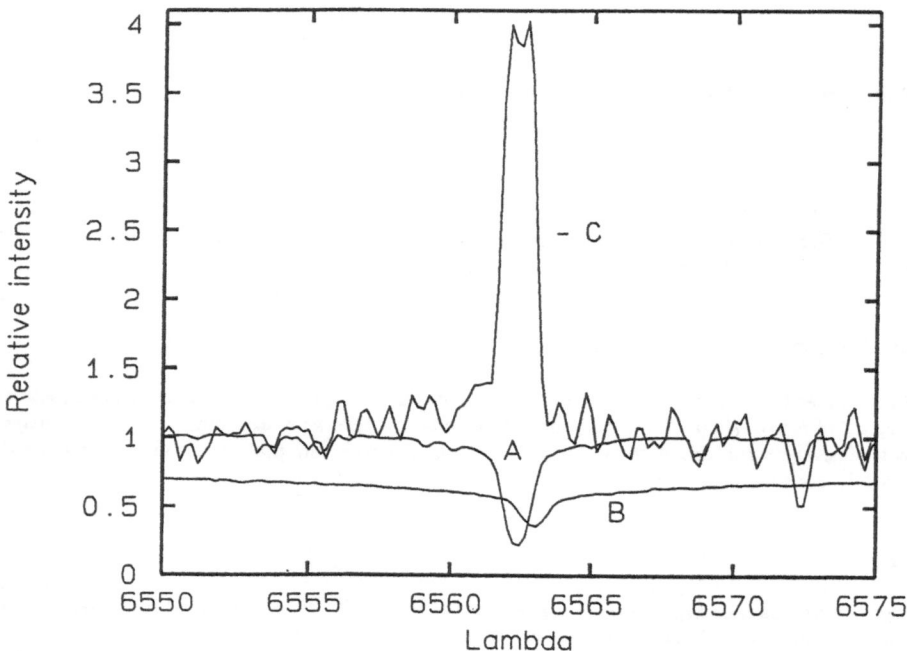

Fig. 1: Hα in the 3 components of the 40 Eri system.

Table 1: Velocity shifts in the 40 Eri system (km s^{-1})

component	A	C (Hα)	C (CaI)	B
measured rel. velocity	-21.48	-22.40	-21.72	+5.94
corr. for grav. shift	-0.60	-0.46	- 0.46	
orbital motion B-C		-1.67	-1.67	+0.62
corrected velocity	-22.08	-21.19	-20.51	+5.32
grav. shift B (Hα)				+26.51
from CaI lines in C				+25.83

and Peterson 1973), 16850 (Wegner 1979), 16900 (Shipman 1979). We will therefore use this value in the following discussion, although it is in conflict with the most recent determination of 16250 K by Bergeron et al. (1990a).

Fig. 2 shows all this information on 40 Eri B in the mass-radius diagram (see the figure caption for explanations). The hatched area is the region allowed by our determinations of redshift and radius, the best value for the mass from this determination is 0.53 ± 0.04 M$_\odot$. The allowed area is consistent with the Hamada-Salpeter relations for C or He, as well as the evolutionary relation for a *thin* hydrogen mass (not shown), but marginally excludes the *thick* hydrogen relation. Both solutions using the Bergeron et al.(1990a) data are inconsistent with the velocity measurement, the solution using their temperature and log g (B1) is also inconsistent with any mass-radius relation. Agreement with our solution could be obtained by increasing their log g by 0.15, and I will come back to this point below.

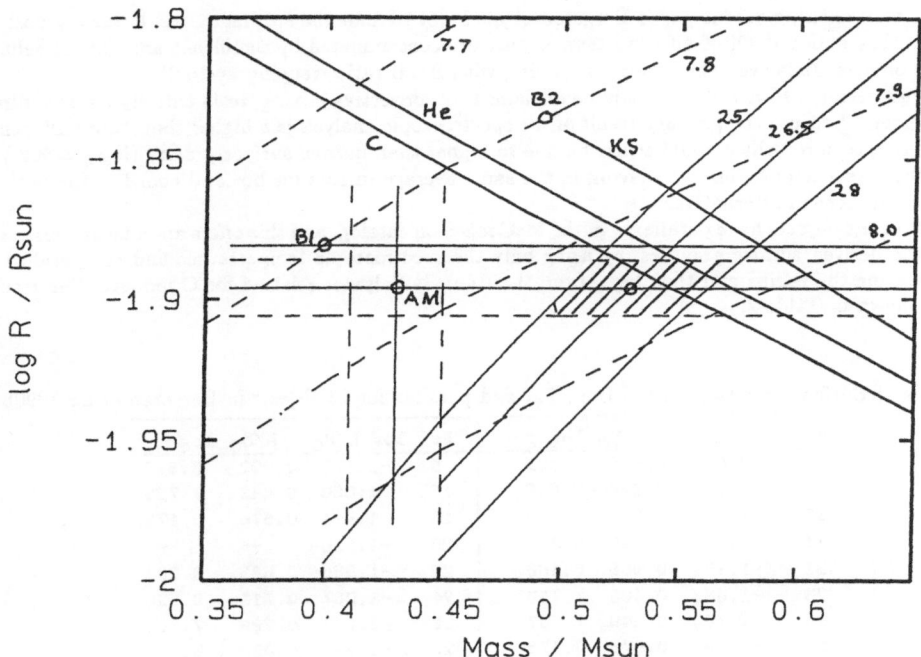

Fig. 2: 40 Eri B in the mass-radius diagram. R (continuous): radius corresponding to T_{eff} = 17000 K with errors (dashed). C, He, KS : theoretical mass-radius relations for C and He (zero temperature, Hamada and Salpeter 1961), and evolutionary relation for T_{eff} = 17000 K (Koester and Schönberner 1986). Continuous diagonal lines from lower left to upper right : lines of constant gravitational redshift, as velocity. Dashed diagonal lines : lines of constant log g. B1: solution for log g and T_{eff} from Bergeron et al. (1990a). B2: solution for log g from Bergeron et al. (1990a) and evolutionary mass-radius relation. AM : astrometric mass solution.

Our data are also inconsistent with the solution corresponding to the astrometric mass determination (AM), and we have therefore taken a look at possible errors of this determination. The maximum error for the mass of component B — neglecting for the moment all considerations of statistical independency — can be written as

$$\Delta M_B = M_B\left(\frac{2\Delta P}{P} + \frac{3\Delta a"}{a"} + \frac{3\Delta \pi}{\pi} + \frac{\Delta f}{f}\right),$$

where P is the period, $a"$ the major axis in arcsec, π the parallax, and f the mass fraction. Taking estimates of the errors from Wielen (1962) and Heintz (1974) the contributions in the brackets are

0.020, 0.017, 0.012, 0.006 in this order. With the most unfavorable constellation this could add up to 0.06 $M_\odot$. The error estimate given by Heintz (0.02) thus seems too optimistic, it seems hardly possible, however, to reconcile our result with the astrometric mass.

3. Spectroscopic Gravity Determinations in Cool DA

In a very exciting recent paper Bergeron et al.(1990b) found the surprising result that almost all DA below about 12000 K effective temperature are contaminated by significant amounts of helium. In 5 objects He is even the dominant species, with He/H ratios ranging up to 25.

This result, which will not easily be explained by convective mixing, rests entirely on an indirect argument, because the primary result of the spectroscopic analysis is a higher than "normal" atmospheric pressure, which could either be due to higher than normal surface gravity (log g = 8.2) or to the presence of helium. Believing in the same average masses for hot and cool DA the authors chose the second alternative.

24 of the objects have parallaxes in the McCook-Sion catalog, and this offers an independent check on the derived surface gravities, by using only the spectroscopic temperatures and the parallax to determine the radius and then log g from the Hamada-Salpeter relation for C models. The results are shown in Table 2.

Table 2: Surface gravities derived from T_{eff} and parallax for 23 objects in Bergeron et al. 1990b

No	log $R/R_\odot$	$M/M_\odot$	log g	No	log $R/R_\odot$	$M/M_\odot$	log g
3	-2.056	0.905	8.509	5	-2.103	0.992	8.642
9	-1.794	0.328	7.545	13	-1.850	0.443	7.786
15	-1.917	0.596	8.050	16	-1.951	0.676	8.172
17	-1.997	0.780	8.326	20	-1.558	--	--
21	-1.955	0.685	8.186	22	-1.960	0.695	8.201
23	-1.833	0.405	7.713	24	-1.967	0.711	8.226
25	-1.750	0.249	7.337	26	-1.777	0.295	7.463
27	-1.952	0.676	8.173	28	-1.930	0.626	8.097
29	-1.920	0.603	8.061	31	-1.952	0.677	8.175
32	-1.855	0.454	7.806	33	-1.878	0.507	7.901
34	-2.073	0.938	8.558	35	-2.001	0.787	8.337
36	-1.808	0.355	7.608	37	-1.920	0.603	8.061

The average log g determined from the 23 objects with reasonable results for the radius is 8.041. This certainly supports the assumption made by Bergeron et al. (1990b) that the stars below 12000 K have the same average masses (and log g) as those above, and that the higher values found spectroscopically could be due to an increased helium abundances.

However, a closer look at their results for individual objects raises some questions. All four of the 5 objects with apparent He/H $\geq$ 1 that have parallaxes, seem indeed to have much higher than average surface gravities. The two objects with the highest He abundances (3, 5) have log g values of 8.51, resp. 8.64 as determined from the parallaxes. Taking this into account would certainly reduce the He abundance necessary to remove the remaining discrepancy with the spectroscopic determination to values corresponding to Δ log g = 0.15 dex, in line with the bulk of the stars.

These stars are therefore extreme cases in this diagram not because of their extreme He abundance, but almost certainly because of a real high surface gravity.

What remains of the result is then a systematic discrepancy between spectroscopic and parallax surface gravities of 0.15 dex, which could be interpreted as a higher He abundance *or* as a "calibration" problem of the spectroscopic g determination, only this time in the opposite sense with spectroscopy giving higher values.

3.1 A CALIBRATION PROBLEM OF SPECTROSCOPIC MASSES?

Comparing the recent spectroscopic gravity determinations with independent results we are faced with a dilemma: at higher T_{eff} the case of 40 Eri B seems to indicate a necessary revision of $+0.15$ dex, while at the lower temperatures a change of -0.15 would bring the spectroscopic results more in line with those derived from parallaxes.

The spectroscopic determinations by Bergeron et al. (1990a,1990b) depend heavily on the occupation number probability formalism developed by Hummer and Mihalas (1988), which describes the "quenching" of the higher Balmer lines due to interaction with perturbing particles. While it is a very elegant mechanism, it necessarily involves several approximations in the physical model used to describe the interactions, and the resulting gravities may have to be "calibrated" for an unknown zero-point, as mentioned by Bergeron et al. (1990b).

With regard to our problem, we want to draw the attention to the fact that the occupation probability is calculated from two completely independent terms in the model for the Free Energy, one describing charged particle interactions and one describing the neutrals. It is easily possible that a calibration demanded by observations could be different (even in sign) for the two contributions, with the dividing line at about 11000 K, where neutral particles start to dominate.

4. The DBA Object GD323

This unique object was originally classified DBp (peculiar) by Greenstein (1969), who noticed weak and sharp He lines. It was found to show broad, shallow hydrogen lines — and thus classified DAB — by Oke et al. (1984) and Liebert et al. (1984).

Using new optical observations and ultraviolet (IUE) spectra, Liebert et al. (1984) demonstrated that this object has an energy distribution from the UV to the red characteristic of a helium-dominated atmosphere around 30000 K, but lines of H and He much too weak for this temperature. They analyzed the spectra using homogenous model atmospheres, but were not able to find a consistent fit. After a discussion of all alternatives (interstellar reddening, rotation, accretion disks, binary system) the only hypothesis that survives — as they put it — is the assumption of a stratified atmosphere, which could not be tested at that time due to the lack of detailed models.

A preliminary analysis using stratified atmospheres was presented in Koester (1989). It indicated the possibility to explain the overall energy distribution as well as the line spectrum with an atmospheric model near the parameters $T_{eff} = 27000$ K and $M_H = 7.5 \ 10^{-18} \ M_\odot$. Since then we (Koester et al. 1990) have used newer and much better optical spectra obtained with the Palomar Double CCD Spectrograph (Greenstein 1986). In addition the IUE spectra used in Liebert et al. (1984) were extracted from the IUE archive, corrected for sensitivity changes with time using a program and data kindly provided by Dr.R.Bohlin and finally corrected to an absolute scale with the white dwarf calibration of Finley et al. (1989).

For the analysis of the observed spectra a new theoretical grid of stratified models and synthetic spectra was calculated for effective temperatures between 26000 and 33000 K, $M_H/M_\odot$ between 10^{-16} and 10^{-19}, and log g = 8 for most models.

4.1 ATMOSPHERE MODELS WITH STRATIFIED ABUNDANCES

The model atmosphere code used is based on that developed for the first systematic study with realistic abundance distributions by Jordan and Koester (1986), but has been improved significantly in the treatment of convection zones. The models are plane parallel atmospheres in LTE, constructed using the Feautrier method with variable Eddington factors as described in more detail in Koester et al. (1979) and Jordan and Koester (1986). Convection is included in the mixing length approximation, using the Schwarzschild criterium for convective stability. The models are iterated until constancy of the total flux is achieved in all layers to an accuracy of usually 0.1%.

The only difference to our normal atmosphere calculations is that the abundance of hydrogen and helium varies with depth. If we assume that the degree of ionization is constant, the equation of diffusion equilibrium can be solved analytically and the solution gives directly the H/He abundance ratio as a function of total pressure (e.g. Dziembowski and Koester 1981). This equation can easily be implemented in atmosphere codes.

Hydrogen is always almost completely ionized in the hot stars considered here. It can also be shown that the abundance profile is not very different for once ore twice ionized helium. Using an effective charge for helium of 2 in the hot models and 1 in the cooler ones is therefore an adequate approximation.

If convection zones occur within the atmosphere, they are treated in the following manner: the analytic solution is used above the convection zone. Within the zone the abundances are held constant. If the bottom of the convection zone lies within the atmosphere, the analytic solution for the deeper layers is fitted continuously to the abundances in the zone. If convection extends down to the bottom of our atmosphere model, we cannot determine the total hydrogen mass present without an envelope solution with the atmosphere as boundary condition. In that case the H masses we calculate from the parameter we use (gas pressure at the depth where H/He = 1) are only lower limits. For the examples I will use in this paper this does not occur, however.

4.2 THE ENERGY DISTRIBUTION

With IUE and optical spectra — which fit together very nicely — the energy distribution is known from 1200 to 7000 Å. This observed distribution can be fitted with theoretical calculations in the range from T_{eff} = 26000 to 31000 K (with different total hydrogen masses), because the slope of the energy distribution becomes steeper if the amount of hydrogen in the atmosphere is increased. These fits to the energy distribution define a curve of possible parameter solutions in a T_{eff} vs. M_H plane that leads from relatively "thick" H layers at 26000 K to a very thin layer at 31000 K.

4.3 THE OPTICAL LINE SPECTRUM

A similar fitting process was carried through with the detailed line spectrum in the optical region from 4000 to 5000 Å, again leading to a sequence of possible solution but with the opposite slope: from "thin" H layers at 26000 K to "thicker" at 31000 K. Both fit curves therefore define a unique solution near 28000 K effective temperature.

Fig. 3 shows the comparison for the energy distribution at this effective temperature, Fig. 4 the optical line spectrum. The agreement between theory and observation is extremely good, the only remaining discrepancy being that the theoretical Balmer lines are slighly narrower than the observed ones.

The theoretical grid was then expended with much finer stepsizes around the expected solution, the fitting procedure repeated and finally an optimum fit determined with the following parameters:

$$T_{eff} = 27500 \pm 500 \text{ K, and } M_H = (5.2\pm1.0) \ 10^{-18} \ M_{\odot}.$$

The model atmosphere with these parameters is stable throughout the whole transition zone. A very thin He convection zone is present in deeper layers, where the composition is almost pure He, and separated from the transition region by at least one pressure scale heigth. We therefore conclude

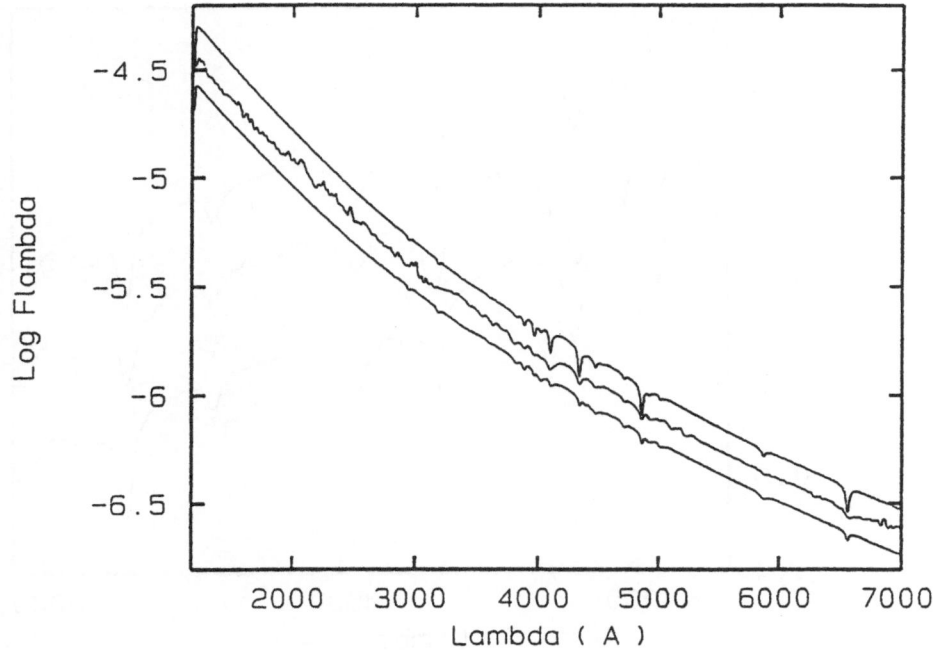

Fig. 3 : Energy distribution of GD323 and models at 28000 K, $M_H/M_\odot = 3.7$ and 9.2 10^{-18}.

that this is a realistic model for this object.

Our hope to get an even better agreement (broader H lines) with a higher surface gravity of log g = 8.5 was disappointed by the calculations: the observed hydrogen lines do only allow a certain number of hydrogen atoms per cm^{-2} on top of the atmosphere. A model with higher surface gravity has higher gas pressure at the same optical depth, but the transition from H to He has to occur at a lower P_g than in the lower gravity model or else the H lines become much too strong. At a given equivalent width therefore the lines cannot be made much broader by using higher log g values.

An alternative explanation for the remaining small discrepancies is the exact structure of the transition profile. As mentioned above, our calculations use a constant degree of ionization for He (Z = 1 in this case) and a ratio of mixing length to pressure scale height of 1. It is possible that variations of these parameters can change the theoretically predicted abundance profile (e.g. widen the transition zone) and lead to an even better agreement as indicated by some numerical experiments we performed. This would, however, introduce new free parameters into the fitting procedure, and we therefore prefer to keep the simple theoretical model.

In any case the excellent fit with stratified models - as opposed to the complete failure with homogeneous atmospheres - lends strong support to the conclusion that in this object we really observe the result of gravitational settling with an extremely thin hydrogen layer.

How does this object fit into our general picture of the evolution of white dwarf spectral types? With such a thin hydrogen layer the object will have to become a DB at slightly lower effective temperatures, when the strong HeI convection develops and mixes and dilutes the hydrogen to undetectability. It is thus tempting to regard GD323 as an object at or very close to the transition

350

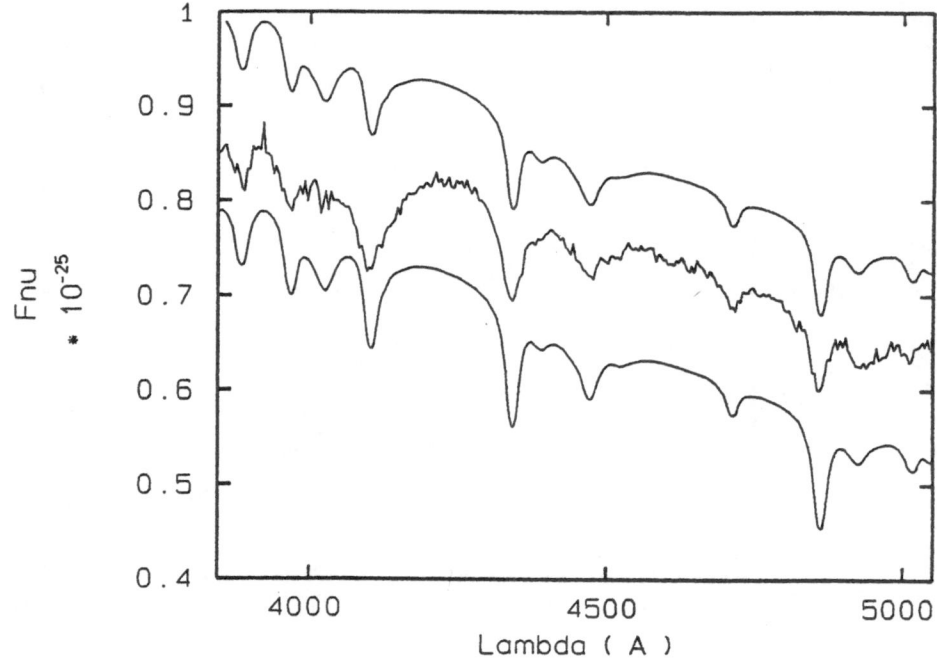

Fig. 4: Optical line spectrum and theoretical models for 28000 K, $M_H/M_\odot = 4.6\ 10^{-18}$ (top) and $5.8\ 10^{-18}$ (bottom).

from a DA to a DB, a transition that has to occur around 30000 K, if our current interpretation of the absence of He spectra between 30000 and 45000 K is correct (Liebert et al. 1987). However, with the same amount of hydrogen as GD323 at higher effective temperatures, a white dwarf would definitely show He lines! A very speculative scenario to solve this puzzle could be the following:

The calculations by MacDonald and Vennes (1990) indicate that a white dwarf with a hydrogen mass of about 10^{-15} $M_\odot$ develops a helium convection zone around $T_{eff} = 28000$ K, which contains of the order of 10^{-10} $M_\odot$. Even if the total amount of hydrogen would be diluted in this zone, the abundance would be only 10^{-5}. Our atmosphere calculations indicate that a hydrogen layer of order 5 10^{-18} $M_\odot$ on top of this envelope would be stable against convection and thus develop the abundance profile predicted by diffusion equilibrium. At the bottom of our atmosphere the H abundance is about 10^{-4}, a convection zone starting much deeper, where this abundance has fallen to e.g. 10^{-5} would not change the observable properties of the star, but could increase the total H mass to 10^{-15} $M_\odot$, which would have made this object a DA between 30000 and 45000 K.

Acknowledgements: This work was supported in part by grants from the National Aeronautics and Space Administration (NAG5-990) and from the National Science Foundation (AST-8814711).

References

Bergeron,P., Saffer,R.A., and Liebert,J.: 1990a, preprint
Bergeron,P., Wesemael,F., Fontaine,G., and Liebert,J.: 1990b, *Astrophys.J. (Letters)* **351**, L21
Dziembowski,W., and Koester,D.: 1981, *Astron. Astrophys.* **97**, 16
Finley,D., Basri,G., and Bowyer,S.: 1989, in *White Dwarfs*, IAU Coll. No.114, ed. G.Wegner
        (Berlin: Springer-Verlag), p.139
Gliese,W., and Jahreiß,H.: 1987, *Catalog of Nearby Stars*, 3rd ed., in preparation, as cited in McCook
        and Sion (1987)
Greenstein,J.L.: 1969, *Astrophys.J.* **158**, 281
Greenstein,J.L.: 1986, *Astrophys.J.* **304**, 334
Greenstein,J.L., and Peterson,D.M. : 1973, *Å* **25**, 29
Hamada,T., and Salpeter,E.E. : 1961, *Astrophys.J.* **134**, 683
Heintz,W.D. : 1974, *Astron.J.* **79**, 819
Hummer,D.G., and Mihalas,D.: 1988, *Astrophys.J.* **331**, 794
Jordan,S., and Koester,D.: 1986, *Astron. Astrophys. Suppl.* **65**, 367
Koester,D.: 1989, in *White Dwarfs*, IAU Coll. No.114, ed. G.Wegner (Berlin: Springer-Verlag),
        p.206
Koester,D., Liebert,J., and Greenstein,J.L.: 1990, in preparation
Koester,D., and Schönberner,D. : 1986, *Astron. Astrophys.* **154**, 125
Koester,D., Schulz,H., and Weidemann,V.: 1979, *Astron. Astrophys.* **76**, 262 Astrophys. 76,262
Koester,D., and Weidemann,V.: 1990, in preparation
Liebert,J., Wesemael,F., Sion,E.M., and Wegner,G.: 1984, *Astrophys.J.* **277**, 692
Liebert,J., Fontaine,G., and Wesemael,F.: 1987, *Mem. Soc. Astron. Italiana* **58**,17
MacDonald,J., and Vennes,S.: 1990, preprint
McCook,G.P., and Sion,E.M.: 1987, *Astrophys.J. Suppl.* **65**, 603
Oke,J.B., Weidemann,V., and Koester,D.: 1984, *Astrophys.J.* **281**, 276
Shipman,H.L.: 1979, *Astrophys.J.* **228**, 240
Wegner,G. : 1979, *Astron.J.* **84**, 650
Wegner,G. : 1980, *Astron.J.* **85**, 1255
Wielen,R. : 1962, *Astron.J.* **67**, 599

THE LYMAN ALPHA LINE WING IN HYDROGEN-RICH WHITE DWARF ATMOSPHERES

N. ALLARD[1] and J. KIELKOPF
University of Louisville
Department of Physics
Louisville
KY 40292, USA

ABSTRACT. We present calculations of the line profile of Lyman alpha, including the very far wing out to 4000 Å. We demonstrate that absorption during the close collision with one or more atoms or ions is probable, and produces significant effects throughout the line wing. New satellites are predicted near 1900 and 2400 Å due to three-body H_3^+ and H_3 interactions. Such effects are identified in some IUE white dwarf spectra.

1. Introduction

In white dwarfs of type DA, diagnostics based on visible spectra have shown atmospheres rich in hydrogen (Shipman, 1989). H and H^+ are the major atmospheric constituents in the temperature range from 10 000 to 16 000 K. Atomic hydrogen is the most significant source of opacity in the Lyman α region under these conditions. The resonance broadened wing and absorption by free atoms in a state of collision produce continua that are particularly prominent in DA white dwarf spectra up to 1800 Å.

There is a maximum, or binary satellite, around 1600 Å, this maximum has been identified as a quasi-molecular absorption of two hydrogen atoms in the singlet state which leads to a satellite band at λ=1623 Å.

Another absorption feature at 1400 Å is attributed to H-H^+ collisions. The 1600 Å atom-atom absorption is less prominent than its 1400 Å atom-ion counterpart at temperatures of the order of 16000 K, but at lower temperatures, of the order of 12000 K, it is quite distinct in a number of stars. Figure 1 exhibits a spectrum of the DA star L481-60 taken with the IUE. Several satellites can be distinguished. The satellite due to H-H collisions and a transition from the X to the B state is apparent at 1600 Å. Two others are also visible. The intensity minimum at about 1250 Å may be the satellite due to H-H collisions and transitions from the X to the C state, which produce a classical singularity at 1269 Å (Sando *et al* 1969). The weak dip near 1900 Å is approximately where a contribution from two neutral atoms and a proton is expected to produce a satellite. Although these two other features are hard to distinguish without filtering, they seem to be present in other

[1]Permanent address:
Observatoire de Paris-Meudon
Département Atomes et Molécules en Astrophysique
92195 Meudon Principal Cedex, France

G. Vauclair and E. Sion (eds.), White Dwarfs, 353–360.
© 1991 *Kluwer Academic Publishers.*

spectra shown by Koester *et al* (1985). Figure 2 shows the spectrum of the DA white dwarf Case 1, based on the data reported by Sion *et al* (1984) that included spectrophotometry at longer wavelengths than available for L481-60. A new weak absorption near 2400 Å appears here. We identify this as resulting from radiative transitions from three hydrogen atoms in collision. For clarity, the four flux minima are labeled with the corresponding molecular designations.

Previous calculations of the wing of Lyman α have been done by Sando *et al* (1969) and Sando and Wormhoudt (1973) in the classical and quantum mechanical theories respectively. Their accurate quantum-mechanical spectrum have been a good test of the validity of the semi-classical approach. Their main results are the existence of a broad satellite band and an absorption which decreases exponentially on the red side ($\lambda \rangle$ 1623 Å) (Sando and Wormhoudt 1973).

Unified calculations valid from the line center to the first satellite have been done by Koester *et al* 1985 in the Unified Franck-Condon treatment of Szudy and Baylis (1975).

While the IUE observations with the long-wave length redundant (LWR) and short-wavelength prime (SWP) cameras yield a spectral coverage $\lambda\lambda$ 1200-3200 (see e.g. Wegner 1982), the one perturber approximation is insufficient to explore the wing farther than the position of the first satellite.

The calculations reported here have been done in the unified theory of Anderson and Talman (1956) which provides the whole profile from the center of the line to the far wing. The quasi-molecular absorption occurs during the collision of two H atoms in the two electronic states $^1\Sigma_g$ and $^1\Sigma_u$ of H_2. The potential difference between them presents a minimum ω_s, corresponding to a wavelength of 1623 Å. Because of this extremum, the unified theory predicts periodic satellites centred at $\omega = k\,\omega_s + \omega_0$, (k=1,2,3,...), where ω_0 is the unperturbed line (see e.g. Allard 1978, Royer 1978, Allard and Kielkopf 1982). This series of satellites corresponds to the simultaneous presence of k perturbers in the collision volume $\mathcal{V}$. Such multiple satellites have been observed experimentally at very high densities (Kielkopf and Allard 1979).

2. Theory

We consider a fixed radiating atom surrounded by moving perturbers. The total power emitted in all directions for a dipole transition of this atom is given by

$$P(\omega) = \frac{4\omega^4}{3c^3} I(\omega) \tag{1}$$

For an isolated line the spectral line shape, I(ω) is the Fourier transform (FT) of the autocorrelation function $\Phi(s)$.

$$I(\omega) = \frac{1}{2\pi} \int_{-\infty}^{+\infty} \Phi(s) e^{i\omega s} ds \tag{2}$$

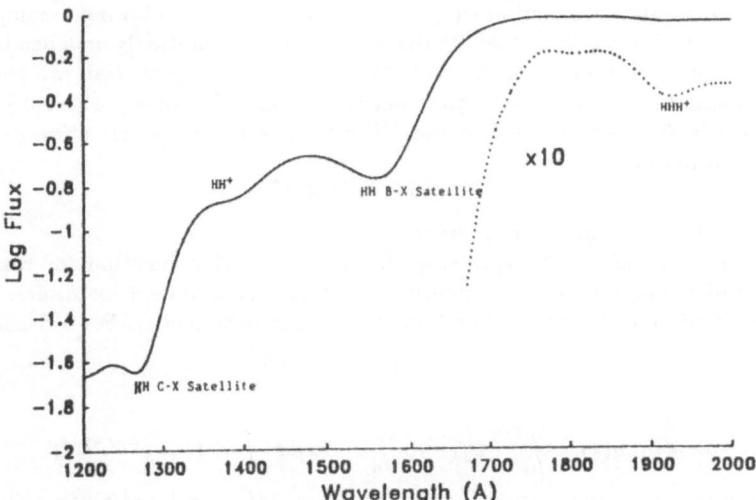

Figure 1. Absorption spectrum of the DA white dwarf L481-60 from an IUE spectrum reproduced in Koester, et al. (1985). The spectrum has been filtered to remove noise, and plotted as $\log_{10}$ to show the range of weak features that make the Lyman α wing. A section expanded by 10x(.....) shows a weak feature at 1900 Å that we attribute to three-body collisions involving an excited H atom, a ground state atom, and H^+.

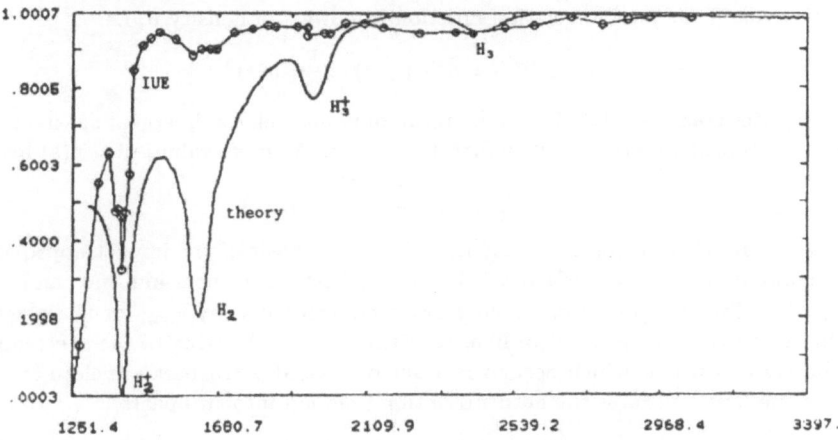

Figure 2. Absorption spectrum of the Lyman α wing for the DA white dwarf Case 1 (WD 1213 + 528, EG 87) from the IUE data of Sion et al. (1984). Satellites due to two- and three-body interactions of H and H^+ are identified.

The autocorrelation function $\Phi(s)$ is calculated with the following assumptions : 1) the radiator is stationary in space; 2) the perturbers are mutually independent; 3) in the adiabatic approach valid in the present case the interaction potentials are scalarly additive. This last simplifying assumption allows us to calculate the total profile $I(\omega)$ – when all the perturbers interact– as the FT of the Nth power of the autocorrelation $\phi(s)$ of a unique atom-perturber pair.

$$\Phi(s) = (\phi(s)^N) \tag{3}$$

Here, N is the total number of perturbers.

The fundamental result expressing the autocorrelation function for many perturbers in terms of a single perturber quantity g(s) was first obtained by Anderson (1952) and Baranger (1958) in the classical and quantum cases respectively. For a density n :

$$\Phi(s) = e^{-ng(s)} \tag{4}$$

where

$$g(s) = \int_0^{+\infty} v f(v) dv \int_0^{+\infty} 2\pi\rho d\rho \int_{-\infty}^{+\infty} (1 - e^{i\eta(t,s)}) dt \tag{5}$$

ρ is the impact parameter and $\eta(t,s)$ is the phase shift calculated along a classical path,

$$\eta(t,s) = \int_t^{t+s} V(t') dt' \tag{6}$$

V represents the difference between the electronic energies of the quasimolecular transition in unit of rad/sec.

The relatively low densities of hydrogen atoms ($n \ll 10^{20}$ atomes/cm^{-3}) involved in the astrophysical conditions allow us to use an expansion of the spectrum $I(\omega)$ in powers of the density (Royer 1978).

We expand the autocorrelation function in powers of density n, i.e.

$$e^{-ng(s)} = 1 - ng(s) + \frac{1}{2}n^2 g(s)^2 + \cdots \tag{7}$$

and Fourier transform Eq. 7 term by term, but this yields a divergent spectrum in the limit of $\omega \longrightarrow 0$ because g(s) is unbounded for large s. At large values of s, g(s) becomes linear in s :

$$g(s) \longrightarrow g_{imp}(s) = \alpha + \beta s \tag{8}$$

where α and β are complex constants. This is the basis of the impact approximation. Let us denote by τ_c the value of s beyond which g(s) becomes linear in s and can be replaced by g_{imp}. The impact appoximation consists in replacing g(s) by g_{imp} for all values of s and is valid in the core of the spectral line. Contributions to the wing of the spectrum arise from radiative transitions which occur when one or several perturbers are close to the radiator. Thus one first separates the autocorrelation function into two parts,

$$g(s) = g_{imp}(s) + \tilde{g}(s) \tag{9}$$

The linear part is usually associated with the line core and the second part $\tilde{g}(s)$ with the line wing. The autocorrelation function becomes:

$$\Phi(s) = e^{-ng_{imp}(s)}(1 - n\tilde{g}(s) + \frac{1}{2}n^2 \tilde{g}(s)^2 + \cdots) \tag{10}$$

We obtain a density expansion for the spectrum by taking the term by term FT of Eq. (10)

$$I(\omega) = I_c(\omega) * (\delta(\omega) - n\tilde{g}(\omega) + \frac{1}{2}n^2\tilde{g}(\omega)^{*2} + \cdots) \tag{11}$$

* symbolizing convolution.

Rather than a direct product, we have now a convolution, where

$$\tilde{g}(\omega) = FT(\tilde{g}(s)) \tag{12}$$

$$I_c(\omega) = FT(e^{-ng_{imp}}) \tag{13}$$

$I(\omega)$ can be expressed as a sum of several different terms. The first two terms of the expansion correspond to the impact and one perturber approximations, but the one perturber spectrum is convolved by the impact distribution. High-order terms of the expansion are successive convolution powers of the wing of the one perturber distribution, all convolved by the impact distribution.

The one perturber distribution is the first term of a sum of contributions each corresponding to the simultaneous presence of k (k=1,2,3,...) perturbers near the radiator in volume $\mathcal{V}$.

This low density approximation is very useful for computing the line profile far in the wing.

3. RESULTS

For a calculation of the line shape following the theory outlined in the previous section, we need the diatomic potentials for the interactions of the colliding atoms in their initial and final states. The difference between the B and X states of H2 is shown in Fig. 3. Lyman α is at 82258 cm-1, and the minimum of this difference potential is -20632 cm-1, producing a wing out to 61626 cm-1 or 1623 Å. For the ion, the relevant molecular curves are for H2+ for which a difference potential minimum is -10667 cm-1, and a binary satellite is at 1397 Å.

To calculate profiles we have represented the potential difference by a square-well potential where $V(R) = \omega_s$ for $R \leq a$, $V(R)=0$ for $R \geq a$, R is the internuclear distance and a the range of the potential. We assume rectilinear trajectories for the atoms of uniform velocity $\bar{v}$, $(\bar{v} = (8kT/\pi m)^{1/2})$. The autocorrelation function is completely analytical leading to very fast calculations. This very simple model allows us to point out the main physical effects.

There are two different expressions for $\phi(s)$ depending on whether the value s is less or greater than $\tau_{max} = 2a/\bar{v}$ where τ_{max} is the maximum collision time. These expressions are given in Allard 1978 (Eqs 4 and 5). The autocorrelation function presents oscillations only during the collision time τ_{max}, and beyond it decreases exponentially as predicted by the impact theory. These oscillations will give rise to satellites.

Fig.2 shows the theoretical profile at 12500 K and 10^{19} cm^{-3}. In the additive approximation the total profile taking into account the collision HH$^+$ and HH is given by

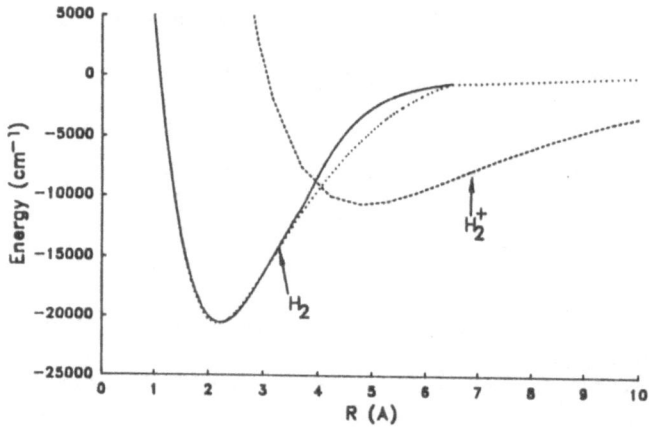

Figure 3. Difference potential energy in cm^{-1} for the B-X branch of the H_2 potential $(\cdots\cdots)$ and the fitted version used to compute line profiles (———). The difference potential for H_2^+ is also shown (---). The molecular potentials are from the compilation of Sharp (1971), and the differences are plotted with respect to the Lyman α line energy.

Table I

A Comparison of Lyman Alpha Satellites

Feature	Wavelengths (Å)				
	Observation		Theory		
	L481-60	Case 1	a priori	unified	additive
Lyman α			1215.7	1215.7	1215.7
H_2 C-X	1265	1268	1268		1268
H_2 B-X	1570	1605	1623	1622	1623
H_3		2375	2610	2432	2439
H_2^+	1391	1395	1397	1397	1397
H_3^+	1927	1950	1978	1959	1962

$$I_{tot}(\omega) = I_{HH^+}(\omega) * I_{HH}(\omega) \tag{14}$$

$*$ symbolizing convolution.

As expected, the binary satellites appear near 1400Å and 1620 Å, and the tertiary satellites are near 1900Å and 2400 Å. The latter is weak at 10^{19} cm-3, and disappears into the wing of Lyman α at densities below about 10^{17} cm^{-3} for a square well potential. The satellite widths are temperature dependent. This occurs because the duration of the collision decreases as the temperature increases, and the reduced interaction time broadens the satellites.

The satellites in the long wavelength wing of Lyman α predicted on the basis of the known potential curves of H_2, H_2^+, H^3, and H_3^+ and these line shape calculations are given in Table I. The unified line shape calculations for the one- and two-perturber satellites are measurements of the peaks of the satellites in the computed spectra. The predicted positions from an additive approximation are based on the diatomic potentials (Allard and Kielkopf 1990). Finally, the a priori wavelengths are derived from quantum calculations of the H_3 and H_3^+ potentials.

The DA white dwarf spectrum shown in Fig. 1 was specifically selected to show both H_2^+ and H_2 features. At somewhat lower temperatures, the H2 satellite at 1623 Å is completely opaque in other DA white dwarfs. This is shown in Fig. 1 of Koester et al (1985) for the star L532-81. Under these circumstances the opacity in the region above 1600 Å is determined by the wing of the satellite, which we have shown depends critically on many-body interactions and the core profile. This effect provides a previously unrecognized source of opacity for the photosphere of a DA white dwarf when it cools below 10000 K.

The effects of the close collisions of many perturbers do not appear at the low densities where the core of Lyman α forms in a typical hot DA white dwarf atmosphere, but the wing of the line is generated at a greater depth and higher density. As a consequence, the effects of multiple perturbers still can be important if there are no competing contributions to the opacity. With increasing depth and temperature, however, the fraction of the hydrogen that is ionized increases, and the perturbation of Lyman α by H^+ adds another dimension to the problem. A theoretical model of the profile shown in Fig. 1 will require a grid of profiles such as we have calculated here that include simultaneous interactions of the radiator with other neutral atoms and ions.

Acknowledgments :
The work done at the University of Louisville has been supported by a grant from Basic Energy Sciences, Chemistry Division, Fundamental Interactions Branch of the United States Department of Energy. Support for Dr. Allard's stay at the University of Louisville was provided by a grant from NATO.

References

Allard, N.F. (1978) 'Alkali-rare-gas line profiles in a square-well potential approximation I.Satellites', J.Phys.B:Atom.Molec.Phys., 11,1383-1392
Allard N.F., Kielkopf J.F. (1982) 'The effect of neutral nonresonant collision on atomic spectral lines', Rev.Mod.Phys.,54,1103-11182

Allard N.F.,Kielkopf J.F. (1990) 'Temperature and Density Dependence of the Lyman α Line Wing in Hydrogen-Rich White Dwarf Atmospheres', in press Astron.Astrophys.

Anderson, P.W. (1952) 'A method of synthesis of the Statistical and Impact Theories of Pressure Broadening', Phys.Rev., 86,809.

Baranger, M. (1958) 'Simplified Quantum Mechanical Theory of Pressure Broadening', Phys.Rev., 111, 481-493.

Anderson, P.W., Talman, J.D. (1956) 'Proc.Conf.Broadening of Spectral Lines', Bell Telephone System Technical Publications,No.3117,29,Murray Hill,N.J.

Baranger, M. (1958) 'Simplified Quantum-Mechanical Theory of Pressure Broadening', Phys.Rev,11,481

Bracewell, R.N. (1965) 'The Fourier Transform and its application', in New York:Mc Graw-Hill

Kielkopf J.F., Allard N.F. (1979) 'Observation of the Simultaneous Effect Additive Effect of Several Xenon Perturbers on the Cs 6s-9p Doublet', Phys. Rev.Lett., 43, 196-199.

Koester, D., Weidemann, E.-M., Zeidler, K.T., Vauclair, G. (1985) 'The Explanation of the 1400 and 1600 Å features in DA White dwarfs', Astron.Astrophys., 142,L5-L8.

Royer, A. (1978) 'Low Density Approximation In the adiabatic theory of pressure broadening ', Acta Phys.Pol. A, 54,805-822.

Sando, K.M., Doyle, R.O., Dalgarno, A. (1969) 'Resonance-broadening Absorption in the Wings of Lyman Alpha', Astrophys.J., 157,L143-L145.

Sando, K.M., Wormhoudt, J.G. (1973) 'Semiclassical Shape of Satellite Bands', Phys.Rev.A, 7, 1889-1898.

Sharp, T.E. (1971) Atomic Data, 2,119

Sion, E.M., Wesemael, F., Guinan, E.F. (1984)' IUE Spectrophotometry of the DA4 Primary in the Short-period White Dwarf-red Dwarf Spectroscopic Binary Case 1', Astrophys.J., 279, 758-762.

Shipman, H.L. (1989) 'Properties and evolution of white dwarf stars in Planetary Nebulae, IAU Symposium 131, ed. S. Torres-Peimbert, Kluwer, Dordrecht, pp. 555-566.

Szudy, J., Baylis, W.E. (1975) 'Unified Franck-Condon treatment of Pressure Broadening of Spectral Lines', JQRST., 15, 641-668.

Wegner, G. (1982) ' Detection of the 1400 Å Absorption in the Ultraviolet Spectrum of the DA White Dwarf LB3303', Astrophys.J., 261, L87-L89.

ATMOSPHERIC PARAMETERS FOR DA WHITE DWARFS IN THE VICINITY OF THE ZZ CETI INSTABILITY STRIP

N. DOLEZ[1,2], G. VAUCLAIR[1], D. KOESTER[3]
[1] Observatoire Midi-Pyrénées, Toulouse, France
[2] CERFACS, Toulouse, France
[3] Louisiana State University, Baton-Rouge, Louisiana, USA

ABSTRACT. A sample of DA white dwarfs whose high S/N spectrophotometry has been obtained at the Palomar 5m telescope is analysed with a grid of model atmospheres. Atmospheric parameters are derived for 38 DA. A tendancy to increased log g with decreasing temperature is found. This does support the conclusion that the apparent increase in gravity is due to the increased He abundance in cool DA produced by convective mixing. The presence of a few non-pulsating DA in the ZZ Ceti instability strip is discussed.

1. Introduction

The main motivation in undertaking this work a few years ago was to search for new ZZ Ceti variable white dwarfs. From the linear non-adiabatic stability analyses of the non-radial g-modes (Winget 1988 and Kawaler and Hansen 1989 for reviews) it is possible to infer important astrophysical informations. The use of asteroseismology for a fine analysis of the internal structure and evolution of white dwarfs has proved to be extremely rewarding (Winget 1990, this workshop). Among the astrophysical quantities which one would like to extract from the observations, the hydrogen content in the DA white dwarfs is of special interest. It is possible in principle to constrain this hydrogen mass content: the mode trapping in the hydrogen outer layers, the location of the instability strip blue and red edges in the H-R diagram, the changes in surface chemical composition due to the convective mixing of the outer hydrogen with the underlying He, do directly depend on the hydrogen mass content. For these reasons we have obtained spectrophotometry of a number of DA white dwarfs from which we have extracted a homogeneous sample of ZZ Ceti candidates. These candidates were observed with a high speed photometer to search for new variable stars. This sample is analysed with a grid of model atmospheres to derive their atmospheric parameters: log g, T_e.

2. Observations

The spectrophotometric observations of DA white dwarfs have been obtained at the Palomar 5m telescope by two of us (N.D. and G.V.) as part of a joint observing

361

G. Vauclair and E. Sion (eds.), White Dwarfs, 361–367.
© 1991 *Kluwer Academic Publishers.*

programme with J.L. Greenstein. The Oke-Gunn double CCD spectrograph (Oke and Gunn, 1982) was used to get high S/N spectrophotometry of about 80 DA white dwarfs. The resolution of the spectra is $\sim 4\overset{\circ}{A}$ in the blue ($\lambda < 5400\overset{\circ}{A}$) and $\sim 6\overset{\circ}{A}$ in the red ($\lambda > 5400\overset{\circ}{A}$). Because of a failure of the blue CCD camera for two nights, the total coverage of the $3100\overset{\circ}{A} - 10000\overset{\circ}{A}$ wavelength interval has been possible for only 59 stars out of the total sample. Some details of these observations and preliminary analysis have been presented by Greenstein (1986) and Greenstein and Liebert (1990).

As far as the search for variability in the sample is concerned, this programme, which was started with great expectation, did result, in a rather frustrating manner, in the discovery of only one new ZZ Ceti star: GD66 (Dolez, Vauclair, Chevreton 1983).

3. Spectra analysis

In the present paper, a sub-sample of DA white dwarfs in the vicinity of the ZZ Ceti stars is choosen (6000 K $\leq T_e \leq$ 15000 K) for detailed analysis. A grid of LTE, pure hydrogen, model atmospheres is used, in the effective temperature range 6000 K -(1000 K) -15000 K and for surface gravities (log g) in the range 7.50 -(0.25) -8.50. This grid of model atmospheres does not use the Hummer-Mihalas formalism in calculating the population of the hydrogen atom levels.

In principle, a fit of the total energy distribution with model atmosphere theoretical flux allows a determination of the effective temperature. In practice, this method alone suffers for some uncertainties: among them, a difficulty in adding the blue and red parts of the spectrum may arise, due to the dichroic filter used to split the stellar light into the two channels of the double spectrograph. For this reason, an automatic χ^2 method is used to fit the Balmer line profiles. By tracing graphically the lines of iso-χ^2 values in a log g-T_e plane, one can derive in principle the log g-T_e values which minimize χ^2. For the high temperature white dwarfs where the line profiles do vary regularly with effective temperature and gravity, this method gives a unique log g-T_e determination. In the vicinity of the ZZ Ceti white dwarfs, however, the determination of log g-T_e values is made more difficult by the well known fact that a given line profile may be equally well reproduced by two different choices of log g-T_e values. The absolute flux distribution (or colors) can then help to choose the correct couple of atmospheric parameters.

The Table 1 gives the results of this exercice for the 38 DA white dwarfs of our sample falling in the temperature interval 15000 K-6000 K and with complete spectrophotometry in the wavelength interval 3100 Å -10 000 Å. It has been frequently found that different Balmer lines lead to different T_e-log g values! At this stage, it is not possible to know whether these uncertainties arise in our model atmosphere grid being too sparse or in difficulties with the Balmer lines broadening theory. The values given in table 1 results from the following choice: in case of discrepancies in the log g-T_e values derived from various Balmer lines, the T_e has

TABLE 1 : Effective temperature and surface gravity of DA white dwarfs

WD	Name	T_e (K)	log g
0009 − 058	G158-39	10000 ± 50	8.45 ± .04
0032 − 175	G266-135	9980 ± 50	8.30 ± .10
0052 + 226	LHS5016	9000 ± 100	8.37 ± .12
0107 − 192	GD685	14360 ± 200	8.25 ± .02
0143 + 216	G94-9	9000 ± 50	8.34 ± .10
0148 + 641	G244-36	9000 ± 50	8.17 ± .15
0231 − 054	GD31	11780 ± 100	8.37 ± .15
0236 + 744	G221-2	9000 ± 50	8.35 ± .15
0239 + 109	G4-34	7000 ± 100	8.35 ± .10
0243 − 026	LHS1442	7000 ± 100	8.35 ± .15
0302 + 621	GD426	11000 ± 150	8.50 ± .03
0339 + 523	Rubin70	12100 ± 100	8.00 ± .10
0348 + 339	GD52	13000 ± 100	8.50 ± .10
0354 + 463	Rubin80	8000 ± 100	8.00 ± .10
0407 + 179	HZ10	14200 ± 100	8.25 ± .10
0408 − 041	GD56	15000 ± 100	8.35 ± .12
1610 + 166	GD196	14500 ± 50	8.11 ± .13
1624 + 477	G202-49	9000 ± 100	8.30 ± .10
1625 + 093	G138-31	7000 ± 100	8.35 ± .10
1636 + 160	GD202	12000 ± 100	8.35 ± .10
1637 + 335	G180-65	10000 ± 100	8.50 ± .10
1647 + 591	G226-29	11000 ± 100	8.40 ± .10
1710 + 683	G240-47	6960 ± 100	8.20 ± .05
1716 + 020	G19-20	14250 ± 200	7.90 ± .10
1811 + 327.1	G206-17	8000 ± 100	8.38 ± .04
1826 − 045	G21-16	9000 ± 100	8.20 ± .15
1827 − 106	G155-19	14800 ± 100	7.85 ± .12
1855 + 338	G207-9	11150 ± 100	8.50 ± .10
2010 + 613	GD544	14500 ± 150	8.12 ± .12
2111 + 261	G187-32	8500 ± 100	8.36 ± .10
2136 + 229	G126-18	10000 ± 100	8.40 ± .10
2139 + 115	GD235	15000 ± 100	8.25 ± .10
2151 − 015	G93-53	9000 ± 100	8.35 ± .10
2207 + 142	G18-34	8000 ± 100	8.30 ± .10
2258 + 406	G216B14B	10000 ± 100	8.25 ± .10
2311 + 552	GD556	11000 ± 100	8.50 ± .10
2326 + 049	G29-38	11000 ± 100	8.37 ± .12
2347 + 128	G30-20	11000 ± 100	8.50 ± .10

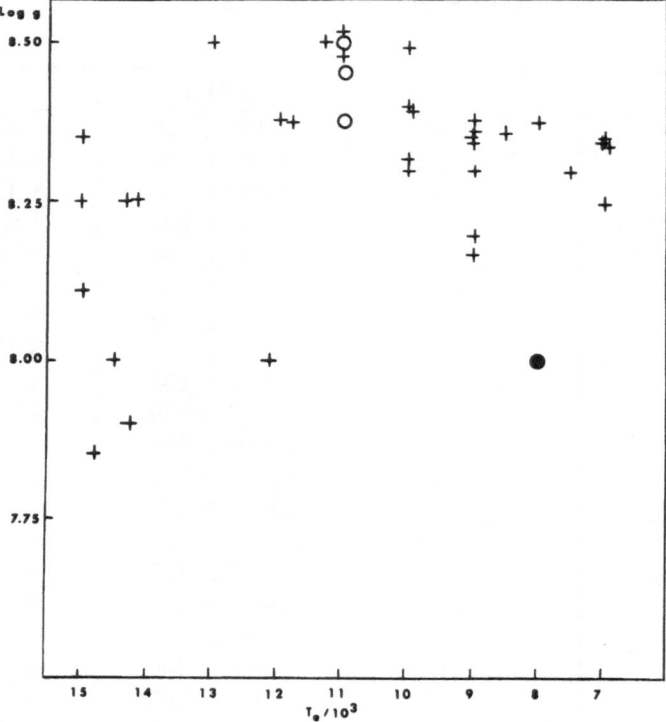

Figure 1: Atmospheric parameters for a sample of DA white dwarfs. Open circles are the ZZ Ceti of this sample: G226-29, G207-9 and G29-38 while the full circle is the composite spectrum (DA+dM) Rubin 80.

been taken from the lower Balmer lines (Hα -Hγ), while the gravity has been derived from higher Balmer lines (Hε, H8), more gravity sensitive.

4. Discussion

The results of our analysis are summarized in figure 1. This analysis clearly shows a tendancy for an increase of $< \log g >$ as T$_e$ decreases. The exception to that tendancy (full circle in Figure 1) is the white dwarf Rubin 80 which has a composite spectrum (DA + dM): this may have contaminated the log g determination or the mass of the white dwarf may be lower than the average if Rubin 80 forms a binary system.

As discussed in Bergeron et al. (1990), it is not possible to separate the effects on the line profiles of an increased gravity from those of an increased helium abundance. Bergeron et al (1990) did choose to maintain the surface gravity of their white dwarf sample at a constant standard log g = 8 value, and fit the Balmer lines profiles by varying the He/H abundance ratio and effective temperature. Here, the other choice made is to maintain a pure hydrogen composition and fit the line profile by varying effective temperature and gravity. If one reasonably assumes that indeed the mean surface gravity of cool DA white dwarfs should be the same as the mean surface gravity of hotter white dwarfs, our results confirm the Bergeron et al (1990) conclusion that the surface composition of cool white dwarfs is polluted by He as a result of the convective mixing. An important implication of this result is that the hydrogen mass content of the DA must be small enough to allow convective mixing to occur at effective temperatures as high as $\sim$ 11000 K.

Another interesting result of this analysis concern the three ZZ Ceti variables of the sample (G226-29, G207-9, G29-38). All three lie at the red edge of the instability strip and do require high gravity ($8.40 \leq \log g \leq 8.50$) to fit their Balmer lines. Does this mean that they are already convectively mixed? If so, what would be the role of the He abundance on the excitation mechanism and the consequence of the disappearance of the H/He transition zone on the mode trapping?

Finally, a few DA white dwarfs falling in the ZZ Ceti instability strip have not been found to pulsate with amplitude larger than a few mmag. These are: GD31, GD426, Rubin 70, GD52, GD202, GD556, G30-20. This conclusion depends of course on the exact limits of the instability strip. Some of these non pulsating DA are in fact close to the edges and could well be outside in view of the uncertainties on their effective temperature. A few candidates close to the red edge (GD426, GD556, G30-20) may be outside of the instability strip if the red edge is at 11320 K as suggested by Daou et al. 1990. Similarly, GD52 is close to the blue edge and may also be outside. Clearly more precise effective temperature for these stars are needed as they are usefull to determine the edges of the instability strip. The remaining stars GD31, Rubin 70 and GD202 do sit in the middle of the instability strip and remain the most serious candidates as non pulsating DA in the ZZ Ceti instability strip.

References

Bergeron, P., Wesemael, F., Fontaine, G. and Liebert, J.: 1990, *Astrophys. J. Letters*, **351**, L21

Daou, D., Wesemael, F., Bergeron, P., Fontaine, G. and Holberg, J.B.: 1990, *Astrophys. J.* **364**, 242

Dolez, N., Vauclair, G. and Chevreton, M.: 1983, *Astron. Astrophys. Letters*, **121**, L23

Greenstein, J.L.: 1986, *Astrophys. J.* **304**, 334

Greenstein, J.L. and Liebert. J.: 1990, *Astrophys. J.* **360**, 662

Kawaler, S.D. and Hansen, C.J.: 1989, in *"White dwarfs"*, IAU Colloquium 114, Ed. G. Wegner, Lecture Notes in Physics, Springer-Verlag, p. 97

Oke, J.B. and Gunn, J.E.: 1982, *Publ. Astron. Soc. Pacific*, **94**, 586

Winget, D.E.: 1988, in IAU Symposium 123, *"Advances in Helio and Asteroseismology"* eds. J. Christensen-Dalsgaard and S. Frandsen, Dordrecht, Reidel, p. 305

Winget, D.E.: 1990, in the Nato Advanced Research Workshop *"7th Europeen Workshop on white dwarfs"*, Ed. G. Vauclair and E. Sion

DISCUSSION

WINGET :

(1) I have a comment and a related question. First, one expects to find a small number of very low amplitude pulsators in the DAV strip based on inclination angle effects but probably not as many as you find. The related question in then, what are the amplitude limits you have for the non-pulsator candidates?

(2) A comment on the importance of finding non-pulsators. The Rochester-Texas-Montréal-Colorado calculations imply that DA stars with $M_H > 10^{-8} M_*$ will not pulsate within the strip, and the Los Alamos calculations imply that all stars should pulsate within the strip.

VAUCLAIR :

All observations were on the 1.9m at Haute Provence and the limits were 1 to 2 mmag.

KEPLER :

Comment : I have observed many non-variable stars in the color range of the ZZ Ceti instability strip, but we lack accurate temperatures, specially for the southern stars, to do any statistical analysis of the ratio of variable to non-variables in the strip.

WEIDEMANN :

Did you compare your results with our 1984 analysis which used multichannel observations and the theoretical dependence of the energy distribution on T_{eff} and g. We did not use line profiles also for the reason that the g dependence in the ZZ Ceti region is much more pronounciated in the energy distribution.

VAUCLAIR :

We did not make such a comparison. As I have explained, some uncertainties may arise when one adds the blue and the red spectra obtained with the double spectrograph, because of the dichroic filter. We choose to fit the high Balmer line profiles to make full use of the information contained in these profiles.

SPACE TELESCOPE OBSERVATIONS OF WHITE DWARF STARS

HARRY L. SHIPMAN
Physics and Astronomy Department
University of Delaware
Newark, Delaware 19711 USA

ABSTRACT. The Hubble Space Telescope will present an opportunity for a dramatic improvement in our understanding of white dwarf stars. A collaboration of researchers has obtained observing time to study white dwarf stars. This paper reviews our proposed observations, the genesis and history of the collaboration, and the current status of this project in light of the telescope's shortcomings.

1. INTRODUCTION

The capabilities of the Hubble Space Telescope (HST) are well known to the astronomical community. Its limitations are becoming better known. For better or worse, it is one of the central instruments for the decade of the 1990s and perhaps beyond. Space Telescope presents the opportunity for a great leap forward in two major areas relating to white dwarf stars. One primary advantage of HST is its considerably improved spectral resolution and sensitivity in comparison with the International Ultraviolet Explorer (IUE). Another, perhaps less appreciated advantage is its higher spatial resolution. Astronomers studying white dwarf stars have been bedeviled for decades by the proximity of Sirius B and Procyon B to two of the brightest stars in the sky. A more recently emerging difficulty is that two more white dwarf stars in binary systems, the two components of G 107-70, are sufficiently close together that obtaining independent observations of the two components is quite difficult. Given the current spherical aberration in HST, use of this second capability may be subject to some limitations.

This paper presents a progress report on our efforts to see that observations of white dwarf stars are included in the Space Telescope observing program. A massive team of researchers has participated, so far, in these efforts. The team includes G. Basri and D. Finley (University of California, Berkeley); H. Bond (Space Telescope Science Institute), F. Bruhweiler (Catholic University of America), F. Cordova (Los Alamos National Laboratory and Penn State University), G. Fontaine and F. Wesemael (Universite de Montreal), P. Hintzen (NASA-Goddard Space Flight Center and University of Nevada-Las Vegas), J. Holberg and J. Liebert (University of Arizona), K. Jensen (NASA-Goddard Space Flight Center and STX Corporation), D. Koester (Louisiana State University), J. Nousek (Penn State University), T. Oswalt (Florida Institute of Technology), E. Sion (Villanova), S. Starrfield (Arizona State), D. Tytler (Columbia

369

G. Vauclair and E. Sion (eds.), White Dwarfs, 369–378.

University and University of California-San Diego), G. Vauclair (Observatoire Pic-du-Midi et de Toulouse), G. Wegner (Dartmouth College), and V. Weidemann (Kiel) in addition to myself, the Principal Investigator.

At this early stage of this project, a paper that is most likely to be of lasting archival value is one that includes some discussion of how this collaboration began and how it has stayed cohesive through the peregrinations, wild oscillations, and general turbulence of the Space Telescope program. As a result, this paper is part history, and part proposed observations. Section 2 discusses the nature and origin of this collaboration. Since no successful HST observations of white dwarf stars have yet been made at the time this is being written (early 1991), Section 3 of this paper can only describe the science which we hope to do with HST. Section 4 describes the developments of 1990: the launch of HST, the revelation of its flawed optical system, and the impact on our project. A consequence of this approach to writing this paper is that it is a bit more personal in style than the usual scientific paper. While fellow team members have contributed enormously to the development of this project, they are in no way responsible for the contents of this article -- in particular my statements about the social and institutional context in which this project evolved.

2. NATURE AND ORIGIN OF THE COLLABORATION

It all began, as so often happens, over a glass of beer. The Space Telescope mission has a long history, dating back to the mid 1940s. A number of us had been aware for some time that this mission offered several opportunities for our field. But we also realized that competition for observing time on Space Telescope was likely to be extremely intense. How were we to ensure that an appropriate amount of time on Space Telescope would be devoted to the study of white dwarf stars?

In the summer of 1984, my own concerns intensified. Would any time on Space Telescope be allocated to the study of white dwarf stars? Malcolm Longair, in a talk for the 1984 Baltimore meeting of the American Astronomical Society (AAS), asserted that stellar evolution was basically fairly well understood and implied that stellar astronomy was a completed field of science. Did this mean that Space Telescope would not observe any target closer than the Andromeda Galaxy, M 31? The process of developing "key projects," in which the study of individual stars was considered by only one of several groups, was another indication that many people associated with the Space Telescope were casting their eyes beyond the neighborly confines of the Milky Way Galaxy. Many committees were worrying about extragalactic objects while all of stellar astronomy, from T Tauri stars to white dwarf stars, was considered by only a single group.

Other clouds on the horizon appeared as the process of allocating time on Space Telescope among various competing projects became more widely discussed. Progress in our field, white dwarf stars, has largely followed the "small science" model of individual investigators each applying for allotments of telescope time on the ground or in space. Would this model work with Space Telescope? Even in the early 1980s there was much discussion of big projects and massive allocations of observing time to "very important" science. Surveys by the Space Telescope Science Institute suggested that the oversubscription rate for Space Telescope proposals could be as much as a factor of ten. Some of us thought that if we each submited short proposals for Space Telescope, with a few targets apiece, none of us might end up with anything. Space Telescope might not ever study white dwarf stars, or at least not for a while until the community adjusted to the new way of doing business.

A number of us talked about these issues in hallways and in other places at the Baltimore AAS meeting and at the European Workshop on White Dwarf Stars held in Kiel later in the summer. A one-hour session at the close of the Kiel workshop revealed broad consensus, if not unanimous agreement, that a collective approach towards competing for time on HST was the way that the community wanted to go. Then and now, we visualize this team effort as being collective only in the early stages of target selection and data gathering. Once the data is in hand we can resume our traditional method of doing science by ones, twos, and threes.

I believe that there were basically two reasons that we formed a team rather than proceeding as individuals. First, many of us shared the thinking discussed above. Only by forming a team could we as a community avoid being swamped by big science, we believed. A second reason was that we felt that we as a community should end up making the decisions as to what targets are going to be observed, rather than leaving the decisions up to a telescope allocation committee which would contain few if any people who knew much about white dwarf stars.

Time will tell whether these thoughts should apply to other missions. We cannot run the experiment over again and see what would have happened had we submitted proposals as individuals or in smaller groups. I do offer some observations which suggest that "small science" can still survive in the space astronomy era of the 1990s. The team approach may not be necessary. After our proposal was submitted, there was another intense competition for observing time on ROSAT. The available observing time on ROSAT was divided up into tiny little pieces, with everyone getting some part of the pie. Another observation is of the fate of other teams which formed in many other areas of stellar astronomy. I have not done a systematic study of what happened to these team proposals; such a study might be interesting. Anecdotal information suggests that in several cases at least the team's proposal for observing time was unsuccessful. As a result little science in these particular fields of astronomy will be done until Cycle 2 of HST observations, if then. I know of at least one other case where the team was not able to develop a consensus approach to using HST and decided in the end to submit individual proposals, at least some of which were accepted.

Our team, at least, stayed together through the turbulent 1980s. In 1986, everyone expected that Space Telescope would be launched at some time that year, and I was busily preparing drafts of a proposal, bringing it to meetings, and revising it. And then, on January 27, 1986. the Space Shuttle Challenger exploded. Leaders at the Space Telescope Science Institute quickly realized that the launch of HST was likely to be delayed for some time. Our proposal, and all others, were put on hold. This delay was to be only the first of many. There had been further discussion of this project at various meetings in 1987 and 1988, and an accretion disk of astronomers had accumulated on this proposal. The next section of this paper states what we hope to do with HST. The thoughts are basically the same as those we had in 1988, since at this writing there is no data on white dwarfs which has been obtained so far with HST.

3. WHAT HST CAN DO TO STUDY WHITE DWARF STARS

Perhaps the most important unifying idea of our project requires only two hours of observing time. We seek to nail down our understanding of the physics of stellar degeneracy by confirming that white dwarf stars do indeed fall on the mass-radius relation for degenerate objects. Virtually everyone believes in the theory of stellar degeneracy, a Nobel-Prize winning idea first proposed fifty years ago. However, a sobering thought is that observational confirmation rests on three objects: Sirius B, 40 Eri B, and Stein 2051 B. A number of us have realized, as Weidemann pointed out at the Kiel workshop six years ago that this is hardly a good scientific test of a basic physical theory. Furthermore, one of the three objects, 40 Eri B, falls two standard deviations away from the predicted mass radius relation (see, e.g., Shipman and Sass 1980; Wegner 1979, 1980; Wegner and Yackovich 1983).

Three other white dwarfs with known, high-precision parallaxes do exist: Procyon B and the two components of G 107-70. (While V 471 Tauri is a binary white dwarf, its parallax has been claimed to be uncertain.) Procyon B's properties are virtually unknown because of its closeness to Procyon A. Usually, the two stars are about 3 arc seconds apart, but as chance would have it, the separation is considerably larger in the late 1990s. Investment of a very few hours of HST time in observations of Procyon B and the two components of G 107-70 can double the number of objects used for this fundamental test of the theory of stellar degeneracy, either placing it on a secure observational footing (at last!) or undermining it completely. The consequences will be profound if this test of relativistic degeneracy fails.

The remainder of the HST observations of white dwarf stars deal with the complex ways in which the chemical compositions of white dwarf photospheres evolve. Several review papers have been published in conferences in the late 1980s (Fontaine and Wesemael 1987; Liebert, Fontaine, and Wesemael 1987; Shipman 1989; Sion 1986). Other papers in this volume discuss these processes in more detail, and a forthcoming paper by MacDonald and Vennes (1991) gives a comprehensive view. Models which deal with the competition between accretion, gravitational settling (also known as diffusion), and convective mixing often make quite specific predictions about the abundances of different elements or ions.

The difficulty we have with IUE is that its limited sensitivity and spectral resolution limits either the variety of chemical elements which can be observed, or the type of star in which they can be observed, or both. Only the most astrophysically common elements, the "usual suspects" of C, N, O, Si, Ca, and Mg, are accessible from IUE. ST can in principle extend the way in which we can probe these processes by being sensitive to ions which are two orders of magnitude less abundant in stellar photospheres.

Specific examples of scientific questions which can be addressed with HST include:

- accretion of material in relatively close, detached binaries like V 471 Tauri

- theories which use accretion to explain the heavy elements in DBZ stars like G 200-39 and GD 40

- circumstellar features and the connection between planetary nebulae and white dwarf stars

- the highly peculiar chemical compositions of the hottest DO stars like H1504+65, PG 1159, and others.

- line profiles of the peculiar absorption features at 1400 and 1600 A which are satellite bands of H_2 and H_2^+.

However, the difficulty we face with using HST is that the observing time is at present considerably more limited than it is for IUE. Even with the fairly substantial allocation of 10.5 hours of observing time during cycle 1, which should roughly correspond to the first calendar year of general observations, choosing targets is a difficult task.

Different approaches can be considered in putting together an interesting program for the first year of HST observations. We must do enough science to demonstrate HST's power in solving problems in our field and to impress future Telescope Allocation Committees. But we must also investigate a broad enough range of white dwarf stars in order to guide future HST work on white dwarf stars, both done by this team and by others. One path through this dilemma was taken by the planners of the IUE and EINSTEIN missions: to simply observe a collection of the brightest objects of all spectral classes. This approach does allow for the possibility of unexpected discoveries. The disadvantage is that the really exciting discoveries -- with both of these spacecraft -- came from a minority of white dwarf stars. Indeed, many of the initial observations of white dwarf stars by IUE and by EINSTEIN were relatively uninteresting.

A second approach, taken with EXOSAT and which is largely but not exclusively the basis for our target selection, is to select a variety of objects in which previous observations suggest that HST has the potential to make an immediate contribution to the solution of a number of outstanding problems. Specifically, the idea is to look for a number of key chemical elements in a number of key targets and hope that the results will offer a clue to the chemical evolution of white dwarf stars.

A potential disadvantage of this second approach is that by focusing on "interesting" objects, where previous work has been used to determine just what is interesting, that we will miss some new phenomenon which can be unveiled by the unprecedented spectral resolution and quality of HST. At its extreme, what happens is you explore in detail the science you understand already, and you completely overlook the new phenomena. My first hope was that we would obtain enough telescope time to include some comprehensive observations of a few bright members of each major class of white dwarf stars, including enough of the broad-brush approach to guide further work. A high priority for the final observing program is to observe at least one target with the highest signal/noise and resolution which is possible with HST in order to guide further work in the field. Choice of targets is critical here; for instance, the brightest DA white dwarf, 40 Eri B, has a much less interesting spectrum at high IUE resolution than one of the next brightest, CoD-38 10980.

Still another consideration for HST observing is that once a target has been selected, there are still a number of choices to be made. What spectrograph? What resolution? What signal/noise should one try to obtain? Previous space observatories have been sufficiently less capable that the spacecraft limitations basically dictated what it meant to observe a target. For IUE, for example, in the early stages one basically observed long enough to get signal/noise of around ten and the target's brightness determined whether the spectral resolution was high or low. In principle, at least one of the spectrographs (the Goddard High Resolution Spectrograph or GHRS) is capable of obtaining spectra with S/N near 100, but for many targets obtaining such resolution takes an excessive amount of telescope time, particularly given its degraded optics.

I have borrowed the terms "reconnaissance," "exploration," and "intensive study" from the planetary science community as guides to thinking about just what to do with a new instrument (COMPLEX 1978, Hinners 1982, Shipman 1987). Their thinking is that exploration of a new solar system object proceeds in three logical evolutionary steps. First you ask broad questions in a reconnaissance phase. Then the questions become gradually narrower as you proceed through exploration to intensive study. When too much early effort is devoted to narrow questions, as was the case with the Viking biology explorations of Mars, it turns out that much effort is wasted. You can ask the wrong question. Or you can pose it in the wrong way. What happened with Viking biology is that the narrow questions characteristic of the intensive study phase were asked before the reconnaissance phase suggested what questions to ask. We don't want to make the same mistake.

Depending on what type of white dwarf is being considered, the IUE experience has taken us through some of these phases. IUE observations of cool white dwarfs have barely begun a reconnaissance, since the number of targets which are bright enough to be observed is extremely small and in many cases useful observations can only be obtained at long wavelengths. We're still in a reconnaissance phase here. Some hotter stars are bright enough so that IUE can obtain usable high dispersion spectra; these are ripe for intensive study with HST. There's a lot in between, where HST is in the exploration phase.

The result of the preceding considerations and proposal writing activity was that prior to launch our team won a reasonable time allocation for HST, consisting of 10 spacecraft hours (cycle 1) and 15 hours (cycle 2). A "spacecraft hour" includes time allocated to instrumental overhead and locking onto the target -- time under the control of the observing program. It excludes time lost to earth occultations and to slewing to the target. The 10 hour allocation was increased to 10.5 hours to allow for an interactive acquisition of Procyon B.

At the moment, the only detailed observing plan which exists is one which fits into the 10.5 hour allocation for cycle 1 and which assumes pre-launch observation capabilities. The bulk of the time is devoted to ultraviolet spectroscopy, primarily in the range shortward of 2000 A. Plans were to observe some targets at very high resolution in limited spectral regions (GD 394, CoD -38 10980, KPD 0005) where IUE data suggest the desirability of intensive study. Some other targets will be observed with more modest resolution and S/N, where reconnaissance has been done with IUE (e.g., V 471 Tauri, H1504+65). Still others (GD 40, L 745-46A) will be observed at relatively low resolution (still 7 times better than IUE) and relatively low S/N of around 10. Those plans can be described in summary form in Table 1 below.

However, it is highly uncertain that the program listed in Table 1 can be implemented in the form which was initially visualized. With the degraded optical efficiencies, it would take approximately 27 hours of spacecraft time to accomplish what we originally had in mind. The two spectrographs referred to are the Goodard High Resolution Spectrograph (GHRS) and the Faint Object Spectrograph (FOS). Broadly speaking the GHRS has higher resolution and covers a narrower spectral range, and the FOS has lower resolution and quite broad spectral coverage. Both instruments have multiple observing modes. The higher resolutions of the GHRS may be particularly interesting for our field.

4. DEVELOPMENTS IN 1990

As is well known, in June 1990 NASA shocked us all by announcing that HST contained a serious flaw in its optical system. As a result we may not be able to do all the observations in the table above. Reports in the media, perhaps accurately reflecting a deep sense of disappointment among those who had devoted substantial parts of their science careers to this mission, created an impression that HST would be almost useless. "Crippled" was a term which appeared quite widely in the American press.

However, HST is still a large telescope and it can be used to do good science. Its collecting area is considerably larger than that of IUE, and its detectors and spectrographs are newer. Its image profile, while blurrier than ideal, can be considered to be quite stable. Those of us who wish to make use of HST will simply have to figure out how to work around its limitations.

The work-arounds are only beginning to be discovered, especially for spectroscopic observations. At the time of the Toulouse conference, there was not a single spectrum available from HST. Even now there are only a few. The highest resolution can still be obtained for very bright targets by using small entrance apertures to sharpen the resolution of the spectrographs, which were designed anticipating an image size of 0.1 arcsecond and how have to live with 2 arcsecond images. There is much discussion of deconvolution, though most real work has been with the imaging instruments. Deconvolution of HST spectra seems possible but has been relatively unexplored.

TABLE 1. The Initially Proposed Cycle 1 Observing Program

Project	Targets and Observations	Observing Time
Mass-Radius Relation	Procyon B, G 107-70 images in several filters	1.3 hr
Binaries	Time resolved GHRS(0.6 A resolution) spectra of V 471 Tauri	0.7 hr
Cool He-dominated Stars	FOS (1 A resolution) spectra of L 745-46A, GD 40; S/N ~ 10	2.1 hr
Hot He-dominated Stars	GHRS (0.6 A) spectra of H1504+65 and KPD0005+5106; 0.6 A and 0.06 A (for selected wavelengths) of PG 1034+001. S/N ~ 30 (high resolution); 40-100 (lower resolution). FOS spectra of K 1-16.	3.1 hr
DA Stars	GHRS (0.06 A and higher resolution) spectra of CoD-38 10980 (done in an "intensive study mode) and GD 394; 0.6 A resolution spectra of 40 Eri B (S/N ~ 120) and PG 0950+139.	3.1 hr

Also not yet clear is the nature of HST's observing program. Some time consuming and exciting programs will undoubtedly have to be deferred until a second generation of scientific instruments are in place several years from now. Those projects which can be done, such as the spectroscopy described here, will have to be modified. The Telescope Allocation Committee will meet once again to reallocate the first cycle of observing time, and just what we will be able to do will depend on the final allocation of cycle 1 observing time which we end up receiving. One purpose of presenting the program at this meeting was to allow members of the HST team, and others in the white dwarf community, to react to plans as they are currently set and suggest for modifications of the observing program.

Both the collaboration in particular and the HST mission more globally have devoted little attention to cycle 2 of HST observations. Currently, the ST proposal selection process has designated this proposal as a "long term" proposal and allocated 15 hours to us for cycle 2. Some discussion of the possibilities for white dwarf stars took place at this meeting. A "strawman" cycle 2 program allocated a considerably greater proportion of the observing time to the warm He-dominated stars. I hope that we can obtain spectra of Procyon B and of G 107-70. There is understandable concern that spacecraft jitter may make it undesirable to put a first magnitude star (namely Procyon A) within 5 arcsec of the spectrograph's entrance aperture, at least in the early stages of the HST mission. Specifically, technical considerations dictated that spectra of Procyon B not be obtained in the first cycle of HST operations. Based on experience so far, it is likely that some target selection for cycle 2 can be based on experience we gather during cycle 1, even though cycle 2 proposals will be gathered before any significant amount of cycle 1 data is gathered.

Thus, in summary, prospects for doing good white dwarf science with HST remain bright, though we will get less data than we had hoped for and will have to work harder in order to make the most of the data that we do get. We also must wait. Recent experience has shown that in a mission this complex, apparently simple things like verifying the proper functioning of particular instruments takes much longer than anyone anticipated. However, even with all the waiting and frustration, I believe that it will have been worth it in the end.

This research has been supported by the NSF and by NASA.

REFERENCES

COMPLEX 1978, Committee on Planetary and Lunar Exploration, Strategy for Exploration of the Inner Planets 1977-1987, National Academy of Sciences, Washington, D.C.

Fontaine, G., and Wesemael, F. 1987, in Proceedings of IAU Colloquium No. 95: The Second Conference on Faint Blue Stars, ed. A.G.D. Philip, J. Liebert, and D.S. Hayes, L. Davis Press, Schenectady, N.Y., p. 319.

Hinners, N. 1982, in J.K. Beatty, B. O'Leary, and A. Chaikin, eds., The New Solar System, 2nd ed., Sky Publishing, Cambridge, Mass., USA, pp. 3-10.

Liebert, J., Fontaine, G., and Wesemael, F. 1987, in Memoria della Societa Astronomica Italiana: Proceedings of the Sixth European Workshop on White Dwarfs 58, 17.

MacDonald, J., and Vennes, S. 1991, Astrophysical Journal, in press.

Shipman, H. 1987, Space 2000: Meeting the Challenge of a New Era, Plenum Press, New York, pp. 210-214.

Shipman, H.L. 1989, in G. Wegner, ed., White Dwarfs, Springer-Verlag, Berlin, Germany, p. 220.

Shipman, H.L., and Sass, C.A. 1980, Astrophysical Journal 235, 177.

Sion, E.M. 1986, Publications of the Astronomical Society of the Pacific 98, 821, 1986.

Wegner, G. 1979, Astronomical Journal 84, 650.

Wegner, G. 1980, Astronomical Journal 85, 1255.

Wegner, G., and Yackovich, F. 1983, Astronomical Journal 275, 240.

A DEEP SPECTROSCOPIC SURVEY OF WHITE DWARFS
IN COMMON PROPER MOTION BINARIES

TERRY D. OSWALT*
Department of Physics and Space Sciences
Florida Institute of Technology
Melbourne, Florida 32901 USA

EDWARD M. SION*
Department of Astronomy and Astrophysics
Villanova University
Villanova, Pennsylvania 19085 USA

PAUL M. HINTZEN*
Department of Physics and Astronomy
University of Nevada
Las Vegas, Nevada 89154 USA

JAMES W. LIEBERT*
Steward Observatory
University of Arizona
Tucson, Arizona 85721 USA

ABSTRACT. We report preliminary results of a comprehensive spectroscopic survey of 511 Luyten-Giclas common proper motion binaries which contain known or suspected white dwarf components. Approximately 75% of the sample, which reaches to $m_{pg} \sim +21$, has now been spectroscopically classified.

1. Introduction

Over the last twenty years it has become evident that wide common proper motion binaries (CPMBs) constitute the most prevalent type of binary system (Luyten 1971). CPMBs constitute fertile ground for studies of WDs. Because they are gravitationally bound, these pairs may be assumed to be coeval. Because they are widely separated, their evolutionary histories are not complicated by the effects of mass exchange, tidal distortion, and other interactions complicating the analysis of close binary systems.

The Proper Motion Survey with the Forty-Eight Inch Schmidt Telescope (the LP survey; Luyten, 1963 *et seq.*) covers the area north of declination -33° included in the original Palomar Sky Survey and contains thousands of such pairs down to the plate limit of $m_{pg} \sim 21$. Luyten (1969, 1974, 1979) has drawn special attention to 407 CPMBs containing suspected white dwarfs (WDs). Another 104 similar pairs have been discovered in the course of surveys conducted by Giclas *et al.* (1971, 1978), Ruiz and Maza (1990), or in a few cases by Oswalt (1981). Most candidates were chosen on the basis of photographic magnitude, color, and proper motion. Nevertheless, as noted by Greenstein (1986b), most known CPMBs remain unstudied for lack of identification charts, a problem which we have begun to address (*cf.* Oswalt, Hintzen, and Luyten 1988).

---

* Visiting Astronomer, Kitt Peak National Observatory and Cerro Tololo Interamerican Observatory, operated by the Association of Universities for Research in Astronomy, Inc., under contract with the National Science Foundation

G. Vauclair and E. Sion (eds.), White Dwarfs, 379–393.
© 1991 Kluwer Academic Publishers.

In 1988 we began a comprehensive investigation of CPMBs containing suspected WD components. This paper summarizes our progress to date as well as plans for work to be conducted over the next three years. Our original two-year objective of spectroscopically classifying the entire sample of over 500 CPMBs is now nearly complete; this reconnaissance phase prepares the way for our remaining major objectives:

(1) to determine gravitational redshifts for several hundred newly identified WDs;
(2) to derive complete space motions for the CPMB sample;
(3) to derive luminosity functions (LFs) for the cool WD and main sequence (MS) components; and
(4) to examine the orbital evolution of CPMBs as a probe of post-MS mass loss and Galactic structure.

2. Observational Strategy

In view of their large number (~500) and the fact that many approach the Palomar Sky Survey limit of $m_{pg} \approx +21$, it is difficult to overstate the observational challenge that the Luyten CPMB sample presents. As outlined in Figure 1, this effort is proceeding in three general phases:

(1) low resolution spectra for the initial classification of WD and MS components;
(2) broad-band photometry of CPMBs for the determination of photometric parallaxes; and
(3) high resolution echelle spectra of CPMBs for the determination of radial velocities.

Phase (1) is nearly complete; phase (2) is now underway; and phase (3) will be undertaken in 1991. In each of these cases stars are observed in completeness zones that are randomly distributed on the sky, to guard against the vagaries of telescope time allocations, weather, and instrument failure. Thus at any given time it is possible to draw reasonably reliable conclusions about the entire sample from the statistics of the observed subset (cf. Oswalt, Hintzen, and Luyten 1988; Oswalt and Sion 1989).

2.1. LOW RESOLUTION CLASSIFICATION SPECTRA

Most of the observations required by phase (1) of our program were obtained at NOAO facilities, where high quantum efficiency, low resolution (~7-14Å), two-dimensional spectrographic systems are available (cf. Phillips and Heathcote, 1986; Seitzer and Reid, 1987). This effort alone has required nearly 50 observing nights on the CTIO and KPNO 4-m telescopes. Spectroscopy of the southern sample was completed in January 1990 at CTIO. The larger northern sample should be completed in late 1990 at KPNO. A few dozen previously observed objects will warrant follow-up spectroscopy during the next year, e.g., pairs with original spectra of low signal-to-noise, suspected composites, suspected variables, or pairs posing special observational constraints. In addition to the spectroscopic classifications, these spectra are being used for the determination of temperatures and gravities for the WD components (Penton 1990).

2.2. BROAD-BAND AND CCD FILTER PHOTOMETRY

Photometric magnitudes and colors are essential to a proper statistical analysis of virtually every physical trait of the CPMBs in our sample. They are also needed for DA stars between $10 < T_3 < 30$ since the line spectra cannot yield unambiguous temperatures; likewise for the DB stars in the temperature range $18 < T_3 < 30$. Color information is also helpful in delineating the various types of hot subdwarfs.

In collaboration with M. Wagner, we are beginning to obtain CCD-based UBVRI photometry of the CPMBs using the 1.8-m telescope at Lowell Observatory. Supplementary UBVRI photometry of CPM components brighter than $m_{pg} = +15$ will be obtained on a time-shared basis with the F.I.T. 0.64-m automated photometric telescope. These instruments are capable of providing good magnitudes and colors for the entire sample in ~40 nights of observing time over the next two years. This photometry will provide photometric parallaxes for the warmer members of our sample using the well-known Mv, B-V and/or Mv, R-I relations for MS stars.

S. Leggett is leading efforts to obtain infrared photometry of the Luyten CPMBs of color class "k" or redder, using the 3.0-m NASA Infrared Telescope Facility (IRTF) at Mauna Kea Observatory. JHK photometry provides accurate effective temperatures, photometric parallaxes, and luminosities for the coolest WD and MS stars (cf. Leggett and Hawkins 1988; Leggett 1989). It also readily identifies the subdwarfs in our sample, as they are blue in J-H. Finally, we anticipate that a few of the WDs have unresolved low mass MS companions, detectable by comparing the JHK and optical data-- a technique pioneered by Zuckerman (1989).

2.3. HIGH RESOLUTION SPECTROSCOPY

The low resolution spectra are used to identify the best candidates for radial velocity determinations in phase (3) to be conducted primarily at NOAO facilities beginning in 1991. Pilachowski and Milkey (1984, 1987) showed that useful echelle spectra can be obtained for WDs as faint as $m_{pg}= +17$; in our sample there are nearly 200 pairs in which both components meet this magnitude limit. Over 100 additional CPMBs have at least one component brighter than $m_{pg}= +17$ that will provide useful kinematical data. Thus even if the actual signal-to-noise of the spectra of the fainter stars limits us to a somewhat brighter limiting magnitude, we are assured of a large number of radial velocity measures.

In a parallel effort, J. Liebert and R. Saffer are using the Multiple Mirror Telescope Observatory to obtain radial velocities of CPMBs selected by our program. The MMT echelle spectrograph + red detector provides useful radial velocities for the Hα region of bright MS components of northern CPMBs. Experience with this system has shown that integration times of 30 minutes for nondegenerate components with Hα emission are sufficient even at $m_v= +19$.

Some of the most astrophysically important CPMBs are approved targets for ultraviolet spectroscopy. Observations with the International Ultraviolet Explorer spacecraft were used by Oswalt *et al.* (1991) to estimate the first pressure shift and gravitational redshift for a cool carbon-rich CPMB WD (LDS 678A). Several of the faintest double degenerates identified in our low resolution survey are also scheduled for UV spectroscopy using the Hubble Space Telescope's Faint Object Spectrograph and the Goddard High Resolution Spectrograph (Shipman *et al.* 1988).

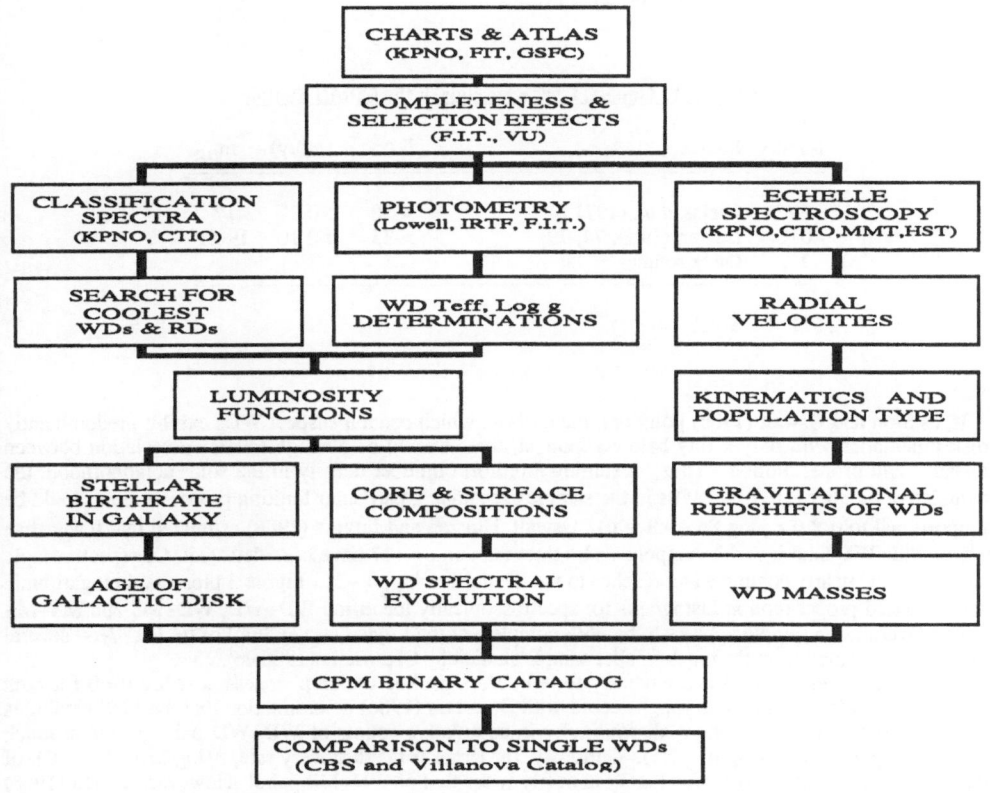

Figure 1. Outline of the Luyten-Giclas CPMB project.

3. Physical Characteristics of the CPMBs

Aside from the spectroscopic identification of the largest sample of binary WDs since the conclusion of the landmark Eggen and Greenstein surveys (1965a,b, 1967), the Luyten CPMB project's actual yield of WDs (~60% of the suspect CPMBs observed to date) is providing critical information on the effectiveness of the screening methods used in the surveys and the selection effects involved in accounting for duplicity in stellar LFs. Although detailed corrections for the completeness and selection effects of the sample cannot be made before photometric magnitudes and colors are available, several interesting properties of the CPMBs are already apparent.

3.1. COMPLETENESS OF THE CPMB SAMPLE

Except for a handful of southern hemisphere pairs from the Luyten-Bruce Survey (Luyten 1941), all of the CPMBs containing suspected WDs published by Luyten were discovered during the course of the LP survey. The sample is the deepest and most complete available; it excludes only the dense regions near the Galactic plane and the southernmost third of the celestial sphere. However Luyten intentionally did not list CPMBs previously published by Giclas, Burnham, and Thomas (1971, 1978), most of which have components brighter than m_{pg}= +17. Thus, to maximize completeness of the CPMB sample, it is necessary to include pairs from both surveys. Table 1 summarizes the basic characteristics of the combined sample. Figure 2 indicates that the typical CPMB has a primary of m_{pg}~ +16 and a companion that is roughly two magnitudes fainter, hence surveys such as those of Giclas which had plate limits near m_{pg}= +17 failed to detect many CPMBs in the solar neighborhood.

TABLE 1. General Characteristics of the CPMB Sample

# Pairs	Source	δ (°)	μ ("/y)	m_{pg}
87	Giclas *et al.* (1971,8)	>-30	>0.27	<17
407	Luyten (1969, 74, 79)	>-33	>0.10	19-21
17	Other sources			
511	Total			

If, as Sion and Oswalt (1988) point out, the CPMBs which contain suspect WDs exhibit predominantly disk kinematics with only a tiny halo component, there should be a rough inverse correlation between distance and proper motion. Thus, assuming a uniform number density in the solar neighborhood, the cumulative number (Σn) of CPMBs in the sample exceeding a particular limiting proper motion should be proportional to μ^{-3} (*i.e.*, log Σn ~ -3log μ). Oswalt, Hintzen and Luyten (1988) estimated that fewer than 1% of wide WD binaries with components brighter than m_{pg}= +17 have been detected. Our spectroscopic survey is now nearly complete and reaches to the LP limit of m_{pg}= +21. Figure 3 presents, in logarithmic form, revised proper motion histograms for spectroscopically identified WD+WD, WD+MS and MS+MS pairs. Overall, the sample appears to be ~6% complete at the Luyten survey limit of μ= 0.1 "/yr-- several times that estimated from the much smaller sample studied by Greenstein (1986a/b).

Also apparent in Figure 3 is a tendency for an abrupt decline in completeness near log μ= 0 for both WD+MS and MS+MS pairs, in rough accord with Dawson's (1986) determination that the LHS catalog is complete to at least μ= 1"/yr. A discontinuity in cumulative counts of WD+WD pairs occurs at much smaller proper motions (log μ~ -0.5). Curiously, for larger μ the discovery rate, Δ(log Σn) / Δ(log μ), of this exceedingly rare type of wide binary is nearly twice that of WD+MS pairs! However, Luyten (1988) noted that the automated measuring machine used in the LP survey more easily distinguished CPMBs of close separation if components had identical magnitudes and colors-- typical traits of WD+WD pairs.

3.2. ORBITAL EVOLUTION IN WIDE BINARIES

The distribution of semimajor axes among CPMBs contains information about the history of encounters within the Galactic disk and/or the effects of post-MS mass loss on the orbits of wide binaries. Retterer and King (1982) note a possible cut-off in the distribution of wide binary semimajor axes greater than ~0.1pc which may be dynamically induced by perturbations of the Galactic disk and/or collisions with giant molecular clouds (*cf.* Bahcall and Soneira, 1981; Bahcall, Hut, and Tremaine (1985)). The predominantly old disk population WD+MS CPMBs provide an excellent test of these theories; it is these pairs for which the cumulative effects of large impact parameter encounters are most severe and the probability of very rare disruptive encounters is highest. Furthermore, the semimajor axis frequency distribution can in principle be fit to models such as those provided by Weinberg, Shapiro, and Wasserman (1987) to derive useful constraints on the birthrate and orbital evolution of wide binaries in the Galactic disk.

TABLE 2. Comparison of CPMB Separation Indices

Sample	Σn	$<\log s/\mu>$	m.e.
MS+MS	130	2.156	0.063
WD+MS	149	2.078	0.051
WD+WD	21	1.593	0.138
DQ (all)	9	1.394	0.108
DQ+MS	6	1.495	0.214
DQ+WD	3	1.193	0.244
DA (best)	73	2.239	0.078
non-DA (best)	49	1.985	0.078

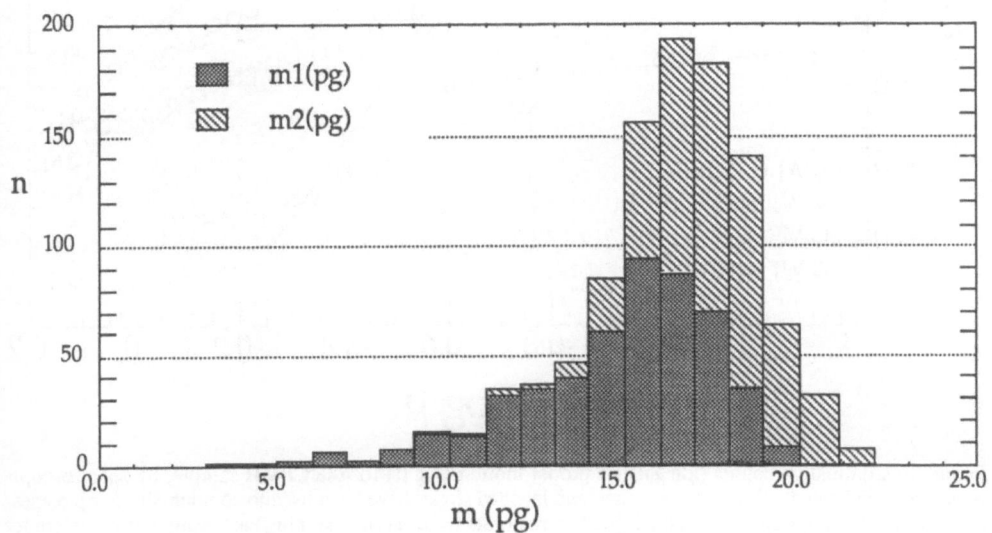

Figure 2. Frequency distribution of photographic magnitudes of Luyten-Giclas CPMB sample.

384

In lieu of spectral type and trigonometric parallaxes for the vast majority of the CPMB sample, Oswalt and Sion (1989) defined a statistical measure of projected semimajor axis, the so-called separation index, log s/μ, where s is the angular separation of the components in seconds of arc and μ is the proper motion in arc seconds per year. Geometrically, one would expect the relation $s/\mu \sim a/V_t$, where a is the projected semimajor axis of the binary in a.u. and V_t is its transverse velocity in km/s. Figure 4 illustrates a test of this assumption using CPMBs with trigonometric parallaxes larger than 0.050" determined by Harrington et al. (1985). Although the exceptionally wide pair G130-43/G34-15 of questionable physical association carries inordinate weight, the empirical relation has nearly the expected unit slope. We therefore feel justified in using log s/μ for a preliminary assessment of the statistical properties of CPMBs which lack either spectra or parallaxes. The y-intercept implies that the mean transverse velocity $<V_t> \sim 25$ km/s. The dispersion in V_t within this sample is responsible for most of the scatter.

Using the separation index to evaluate a large sample of CPMBs culled from the Luyten lists, Oswalt and Sion (1989) showed that those pairs containing suspected WD components had a mean separation implied by $<\log s/\mu>$ is nearly a factor of two larger than those pairs likely to contain only MS components. Spectral types are now available for most of this sample, allowing us to re-examine the question of orbital expansion. Figure 5 displays the frequency histograms of log s/μ for the CPMBs observed to date.

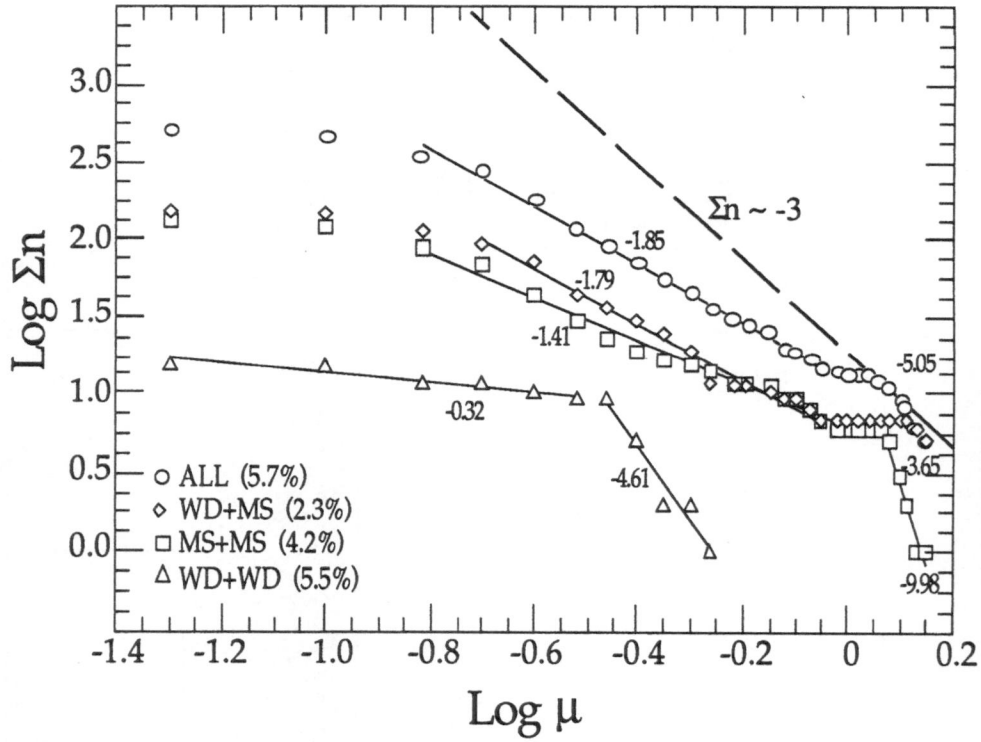

Figure 3. Cumulative counts (log Σn) vs. proper motion (log μ) for the CPMB sample, by spectroscopic subgroup as indicated. The regression lines and labelled slopes have been used to compute the completeness of each CPMB subgroup given in the legend; zero points were set by assuming each sample is complete for log $\mu \geq 0$. The expected detection rate for a uniform number density of CPMBs in the solar neighborhood is indicated by the dashed line.

Surprisingly, there appears to be no significant difference in <log s/μ> between the spectroscopically identified WD+MS and MS+MS pairs (see also Table 2). The straightforward geometrical relation between log s/μ and log a, combined with Greenstein's (1986a) finding that significant orbital expansion has in fact occurred among a smaller but very well-observed group of CPMBs containing WDs, leads to the conclusion that the similar <log s/μ> shown in Figure 5 must result from a difference in mean transverse velocity. It also implies that the crude proper motion, magnitude and color class criteria used in preliminary studies do not completely discriminate the two groups. As evidence of this we note that, contrary to the results of Oswalt and Sion (1989), the fraction of MS+MS pairs is constant with respect to separation index.

Oswalt and Sion (1989) raised the possibility that CPMBs which contain a spectroscopically identified DQ WD have significantly smaller <log s/μ> than other WD+MS pairs. This finding appears to be corroborated by Figure 5 and Table 2. However, since there is no *a priori* reason to expect separations of CPMBs to follow a normal distribution, a nonparametric test of significance is more appropriate. A Wilcoxon rank sum test (*cf.* Wonnacott and Wonnacott 1972; Wilcoxon, Katti, and Wilcox 1973) applied to the DQ+MS and WD+MS groups indicates a less than 1% probability that the observed differences in <log s/μ> are due to random sampling of the same parent population. A similar test of the three WD+WD pairs which contain a DQ indicated that there is roughly a 6% probability that they share the same parent population as the other WD+WD pairs. Thus, the smaller <log s/μ> exhibited by CPMBs containing a DQ WD may be real. Heber *et al.* (this volume) suggest that the lower than expected helium abundances in the PG 1159 stars make it unlikely that the DQ degenerates are their progeny. Clearly the DQ stars in our CPMB sample may shed additional light on their origin. It remains to be seen whether intrinsic differences in <log a> and/or <V$_t$> are responsible, although the latter is suspected (see Section 5.2).

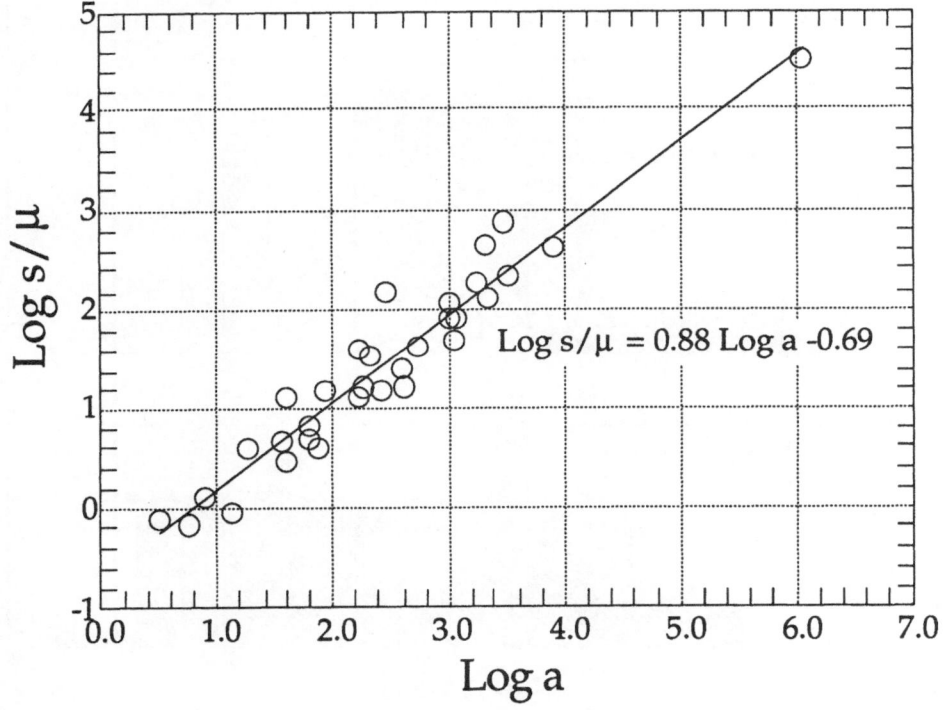

Figure 4. Correlation between separation index (log s/μ) and projected semimajor axis (log a) for CPMBs with trig parallaxes Π≥ 0".05.

Among CPMBs with the highest quality spectra (see Table 2), <log s/μ> for CPMBs which contain DA components implies separations nearly twice that of the nonDA sample. However, the latter sample includes pairs with DQ components which, as shown above, exhibit smaller <log s/μ>. Also, the distribution of DA separation indices seems to have a second maximum near log s/μ= 3.0. However, with the exception of the WD+WD pairs discussed below, Wilcoxon tests of all other possible data pairs in Figure 5 did *not* reject the hypothesis that they could have arisen from the same parent population.

We also find in Table 2 and Figure 5 that <log s/μ> for the WD+WD sample is significantly smaller than either WD+MS or MS+MS pairs (the formal Wilcoxon probability that this is a chance result of data drawn from the same parent population is ~1%). Using parallax estimates derived from our spectra Sion *et al.* (this volume) show that the projected semimajor axes of the WD+WD pairs are also smaller than those of WD+MS and MS+MS pairs observed by Greenstein (1986a), ranging between 7 a.u. to 8500 a.u. with <log a> = 3.1. Although this difference could arise from smaller original separations among double degenerate progenitors, Sion *et al.* (this volume) point out that nonexplosive mass loss in pairs of small

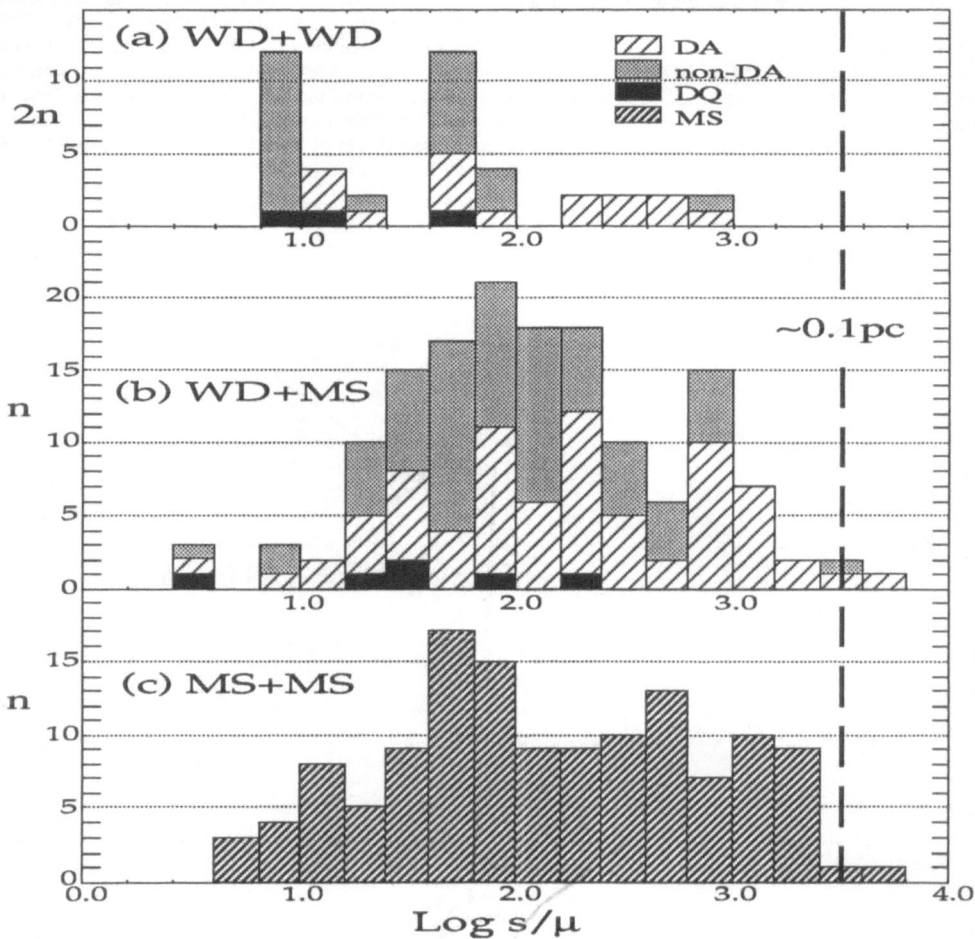

Figure 5. Comparison of Log s/μ separation indices for spectroscopically identified (a) WD+WD, (b) WD+MS and (c) MS+MS CPMBs. Note that each component is included in the histogram for WD+WD pairs (a). Vertical dashed line is approximate cut-off in binary separations proposed by Retterer and King (1982), assuming <V_t>≃ 25 km/s for all CPMBs.

initial mass ratio would lead to less efficient orbital amplification. The low value of <log s/μ> for double degenerates relative to the other two groups of CPMBs can also be due to a higher mean transverse velocity, in accord with the great ages implied by their cooling times. Some of these effects were predicted by Greenstein (1986b) but could not be demonstrated with his sample of only 6 WD+WD pairs. Furthermore, the response of the wide orbits to mass loss is not a straightforward problem and requires further dynamical exploration (cf. Valls-Gabaud 1988).

The photometric parallaxes of the MS components obtained by our program will allow us to determine, independent of transverse velocity influences, the physical separations of both MS+MS and WD+MS pairs and also any differences between DA and non-DA systems. In addition, MS components of CPMBs provide firm lower limits to the progenitor masses of WD companions. *Thus, the CPMB sample may ultimately provide useful constraints on wide binary orbital evolution and post-MS mass loss as a function of original MS spectral type.*

4. Luminosity Functions of CPMB Components

The factor-of-two discrepancy between the dynamically inferred mass in the solar neighborhood (Bahcall, 1984) and the mass that can be ascribed to the observable stellar LF is widely suspected to be due to incomplete detection of binary and multiple stars (cf. Greenstein, 1986a; Wasserman and Weinberg 1987). Moreover, Luyten (1969, 1971, 1987) argues persuasively that CPM pairs are in fact the dominant type of binary in space.

Low luminosity WD or MS stars are also prime suspects in the search for hidden mass in the Galactic disk. For instance, Larson (1986) proposed a bimodal star formation hypothesis which would account for much of the missing mass as cool degenerate stars born early in the history of the Galactic disk during a "high mass" star formation epoch. Since many of the lowest luminosity stars known in the solar neighborhood were originally discovered as components of wide binaries, (e.g. vB3, vB8, vB10, vB11; van Biesbrock, 1961), the Luyten CPMB sample, with an apparent magnitude limit near m_{pg}=+21, has the strongest potential of any existing survey to refine and extend the LFs of both WDs and MS stars.

4.1. THE LUMINOSITY FUNCTION OF COOL WDs AND THE AGE OF THE GALACTIC DISK

The WD LF essentially terminates at Mv≈ +16 according to Liebert, Dahn, and Monet (1988). Using model cooling ages, Winget *et al.* (1987) and Wood (1990) used this result to estimate the earliest epoch of star formation in the Galactic disk. While an observational terminus to the WD LF would explain the low yield of several earlier searches for cool degenerates (cf. Hintzen 1986, and references therein), the issue remains open because none have utilized a sample as deep or as complete as the Luyten proper motion survey. The Luyten CPMB sample also provides a convenient test of whether the LF of degenerate companions is the same as that of the field.

We observed our entire Luyten "bright list" (both components brighter than m_{pg}= +17; Oswalt 1981; Oswalt, Hintzen, and Luyten 1988) without identifying any new degenerates of Luyten color class "m", a result which is not surprising in view of Greenstein's (1986a/b) results. However, our very first KPNO 4-m telescope run yielded three cool degenerate stars with Mv ≈ +16 (cf. Hintzen, *et al.* 1989). *Our preliminary results are consistent with an abrupt truncation of the degenerate star LF near Mv = +16.2 (Liebert, Dahn, and Monet 1988) and with Winget et al.'s (1987) conclusion that the age of the Galactic disk is 9.3 ± 2.0 Gyr.*

4.2. THE LUMINOSITY FUNCTION OF LOWER MS STARS

The MS LF derived from the Luyten CPMB sample has been shown to compare favorably with that derived from a volume-limited "5.2 pc" sample deemed complete by Dahn, Liebert, and Harrington (1986). Dawson (1986) also argued that the Luyten LHS sample is 90% complete for μ> 0.5 "/yr and m_R <18 over the portion of the sky covered by the Palomar sky survey. Thus, one might expect the subset of Luyten CPMBs to exhibit a comparable level of completeness. However, as mentioned above, only about 6% of such pairs appear to have been detected. At present it is not clear why the detection rate is so low. Ruiz *et al.* (1990) have extended the search for additional CPMBs to the southern hemisphere regions not covered by the Luyten survey and this should eventually increase the completeness of the CPMB sample. For the present, more than 60 Luyten CPMBs contain red MS components potentially fainter than Mv=+16-- *an important sampling of the faint end of the MS LF which may also provide clues to the cause(s) of incompleteness.*

4.3. DOUBLE DEGENERATE CPMBs

Nearly two thirds (14/24) of all known CPMBs containing two degenerate components have been identified in the course of our spectroscopic survey. This subset of the CPMBs is now large enough to consider as a separate statistical group (for a more detailed discussion of these pairs see the article by Sion *et al.* elsewhere in this volume). Most consist of extremely cool DC stars. Therefore photometric parallaxes are derived from our spectra by using the Greenstein (1976) MCSP index b-v where b = -2.5 log Fv (4255Å) and v = -2.5 log Fv (5405Å). JHK photometry will greatly improve our temperatures and photometric parallaxes for the extremely old double degenerate pairs. By virtue of a decrement in the two color J-H vs. H-K diagram, it will also reveal the presence of photospheric carbon too weak for spectroscopic detection.

The three faintest WDs our survey has yielded to date are all components of double degenerate pairs (Hintzen, *et al.* 1989). Each pair consists of a yellow degenerate primary and a red DC13+ secondary 1.4 to 2.3 magnitudes fainter. One of these systems, LP701-69/70, has survived over 8 billion years without disruption despite its 1500 a.u. orbital major axis. These three binaries comprise nearly half the available sample of stars at the low luminosity terminus of the WD cooling sequence.

One southern pair identified in our survey, L151-81a/b, is the hottest known WD+WD and the first to contain both a DA (hydrogen-rich) and DB (helium-rich) WD (*cf.* Oswalt, *et al.* 1988). This system sets important new constraints on accretion models of WD spectral evolution (*cf.* Sion, 1986) as it is difficult for conventional interstellar accretion models to explain the observed difference in atmospheric composition.

5. Radial Velocities and Space Motions of the CPMBs

Many of the remaining problems in late stages of stellar evolution (e.g. progenitor mass loss, identification of progenitor channels feeding into the WD sequences, remnant masses and mass dispersion, stellar population subcomponents, and space densities) may be solved with the acquisition of radial velocities and masses for WDs. Particularly in the case of the WDs the kinematical data are indispensable-- processes such as gravitational and thermal diffusion, convective dredge-up, dilution and mixing, accretion, late thermal pulses, mass loss, diffusion-induced burning, etc. erases or alters any compositional clue to their evolutionary history. *The individual radial velocities and gravitational redshifts (and therefore complete space motions and masses) are the only fundamental parameters not accurately known for individual WDs.*

The most recent compilation of spectroscopically identified WDs by McCook and Sion (1987) catalogs proper motions, colors, parallaxes, radial velocities, and luminosities for over 1300 degenerate stars among the various spectroscopic subgroups. A kinematical analysis of this sample was carried out by McMullin, Fritz, and Sion (1987). In still another kinematical investigation of a catalog subset, assuming a zero radial velocity for wide binaries containing a WD, Sion and Oswalt (1988) found that the sample is characterized by both young disk and (predominantly) old disk (50 km/s) motions with an apparent paucity of halo members compared to the few percent present among single degenerates. Oswalt and Sion (1989) have shown that significant orbital expansion has occurred during the post-MS evolution of the present WD; whether this is a function of kinematical subgroup membership remains unknown.

The question of whether or not there exist kinematically distinct subgroups of WDs has never been conclusively answered primarily because of the lack of WD radial velocities. Yet a knowledge of the nature of WD progenitors, of their stellar population subcomponents, of whether there are groups having different mean masses or interior compositions, must be inferred from their kinematics because, with few exceptions, the analysis of WD atmospheres yields little information on the underlying envelope, core, or other interior properties. The radial velocities of CPMBs provide the basis for the following investigations.

5.1. GRAVITATIONAL REDSHIFT MASSES OF WDs IN CPMBs

White dwarf masses determined from the gravitational redshifts are nearly independent of the uncertainties inherent to stellar atmosphere analysis. As there are only a handful of WDs for which astrometric masses can be determined, it is not much of an exaggeration to assert that gravitational redshifts offer perhaps the only reasonable hope for a reliable test of the Chandrasekhar mass-radius relation for degenerate matter. The pioneering efforts to measure the radial velocities and mean gravitational redshifts of a large sample of WDs were presented in two classic papers by Greenstein and Trimble (1967) and Trimble and Greenstein (1972). Their determination of mean gravitational redshifts utilized the assumption that the intrinsic space motions are randomly distributed, and hence the derived masses are also of statistical use only.

Orbital velocities for CPMBs are typically less than ~1 km/s, well below the precision of faint star radial velocity measurements. Thus, a WD gravitational redshift can be directly measured relative to the systemic velocity provided by a MS companion. The efficacy of this technique for hydrogen rich DA WDs has been amply demonstrated by Eggen and Greenstein (1965 et seq.), Wegner (1973, 1974, 1979, 1981, 1989); Koester (1987); Wegner, Reid, and McMahan (1989); and Marcum (1989). Accurate mass determinations for DA WDs are possible from measurements of the gravitational redshift of their resolved non-LTE sharp Balmer line cores at high spectral resolution and signal-to-noise. In the cores of these lines, particularly Hα, the pressure shifts have been shown to be utterly negligible (Grabowski et al. 1987). In fact Koester (1987) and Wegner (1989) presented gravitational redshift masses for roughly two dozen DA stars that rival in accuracy the few precision masses of WDs in astrometric binaries like 40 Eri B and Sirius B.

The outlook for gravitational redshift determinations of DB stars and other helium rich WDs is less encouraging because of the poorly understood line asymmetries and pressure shifts which, depending upon instrumental resolution, yield different velocities for each line. However, Oswalt et al. (1991) have shown that CPMBs are capable of setting empirical constraints on computed pressure shifts, and hence gravitational redshifts, for helium-rich WDs. CPMBs containing DB, DC or DQ WDs (and pairs with WDs too close to bright primaries to observe) can at least provide accurate intrinsic radial velocities via the MS companions. Such pairs therefore can be included in kinematical analyses. For the cool helium-rich metallic line degenerates (DZ stars) Hammond (1988) has found excellent agreement between his width and shift measurements of van Maanen 2 (Hammond, 1975) and recent theoretical work (Monteiro et al., 1986,8) on the van der Waals broadening of CaII H&K lines by helium. This work will greatly contribute to our extraction of gravitational redshifts in the CPM binary DZ stars.

In summary, the CPMBs provide a far larger and deeper prospective source of WD masses than any other sample now available.

5.2. KINEMATICAL LINKS BETWEEN WDs AND MS PROGENITORS

The most recent and largest compilation of space motions for WDs by Sion et al. (1988) confirms the earlier conclusions by Eggen and Greenstein (1967) and Sion and Liebert (1977), that the motions of the local WDs represent a mixture of stellar population subcomponents. The majority of WDs belong to the old disk population subcomponent with typical total space motions V_{sp}= 50-60 km/sec with respect to the sun, 4-5% have $V_{sp} \geq 150$ km/sec characteristic of the halo and extreme population II, and several per cent have motions associated with young, fairly massive progenitor stars. The long total stellar ages of WDs and perturbative encounters they suffer during their galactic orbital motions tend to smear out kinematical distinctions among the different types of WDs, (e.g. increase their velocity dispersions with age; cf. Wielen 1977). Despite this expectation, Sion et al. (1988) presented evidence that the DQ (carbon-band) degenerates and the magnetic WDs have higher than average and lower than average space motions, respectively. The higher than average space motions of the DQ degenerates may be indicative of lower mass progenitors. Such objects would be expected to produce lower mass progeny accompanied by deeper convection zones which favor the dredgeup of carbon from its equilibrium diffusion tail (cf. Fontaine et al. 1984; Koester et al. 1982). These investigations strengthen our contention that kinematical distinctions can still be isolated among the WDs-- all the more so with the complete space motions provided by the MS companions to the hundreds of degenerates in the CPMB sample!

Degenerate remnants of progenitors in the mass range 3 to 8 M_{sun} are theoretically expected to leave cores of higher than average mass (cf. Weidemann 1987). According to Renzini (1988) the luminosity at the peak of a thermal pulse should reach the Eddington limit, leading to the hydrodynamic expulsion of the entire hydrogen-rich envelope after only one thermal pulse. Massive WDs have been been identified in open clusters whose ages are so young that only the massive upper MS members of the cluster have had sufficient time to evolve toward the red giant branch (cf. Wegner 1989; Romanishin and Angel 1980). The red giant turnoff mass is 6-8 M_{sun} in some of these clusters, thus implying that the parents of the cluster WDs were at least that massive. In fact, the initial MS masses of two cluster WDs (NGC 2451-1 and NGC 2451-5) must have been ~8-9 M_{sun} (Koester and Reimers 1985, 1990)! Unfortunately, few massive WDs with well-determined parameters are known and the observational verification of the initial-final mass relation (*cf.* Weidemann and Koester 1984) is accordingly weak.

Greenstein (1986a/b) has shown that most CPMBs containing a WD component must have had relatively high original mass ratios. For example, though the average mass of Koester's (1987) small CPMB sample agrees nicely with the average value, $0.6\,M_{sun}$, of single field degenerates, among his sample are three stars with mass values above $0.7\,M_{sun}$ that are likely to be the descendants of more massive upper MS progenitors. There are nearly three dozen MS components of spectral type G or earlier in the Luyten CPM binary sample. Is there a correlation between the "earliness" of the MS spectral type and the mass of its coeval degenerate companion?

5.3. STELLAR SUPERCLUSTER AND MOVING GROUP MEMBERSHIP

Eggen (1987) estimated that 20 to 30 percent of the stars in the solar neighborhood with proper motions larger than 0.04 "/yr, are members of a proposed Hyades supercluster. Ruiz *et al.* (1990) determined that the overwhelming majority of the WDs in the solar neighborhood may be members of the Hyades or Sirius superclusters. Do similar moving groups exist within the CPMB sample? Displayed in Figure 6 is the frequency histogram of proper motion position angles for the CPMB sample broken down by our survey's spectroscopic classifications. The general dearth of position angles between 325 - 50° can be understood in terms of the higher space velocities typical of old proper motion selected samples (for an excellent discussion on this effect see Chandrasekhar 1960). For reference, the approximate position angles of the solar reflex motion and the Sirius and Hyades moving groups given by Ruiz *et al.* (1990) are labelled. The

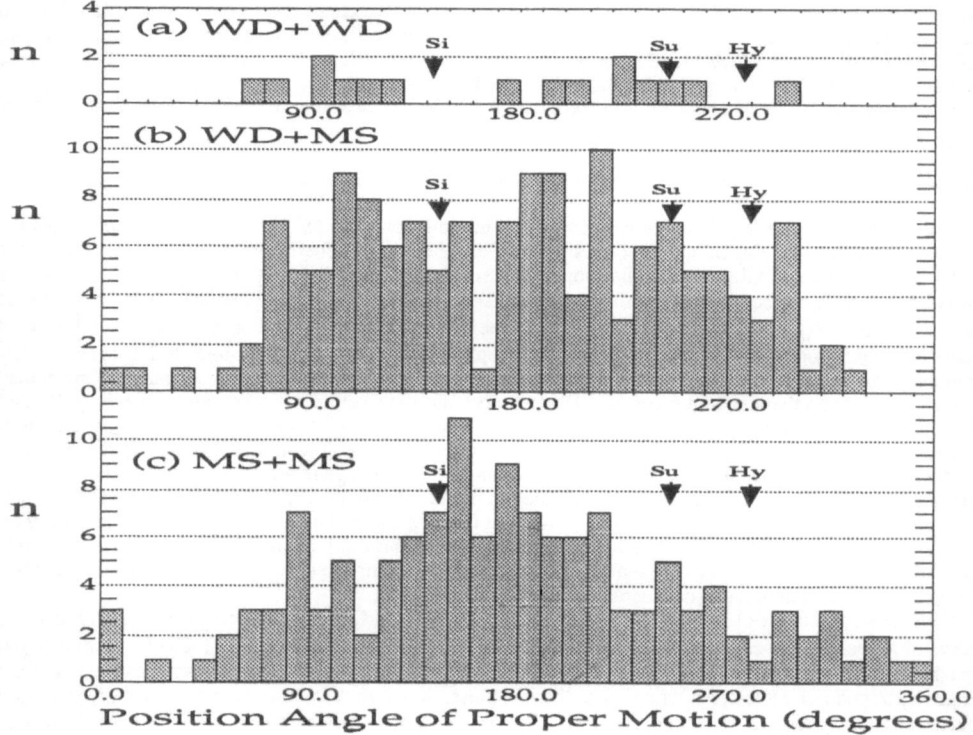

Figure 6. Frequency distribution of spectroscopically classified CPMB sample: (a) WD+WD pairs; (b) WD+MS pairs; and (c) MS+MS pairs. Arrows indicate the solar reflex motion (Su), and the convergent points of the Sirius (Si) and Hyades (Hy) moving groups.

CPMBs appear to exhibit *no* predilection for any of these position angles. However, some dramatic differences do exist between the three subgroups. MS+MS pairs exhibit the most symmetrical distribution. WD+WD pairs, and perhaps WD+MS pairs, may have a bimodal distribution. Do these differences imply kinematical groups within the CPMB sample, or are they related to the bimodal distribution of semimajor axes seen in WD+WD pairs (see Sion *et al.*, this volume)? *Radial velocities are required to assess the frequency of supercluster membership among our CPM pairs, and to determine whether their formation as wide binaries is associated with the supercluster aggregation of stars.*

5.4. WD ROTATION RATES AND MAGNETIC FIELDS

Greenstein (1976) recognized that the sharp non-LTE line cores of Hα observed at high resolution constrain the rotation rates of DA WDs to v sini < 60 km/s, indicating that major angular momentum loss accompanies post-MS evolution. More recently, Pilachowski and Milkey (1984, 1987) have assessed the rotational characteristics of 15 additional single DA WDs with similar results. The same high resolution spectra used to derive radial velocities provide a unique opportunity to assay both the rotation rates and the magnetic field characteristics of WDs in CPMBs and to establish whether they differ significantly from those of single WDs. It is possible that the angular momentum and/or mass-loss history of the wide binary WDs is different from that of single WDs.

6. Summary

Our primary motive for constructing the first (nearly) complete set of identification charts for the Luyten CPMBs was to facilitate a comprehensive spectroscopic survey of these long-neglected but astrophysically important binaries. Based upon the fraction of the CPMB sample observed thus far, we can report several preliminary conclusions:

1. All but 39 of the 511 Luyten and Giclas CPMBs which contain known or suspected WD components have been firmly identified. A complete atlas of finding charts is being published in installments.
2. The Luyten-Giclas sample is ~6% complete to the survey limit of μ~ 0.1 "/yr.
3. About 75% of the entire CPMB sample has now been spectroscopically classified. The yield of WDs is quite high; ~60% of the observed pairs have at least one WD component.
4. The DA:nonDA ratio of WDs in CPMBs is ~2-3 times that of single WDs.
5. WD+MS pairs appear to have significantly larger projected semimajor axes than MS+MS pairs indicative of substantial systemic mass loss during post-MS evolution.
6. The separation indices of WD+WD, and possibly WD+MS pairs, appear to be bimodal.
7. At least in one case, L151-81A/B, accretion of material from the interstellar medium cannot easily explain the dissimilar atmospheric abundances of the two WD components.
8. The Luyten-Giclas sample has yielded 14 of the 21 known double degenerate CPMBs.
9. Several of the faintest known WDs and MS stars have been identified in the CPMB sample. The faintest three WDs have cooling ages consistent with an age of ~9.3 x 10^9 yrs. for the Galactic disk.
10. The distribution of proper motion position angles for CPMBs which contain at least one WD differs substantially from that of MS+MS pairs, suggesting that CPMBs are a mix of kinematical groups.

Despite an encouraging start, it is clear that much work on the CPMBs remains to be done. We hope to draw attention to the CPMBs as a powerful testing ground for a large array of astrophysical problems, including but not limited to: binary origins and orbital expansion, Galactic structure, WD masses and spectral evolution, and the LFs of both WDs and MS stars. Luyten (1971) best describes the spirit of our investigation of the CPMBs:

> "Observationally speaking, therefore, these proper-motion doubles are the most common type of binaries in space. Nevertheless, double star observers and theoreticians alike continue to ignore them... Perhaps this is part of the reason we have progressed so little in our attempts at understanding the processes of double star genesis: we have concentrated our efforts on the exceptional freaks, and neglected the common man in space."

392

7. Acknowledgements

We are especially indebted to W.J. Luyten, who has spent decades identifying the CPMBs, and to KPNO and CTIO for generous amounts of observing time. One of us (TDO) expresses appreciation to L. Roberts for maintaining our extensive database on CPMBs, to S. Penton for much of the spectroscopic reductions, to W. Gabrenya for helpful discussions on the statistics of CPMB separations, and for NATO travel support to the "7th European Workshop on White Dwarfs". The authors acknowledge support from NSF AST88-02687 (TDO), NSF AST88-02689 (EMS), NASA 188-4151-07 (PMH), and NSF AST88-40482 (JL).

8. References

Bahcall, J.N. 1984, *Ap.J.*, 276, 169.

Bahcall, J.N., Hut, P., and Tremaine, S. 1985, *Ap.J.*, 290, 15.

Bahcall, J.N., and Soneira 1981, *Ap.J.*, 246, 122.

Chandrasekhar, S. 1960, "Principles of Stellar Dynamics", (Dover: New York).

Dahn, C.C., Liebert, J., and Harrington, R.S. 1986, *A.J.*, 91, 621.

Dawson, P.C. 1986, *Ap.J.*, 311, 984.

Eggen, O.J. 1987, in "The Galaxy", ed. G. Gilmore and B. Carswell (Reidel: Dordrecht) p211.

Eggen, O.J., and Greenstein, J.L. 1965a, *Ap.J.*, 141, 83.

Eggen, O.J., and Greenstein, J.L. 1965b, *Ap.J.*, 142, 925.

Eggen, O.J., and Greenstein, J. 1967, *Ap.J.*, 146, 850.

Fontaine, G., Villenueve, B., Wesemael, F., and Wegner, G. 1984, *Ap.J. (Letters)*, 277, L51.

Giclas, H.L., Burnham, R., and Thomas, N.G. 1971, "Lowell Proper Motion Survey, Northern Hemisphere: The G-Numbered Stars" (Flagstaff: Lowell Observatory).

Giclas, H.L., Burnham, R., and Thomas, N.G. 1978, "Lowell Proper Motion Survey, Southern Hemisphere Catalog". Lowell Obs. Bull. No. 164, Vol. VIII, p.89

Grabowski, B., Madej, J., and Halenka, J. 1987, *Ap.J.*, 313, 750.

Greenstein, J.L. 1976, *A.J.* 81, 323.

Greenstein, J.L. 1986a, *A.J.*, 92, 859.

Greenstein, J.L. 1986b, *A.J.*, 92, 867.

Greenstein, J.L., and Trimble, V. 1967, *Ap.J.*, 149, 283.

Hammond, G.L. 1975, *Ap.J.*, 196, 291.

Hammond, G.L. 1988, in "White Dwarfs", ed. G. Wegner, (Springer-Verlag: Berlin), p346.

Harrington, R., Dahn, C., Kallarakal, V., Riepe, B., Christy, J., Guetter, H., Ables, H., Hewitt, F., Vrba, F., and Walker, R. 1985, *A.J.*, 90, 123.

Hintzen, P.M. 1986, *A.J.*, 92, 431.

Hintzen, P., Oswalt, T., Liebert, J., Sion, E. 1989, *Ap.J.*, 346, 454.

Koester, D. 1987, *Mem. Soc. Astr. Ital.*, 58, 45.

Koester, D. and Reimers, D. 1985, *A.Ap.* 153, 260.

Koester, D. and Reimers, D. 1990, *A.Ap. (Letters)*, 217, L1.

Koester, D., Weidemann, V. and Zeidler, T. 1982, *A.Ap.*, 116, 147.

Larson, J. 1986, *M.N.R.A.S.*, 218, 409.

Leggett, S.K. 1989, *A.Ap.* 208, 141.

Leggett S.K. and Hawkins, M.R.S. 1988, *M.N.R.A.S.*, 234, 1090.

Liebert, J., Dahn, C., & Monet, D. 1988, *Ap.J.* 332, 891.

Luyten, W.J. 1941, "The Bruce Proper Motion Survey", (Minneapolis: Univ. Minn. Press).

Luyten, W.J. 1963, "Proper Motion Survey with the 48-Inch Schmidt Telescope" I, (Univ. Minn. Press).

Luyten, W.J. 1969, "Proper Motion Survey with the 48-Inch Schmidt Telescope" XXI. (Univ. Minn. Press).

Luyten, W.J. 1971, *Ap.Sp.Sci.*, 11, 49.

Luyten, W.J. 1974, "Proper Motion Survey with the 48-Inch Schmidt Telescope" XXXVIII. (Univ.Minn.Press).

Luyten, W.J. 1979, "Proper Motion Survey with the 48-Inch Schmidt Telescope" LII. (Univ. Minn. Press).

Luyten, W.J. 1988, *A.Sp.Sci.*, 142, 17.

Marcum, P.M. 1989, M.S. Thesis, Florida Institute of Technology.

McMullin, J., Fritz, M., Sion, E. 1987, *IAU Coll.No. 95*, "Faint Blue Stars" (Schenectady: L. Davis Press).

McCook, G., and Sion, E.M. 1987, *Ap.J.Suppl.*, 65, 603.

Monteiro, T., *Cooper, I., Dickinson, A., and Lewis, E.* 1986, *J.Phys.B.*, 19, 4087.

Monteiro, T., Danby, G., Cooper, I., Dickinson, A., Lewis, E. 1988, *J.Phys.B.*, 21, 4165.

Oswalt, T.D. 1981, Ph.D. Thesis, The Ohio State University.

Oswalt, T.D., Hintzen, P.M. and Luyten, W.J. 1988, *Ap.J. Suppl.* 66, 391.

Oswalt, T.D., Hintzen, P.M., Liebert, J. and Sion, E.M. 1988, *Ap.J. (Letters)*, 333, L87.

Oswalt, T.D. and Sion, E.M. 1989, in "White Dwarfs", ed. G. Wegner, (Springer-Verlag: Berlin), p454.

Oswalt, T., Sion, E., Vauclair, G., Hammond, G., Liebert, J., Wegner, G., Marcum, P. 1991, *Ap.J.* (in press for February issue).

Penton, S.V. 1990, M.S. Thesis, Florida Institute of Technology.

Phillips, M. and Heathcote, S. 1986, "Digital Spectroscopy at CTIO", CTIO Facilities Manual.

Pilachowski, C. and Milkey, R. 1984, *Pub. A.S.P.*, 96, 821.

Pilachowski, C. and Milkey, R. 1987, *Pub. A.S.P.*, 99, 836.

Renzini, A. 1988, in *IAU Symposium No. 131*, ed. S. Torres-Peimbert, (Dordrecht: Reidel).

Retterer, J.M. and King, I. 1982, *Ap.J.*, 254, 214.

Romanishin, R. and Angel, J.R.P. 1980, *Ap.J.*, 235, 992.

Ruiz, M.T., Anguita, P., Maza, J. and Roth, M. 1990, *A.J.* (in press).

Ruiz, M.T., and Maza, J. 1990, *A.J.*, 99, 995.

Seitzer, P., and Reid, M. 1987, *NOAO Newsletter* 9, 39.

Shipman, H.L., Basri, G., Bond, H., Bruhweiler, F., Cordova, F., Finley, D., Fontaine, G., Hintzen, P., Holberg, J., Jenson, K., Koester, D., Liebert, J., Nousek, J., Oswalt, T., Sion, E., Starrfield, S., Tytler, D., Vauclair, G., Wegner, G., Wesemael, F., Weidemann, V. 1988, "White Dwarf Stars", Guest Observer proposal to the Space Telescope Science Institute.

Sion, E.M. 1986, *PASP* 98, 821.

Sion, E.M., Fritz, M., McMullin, J., and Lallo, M. 1988, *A.J.*, 96, 251.

Sion, E.M. and Liebert, J., 1977, *Ap.J.*, 213, 468.

Sion, E.M. and Oswalt, T.D. 1988, *Ap.J.*, 326, 249.

Trimble, V. and Greenstein, J.L. 1972, *Ap.J.*, 177, 441.

Valls-Gabaud, D. 1988, *Ap.Sp.Sci.*, 142, 289.

Van Biesbrock, G. 1961, *A.J.*, 66, 528.

Wasserman, I. and Weinberg, M.D. 1987, *Ap.J.*, 312, 390.

Wegner, G. 1973, *M.N.R.A.S.*, 165, 271.

Wegner, G. 1974, *M.N.R.A.S.*, 166, 271.

Wegner, G. 1979, *A.J.*, 84, 1384.

Wegner, G. 1981, *A.J.*, 86, 264.

Wegner, G. 1989, in "White Dwarfs", ed. G. Wegner, (Springer-Verlag: Berlin), p401.

Wegner, G., Reid, N. and McMahan, R. 1989, "White Dwarfs", ed. G. Wegner, (Springer-Verlag: Berlin), p378.

Weidemann, V. 1987, *A.Ap.*, 132, 367.

Weidemann, V. and Koester, D. 1984, *A.Ap.*, 132, 195.

Weinberg, M., Shapiro, S.L. and Wasserman, I. 1987, *Ap.J.*, 312, 367.

Wielen, R. 1977, *A.Ap.*, 60, 263.

Wilcoxon, F., Katti, S.K. and Wilcox, R.A. 1973, *Selected Tables in Mathematical Statistics*, p171.

Winget, D., Hansen, C., Liebert, J., Van Horn, H., Fontaine, G., Nather, R., Kepler, S. Lamb, D. 1987, *Ap.J.*, 315, L77.

Wonnacutt, T.H. and Wonnacutt, R.J. 1972, "Introductory Statistics", (Wiley and Sons: New York), p404.

Wood, M. 1990, Ph.D. Thesis, University of Texas.

Zuckerman, B. 1989, *Highlights of Astronomy* 8, 119.

DOUBLE DEGENERATE COMMON PROPER MOTION BINARIES

Edward M. Sion
Observatoire Midi-Pyrenees
Centre National De La Recherche Scientifique
Toulouse
and
Department of Astronomy and Astrophysics
Villanova University

Terry D. Oswalt
Department of Physics and Space Sciences
Florida Institute of Technology

James Liebert
Steward Observatory
University of Arizona

Paul Hintzen
Department of Physics and Astronomy
University of Nevada, Las Vegas

ABSTRACT. We present spectral types and spectrophotometry for 21 double degenerate (DD) common proper motion binaries and provide estimates of their colors, absolute visual and bolometric magnitudes and cooling ages. The most typical double degenerate pair contains stars of similar color and spectral type, with an average $M_v = 14.20$, and average cooling age of $\approx 3 \times 10^9$ years, thus typically belonging to the extremely old disk population subcomponent. The oldest pairs in the sample are 9×10^9 years. The median and mean separations of the DD pairs are 2.63 A.U. and 2.61 A.U., respectively, both smaller than the values for main sequence plus white dwarf pairs (WD+MS). Their distribution of separations N (log a) may be skewed toward smaller bins of log a than the distribution characterizing the MS+WD pairs but is larger than the distribution characterizing the MS+MS pairs. The latter comparison, once the role of selection effects in the two samples are properly evaluated, may be indicating orbital amplification of the DD pairs due to non-explosive, isotropic, mass loss on a time scale longer than the orbital period, while the product of the total systemic mass and the semi-major axis was conserved. Their smaller separations relative to MS+WD pairs may be attributed to the smaller dynamical effect of mass loss on their orbits, due to their lower initial mass ratios. The DD sample, even in the absence of radial velocities, appears to have velocity dispersions, 2 to 7 times larger than selected samples of single white dwarfs and MS+WD pairs, chosen to be in the same color/absolute magnitude range as the DD components. This property may be a result of the dynamical inflation of their velocity dispersions due to their extremely ancient total stellar ages.

395

G. Vauclair and E. Sion (eds.), White Dwarfs, 395–408.
© 1991 *Kluwer Academic Publishers.*

1. Introduction

There are over 500 wide, non-interacting, common proper motion (CPM) binaries identified in the Luyten and Lowell-Giclas proper motion surveys as containing at least one degenerate star. Spectroscopic observations demonstrate that these pairs consist predominately of two main sequence components (MS+MS) or of a main sequence star and white dwarf pair (WD+MS). A rare subset of the CPM sample are pairs of white dwarfs which comprise roughly 5 % of the total CPM sample. These double degenerates (DD) offer critical astrophysical insight about stellar evolution since the two degenerate stars presumably share the same total stellar age and the same initial chemical composition but provide a differential cooling clock since their birth as white dwarfs, which implies the masses (mass ratio) of their progenitors. Because the total progenitor masses of the DD binaries are expected to be higher, in general, than those of the main sequence-white dwarf (WD+MS) pairs, the white dwarf masses themselves should be somewhat higher, to be consistent with the initial-to-final mass relation obtained from data for the asymptotic giant branch and white dwarfs (Weidemann 1987). Since more massive degenerates cool more slowly during the first three billion years but after that time the more massive degenerates cool faster, it is possible that the faintest observed degenerates could actually be less massive. The faintest double degenerate pairs should be among the oldest stars in the disk of the Galaxy. This point has been emphasized recently with the discovery of three of the lowest luminosity degenerates known, all three being in double degenerate common proper motion pairs (Hintzen *et al.* 1989).

The double degenerate pairs provide a useful test of whether mass loss prior to the white dwarf phase leads to orbital amplification. This bears directly on the question of whether mass loss influences the distribution of separations among the wide binaries and the possibility that wide binaries can be produced from close pairs (cf. Valls-Gabaud 1988). If mass loss is a factor in orbital amplification, it follows that one would expect the double degenerates to have suffered the most extensive mass loss episodes compared to the main sequence pairs or main sequence-white dwarf pairs. Therefore the average separation of DD pairs should be larger if this hypothesis is correct. The distribution of separations N (log a) for the DD pairs has never been explored in detail prior to this work and also holds implications for the stability of binary orbits toward disruption by various perturbers and can be used to test for the existence of large numbers of undetected binaries.

The physical properties of six well-studied systems were presented by Greenstein (1986b) who used stellar evolution theory to derive estimates of their initial mass ratio, progenitor masses, total stellar ages, and group age. Greenstein noted that the two degenerates in each pair are closely similar in color and luminosity, which is a result of their cooling times being much longer than the nuclear lifetimes of their progenitors and the fact that the least massive white dwarf in a DD pair cools faster than the primary. Our spectroscopic survey of the Luyten CPM binaries has now more than tripled the size of the DD sample analyzed by Greenstein (1986b).

2. Observations and Data Reduction

Spectrocopic Observations of the Luyten DD CPM binaries, covering the 3600Å-6300Å range were obtained in a number of observing runs with the 4 m telescopes at Kitt Peak

National Observatory and at Cerro Tololo Inter-American Observatory. At KPNO the instrumental system consisted of a UV-flooded TI 800 x 800 CCD with the Ritchey-Chretien spectrograph. Integration times of 2400 seconds were typically used for 18th magnitude program stars. Exposures of each pair were usually obtained with a 2" slit rotated to allow observation of the two components simultaneously. The BL250 grating, ruled with 158 lines/mm and blazed in first order at 4000Å, provided coverage from 3600Å to 6300Å, with a resolution of 14Å FWHM (see De Veny 1988 for a complete description of the instrument). Observations at CTIO were obtained with the 2-D Frutti. The instrumental description and setup for the CTIO observations are found in Oswalt et al. (1988).

Observations of the Massey et al. (1988) standard stars were obtained with a 7" slit to determine the flux calibration for the program stars. All data reductions including pixel-to-pixel response corrections, sky subtraction, wavelength calibrations and flux calibrations were done using standard IRAF reduction routines.

3. Physical Properties of Wide Double Degenerate Binaries

3.1 Colors, Magnitudes and Photometric Parallaxes

The present sample of known wide (non-interacting) double degenerate binaries comprises 21 systems, 15 more than were available to Greenstein (1986b) in his landmark papers. Estimates of monochromatic colors and magnitudes for the Luyten pairs were obtained from the classification spectra in the following manner. The V-magnitudes were computed from the observed flux at 5500Å as an approximation to the effective wavelength of the Johnson V-magnitude for late type stars. For each star we have calculated the $b - v$ color from the spectrum as defined by Greenstein (1984), where b and v are the monochromatic magnitudes at 4255Å and 5405Å respectively. Then

$$(b - v) = -2.5 Log[Fnu(4255)/Fnu(5405)].$$ (1)

The MCSP data tabulated by Greenstein (1984), specifically the M_v versus $(b - v)$ data for single degenerates, was used to construct color-magnitude calibrations separately for the DC stars and cool DA stars in the double degenerate sample. For the DC stars we employed the following polynomial best fit to the Greenstein (1984) M_v versus $(b-v)$ data:

$$M_v = 12.738 + 4.3853(b - v) - 2.2134(b - v)^2 + 0.45432(b - v)^3.$$ (2)

This relation has a dispersion of 0.964 magnitudes and was utilized for all helium-rich degenerates with $M_v > 12.00$. In order to determine bolometric corrections for the non-DA stars we followed the procedure described in Liebert et al. (1988) where bolometric corrections for black bodies of the same effective temperature were applied to each non-DA star in their sample. We used the following quadratic least squares fit to the non-DA data points in the M_{bol} versus M_v diagram of Liebert et al. (1988):

$$M_{bol} = -19.55 + 3.847M_v - 0.1042M_v^2.$$ (3)

For the DA stars, most of which are cooler than spectral type DA4, we also used the MCSP data in Greenstein (1984) to construct a color-absolute magnitude relation (M_v vs (B-V)). Bolometric corrections were obtained from the cool DA model atmospheres of

Shipman (1972) as tabulated in Table 5 of Liebert *et al.* (1988). For the DA stars hotter than 20,000K the bolometric corrections were adopted from Wesemael (1979).

The general properties of the double degenerate cpm sample is shown in Table 1 where column (1) lists the proper motion discovery name, column (2) the WD number from the catalog of McCook and Sion (1987), column (3) the apparent photographic magnitude or MCSP visual magnitude of each component and below these two magnitude values, the common proper motion is listed; column (4) gives the absolute visual magnitude estimate for each component and below these absolute magnitude values is listed the position angle of the proper motion vectors; column (5) tabulates the absolute bolometric magnitude in the manner described above; column (6) gives the best available approximate spectral type of each component on the system of Sion *et al.* (1983) and column (7) also give the angular separation and below that, values of the actual linear separation in AU, based upon the photometric parallax generated from columns (3) and (4).

The distance moduli for the double degenerate sample reveals that 45% of the pairs are closer than 40 parsecs, 25% have distances between 47 and 60 parsecs, and 30% of the sample are more distant than 85 parsecs. There is no apparent correlation of the separations in AU with grouping of the pairs into these intervals of distance. Both nearby systems and systems with large distances (> 85 parsecs) reveal a large range of separations.

3.2 The DA/non-DA Ratio Among the Double Degenerates

Among the DA stars included in Table 1, the coolest DA white dwarfs in the sample have spectral type DA8 with M_{bol} = 13.60-13.70. It is therefore not unreasonable to regard M_{bol} = 14.0 as the limiting absolute magnitude fainter than which the Balmer lines would not be detectable at moderate resolution spectra. The choice of this limit is corroborated by the data presented in Liebert *et al.* (1988) where the coolest DA star in their Figure (1) has T_{eff} =5800K which corresponds to M_v = 14.0. For a DA degenerate M_v =14 corresponds to M_{bol} = 13.60.

In Table 1, 11 of the 21 total systems have one or both components with M_v < 14.00. Of these 11 systems, there are 10 DA stars and 7 non-DA stars, but only 4 of the non-DA stars are hot enough to have shown Balmer lines. Therefore among the double degenerate pairs the DA/non-DA ratio is nearly 3:1, which is close to the ratio one derives for white dwarfs hotter than 15000K. At lower temperatures convective mixing (cf. Sion 1984; Greenstein 1986c), or possible diffusion-induced burning (Michaud *et al.* 1984) apparently transforms hydrogen-rich atmospheres into non-DA atmospheres. Although the sample of double degenerates is admittedly small, it is interesting that the DA/non-DA ratio is larger than the value corresponding to the common proper motion pairs containing a main-sequence component and a white dwarf: Among MS+WD pairs this same ratio is roughly 1:1. This apparently higher frequency of DA degenerates, is statistically very weak: even if the true DA/non-DA ratio among DD pairs were 1:1 there would be a 9% chance of 10 or more DA components in a sample of 14 stars, and, of course, a 9% probability of 10 or more non-DAs among the 14 stars. Therefore, there is an 18% probability that our small sample would have 10 or more stars of the same class even if the parent population contains equal numbers of the two types. Nevertheless, it would not be surprising if further data confirm this statistically weak "excess" of DA stars among the DD pairs, given the observational evidence that DA white dwarfs, not non-DA spectral types, appear to be the progeny of the most massive single parent stars (cf. Sion 1990 and references therein). Since the double degenerate wide binaries are thought to have been initially more massive and to have had a smaller mass ratio than typical WD+MS pairs (Greenstein 1986a,b), it is possible that they could indeed have a differing mix of DA and non-DA atmospheres. Unfortunately the sample of double degenerates is still too small to derive a definitive conclusion.

3.3 Differential Cooling Times of Double Degenerate Pairs

The components of a DD pair tend to be similar in color as seen in Table 1: typically DD pairs differ by only one or two temperature subclasses. For such stars of similar temperatures we expect that

$$\Delta M_{bol} = \Delta V. \tag{4}$$

Therefore the average $\overline{\Delta V}$ in Table 1 ($\overline{\Delta V} = 1.04$), implies $\Delta M_{bol} = 1.04$, which, by simple Mestel cooling theory (for an assumed mass and core composition) yields

$$\Delta \log \tau_c = 2/7 \Delta M_{bol}. \tag{5}$$

Thus for the overall sample in Table 1 we have

$$\overline{\Delta \log \tau_c} = 0.30. \tag{6}$$

Thus for the typical DD pair in our sample, the components differ in cooling age by a factor of two. In Table 2 the apparent magnitude difference, absolute bolometric magnitude difference and the differential cooling time measured by $\Delta \log \tau_c$, are listed for each individual system. The oldest systems in Table 2 are represented by the three DD systems observed by Hintzen *et al.* (1989). The three systems, L197-5/6, L701-69/70, and L77-57/56, have cooling ages of $8-9^9$ years and differential cooling times in Table 2 of 0.49, 0.52 and 0.26 dex respectively. Note that seven of the DD systems in Table 2 have $\Delta \log \tau_c >$ 0.3, with the largest estimated $\Delta \log \tau_c = 0.84$, for the DD system LP707-8/9. This system has a difference in cooling of a factor of 7 and would be the progeny of an initial binary having a large initial mass ratio, which is atypical of these archaic double systems.

If accurate differential photometry (i.e. $\Delta(b - v)$ and ΔV) were available for the components of the DD pairs, then one could derive $\Delta \log M$ and $\Delta \log \tau$ independently of distance, from the relative displacement of the two components with respect to a white dwarf cooling track in the $(M_{bol}, \log T_{eff})$-plane, a point emphasized to us by an anonymous referee. Unfortunately, our errors in ΔV and $\Delta(b - v)$ are quite significant, amounting to no less than 0.1 magnitude (rms), making an assumption of 10% errors in temperatures and luminosities optimistic. Our principal drawback is that we frequently observe the stars with the slit oriented along their position angle to economize observing time instead of having the slit oriented along the direction of atmospheric dispersion. It is therefore prudent that we defer the derivation of mass information until we are able to secure accurate photometry or even better, to re-observe the DD sample spectrophotometrically, taking care to position and widen the slit for maximum accuracy and extracting colors from the 2D CCD arrays.

3.4. The Frequency Distribution Function of Separations, N (Log a)

The increased size of the double degenerate sample reported here allows an initial comparative exploration of the distribution of linear separations relative to MS+MS, MS+WD pairs and wide binaries with orbit solutions. A most important question is whether the large scale mass loss by the progenitors of the white dwarfs in the cpm binaries had a measurable dynamical effect on the distribution of separations. Mass loss as extensive as 80% of the initial binary mass must have a dynamical effect on the orbital evolution (cf. Kopal 1984). Non-explosive, isotropic mass loss by white dwarf progenitors during the giant stages and post-AGB phase may be the primary mechanism for orbital amplification in these systems (cf. Greenstein 1986a, Valls-Gabaud 1987).

If two stars have separations, a, so large that they evolve independently of each other, and for which the mass loss is non-explosive, isotropic and occurs steadily or quasi-steadily on a time scale $t_{wind} \gg P_{orb}$, then the integral

$$(M_1 + M_2)a = const \tag{7}$$

should hold, a property first derived by Jeans (1924) with the method of adiabatic invariants. It can be shown that this principle has the consequence that the angular momentum per unit of reduced mass is conserved (Dommanget 1963). We shall employ equation (7) in the analysis which follows.

Oswalt and Sion (1989) have presented statistical evidence that orbital expansion has indeed occurred in wide binaries which contain a white dwarf. In a sample of over 300 CPM binaries, the physical separation index, $\log s/\mu$ (where s is the angular separation of the pair and μ is the proper motion), for MS+WD pairs ($< \log s/\mu >= 2.22 \pm 0.04$ m.e.) was shown to be about a factor of two larger than for MS+MS pairs ($< \log s/\mu >= 1.91 \pm 0.07$). Our present sample of DD pairs yields $< \log s/\mu >= 1.58 \pm 0.14$ m.e. Thus contrary to expectations, the DD pairs appear to have less than half the average separation of even MS+MS pairs!

Our flux-calibrated spectra provide a means of deriving projected semi-major axes for individual DD pairs. The distance moduli for the 21 double degenerate binaries in Table 1 were derived from the absolute visual magnitudes estimated for each system. Then, from the observed separation in seconds of arc, the linear separation in A.U. was derived. The median separation for the sample in A.U. is $\log a = 2.63$ and the mean separation is $\log a = 2.61$.

The frequency distribution of separations, N ($\log a$) exhibits a relatively small but possibly significant departure from the distribution in intervals of $\log a$, shown by Greenstein's (1986a) sample of 68 MS+WD pairs. The observed frequency distribution for the DD pairs is compared in Table 3 with the observed and normalized distribution functions N ($\log a$) for the observed MS+WD sample of Greenstein (1986a) and the distribution of Kuiper (1935a,b) normalized to Greenstein's sample size of 68 WD+MS systems. While the DD sample is admittedly small, it is apparent that 45% of the DD pairs lie in the range $2.5 \leq \log a \leq 3.5$ while 70% of the WD+MS pairs fall in that range. In the bin $1.5 \leq \log a = 2.5$, however, only 16% of the MS+WD pairs are found there while 45% of the DD pairs are found in this bin of separations. Only one DD system (or 5%) falls into the bin of widest separation ($3.5 \leq \log a \leq 4.5$) while in the WD+MS sample of Greenstein (1986a) 13% fall into this same range. Although the difference between the the two groups (DD versus WD+MS) is not large, it is possible to conclude on the basis of these results that the DD pairs tend to have a higher frequency of smaller separations and fewer systems at the largest separations than the WD+MS pairs. The near absence of very wide DD pairs is very likely a consequence of their greater mean age (see Section 3.5 below) since the fraction of wide binaries which get disrupted by encounters with passing perturbers must increase with systemic age.

It is illuminating to compare the mean and median separations of the present DD sample, with the smaller sample of seven DD pairs in Greenstein (1986b) and the 68 MS+WD pairs in Greenstein (1986a). The mean and median separation of the 20 DD pairs in this paper are $\overline{\log a} = 2.61$ and $\overline{\log a} = 2.63$ respectively, both larger than the respective values quoted in G86b and in Valls-Gabaud (1987). However, the mean separation of the DD pairs of Table 1 is nevertheless slightly smaller than $\overline{\log a} = 2.98 \pm 0.07$ for the 68 WD+MS pairs in G86a, a difference significant at the 5% level, assuming Gaussian statistics. This difference, if real, could be attributed to a selection effect since DD pairs of smaller separation may be easier to detect because their magnitude difference is much smaller than the magnitude difference between members of a WD+MS pair. If, however, they have the same detection probability (cf. Oswalt and Sion 1989), then it is tempting to derive the preliminary conclusion that the DD pairs tend to have smaller separations than the MS+WD pairs but significantly larger separations than Kuiper's (1935a,b) MS+MS most probable separation of log $a = 1.27$. Kuiper's sample, however, is based upon visual surveys with the Lick long focus refractor in which visually resolvable companions could be detected, up to several magnitudes fainter, as close as several seconds of arc, and companions of comparable magnitude could be detected at separations under 1 arc second. The Luyten and Lowell proper motion surveys could not possibly detect companions this close since they would have been lost in the photographic halo of the bright object on the Palomar-Schmidt and Lowell plates. Therefore, due to the selection of the Kuiper sample toward genuinely close (in A.U.) pairs, a comparison of the average separation of the cpm pairs with the Kuiper sample cannot be a reliable indicator of a true difference in their respective separations. The difference in average distance of the two groups for example, must be assessed. When the magnitude of the selection effects inherent in the two samples are eventually addressed in a quantitative manner, it is possible that the average separation of the DD pairs will persist in being genuinely larger than the average separations of the MS+MS pairs. Until then, caution is advised in using the difference in separation between the two samples as any indicator of orbital evolution.

Concerning the DD pairs of the widest separations, the DD system, LP128-254/255, has the largest separation, with $a = 8562$ parsecs, a value well below the maximum separation of 10^4 A.U. discussed in Greenstein (1986a) and confirmed by Abt (1988) for pairs containing a white dwarf. The possibility should be noted that the DD pairs of widest separation may have been disrupted by numerous encounters with perturbers over the long total stellar lifetimes of these systems.

The MS+WD pairs are believed to originate from binaries with initial mass ratio $Q \geq 5$ while the DD pairs are predominately descended from from binaries with $Q = 1$ (cf. Greenstein 1986a,b and references therein). If we consider a DD pair with an initial total mass of, say, 9 $M_\odot$, and with degenerate progeny of combined total mass 1.2 $M_\odot$ we should expect that isotropic, non-explosive mass loss with $M_{wind} \gg P_{orb}$, should lead to an orbital amplification of 0.9 in log a, corresponding to a loss of over 86% of the mass of the system. It is therefore possible for a MS+MS binary of average separation Log $a = +1.27$ to undergo orbital expansion due to mass loss and produce a DD pair with Log $a = +2.17$, which lies in the well-populated bin $1.5 < \text{Log } a < 2.5$. However, the median separation of the DD pairs (Log $a = 2.5$) determined in this paper, while less than the WD+MS pairs, requires an amplification in Log a of a factor of 17 larger than the most probable separation of MS+MS pairs. The larger mean separations of the DD and WD+MS pairs relative to the proper motion-selected MS+MS pairs must in some measure be due to the presence in the white dwarf pair groups of at least one star exceeding one solar mass or much greater whereas the MS+MS pairs are low mass dwarfs. It is also possible that the low mass MS+MS pairs must form closer together in order to be gravitationally bound at the outset. Despite the smaller log a of the Kuiper sample, more very wide MS+MS pairs should exist because they are younger and perhaps because they are spared disruption by post-main sequence, impulse-like mass ejection.

The apparent tendency of DD pairs to have smaller separations than the MS+WD pairs would be consistent with the expectation that their initial mass ratio(s) Q were close to unity as compared with the larger Q $\sim$ 5-10, characterizing the initial mass ratio of the WD+MS pairs. This can be understood by considering that many progenitor binaries of the MS+WD pairs have large initial mass ratio with a main sequence star of very low mass. Thus, by equation (7), the resulting orbital amplification due to non-explosive mass loss by the WD progenitor would lead to a greater orbital amplification.

Finally, we note the possibility of a bi-modal distribution of separations for the double degenerate binaries with ten systems having $\log a < 2.61$, and ten systems having $\log a > 2.61$ with peaks at $\log a = 2.2$ and $\log a = 3.2$. While the sample size is small and requires statistical confirmation, this possibility merits further investigation.

3.5. Kinematics of the Double Degenerate Sample

For most of the double degenerate systems(DD) in Table 1 it is unlikely that complete space motions will ever become available because of either their DC spectral type or their lack of a companion with accurately measurable radial velocity. It is therefore appropriate and interesting to examine their kinematical properties, with the assumption of zero radial velocity, and compare their space motions with earlier analyses of single degenerates (Sion et al. 1988) and MS+WD pairs (Wegner 1981; Sion and Oswalt 1988).

We have used spectrophotometric parallaxes and proper motions in Table 1 to compute the vector components of the galactic space motion, U, V, W, relative to the sun in a right handed system following Wooley et al. (1970), where U is counted positive in the direction of the the galactic anti-center, V is counted positive in the direction of the galactic rotation, and W is counted positive in the direction of the north galactic pole. The mean velocity, mean tangential velocity, and velocity dispersion in each component were computed and the results are presented in Table 4 where Column (1) lists the WD number, Column (2) the U velocity, column (3) the V velocity, column (4) the W velocity and Column (5) the transverse velocity defined by

$$V_T = \sqrt{U^2 + V^2 + W^2}. \tag{8}$$

The velocity dispersions in each component, the average velocity in each component and the average transverse velocity are listed at the bottom of Table 4.

The average motions in the U and V component for double degenerates is 2-3 times larger than the average motions given by Sion et al. (1988) for single DAs, single DBs, single DC stars, and single DZ stars, and a factor of 2-3 times larger than the motions of main sequence and white dwarf pairs, presented by Sion and Oswalt (1988) (with V_r =0), and a factor of 2-3 times larger than the sample of 24 main sequence plus white dwarf pairs with complete space motions presented by Wegner (1981).

Eight of the 21 DD pairs (37%) have total motions > 100 km/s and 6 out of 21 (32%) have motions larger than 120 km/s, even assuming zero radial velocities. It is significant that these frequencies of large space motions greatly exceed the same statistics for DA, DQ and DZ stars in the same interval of M_v (12.00 $\leq M_v \leq$ 15.5) as 90% of the DD binaries.

It is also interesting that the velocity dispersions σ_u, σ_v and σ_w for the double degenerates are a factor of 2 to 7 times larger than the dispersions for single white dwarf subgroups (see Table IIIa in Sion *et al.* 1988), or the velocity dispersions of previously analyzed binaries containing a white dwarf (see Table 2 in Sion and Oswalt 1988).

There may be at least two reasons why the double degenerates as a group have larger mean velocities, larger velocity dispersions, and a higher frequency of large space motions, relative to the white dwarf subgroups in the same interval of M_v. First, the double degenerates as a group are extremely old, an overall property already pointed out by Greenstein (1986b) and corroborated with our factor of three larger sample. We find that the sample in Table 1 implies an average $\Delta \log \tau_c = 0.33$ while Greenstein's sample of 6 systems yielded $\Delta \tau_a = 0.11$. The average cooling age for the sample in Table 1 is 3 x 10^9 years. Ten of the systems in Table 1 are as intrinsically faint or fainter than the oldest double degenerate (LP543-33/32) in Greenstein's (1986b) Table 1. Their cooling ages and total stellar ages, in some cases may exceed 9 × 10^9 years. Therefore the high velocity kinematical characteristics of the group is due in part to diffusion of their galactic orbits with an increase in velocity dispersion with time (Wielen 1977). The rate of increase in velocity dispersion as a function of total stellar age could even be larger than predicted by the analysis of Wielen (1977). A second factor may stem from inaccuracies in our estimates of the photometric parallaxes. For instance, a cool degenerate more luminous than our assigned photometric parallax would appear to have a higher space motion.

4. Conclusions

The double degenerate common proper motion binaries are an extremely old ($\tau_c \approx$ 3 × 10^9 yrs.), very rare (< 5%) subset of the CPM systems. They represent a subset in the overall distribution of double white dwarfs, having the widest separations. They contain the oldest known stars in the solar neighborhood, with cooling ages in some cases approaching 9 x 10^9 years, as old as the Galactic disk.

A typical double degenerate pair contains stars of similar color and spectral type, with an average $M_v = 14.20$ and average cooling age of $\approx 3 \times 10^9$ years, thus typically belonging to the extremely old disk population subcomponent. The differential cooling ages, $\Delta \log \tau_c$, range between 0.01 and 0.84. A typical pair has $\Delta \log \tau_c = 0.30$, corresponding to a factor of two difference in cooling age of its components. One system differs in component cooling age by a factor of seven. The median and mean separations of the DD pairs is 426 A.U. and 407 A.U., respectively, both apparently smaller than the WD+MS values. Their distribution of separations N (log a) is skewed toward smaller bins of log a than the distribution characterizing the MS+WD pairs but is considerably larger than the MS+MS distribution. If selection effects are shown to be unimportant in this latter comparison, then the larger mean separation may be understood if orbital amplification of the DD pairs occurred, due to non-explosive, isotropic, mass loss on a time scale longer than the orbital period, while the product of the total systemic mass and the semi-major axis was conserved. Their smaller separations relative to MS+WD pairs may be attributed to the smaller dynamical effect of mass loss on their orbits, due to their lower initial mass ratios. The DA/non-DA ratio, for the pairs hot enough to show Balmer lines (*i.e.* $M_v < 14.0$) is 3:1, similar to the ratio for non-binary degenerates and apparently higher than the anomalous 1:1 found for the MS+WD pairs, though the present DD sample is too small for this difference to be statistically significant.

Their kinematical properties are extreme: the average UVW motions and velocity dispersions are significantly larger than the average velocities and dispersions associated with selected samples of single white dwarfs and MS+WD binaries, when the latter are restricted to the same color/M_v range as the DD systems. The DD sample, even in the absence of radial velocities appears to have velocity dispersions 2 to 7 times larger than selected samples of single white dwarfs and MS+WD pairs chosen to be in the same color/absolute magnitude range as the DD components. This property may be a result of the dynamical inflation of their velocity dispersions due to their extremely ancient total stellar ages.

It is a pleasure to thank Beth Jewell for assisting with the camera ready manuscript. This research was supported by NSF grants AST88-02689 (E.M.S.) to Villanova University, The Centre National De La Recherche Scientifique, URA285 (E.M.S.), AST89-18471 (J.L.) to the University of Arizona, AST88-02687 (T.D.O.) to the Florida Institute of Technology and by NASA grant 188-4151-07 (P.H.).

References

Abt, H. 1988, *Ap.Sp.Sci.*, **142**, 111.
Dommanget, J. 1963, *Ann. Obs. Roy. Belgique, 3eme Serie*, **IX** (5), 216.
Greenstein, J. L. 1984, *Ap.J.*, **276**, 622.
Greenstein, J. L. 1986a, *A.J.*, **92**, 859 (G86a).
Greenstein, J. L. 1986b, *A.J.*, **92**, 867 (G86b).
Greenstein, J. L. 1986c, *Ap.J.*, **304**, 334.
Hintzen, P., Oswalt, T. D., Liebert, J., and Sion, E. M. 1989, *Ap.J.*, **346**, 454.
Jeans, J. H. 1924, *MNRAS*, **85**, 2.
Kopal, Z. 1984, *Ap.Sp.Sci.*, **99**, 3.
Kuiper, G. P. 1935a, *PASP*, **47**, 15.
Kuiper, G. P. 1935b, *PASP*, **47**, 121.
Liebert, J., Dahn, C., and Monet, D. 1988, *Ap.J.*, **322**, 891.
McCook, G. P., and Sion, E. M. 1987, *Ap.J. Suppl.*, **65**, 603.
Oswalt, T., and Sion, E. M. 1989, in *White Dwarfs*, ed. G. Wegner (Berlin: Springer Verlag), p. 454.
Oswalt, T. D., Hintzen, P. M., Liebert, J. L., and Sion, E. M. 1988, *Ap.J. (Letters)*, **333**, L87.
Sion, E. M. 1990, in *Mass Loss and Angular Momentum Loss from Hot Stars*, ed.L.A. Willson and R. Stalio (Kluwer: Dordrecht), in press.
Sion, E. M., and Oswalt, T. D. 1988, *Ap.J.*, **326**, 249.
Sion, E. M., Greenstein, J. L., Landstreet, J. D., Liebert, J., Shipman, H. L., and Wegner, G. A. 1983, *Ap.J.*, **269**, 253.
Sion, E. M., Fritz, M. L., McMullin, J. P., and Lallo, M. D. 1988, *A.J.*, **96**, 251.
Valls-Gabaud, J. 1988, *Ap.Sp.Sci.*, **142**, 289.
Wegner, G. 1981, *A.J.*, **86**, 264.
Weidemann, V. 1987, *Astr.Ap.*, **188**, 74.
Weilen, R. 1977, *Astr.Ap.*, **60**, 263.
Wooley, R., Epps, E. A., Penston, M. J., and Pocock, S. B. 1970, *Ann. R. Obs.*, No. 5.

TABLE 1: PROPERTIES OF DOUBLE DEGENERATE BINARIES						
DESIG	WD NO.	V/μ	M_v/ϕ	M_{bol}	SPCTRL TP	(a)/a(AU)
LP406-62/63	0102+210	17.94	14.96	14.68	DC 9	(28")
		18.14	15.48	15.03	DQ 9/DC 9	1104
		0.48	208			
LP647-33/34	0114-049	18.50	15.18	14.83	DC 9	(2")
		18.62	15.12	14.83	DC 9	100
		0.374	226			
LD197-5/6	0222+422	16.8	14.1	13.70	DA 8	(7")
		18.9	16.2	15.42	DC 14	240
		0.42	117			
Gr566/567	0727+242	15.48	15.32	14.94	DC 9	(0.66")
		14.73			DC 9	
		15.48	15.32	14.94	DC 9	7
LP543-33/32	0747+073	16.67	15.40	14.98	DC 9	(16")
		17.00	15.83	15.23	DC 9	275
		1.78	173			
LP535-288/287	0841+69	18.62	15.18	14.83	DC	(3")
		17.93	14.55	14.36	DC/DZ	153
		0.34	223			
LP462-56A/B	0935-371	14.30	12.58	12.36	DQ 6	(4")
		14.88	13.27	12.86	DA 6	88
		0.37	2.95			
LP370-50/51	0942+236	17.33	13.57	13.30	DA 7	(13")
		17.01	13.68	13.30	DA 7	734
		0.22	244			
LP549-33/32	1012+083	16.16	13.84	13.74	DZ ?	(26")
		17.7	15.09	14.78	DA ?	757
		0.32	301			
LP128-254/255	1051+533	18.00	11.58	10.37	DA 4	(44")
		18.30	12.00	11.18	DA 5	8462
		0.112	257			
ES0439-162/163	1127-311	19.84	14.28	14.14	DC 9	(23")
		18.77	–	–	DQ	2976
		0.38	2.33			
LP322-500A/B	1302+313	17.7			DA	(12")
		19.2				
		0.03	147		DA	
LP96-66/65	1308+61	19.4	14.73	14.51	DC	(18")
		21.2(B)	15.52	15.06	DC	1546
		0.29	284			
L151-81A/B	1454-630	16.6	11.8	11.2	DB 3	(2")
		17.0	12.00	11.18	DA 5	219
		0.05	191			
Gr576/577	1704+481	14.48	11.58	10.37	DA 4	(4.5")
		14.45	12.97	12.52	DA 5	171
LP567-39/38	1726+074	18.50	14.76	14.53	DC	(2")
		17.4	15.17	14.83	DB ?	93
		0.3	198			
G206-17/18	1811+327	16.36	13.59	13.20	DA 7	(55")
		17.04	14.00	13.60	DA 8	2120
		0.27	220			
LP701-69/70	2301-072	17.3	13.5	13.39	DC 7	(26")
		19.6	15.8	15.22	DC 17	1500
		0.338	130			
LP77-5457	2351+655	16.74	14.6	14.40	DQ 9	(3")
		18.6	16.0	15.32	DC 13	80
		0.409	61			
LP472-70/69	0322+145	18.455	13.59	13.20	DA	(3")
		21.01	15.633	15.12	DC	914
LP707-819	0106-10.9	17.51	12.30	11.80	DA	(12")
		19.94	15.03	14.73	DC	1320
		0.198	98			

TABLE 2: DIFFERENTIAL COOLING TIMES				
DESIG	WD NO.	ΔV	ΔM_{bol}	$\Delta \log \tau_c$
LP406-62/63	0102+210	-0.20	-0.35	0.1
LP647-33/34	0114-049	+0.12	0	0
LD197-5/6	0222+422	-2.1	-1.72	0.49
Gr566/567	0727+242	0.0	0	0
LP543-33/32	0747+073	-0.33	-0.25	0.07
LP535-288/287	0841+69	-0.69	-0.47	0.136
LP462-56A/B	0935-371	0.33	-0.50	0.14
LP370-50/51	0942+236	+0.32	0	0
LP549-33/32	1012+083	-1.54	-1.04	0.30
LP128-254/255	1051+533	-0.30	-0.81	0.23
ES0439-162/163	1127-311	-1.07	–	–
LP322-500A/B	1302+313	–	–	–
LP96-66/65	1308+61	-1.8	-0.55	0.16
L151-81A/B	1454-630	-0.4	+0.02	0.006
Gr576/577	1704+481	+0.3	-2.15	0.61
LP567-39/38	1726+074	+1.1	-0.30	0.09
G206-17/18	1811+327	-0.68	-0.40	0.114
LP701-69/70	2301-072	-2.3	-1.83	0.52
LP77-5457	2351+655	-1.86	-0.92	0.26
LP472-70/69	0322+145	-2.56	-1.92	0.55
LP707-819	0106-10.9	-2.43	-2.93	0.84

Table 3 Frequency Distribution of Separations N (Log a) Log a (A.U.)						
Sample	-1.5 to 0.5	-0.5 to 0.5	0.5-1.5	1.5-2.5	2.5-3.5	3.5-4.5
G86a	0	0	1	11	47	9
Kuiper	5	13	19	17	9	3
DD Pairs	0	0	1	9	9	1

TABLE 4: KINEMATICAL DATA FOR DOUBLE DEGENERATE BINARIES				
WD No.	U	V	W	V_T
0102+210	-53	-33	-68.	93
0106-109	85	-82	-4	118
0114-049	-73	-22	-37	86
0222+422	51	-67	-13	85
0322+145	205	-148	244	352
0747+073	-57	-139	-55	160
0841+073	68	-60	-37	98
0935-371	45	-11	-18	50
0942+236	42	-41	-45	75
1012+083	49	0	-20	53
1051+533	90	-61	-42	118
1127-311	109	-131	-180	248
1308+61	117	-59	-14	132
1454-630	21	-22	-27	41
1726+074	-13	-73	-21	77
1811+327	-25	-39	8	47
2301-072	43	-82	-56	108
2351+655	53	-35	8	64

CLOSE BINARY WHITE DWARFS

JAMES LIEBERT, P. BERGERON and REX A. SAFFER
Steward Observatory
University of Arizona
Tucson, Arizona 85721
USA

ABSTRACT. Previously undetected binaries consisting solely of degenerate stellar components have been discovered or their existence inferred by three different techniques, the results of which are discussed in this paper. The first and most useful method is the successful detection of periodic radial velocity variations due to orbital motion. Secondly, close binary evolution may be the only way of explaining the existence of stars having derived masses safely below the mass limit for core helium ignition. Finally, at least three stars exhibit inconsistencies between their line spectra and colors which can be explained if the spectra are composite.

1. Introduction

The discovery that close, detached double degenerates (DDs) exist which must have undergone prior phases of common envelope evolution has added a new dimension to the study of these objects. Since the frequency of duplicity is high among other types of stars, perhaps it should not be surprising that at least a few apparently single degenerate stars have now been found to be DDs. Recent theoretical calculations of the evolution of close binaries predict that many Algol-like binaries will turn into short-period, detached binary white dwarfs after passing through at least one stage of common envelope evolution (Webbink 1984; Iben and Tutukov 1984b). Paczynski (1985), Iben and Tutukov (1984a, 1986, 1987), and Tutukov and Yungelson (1987a,b) have estimated that the fraction of all white dwarfs which are binary could be as high as 10–20%. These authors have also suggested that mergers of such close DDs, following a long period of orbital angular momentum loss due to gravitational radiation, could be responsible for the production of some subdwarf O or B stars, and some Type I supernovae. It should also be noted that a class of interacting binary white dwarfs has been known for many years: these are cataclysmic variable-like systems of very short orbital periods observed to be transferring almost pure helium gas, for which the prototypes are AM CVn (HZ 29) and G61–29 (cf. Nather, Robinson and Stover 1981; Solheim et al. 1984). The secondary in these systems is almost certainly a low mass helium degenerate, while the primary could have a core composed of carbon and oxygen.

I will discuss here results of three observational methods by which these close DDs have been discovered, or their existence inferred. These include (1) the discovery of white dwarf spectroscopic binaries, (2) the implication of close binary evolution

G. Vauclair and E. Sion (eds.), White Dwarfs, 409–416.
© 1991 Kluwer Academic Publishers.

for stars having very low masses derived from the atmospheric analyses, and (3) stars showing peculiar, composite spectra and/or colors. DDs inferred from the last approach need not be close enough in separation that they must have undergone common envelope evolution. Here this paper overlaps with the results of Greenstein (1986a), Sion *et al.* (1991, this conference) and Oswalt *et al.* (1991, this conference) who discuss studies of double degenerates found as common proper motion pairs.

2. Radial Velocity Variables

The first comprehensive search for short-period, detached DDs by trying to detect radial velocity variations was that of Robinson and Shafter (1987). These authors found no variables having very short periods between about 30 seconds and 3 hours among 44 DA and DB white dwarfs in which H or He I line variations were sought. The purpose of this investigation was to discover or put limits on systems which might be progenitors of Type I supernovae (*i.e.*, those which would merge within the next several billion years, hopefully with enough mass to exceed the Chandrasekhar limit). However, the modest limits on radial velocity variations of this study meant that little could be inferred about the existence of systems with periods longer than 3 hours.

Saffer, Liebert and Olszewski (1988), following up the suggestion of Greenstein (1983), discovered that L870-2 (EG 11, WD0135-052) is a double-lined spectroscopic binary consisting of cool DA stars but with a much longer period of 1.55 days. Considerably more is now known about this system than any of the others discussed below. Due to its excellent trigonometric parallax (*i.e.*, proximity to Earth), Greenstein (1985) had noted its overluminosity in an HR Diagram with respect to other white dwarfs of similar colors. Using this and the available spectrophotometry and photometry, Bergeron *et al.* (1989) determined the masses and atmospheric parameters of the components. They found masses, effective temperatures and surface gravities of $M_1 = 0.47 \pm 0.05\ M_\odot$, $T_1 = 7470 \pm 500\ K$, $\log g_1 = 7.80 \pm 0.10$, and $M_2 = 0.52 \pm 0.05\ M_\odot$, $T_2 = 6920 \pm 500\ K$ and $\log g_2 = 7.89 \pm 0.10$, respectively. Tutukov and Yungelson (1988) find that the mass ratio of 0.86 is consistent with the conclusion that this is a post-Algol binary consisting of two degenerate helium cores, for which the masses quoted above are marginally consistent. In any case, the masses are clearly too low (and the separation too wide) for this system to be a potential progenitor of a supernova.

As a result of a comprehensive survey similar to that of Robinson and Shafter (1987), Bragaglia *et al.* (1990) found a second DA radial velocity variable, WD0957-066, again with a relatively long period of 1.15 days. While this is a single-lined spectroscopic binary, the derived mass for the companion and failure to detect it imply that it is a cooler degenerate object. Given their failure to find any short-period pairs, the authors argued that DDs are unlikely to be the dominant progenitors of Type I supernovae. Moreover, they found that WD0957-666 and their other suspects for radial velocity variability generally have derived surface gravities much lower than the average for white dwarfs, suggesting that the visible components, at least, will turn out to be low-mass helium degenerate stars. These derived surface gravities, however, were taken from compilations by Guseinov, Novruzova and Rustamov (1983) of a heterogeneous literature. Bragalia and Renzini invited us to fit their optical spectrum of WD 0957-066, using the fitting techniques described by Bergeron, Saffer and Liebert (1991, these proceedings).

Our preliminary fit confirms their conclusion that the mass of the visible DA star is well below 0.4 $M_\odot$.

In a study consisting of multiple radial velocity measurements primarily of hot DA white dwarfs, Foss, Wade and Green (1991) again report no discoveries of short-period binaries among 25 stars. These authors calculated carefully the detection probabilities for detecting periods between three and ten hours for the different velocity samplings of each separate star. For three different assumed distributions of initial binary period, these authors find that the upper limits on the frequency of binaries massive enough to explode is well below the observed rate of Galactic Type Ia supernovae.

Foss, Wade and Green (1991) and Foss and Saffer (1991, in preparation) did discover one subdwarf B star to be a single-lined spectroscopic binary with a period of 2.5 days, outside of their optimum period range; there may be insufficient observations at longer wavelengths, however, to rule out the possibility that the companion is a low mass main sequence star. However, LB3459 is a known binary with a period near 6 hours believed to consist of a subdwarf O star and probably a hot white dwarf (Kilkenny, Penfold and Hilditch 1979, and references therein). Finally, Saffer and Liebert (1989) found that PG 1102+499 apparently is a binary subdwarf O star showing doubled helium lines with an indicated velocity amplitude larger than 100 $km\ s^{-1}$; this amplitude implies an orbital period of the order days or less.

3. An Apparent Sequence of Low Mass Helium Degenerates

One of the clear results of the mass distribution of DA white dwarfs determined by Bergeron, Saffer and Liebert (1991) is the existence of white dwarfs having masses which are apparently below the $0.40-0.45 M_\odot$ value at which theoretical calculations show that helium ignition takes place in the evolution of single stars of solar metal abundance. An excellent example of a spectrum with unusually narrow high Balmer absorption lines fitted to a low surface gravity is PG 1101+364, illustrated in Figure 6 of Bergeron, Saffer and Liebert (1991, these proceedings). The percentage of such stars in the overall distribution depends on uncertainties in our mass determinations and in the core mass for helium ignition, but appears to include some 5–10% of the DA stars.

There is no known way white dwarfs with only helium cores can finish their nuclear evolution as single stars within the lifetime of the Galaxy. The only logical explanation for the sequence of DA stars having masses below $0.4 M_\odot$ is to assume that they have been stripped of their envelopes during evolution from the main sequence to the tip of the red giant branch, i.e. before the helium core could build up the mass and central temperature necessary for helium ignition. This is the so-called "Case B" mass transfer of Kippenhahn. We thus predict that these are close DD binaries. They could show double-lined spectra like L870–2. However, the spectra will be single-lined but variable if the original primary is enough older (cooler) or smaller in radius. It is conceivable also that these low mass stars could be post-merger objects, or represent the end state of some AM CVn interacting systems. It is likely, however, that at least the majority of these objects should still be binary, and the Bragaglia et al. (1990a) system WD 0957–666 may be typical of this group.

We are attempting to prove that the stars having very low derived masses in Bergeron, Saffer and Liebert (1991) are DD binaries by attempting to detect radial velocity variations. Since only the DA white dwarfs with $T_{\text{eff}} \leq 20,000$ K have sharp, non-LTE Hα cores, definitive observations may be possible only for the cooler stars in this sample. Such observations are also expensive in terms of large telescope time; reasonable allocations of time are generally available only on smaller telescopes, and the necessary observations may be obtained only for those stars which are bright in apparent magnitude. Thus, progress in testing these candidates may be slow.

4. Unresolved Double Degenerate Spectra

A third method by which the existence of unresolved DD binaries has recently been inferred is by comparison of the line spectrum and the continuum colors. Three DA stars in the sample observed by Greenstein (1986b) with the Palomar 5–meter Hale reflector using a CCD spectrophotometric detector showed line spectra which were markedly inconsistent with the continuum colors (Bergeron 1988; Greenstein and Liebert 1990). The only interpretation by which the sets of data could be reconciled was to assume that the visible spectrum included contributions from both a DA star and a line-free DC companion. This hypothesis was quantified by Bergeron, Greenstein and Liebert (1990), who derived T_{eff} and crude log g values for each component of the proposed binaries, demonstrating in each case a satisfactory fit to both the set of observed Balmer lines and to the optical continuum.

These stars have not yet been observed for radial velocity variations. Unfortunately, each is fainter than V = 16, posing the same difficulties encountered previously in obtaining the required observations. Moreover, the predicted parallaxes from Bergeron et al. (1990) indicate maximum separations of some 100–300 a.u., assuming a maximum angular separation of 2 arc seconds. They could well overlap in separation some of the nearby common proper motion DD pairs discussed by Greenstein (1986a) and Sion et al. (1991). Relative orbital velocities in these systems might be too small to detect, even at very high spectral dispersion.

These three pairs of apparent DA+DC components were in many ways optimal for the discovery of unresolved DDs. Bergeron, Greenstein and Liebert (1990) noted that these pairs include a DA component not far from the peak of its hydrogen line strengths. For much hotter and cooler stars, the line profiles are much more sensitive to T_{eff} and the interpretation is more sensitive to the accuracy of color measurements. Moreover, a very cool companion could obviously not be found by this technique.

Finally, it seems likely based on the statistics of both common proper motion DD pairs (Sion et al. 1991) and the field white dwarfs that many more unresolved DA+DA stars should exist among known samples. In attempts to determine the temperatures and gravities of the double DA components of L870-2 (Bergeron et al. 1989), it was found that it was generally possible to fit the summed spectra of two DA models having different parameters with a model for a single star. Since the L870–2 components have temperatures near 7,000 K, only temperatures up to about 12,000 K were considered.

This result has now been generalized using pure hydrogen models at hotter temperatures, as described below. We coadded all possible combinations of models with $15,000 \leq T_{\text{eff}} \leq 45,000$ K in intervals of 10,000 K, and log g values of 7, 8 and 9. We then found the best fit to the derived composite theoretical spectrum with

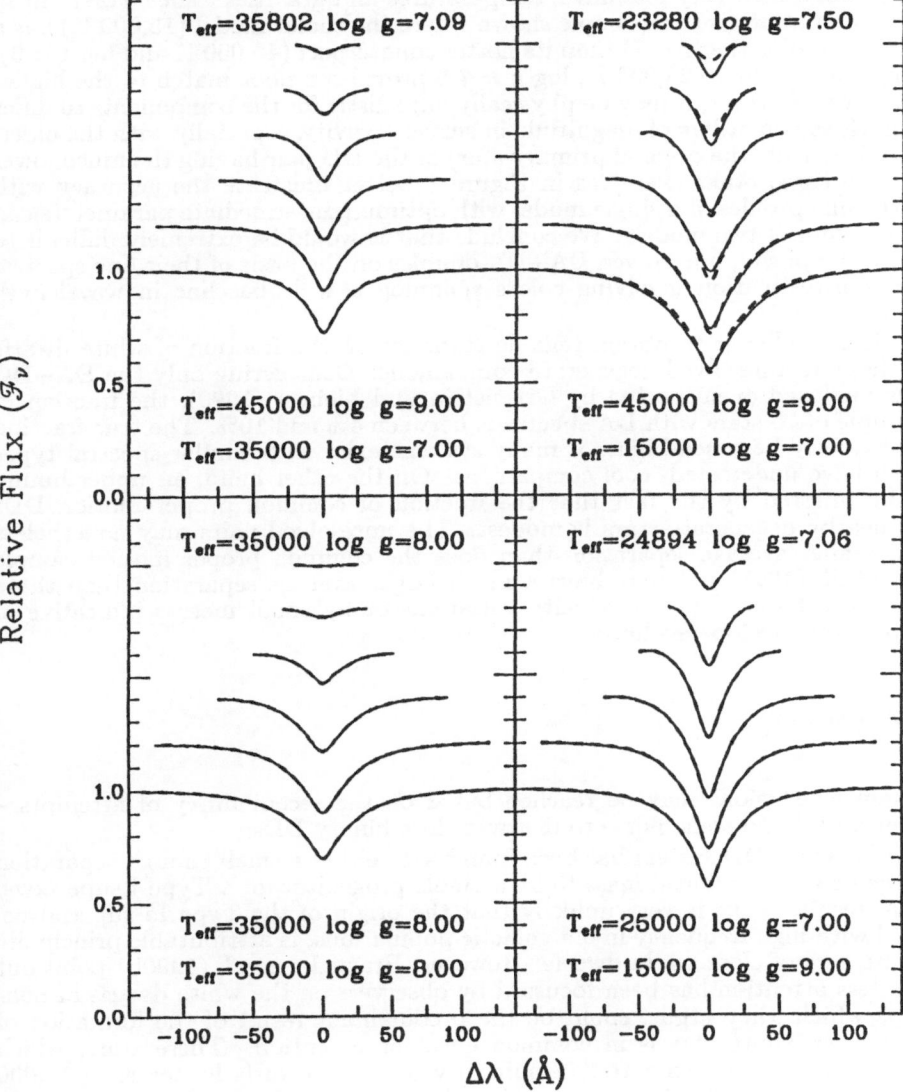

Figure 1. Four comparisons of coadded line profiles of two model spectra with every possible pairing of temperatures and gravities over the ranges stated in the text with the profiles of a single model having intermediate values of these parameters. Balmer line profiles are offset vertically with Hβ at the bottom and H9 at the top. The parameters of the two models are listed at the bottom of each panel, and the profiles are plotted as solid curves; the parameters of the best-fitting single model are listed at the top with the profiles plotted as dashed curves.

a model for a single T_{eff} and gravity. Four examples are illustrated in Figure 1, including some with very dissimilar temperatures and gravities. The "worst" fit of all the combinations is the example shown where the cooler model (15,000 K) has a much larger radius (log g = 7) than its hotter counterpart (45,000 K and log g = 9). The single model fit to 23,000 K, log g = 7.5 provides a poor match to the higher Balmer lines. However, it may be physically unrealistic for the components to differ by as much as two orders of magnitude in surface gravity, especially with the older, cooler component (the original primary star) of the DD pair having the much lower mass. The three other examples in Figure 1 better illustrate the accuracy with which the line profiles of a single model with optimum intermediate parameters can match the sum of two models. We conclude that it would be extremely difficult to recognize composite, unresolved DA+DA doubles on the basis of their line spectra. A similar investigation involving colors spanning a wide baseline in wavelength should be pursued.

It is clearly difficult to obtain reliable statistics on the fraction of white dwarfs having fainter, unresolved degenerate companions. Considering only the DA+DC systems analyzed or suspected by Greenstein and Liebert (1990), the fraction in that sample of 75 stars with DA spectra is between 4% and 10%. The true fraction could obviously be much higher if many systems exist with similar spectral types or which have undetected, cool companions. On the other hand, an upper bound might be inferred by the fact that the fraction of common proper motion DDs among nearby degenerate stars is modest. The unresolved pairs may nonetheless have a smaller average separation than does the common proper motion sample of Sion *et al.* (1991) but may have a much larger average separation than those found on the basis of radial velocity variations or by small masses indicative of prior common envelope evolution.

5. Implications

Two main conclusions may be reached based on the recent flurry of attempts – some successful and some not – to discover close binary DDs:

(1) Not a single DD system has been found with either a small enough separation or a large enough combined mass to be a viable progenitor for a Type I supernova. The null results make it very unlikely that the origin of the Type Ia supernovae, observed with high frequency in old galactic populations, is attributable principally to the mergers of close white dwarfs. However, Bragaglia *et al.* (1990b) point out that far less attention has been focussed by observers on the white dwarfs of non-DA type which, they argue, could be the predominant result of the formation of carbon-oxygen white dwarfs in common envelope evolution. These stars, which represent only an approximate 25% minority of white dwarfs hotter than 10,000 K, are much more difficult to observe since the observed lines – if any – generally suffer greater pressure broadening in a helium-rich atmosphere.

(2) Despite the negative conclusion with respect to finding massive and very close DD binaries, the frequency of close DD pairs may in fact be substantial. It is possible that the majority of these include at least one component of low mass composed of helium. Thus, recent theoretical studies focussing on the consequences of merging low mass, helium white dwarfs (Iben 1990) are timely. Such an event can result in a core-helium-burning stage lasting of the order 10^8 years; it has been

suggested (cf. Tutukov and Yungelson 1987; Iben 1990, and references therein) that some or all hot subdwarf (sdB, sdO) stars originated as mergers of low mass DDs. The observational results suggest that the great majority of the DDs may nevertheless lie at separations too large for merger to be expected within a Hubble time, especially given the strong dependence of the angular momentum loss to to gravitational radiation on the component masses. Many wider pairs may exist whose core compositions may not have been influenced by common envelope evolution.

This research was supported by the National Science Foundation through grant 89–18941, by an NSERC postdoctoral fellowship to P.B., and by NATO.

REFERENCES

Bergeron, P. 1988 PhD Thesis, Université de Montréal.
Bergeron, P., Greenstein, J. L. and Liebert, J. 1990, *Ap. J.*, **361**, 190.
Bergeron, P., Wesemael, F., Liebert, J. and Fontaine, G. 1989, *Ap. J. (Letters)*, **345**, L94.
Bergeron, P., Saffer, R. A. and Liebert, J. 1991, these Proceedings.
Bragaglia, A., Greggio, L., Renzini, A. and D'Odorico, S. 1990a, *Ap. J. (Let.)*, in press.
Bragaglia, A., Greggio, L., Renzini, A. and D'Odorico, S. 1990b, in *Supernovae*, ed. S. E. Woosley (Berlin: Springer-Verlag), in press.
Foss, D., Wade, R. A. and Green, R. F. 1991, *Ap. J.*, in press.
Greenstein, J. L. 1983, *M. N. R. A. S.*, **203**, 1213.
Greenstein, J. L. 1985, *Pub. A. S. P.*, **97**, 827.
Greenstein, J. L. 1986a, *A. J.*, **92**, 867.
Greenstein, J. L. 1986b, *Ap. J.*, **304**, 344.
Greenstein, J. L. and Liebert, J. 1990, *Ap. J.*, **360**, 662.
Iben, I. Jr. 1990, *Ap. J.*, **353**, 215.
Iben, I. Jr. and Tutukov, A. V. 1984a, *Ap. J.*, **282**, 615.
Iben, I. Jr. and Tutukov, A. V. 1984b, *Ap. J. Suppl.*, **54**, 335.
Iben, I. Jr. and Tutukov, A. V. 1985, *Ap. J. Suppl.*, **58**, 661.
Iben, I. Jr. and Tutukov, A. V. 1986, *Ap. J.*, **311**, 753.
Iben, I. Jr. and Tutukov, A. V. 1987, *Ap. J.*, **313**, 727.
Kilkenny, D., Penfold, J. E. and Hilditch, R. W. 1979, *M. N. R. A. S.*, **187**, 1.
Nather, R. E., Robinson, E. L. and Stover, R. J. 1981, *Ap. J.*, **244**, 269.
Oswalt, T.D., Sion, E.M., Hintzen, P.M. and Liebert, J. 1991, these Proceedings.
Paczynski, B. in *Cataclysmic Variables and Low Mass X-ray Binaries*, eds. D.Q. Lamb and J. Patterson (Dordrecht: Reidel), p. 1.
Robinson, E. L. and Shafter, A. W. 1987, *Ap. J.*, **322**, 296.
Saffer, R. A., Liebert, J. and Olszewski, E. M. 1988, *Ap. J.*, **334**, 947.
Saffer, R. A. and Liebert, J. 1989, in *White Dwarfs*, Proc. IAU Coll. 114, ed. G. Wegner (New York: Springer Verlag), p. 408.
Sion, E.M., Oswalt, T.D., Hintzen, P.M. and Liebert, J. 1991, these Proceedings.
Solheim, J. E., Robinson, E. L., Nather, R. E. and Kepler, S. O. 1984, *Astr. Ap.*, **135**, 1.
Tutukov, A. V. and Yungelson, L. R. 1987a, in *I.A.U Colloquium No. 95, The Second Conference on Faint Blue Stars*, eds. A. G. Davis Philip, D. S. Hayes and J. W. Liebert, (Schenectady: L. Davis Press), p. 435.
Tutukov, A. V. and Yungelson, L. R. 1987b, *Comments in Astrophysics*, **XII**, 51.

Tutukov, A. V. and Yungelson, L. R. 1988, *Pis'ma Astron. Zh.*, **14**, 623.
Webbink, R. F. 1984, *Ap. J.*, **277**, 355.

NEW RESULTS ON CATACLYSMIC VARIABLE WHITE DWARFS

Edward M. Sion
Observatoire Midi-Pyrenees,
Centre National De La Recherche Scientifique, URA285,
14, Avenue Edouard Belin,
31400 Toulouse, France
and
Department of Astronomy and Astrophysics,
Villanova University,
Villanova, PA 19085 USA

ABSTRACT: The effective temperatures derived for cataclysmic variable white dwarfs are reviewed and updated. Luminosities are derived, using white dwarf masses determined by Webbink (1990). Lower limit cooling ages are derived by comparison with cooling curves for single degenerates, on the assumption that erosion of the core mass due to dredgeup by repeated nova explosions is negligible. The distribution of CV white dwarfs in the Log P_{orb} - Log T_{eff} plane is compared with the theoretical evolution of T_{eff} predicted by the mass transfer rate versus orbital period-correlation modelled by McDermott and Taam (1989). The distribution and observed range of CV degenerate T_{eff} values above the period gap is quite different from those below the period gap. The magnetic CV degenerates may have a distribution differing from non-magnetic CVs. The distribution and lower limit cooling ages of non-magnetic CVs are not inconsistent with the McDermott and Taam theory. The implied age of the VV Puppis degenerate may be anomalous especially if it is massive. The presence of substantial carbon in the atmosphere of the WZ Sge degenerate is briefly discussed.

1. Introduction

The far ultraviolet (and in some cases the optical) spectroscopic detection of the underlying white dwarfs in cataclysmic variables (CVs) has been possible thus far in three types of CV systems: (1) some dwarf novae with low mass transfer rates during quiescent intervals (between outbursts); (2) some nova-like variables with high mass transfer rates during *low* brightness states of markedly lower mass transfer and; (3) some magnetic cataclysmics during *low* states when accretion appears to shut off and the heated polar accretion caps have cooled. The observational efforts documenting these detections have been reviewed by Smak (1984), Sion (1985, 1986, 1987), Sion and Starrfield (1986), Szkody, Downes amd Mateo (1988), Szkody and Sion (1989) and references therein. These detections of the white dwarf line and continuum radiation are in most cases firmly established and offer several intriguing possibilities for new insights into the accretion physics of CVs: the depth of heat diffusion into the white dwarf (cf. Sion 1985, Pringle 1988; Szkody and Sion 1990; Szkody *et al.* 1988), the role of diffusion and elemental abundances of a freshly

G. Vauclair and E. Sion (eds.), White Dwarfs, 417–430.
© 1991 *Kluwer Academic Publishers.*

accreted white dwarf photosphere (cf. Holm 1982; Shafter *et al.* 1985; Kiplinger, Sion and Szkody 1991; Sion, Leckenby and Szkody 1990), the accretion of disk matter with angular momentum (shear mixing; cf. Sparks *et al.* 1990; Kutter and Sparks 1989), the extent of boundary layer heating and the energy sources which govern the luminosity/effective temperature of a CV white dwarf between nova explosions (cf. Iben 1982; Sion 1985; Prialnik 1986; Regev and Shara 1989; Shaviv and Starrfield 1987) and the possibility of important constraints/implications for the lifetimes of CVs, and their long term evolution above, below and possibly through the period gap.

First, an update of CV white dwarf effective temperature determinations is presented, followed by an initial exploration of the possible implications/constraints on CV lifetimes and evolution, based upon the ensemble of white dwarf effective temperatures as a function of orbital period. CV white dwarf luminosities are derived by using the T_{eff} data and adopting the masses (radii) of individual CV white dwarfs determined by Webbink (1990). Their positions in the (Log ($L/L_\odot$)-Log T_{eff}) plane are compared with cooling curves for single degenerates and lower limit cooling ages are determined. These lower limit cooling ages are subject to possible caveats relating to core mass erosion following repeated nova outbursts (Webbink 1990b) and to the uncertain time interval as a pre-CV white dwarf, preceding the onset of CV mass transfer. The distribution of CV white dwarf T_{eff} in the (Log P_{orb} - Log T_{eff}) plane is compared with the theoretically predicted evolution of T_{eff} according to the correlation between mass transfer rate and orbital period, calculated by McDermott and Taam (1989). Finally some very recent results on the atmospheric composition of the white dwarf in WZ Sagittæ are presented.

2. The Effective Temperature Data

The effective temperatures of cataclysmic variable white dwarfs (CVWDs) are compiled from their original sources in the appendix of the review article by Sion (1987) along with their references and will not be repeated here unless a revised temperature determination has appeared since 1987 or additional CV systems with temperature information have become available. Temperature determinations exist for 14 objects, from actual line profile/continuum fits using grids of white dwarf model atmospheres. These should be regarded as being the most accurate values (see Table 1 below). The remaining 13 CV white dwarfs have temperatures which in most cases have been determined indirectly without actual line profile fits, for example, from decreased UV emission during low accretion states (cf. Szkody *et al.* 1988), from disk emission line strengths, assuming a $1/r^2$ photoionizing flux for the white dwarf (cf. Smak 1984) and other methods. The temperature data for this latter group is not nearly as secure so that any given value of T_{eff} must be regarded as an upper limit only.

There are several recent determinations of exceptional importance. The T_{eff} values for the magnetic nova V1500 Cygni (Stockman, Schmidt and Lamb 1988) and WZ Sge have been measurably decreasing since their respective outbursts in 1975 and 1978; their values at present are respectively 110,000K (see Sec.4.2 below) and 12,500K (LaDous 1990). A new analysis of the T_{eff} of the white dwarf in OY Car has yielded 15,000K ± 2000K (Hessmann *et al.* 1989). A refined *IUE* analysis of the white dwarf in U Gem during quiescence has yielded 30,000K and a possible temperature change from 40,OOOK immediately after the dwarf nova outburst to 30,000K, 103 days into quiescence and just before the next outburst

(Kiplinger, Sion and Szkody 1991). An new analysis by Beuermann *et al.* (1990) of the eclipsing magnetic CV, V834 Cen (E1405-451), has yielded an actual temperature from the synthetic spectrum fit to the optical/near infrared spectrum (including the Zeeman absorption components of Hα) of the magnetic white dwarf, after correction for the cool star. The resulting temperature is 15,000K which replaces the earlier limit of $T_{eff} < 26,500$K by Maraschi *et al.* (1984), cited in Sion (1987). An analysis of the AM Her system UZ For (EXO00333-2554.2) by Beuermann, Thomas and Schwope (1988) has yielded $T_{eff} < 20,000$K for its massive magnetic degenerate. The location of this system at the lower boundary of the period gap is discussed in below.

3. Implications of the Cataclysmic Variable White Dwarf Temperature Distribution for the Lifetimes of Cataclysmic Variables

3.1. MASSES AND LUMINOSITIES OF CVWDs

The most secure values of T_{eff} for CVWDs at present are listed in Table 1 and are employed to determine white dwarf luminosities, by adopting the CV white dwarf mass determinations from Webbink (1990a). These masses were derived in a consistent manner, based upon empirical calibrations derived from several well-known relations between observable quantities of CVs (*e.g.*, parameters of emission line profiles, velocity width at half or mean intensity, rms line widths and velocity separation of doubled emission peaks), and the underlying physical parameters of the CV binaries. The observable quantities were first calibrated against a set of well observed doubled-emission lined CV and Algol binaries from which mass ratios were determined. Using these results, Webbink (1990a) calibrated an empirical mass-radius relation for the CV secondaries, from which the white dwarf masses were estimated. Webbink's data set is used in the analysis below. Nine of the mass values listed in Table 1 (for VV Pup, Z Cha, V834 Cen, AM Her, TT Ari, V1500 Cyg, H0139, MV Lyr, and CW1103) are designated as most reliable by Webbink (1990) because they depend only weakly on any assumptions inherent in their solutions.

For each adopted value of mass, a radius was obtained graphically from the zero temperature model grid of Hamada and Salpeter (1961). The luminosity was calculated using the T_{eff} values and the zero temperature radii corresponding to the masses listed in Table 1. Since 90% of the CVWDS in Table 1 have log $T_{eff} < 4.4$, the departures from the zero temperature radii should amount to only a few per cent. Note however that small departures from the Hamada-Salpeter relation due to *bloating* of the white dwarf in response to long term accretion (regardless of the details of the accretion process) have been quantitatively determined from long term quasi-static model sequences with radial accretion. This long term adjustment (response) of the white dwarf due to structure changes and compressional heating associated with the accretion process should be the dominant source of time-averaged surface luminosity of a CV white dwarf between classical nova outbursts (Sion 1985). It is expressed by

$$L_{g,acc} = f[\frac{GM_{wd}\dot{M}}{R_{hs}}) - (\frac{GM_{wd}\dot{M}}{R_{wd}})] \tag{1}$$

where f is that fraction of the total accretion energy which is liberated by the white dwarf in response to long term accretion, regardless of the details of the accretion process, M_{wd} is the white dwarf mass, R_{wd} is its radius and R_{hs} is the Hamada-Salpeter zero temperature radius. The coefficient f, evaluated from long term model sequences with accretion, lies in the interval 0.15 to 0.25, over a range of accretion rates and white dwarf masses (Iben 1982).

All but five of the luminosities fall in the range $-3.98 < \text{Log} (L/L_{\odot}) < -1.5$. The highest luminosity case is the magnetic CVWD in V1500 Cygni and the lowest luminosity

Table 1
Effective Temperatures, Masses and Luminosities

#	CV Name	T_{eff} (K)	$M_{wd}(M_{\odot})$	Log $(L/L_{\odot})$
1	VV Puppis	9000	1.29	-3.98
2	HT Cas	12000	0.76	-2.69
3	Z Cha	13000	0.84	-2.38
4	OY Car	15000	1.24	-2.94
5	U Gem	30000	1.12	-1.45
6	MV Lyr	50000	0.76	-0.21
7	AM Her	20000	0.86	-2.90
8	TT Ari	50000	0.98	-0.44
9	VW Hyi	18000	0.23	-2.49
10	WZ Sge	12500	0.93	-2.67
11	CW1103	11000	0.43	-2.53
12	V794 Aql	50000	0.98	-0.44
13	H0139-68	20000	0.28	-1.38
14	V834 Cen	15000	0.65	-2.18
15	UZ For	<20000	1.33	-2.64
16	V1500 Cyg	110000	1.28	—

case is the CVWD in VV Puppis. A discussion of these luminosities in terms of cooling ages is complicated by at least three factors: (a) part of the white dwarf surface luminosity results from heating due to the accretion process, then deriving a cooling age for a CV white dwarf from cooling sequences of single non-accreting degenerates is strictly a lower limit to the white dwarf's lifetime; (b) the cooling age of a CV degenerate must include the uncertain time during which the white dwarf was cooling in the pre-cataclysmic detached binary phase prior to the onset of Roche lobe overflow; (c) the mass of a CV white dwarf may decrease with time due to erosion of its mass by repeated nova episodes. The first two complications muddy the derivation of a lower limit cooling age from a CVWD T_{eff} but factor (c) casts doubt even upon a lower limit and is the least predictable, as discussed below.

3.2 DO CVWDs UNDERGO ACCELERATED COOLING DUE TO CORE MASS EROSION CAUSED BY MULTIPLE NOVA EXPLOSIONS?

Any interpretation of the distribution of CVWD effective temperatures may be complicated by the possibility that a CVWD undergoes accelerated cooling due the gradual erosion over time of its core mass by successive nova explosions, as suggested by Webbink (1990b). If the total white dwarf mass decreases with time, it responds by increasing its radius, causing the non-degenerate ion gas in the core to expand and cool adiabatically. Thus, when a higher mass CVWD is eventually *whittled down* to a lower mass degenerate, it could appear considerably older (cooler) than its true chronological age. The observational evidence for core mass erosion is primarily derived from the heavy element abundances in nova ejecta where the degree of core dredgeup is estimated by abundance analyses of nova ejecta (cf. Truran and Livio 1986). There are estimates that 25% of the ejecta are heavy elements that could not be produced in the thermonuclear runaway and must have been dredged up from the core. The specific dredgeup mechanism is not known. Note however that even if core mass erosion occurs, the cooling effect could be offset by re-heating of the upper core due to the nova outbursts. In the absence of detailed long term calculations and the unknown nature of the dredgeup mechanism, this effect cannot be predicted at present. There is one theoretical calculation of a full nova cycle by Prialnik (1986) which does demonstrate that the white dwarf mass has been eroded after two nova explosions. But the amount of mass lost is small; after 10^4 nova explosions, the total amount of erosion is only 0.015 $M_\odot$, according to Prialnik (1986). In the following discussion we assume that the white dwarf mass remains constant during its long term evolution.

3.3 LOWER LIMIT COOLING AGES

Let us assume that the CVWD mass erosion due to nova outbursts is insignificant, as indicated by Prialnik's (1986) models. According to the white dwarf cooling calculations of Winget, Lamb and Van Horn (1990), the cooling time to Log $(L/L_\odot)$ = -1.5 is 7.5 x 10^7 years while the cooling time to Log $(L/L_\odot)$ = -3.0 (*e.g.* OY Car) is 1.0 x 10^9 years, if these CV degenerates underwent core cooling like single non-accreting degenerates. Since the T_{eff} values are strict upper limits, these cooling ages provide lower limits to the lifetime of a CV. Over the luminosity range indicated in Table 2, and if, at CV birth, the white dwarf had Log T_{eff} = 4.4, then the lower limit evolutionary timescale for a CV is 9.25 x 10^8 years. Note that by these same arguments, the magnetic degenerate in VV Puppis may have a lower limit evolutionary lifetime of approx. 5.0 x 10^9 years, suggesting that something may be amiss with this magnetic CV system.

In Figure 1, the CVWDS are shown plotted, along with the cooling tracks of Winget, Lamb and Van Horn (1990), in an H-R diagram. The cooling tracks represent masses of 0.4, 0.6, 0.8 and 1.0 $M_\odot$. Isochrones (the dashed lines) are labelled in units of 1.0 x 10^8 years and the boundary for onset of core crystallization ($T_c < T_{debye}$) is the solid line running diagonally from upper left to lower right. The absence of CVWDs with Log $(L/L_\odot)$ < -4.0, is most likely due to the observational difficulty of detecting such objects since their accretion rates would be extremely low ($\dot{M} < 1$ x 10^{-11} $M_\odot$/yr). The sparsity of very hot (log $T_{eff} > 4.4$) CVWDS is most likely due to either (1) the fact that young CVWDS, simply born hot and remaining so at the onset of mass transfer, undergo rapid (mostly neutrino) cooling down to Log $(L/L_\odot)$ = -1, and thus evade observational detection due to their short-lived cooling phase or (2) the few (four) CV degenerates hotter than log T_{eff} = 4.4, have been heated down to large envelope depth by high rates of accretion or a recent thermonuclear runaway and their envelopes are still cooling down on their respective thermal timescales. Three of the four objects are UX UMa nova-like systems and one object, the magnetic CVWD in V1500 Cygni, is still undergoing cooling to recover its long

422

term pre-nova photospheric temperature (see Sec. 4.2 below).

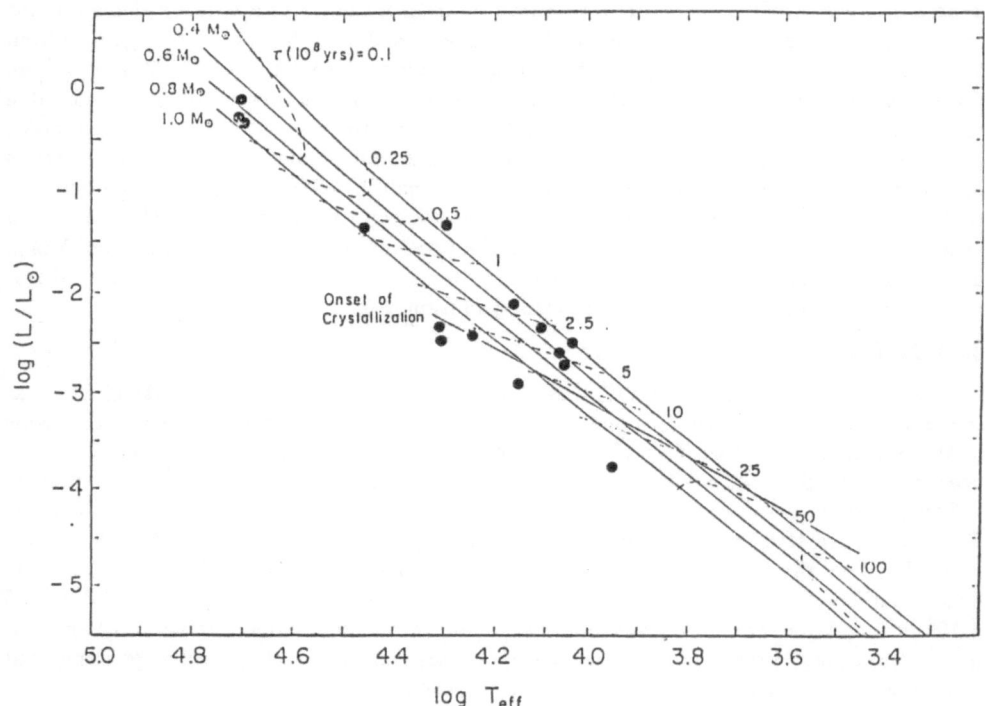

Fig. 1: The CVWDS (filled circles) shown plotted, along with the cooling tracks of Winget, Lamb and Van Horn (1990), in an H-R diagram. The cooling tracks represent masses of 0.4, 0.6, 0.8 and 1.0 $M_\odot$. Isochrones (the dashed lines) are labelled in units of 1.0 x 10^8 years and the boundary for onset of core crystallization ($T_c < T_{debye}$) is the solid line running diagonally from upper left to lower right.

4. The Observed Distribution of CVWD Effective Temperatures as a Function of Orbital Period: A Comparison Between Theory and Observation

If there is a true correlation between mass transfer rate in CVs and their orbital period, then it is not unreasonable to expect that this relationship will be manifested at some level in the distribution function of white dwarf T_{eff} since some fraction of the total accretion energy goes into heating the underlying degenerate star, whether by the disk boundary layer, a dwarf accretion event, or a polar accretion column. The widely held assumptions underlying a correlation between mass transfer rate and CV orbital period are: (1) mass transfer above the top of the CV period gap (*i.e.* above P_{orb} = 2.8 hours) is driven by angular momentum loss due to the braking by the magnetic stellar wind of the Roche lobe-filling secondary star, (2) that the period gap itself results when the secondary star, of diminishing mass, becomes fully convective, which either suppresses the magnetic stellar wind or destroys (or rearranges) the field and (3) that following a detached state of no mass transfer in the period gap, increasing angular momentum losses due to gravitational wave radiation, shrinks the orbit to the extent that Roche lobe contact is re-established by the secondary at the lower edge of the gap (P_{orb} = 2.1 hours), and mass transfer resumes on a longer time scale, driven by gravitational radiation. The quantitative aspects of this scenario have been explored with detailed model calculations by McDermott and Taam (1989), whose numerical results confirm the earlier ideas and model computations of Rappaport, Verbunt and Joss (1983), Spruit and Ritter (1983) and Mestel and Spruit (1987). This work reproduces the observed width and location of the gap, provided that the average mass transfer rate above the upper edge of the gap is $\dot{M} = 1.9 \times 10^{-9} M_{\odot}$/yr, and the average rate below the lower edge of the gap, is $\dot{M} = 1 \times 10^{-10} M_{\odot}$/yr. Their calculations constrain the mass of the secondary at P_{orb} = 2.8 hours to lie in the range $0.25 \ M_{\odot} < M_2 < 0.28 M_{\odot}$. Given the empirical success of the McDermott and Taam (1989) models, it is interesting to compare, as a further test of CV evolution, the observed distribution of CVWD T_{eff} with CV evolution theory.

The effective temperatures (including upper limits) for the CV white dwarfs determined to date, are plotted versus the orbital period of the CV and displayed in Figure 2. Below the period gap, there is a clear concentration of CVWDS at low temperatures with all of them having $T_{eff} < 25,000$K. In contrast, only four systems above the period gap are cooler than 25,000K while 10 systems above the gap contain white dwarfs with $T_{eff} > 30,000$K. The hottest CVWD, V1500 Cygni, lies at $T_{eff} = 110,000$K, above the gap, while the coolest CVWD, VV Pup, lies at 9000K, below the gap. Among the CV sub-types indicated in Figure 2, the the UX UMa and VY Scl-type nova-like, non-magnetic CVs (and the classical nova V1500 Cygni) reveal the hottest degenerates while the AM Her magnetic CVs reveal a tendency to have the coolest degenerates. The former systems tend to have thick disks and winds in their *high states* and are associated with higher mass transfer rates (Warner 1971; Sion and Guinan 1982; Guinan and Sion 1982) while the latter are accreting radially only at their magnetic poles but are associated with lower mass transfer rates, of the same order roughly as low accretion rate dwarf novae in quiescence.

4.1 ARE THE LOWER LIMIT COOLING TIMES OF CV DEGENERATES CONSISTENT WITH THEORETICALLY PREDICTED CV LIFETIMES FOR EVOLUTION THROUGH THE PERIOD GAP?

As a first exercise it is of interest to ask whether the theoretically predicted timescales for CV evolution are consistent with the distribution of T_{eff} in Figure 2, using the T_{eff} values (or upper limits thereof) as LOWER limit cooling times (the cooling time of the white dwarf since the birth of the CV, *i.e.* its combined age as a post-common envelope pre-CV + CV lifetime). This exercise implicitly assumes that: (a) the CVs represented in Figure 2 were born above the period gap; (b) subsequent evolution with decreasing orbital

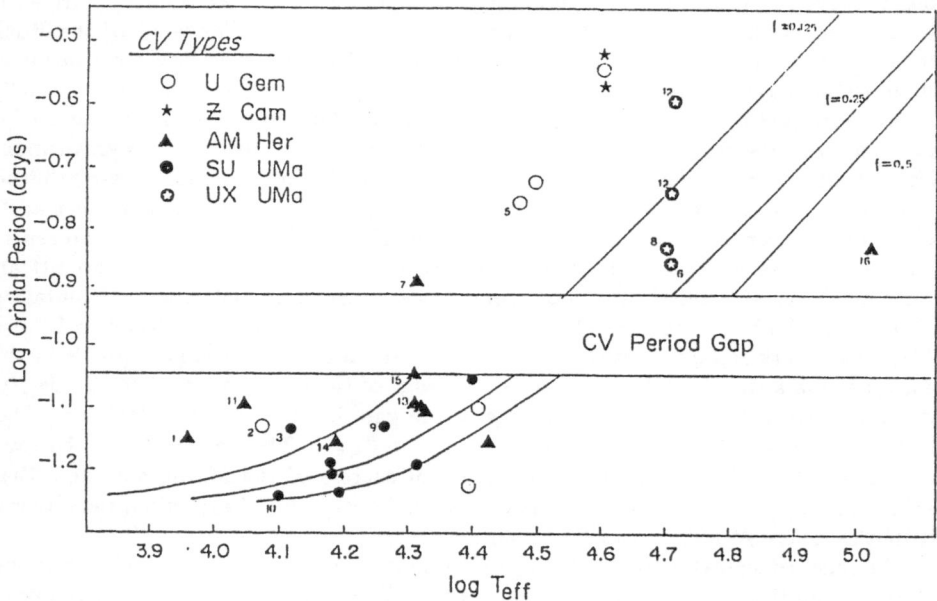

Fig. 2: The CVWDs (labelled by CV subtype with a filled triangle for AM Her systems, a filled circle for SU UMa systems, an open circle for U Gem systems, a filled star for Z Cam systems, and a circle with an embedded open star for UX UMa systems) are shown in the (Log P_{orb}-Log T_{eff}) plane with the boundaries of the CV period gap indicated by the horizontal lines. The numbers next to individual CV degenerates indicate stars with the most accurate values of T_{eff} and correspond to the entry numbers in Table 1. Plotted for comparison with the observed CVWD T_{eff} distribution is the evolution of T_{eff} versus orbital period based upon the correlation between mass transfer rate and orbital period calculated by McDermott and Taam (1989), where the predicted $\dot{M}$ has been converted to T_{eff} for three fractions of the total accretion energy; $f = 0.5$, 0.25, and 0.125.

period has carried them below the gap and; (c) during the long term evolution of the CV, the white dwarf mass has remained constant.

First, the timescale for a CV with initial $P_{orb} = 4$ hours to evolve to the upper edge of the gap is 1.7×10^8 years while the timescale during the detached state in the gap is 2×10^8 years, giving a total timescale to the bottom edge of the gap of 3.7×10^8 years. Then gravitational radiation losses take over, yielding a timescale of a few times 10×10^9 years. However the minimum P_{orb} for a hydrogen CV has been shown by Pacynzski and Sienkiewicz (1981) to be 80 minutes and indeed the dwarf nova WZ Sge is very close (81 min.) to this period minimum. The timescale of interest is that required by a CV to evolve from $P_{orb} = 2.1$ hours (lower edge of gap) to the minimum $P_{orb} = 80$ minutes. I find graphically from the plots of Rappaport, Verbunt and Joss (1983) that this timescale is roughly 1.2×10^9 years.

Returning to Figure 2, these timescales imply that above the gap, if the average accretion rate is 1.91×10^{-9} $M_{\odot}$/yr, the amount of matter accreted by the white dwarf is 0.32 $M_{\odot}$ (!) before it reaches the top of the gap (already in excess of the range of secondary masses predicted by McDermott and Taam 1989). Below the gap, if the average accretion rate is 10^{10} $M_{\odot}$/yr, then one can compute how much mass has been accreted by the white dwarf if one knows the cooling age (lower limit) of the white dwarf. For a typical cool CVWD with $L = 1 \times 10^{-3}$ $L_{\odot}$, a lower limit cooling age of 10×10^9 years is implied, not inconsistent with the timescales asociated with the McDermott and Taam (1989) CV evolutionary calculations.

If the CVWD for example began its CV evolution above the gap (with $P_{orb} = 4$ hours initially) then it has been below the gap for 10×10^9 - 3.7×10^8 years or 6.3×10^8 years. During this time below the gap it would have accreted 0.06 $M_{\odot}$, enough for 300 nova explosions (!), based upon a critical envelope mass of 2×10^{-4} $M_{\odot}$ for a thermonuclear runaway on a 0.8 $M_{\odot}$ white dwarf. If one assumes that the heating of the VV Puppis magnetic degenerate by the accretion process (including nova outbursts) has not retarded the cooling of its core (or core mass erosion has not accelerated its cooling clock) in any way, then its cooling age by reference to the cooling curves of Winget, Lamb and Van Horn (1990), is 2-5×10^9 years, depending upon its mass. This cooling age is not consistent with the McDermott and Taam (1989) evolutionary timescales, for evolution from above the gap to below the gap.

4.2 HOW CONSISTENT IS THE OBSERVED DISTRIBUTION CV WHITE DWARF EFFECTIVE TEMPERATURES WITH THE THEORETICAL MASS TRANSFER RATE VERSUS ORBITAL PERIOD RELATION?

Suppose we adopt the theoretical mass transfer rate versus orbital period relation, as determined by the the model calculations of McDermott and Taam (1989). For each orbital period we calculate the mass transfer rate for all systems with $P > 2.8$ hours by

$$\dot{M} = \frac{1.17 \times 10^{-9} M_{\odot} yr^{-1}}{[P_{orb}/(3hr)]^{3.7}} \tag{2}$$

then the total accretion energy (luminosity) by

$$L_{acc} = \frac{fGM_{wd}\dot{M}}{R_{wd}} \tag{3}$$

at the predicted $\dot{M}$ for each orbital period, and compute the effective temperature of the white dwarf from the expression

$$T_{eff} = (\frac{L_{acc}}{4\pi R_{wd}^2 \sigma})^{0.25} \tag{4}$$

for fractions f of the total accretion energy, $f = 0.5, 0.25$, and 0.125. Then for a 0.8 $M_{\odot}$ white dwarf at the Hamada-Salpeter (1961) zero-temperature radius, in a CV of initial period 4 hours (the initial parameters of McDermott and Taam), the white dwarf should have a surface temperature (between nova explosions) which changes with time along the paths indicated in Figure 2 for the three different fractions of the total accretion energy. These tracks are compared to the actual (observed) distribution of effective temperatures (or upper limits) as shown in Figure 2.

Above the period gap the hottest CV degenerate, V1500 Cygni with $T_{eff} = 110,000K$, is still cooling from its recent explosion and will be sliding to the left in Figure 2 on the thermal timescale of its remaining envelope, to approximately the location of AM Her, which is the coolest object above the gap. This interpretation is supported by the fact that V1500 Cygni appears to contain a massive white dwarf, ($M_{wd} > 0.9$; Horne and Schneider 1989) and falls almost precisely on the theretical light curve for a 1.25 $M_{\odot}$ degenerate (see Figure 8 in Prialnik 1986) 15 years after its 1975 explosion. It follows that the restoration of thermal equilibrium in the V1500 Cygni white dwarf, according to the full nova cycle computed by Prialnik (1986), should be achieved at roughly 20,000K – 30,000K. Two other CV degenerates, the UX UMa systems MV Lyrae and TT Ari fall close to the MT tracks for a 0.8 $M_{\odot}$ white dwarf whose surface luminosity a fourth to a half of the total accretion energy, a required fraction too large to be consistent with the temperatures of the majority of CV degenerates (cf. Sion 1985).

It is possible that MV Lyrae and TT Ari may themselves be still recovering from *recent* nova explosions and their surface luminosities have yet to reach their inter-outburst thermal equilibrium values. However, the majority of CV white dwarfs above the gap seem to fall well to the left of the MT relation, and since their temperatures are upper limits, they are likely to deviate still further to the left. Despite the small number of data points and the observational uncertainties, it is worth noting that the rough slope defined by distribution of these objects appears to be close to the MT slope above the gap but appears shifted significantly to the left relative to the MT curves for a 0.8 $M_{\odot}$ degenerate. This would suggest three possibilities: (1) the masses of the CV degenerates are lower than 0.8 $M_{\odot}$ (*e.g.* 0.6 $M_{\odot}$) since the MT cooling paths would shift to the left in better agreement with the temperature observations or; (2) the CV degenerates have cooled substantially prior to onset of mass transfer above the period gap and their subsequent evolution of surface temperature versus time essentially follows that of a single non-accreting degenerate despite occasional excursions to the right in Figure 2 due to (upper?) envelope heating events/processes induced by accretion. It also cannot be entirely ruled out the CV degenerates above the gap were all very cool (1×10^4 K) at the onset of actual CV evolution (due to a long pre-CV waiting period) but subsequent envelope heating to great depths by the accretion process (*e.g.* downward heat diffusion due to spherical accretion or shear mixing) with envelope thermal timescales of thousands of years has produced the snapshot of the CV white dwarf T_{eff} distribution shown in Figure 2. Below the gap there are observations of white dwarf cooling following dwarf nova events which indicate shallow heating depths in the white dwraf envelope (cf. Sion 1987 and references therein; Pringle 1988; Verbunt 1987; Sion and Szkody 1990) but no such clear detections of white dwarf cooling have been made in systems above the period gap.

Below the gap, where mass transfer is presumably driven by gravitational wave radiation and the accretion rates are correspondingly smaller, the distribution of CV degenerate temperatures is quite different. All of their surface temperatures are below 25,000K. It is curious that if these objects were born below the gap, as opposed to evolving from systems above the gap, none of them appear to contain (with the possible exception of the two known classical novae below the gap) white dwarfs within 50 million years (the cooling age of a non-accreting white dwarf to $T_{eff} = 25,000K$) of its emergence from a common envelope, at which time most such white dwarfs are hotter than 25,000K. In Figure 2 reasonable agreement is seen between the observed distribution of T_{eff} and the accretion luminosity predicted for mass transfer rates driven by gravitational wave losses. Where are the hot white dwarf counterparts to those above the gap if CV systems are born below the gap? A preliminary conclusion would be that they are cooler because they have evolved from younger CV systems with higher mass transfer rates above the gap. This could therefore be taken as evidence (though far from conclusive evidence) that CV systems do indeed evolve across the gap.

4.3 DOES THE LONG TERM COOLING OF AM HERCULIS MAGNETIC DEGENERATES DIFFER FROM OTHER TYPES OF CVs?

One final point about Figure 2 is worth noting: the several AM Herculis systems are known to be radially accreting via magnetically channeled accretion columns and the depth of downward heat diffusion may be quite shallow given how quickly the accretion polar caps appear to cool. Of all the CV systems these magnetic degenerate objects would be expected to cool in a manner most similar to single white dwarfs, in a time-averaged sense, given recovery from periodic nova explosions back to the quiescent, long, inter-outburst timescales. In Figure 2 if one allows for the cooling of V1500 Cygni to the vicinity of AM Her, above the gap, and if one allows for the T_{eff} values of UZ For (located at the gap's lower boundary) and V834 Cen to be overestimated (they are upper limits), then the AM Herculis white dwarf primaries may be indicating a sequence of cooling, together with VV Puppis and CW1103, which is independent of the MT cooling paths. This is further suggested by the fact that VV Puppis, UZ For and V1500 Cygni are all thought to have high white dwarf mass (Webbink 1990; Horne and Schneider 1990). By comparison with the CV evolutionary tracks of MT and the empirical orbital period-average mass transfer rate relation of Patterson (1984), the AM Herculis systems for the most part seem to lie suspiciously far to the left of the tracks for a 0.8 $M_{\odot}$ white dwarf primary. This could be suggestive of their cooling evolution occuring with accretion NOT being a factor in the time-averaged evolution of their surface luminosity as the system age increases.

5. Interpretation of the Metallic Line Spectrum Associated With The White Dwarf in the Ultra-Short Period Dwarf Nova WZ Sagittae During Quiescence

Sion, Leckenby, and Szkody (1990) have recently reported the identification of strong neutral carbon absorption features in the photosphere of the white dwarf exposed during the quiescent state of the ultra-short period dwarf nova WZ Sagittae. The strong features appearing at 1657 Å, 1463 Å, 1561 Å, and 1431 Å, are identified as C I resonance absorption features which appear in all six individual spectra. On the basis of its carbon in a hydrogen-dominated photosphere with Si II also present, they assigned the white dwarf in WZ Sge the spectroscopic classification DAQZ5. The presence of carbon in single cool helium-rich degenerates is strongly associated with convective dredgeup of core carbon by a deepening helium convection zone as cooling proceeds.

The presence of carbon in a DA star is extremely rare with only one confirmed case, that being the single DAQZ star G35-26, which has $T_{eff} > 15,000$ K. Assuming these identifications are correct, for the DAQZ star in WZ Sge, the presence of the metals in the photosphere must be strongly associated with accretion and possibly subsequent convective dilution. If the metals were accreted or dredged-up as a consequence of the 1978 accretion event or subsequent accretion at a low rate, they have not had sufficient time in the 12 years of quiescence to diffuse out the bottom of the putative convective zone. Since the mass-transferring companion in WZ Sge has been shown to be sub-stellar in mass, it is unlikely that the accretion flow is enhanced in Triple α processed material. It is therefore more likely that the carbon was brought to the photosphere by dredgeup associated with or induced by accretion. It is tempting to speculate that the mechanism responsible for the carbon could be the same as (or be related to) the way in which the required carbon enhancements are produced to power the classical nova outburst. Further quantitative analysis of this unique CV white dwarf and of other CV degenerates that reveal metal/helium line spectra during dwarf nova quiescence (*e.g.* U Gem) or low brightness states (*e.g.* TT Ari) should have the highest priority.

I am grateful for the financial support of this research by the Centre National De La Recherche Scientifique URA285, the U.S. National Science Foundation grant AST88-02689 and the NASA grant NAG5-1284.

6. References

Beuermann, K., Thomas, H-C., and Schwope, A. 1988, Astr.Ap. **195**, L15.

Guinan, E.F., and Sion, E.M. 1982, Ap.J. **258**, 217.

Hessman, F.V., Koester, D., Schoembs, R., and Barwig, H. 1989, Astr.Ap., **213**, 167.

Holm, A. 1988, A Decade of Ultraviolet Astronomy with the IUE Satellite, Vol. 1 (ESA SP-281),p. 229.

Horne, K., and Schneider, D. 1989, Ap.J. **343**, 888.

Iben, I. 1982, 1982, Ap.J **259**, 244.

Kiplinger, A., Sion, E.M., and Szkody, P.1991, Ap.J. **366**, 569.

Kutter, S.G., and Sparks, W. 1989, Ap.J. **340**, 985.

LaDous, C. 1990, private communication.

Mc Dermott, P.N., and Taam, R.E. 1989, Ap.J. **342**, 1019.

Mestel, L., and Spruit, R. 1987, MNRAS, **226**, 57.

Patterson, J.1984, Ap.J. Suppl. **54**, 443.

Prialnik, D. 1986, Ap.J. **310**, 222.

Pringle, J. 1988, MNRAS, **230**, 587.

Rappaport, S., Verbunt, F., and Joss, P. 1983, Ap.J. **275**, 713.

Regev, O., and Shara, M.M. 1989, Ap.J. **340**, 1006.

Ritter, H.1990, Astr.Ap.Suppl., in press, (5th ed. CV Catalog).

Shafter, A., Szkody, P., Liebert, J., Penning, W.R., Bond, H.E. and Grauer, A.D. 1985, Ap.J. **290**, 707.

Shaviv, G., and Starrfield, S.G. 1987, Ap.J. **321**, L51.

Sion, E.M. 1986, PASP **98**, 821.

Sion, E.M. 1987, Proceedings of the Second Conference on Faint Blue Stars, IAU Colloquium No. **95**, ed. A.G.D. Philip, J. Liebert, and D. Hayes, (Schenectady:

L.Davis Press), p. 413.

Sion, E.M. 1985, Ap.J. **297**, 538.

Sion, E.M., and Starrfield, S.G. 1986, Ap.J. **303**, 186.

Sion, E.M., and Szkody, P. 1991, Proceedings of IAU Colloquium No.122.

Sion, E.M., and Guinan, E.F. 1983, in Cataclysmic Variables and related Objects, IAU Colloquium No. 72, ed. M. Livio and G. Shaviv (Reidel: Dordrecht), p.414

Sion, E.M., Leckenby, H., and Szkody, P.1990, Ap.J. (Letters) **364**, L41.

Smak, J.1984, Acta Astr. **34**, 317.

Sparks, W., Sion, E.M., Kutter, S.G., and Starrfield, S.G. 1990, BAAS **22**, 843.

Spruit, F., and Ritter, H. 1983, Astr.Ap. **124**, 267.

Stockman, H.S., Schmidt, G.D., and Lamb, D.Q. 1988, Ap.J. **332**, 282.

Szkody, P., and Sion, E.M. 1990, Proceedings of IAU Colloquium No.122, The Physics of Classical Novae, ed. A. Cassatella, (Kluwer: Dordrecht), in press.

Szkody, P., Downes, R., and Mateo, M. 1988, PASP **100**, 362.

Truran, J., and Livio, M. 1986, Ap.J. **308**, 721.

Verbunt, F. 1987, Astr. Ap. Suppl. **71**, 330.

Warner, B. 1976, in IAU Colloquium No.73, Structure and Evolution of Close Binary Systems, ed. P.E. Eggleton, S. Mitton and J. Whelan (Dordrecht: Reidel), pp. 85-140.

Webbink, R.E. 1990a, in Accretion-Powered Compact Binaries, ed. C. Mauche, (Cambridge: Cambridge Univ. Press), p.177.

Webbink, R.E. 1990b, private communication.

Winget, D., Lamb, D.Q. and Van Horn, H.M. 1990, private communication.

DISCUSSION

SHIPMAN :

In two spectra of U Gem and WZ Sge there are indications of heavy element lines, namely silicon and carbon. (My chemistry students would be very upset if I called carbon a metal). Have you checked out the abundances of these elements, which I would expect to be solar?

SION :

Without published model below $T_{eff} = 15000K$ it is difficult to comment. The SiII λ 1260 line in WZ Sge is in a noisy region of declining flux and SiII λ 1530 is not included in the Henry, Shipman and Wesemael (1984) tables of UV metal line abundances.

WHOLE EARTH TELESCOPE OBSERVATIONS OF THE INTERACTING WHITE DWARF BINARY SYSTEM AM CVn: FIRST RESULTS

J.-E. SOLHEIM and P.-I. EMANUELSEN
Institute of Mathematical and Physical Sciences
University of Tromsø, N-9000 Tromsø, Norway

G. VAUCLAIR and N. DOLEZ
Observatioire Midi-Pyrenees
14 Avenue E. Berlin, F-31400 Toulouse, France

M. CHEVRETON
Observatoire de Paris-Meudon
F-92195 Meudon Principal Cedex, France

M. BARSTOW, A.E. SANSOM and R.W. TWEEDY
Department of Physics and Astronomy
University of Leicester, LE1 7RH, U.K

S.O. KEPLER and A. KANAAN
Instituto de Fisica
Universidade Federal do Rio Grande do Sul
91500 Porto Alegre - RS, Brazil

G. FONTAINE and P. BERGERON
Départment de Physique, Université de Montréal
C.P 6128, Succ A., Montréal, PQ H3C 3J7, Canada

A.D. GRAUER
Department of Physics and Astronomy
University of Arkansas, 2801 S.University Ave
Little Rock, AR 72204, U.S.A

J.L. PROVENCAL, D.E. WINGET, R.E. NATHER,
P.A. BRADLEY, C.F. CLAVER, J.C. CLEMENS,
and S.J. KLEINMAN
McDonald Observatory and Dept. of Astronomy
The University of Texas at Austin, RLM 15.308
Austin, Tx 78712, U.S.A.

B.P. HINE
NASA Ames Research Center
M.S 244-4, Moffett Field, Ca 94035, U.S.A

T.M.K. MARAR, S. SEETHA and B.N. ASHOKA
Technical Physics Division, ISRO Satellite Centre
Airport Rd, Bangalore, 560 017 India

E.M. LEIBOWITZ and T. MAZEH
Wise Observatory, The Sackler Faculty
of Exact Sciences, Tel Aviv University
Tel Aviv 69978, Israel

ABSTRACT. We report the first results of the Whole Earth Telescope observations of AM CVn in March/April 1990. The Fourier Spectrum of the light curve shows harmonically related peaks. High frequency sidebands with the fine-splitting of 21 μHz are observed for the fundamental period of 1051 s and its 4 lowest harmonics. These have not been observed before. The fundamental period itself is not detected.

431

G. Vauclair and E. Sion (eds.), White Dwarfs, 431–439.
© 1991 *Kluwer Academic Publishers.*

1.Introduction

1.1 THE WHOLE EARTH TELESCOPE.

The Whole Earth Telescope (WET) is a multi-mirror ground-based telescope for time-series photometry of rapid variable stars. It is designed to minimize or eliminate gaps in the brightness record caused by the rotation of the Earth (Nather *et al.* 1990). The telescope makes use of a sequence of existing telescopes distributed in longitude, coordinated from a single control center. Data are returned to the control center by electronic mail and analyzed in real time. This way it is possible to provide data of continuity and quality that permit true high resolution power spectroscopy. A review of some of the results obtained so far with WET is given by D. Winget (1991).

1.2 AM CVn BEFORE THE WET RUN.

AM CVn (=HZ 29) is believed to be an interacting binary white dwarf system (IBWD), where the mass losing companion is a low mass white dwarf star with a helium atmosphere. The system shows a spectrum with broad, shallow absorption lines of neutral helium. The continuum spectrum has a slope like an optically dense disk, and the UV shows narrow blue-shifted absorption lines of NV, SiIV, CIV and HeII from a wind blowing off the disk. The shapes of the line-profiles indicate a disk-inclination of less than 30 degrees (Solheim and Kjeldseth-Moe, 1987). The light curve shows considerable flickering which is a sign of mass transfer. Figure 1 shows a portion of the lightcurve observed with a large telescope.

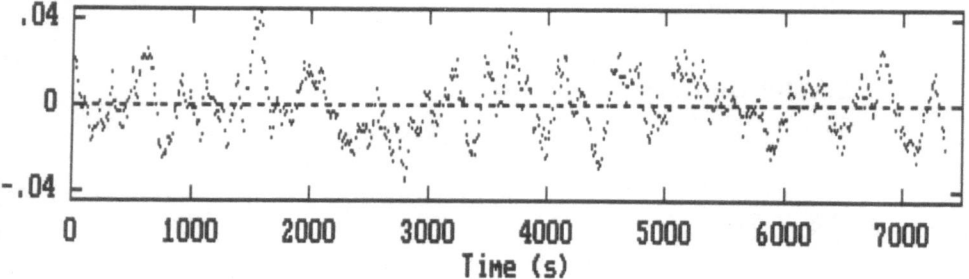

Figure 1: Part of the light curve of AM CVn observed with the CFHT 3.6m telescope during the WET run. The ordinate shows the flux relative to a mean value.

A photometric period of 1051 s, with amplitude 1-2 per cent has been interpreted as the orbital period in a close binary system. The light modulation should then come from the changing aspects of a bright spot created by mass transfer (Warner and Robinson, 1972). A model (the IBWD model) where both stars are white dwarf stars, representing the final stage of binary star evolution, was proposed for the system by Faulkner, Flannery and Warner (1972). The light curve is peculiar in the sense that it sometimes shows one hump per period and sometimes two humps per period. The arrival times of minima show a considerable lack of precision of up to ±0.2 periods variations.

A careful study of the minima in the lightcurve observed since 1962, avoiding cycle count errors, showed that the period changed with a rate of $\dot{P} = (-3.2 \pm 0.6) \times 10^{-12}\,\mathrm{s\,s^{-1}}$ (Solheim *et al.* 1984), and the observed period of 1051 s was interpreted as the period of rotation for a magnetized white dwarf accreting mass from the disk and spinning up. The mass accretion rate calculated from the spinup was of the order 10^{-10} solar masses per year. At the same time it was proposed that the usually double humped light curve might be single humped with a period of 525.5 s, since it was not possible to detect any power in the power spectrum at the the period of 1051 s.

A new period of 1011.4 s was discovered in data obtained in 1978 (Fig 2). It was suggested that this period might be related to the orbital period of the system (Solheim *et al.* 1984), but due to aliasing in the power spectrum, it is not possible to investigate the relation between this period and the possible 1051 s period in any detail in data from one observing site.

Kepler (1984, *private communication*) showed that in data from 1976 and 1978 there was also a period of 350.4 s, exactly 1/3 of the possible 1051 s period. This is seen in figure 2. The system was also observed in 1987. Figure 2 shows an amplitude spectrum with a peak at 1011 s, in addition to the 525 s period and related harmonics, from that run (Emanuelsen 1990).

1.3 OBSERVATIONS OF SIMILAR OBJECTS

Observations of two other IBWDs have shown up to 4 harmonic periods in V803 Cen (O'Donoghue *et al.* 1987) and up to 13 harmonic periods in PG1346+082 (Provencal *et al.* 1989). In both cases the fundamental period is observed and is the strongest, which is not the case for AM CVn. V803 Cen and PG1346+082 vary between a low state at magnitude 17 to a high state at magnitude 13.5, while AM CVn is at a standstill at magnitude 14.2.

The other two systems have been observed in WET runs in 1988. In V803 Cen the harmonics were not coherent through the observing run, and disappeared in the low state (O'Donoghue *et al.* 1990). For PG1346+082 two periods, 1472 s and 1490 s with harmonics, have been investigated (Provencal *et al.* 1989). The pulse shape of the 1493 s period is a pure sinusoid. This period has non-coherent harmonics and is tentatively interpreted as related to the orbital or rotational period. The 1471 s period is non-sinusoidal with coherent lower harmonics throughout the run. This period is interpreted as a possible *g*-mode pulsation on the mass-accreting white dwarf star (Provencal *et al.* 1989, 1991).

1.4 JUSTIFICATION OF A WET RUN

Since the Whole Earth Telescope requires quite some effort to organize and run, only well justified cases can be incorporated into the observing program. In the case of AM CVn we found many reasons to use WET:

-The origin of the periods present in the FT: are they due to rotation, orbital motion or pulsations?

-The identification of possible additional periodicities.

-The existence or not of the 1051s period which may be the fundamental period for the harmonics observed.

-The coherency of the harmonics.

-The variations in the 1011s period and the possible non-linear interaction between the 1011 and the 525s periods.

-The resolution of the phase variations into its various components.

Finally the study of AM CVn is a valuable supplement to the study of the variable IBWDs where the variability limits the length of data strings during which the stars have constant

434

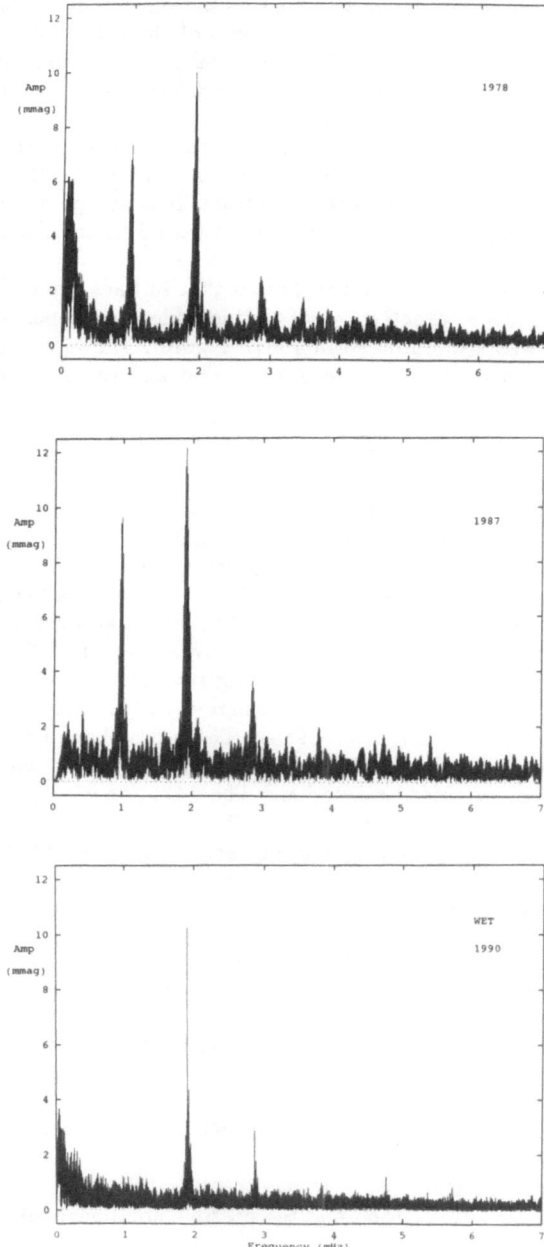

Figure 2, The amplitude-spectra of AM CVn for 1978, 1987 and 1990 in the frequency range 0-7 mHz. The two upper panels show single site data.

magnitude. Because it has a FT with fewer peaks, the signature of non radial g-mode pulsations might be easier to discover than in the complicated FT of PG1346+082 (Provencal *et al.* 1991).

2. Observations

2.1 THE WET RUN

AM CVn was scheduled as the primary object in the first part of the 4th WET-run which took place in March/April 1990. Six Northern hemisphere observatories participated. Five of them got 2 or 3 channel photometric data which were transferred to a command center in Austin, Texas to be reduced and analyzed. The total coverage added up to 141 hrs of photometry in 12 days, or about 50 percent duty cycle. In the maxium coverage part of the run, March 24-27, the duty cycle was 65 percent.

2.2 THE FUNDAMENTAL PERIOD AND ITS HARMONICS

Figure 2 shows the frequency spectrum we observed in 1990, compared with the 1978 and 1987 spectra. The strongest peak is the 525 s peak, with an amplitude about 11 mmags. The peak previously (1978 and 1987) seen at 1011 s is reduced to under 2 mmags amplitude. The FT spectrum is dominated by harmonically related peaks. The precise periods and amplitudes are determined by non linear least square fits and are shown in table 1. All the harmonics are related to a fundamental period P=1051.23 s. In our FT there is a peak at this period with an amplitude of 0.9 mmags, which is not significantly above the average noise in this part of the spectrum to be considered a detection.

TABLE 1. The harmonic peaks observed

	Period (s)	Amplitude(mmag)
1. harm.	$525.618 \pm .003$	$11.6 \pm .5$
2. harm.	$350.407 \pm .006$	$3.2 \pm .4$
3. harm.	$262.799 \pm .009$	$1.1 \pm .25$
4.harm.	$210.244 \pm .005$	$1.4 \pm .25$
5. harm.	$175.200 \pm .005$	$1.0 \pm .25$

The errors given for the amplitudes refer to the mean error
or average amplitude of the noise in the region of the peak.

If we generate an artificial light curve based on the amplitudes of the five harmonics observed, we get the curve displayed in Figure 4. We are able to reproduce the different depths of the minima and the variable shapes of the maxima, which have been reported by many observers, *without* a 1051 s period present. The real light curve (fig 1) shows in addition a considerable flickering. If the other strong period (1011 s) is present, the light

436

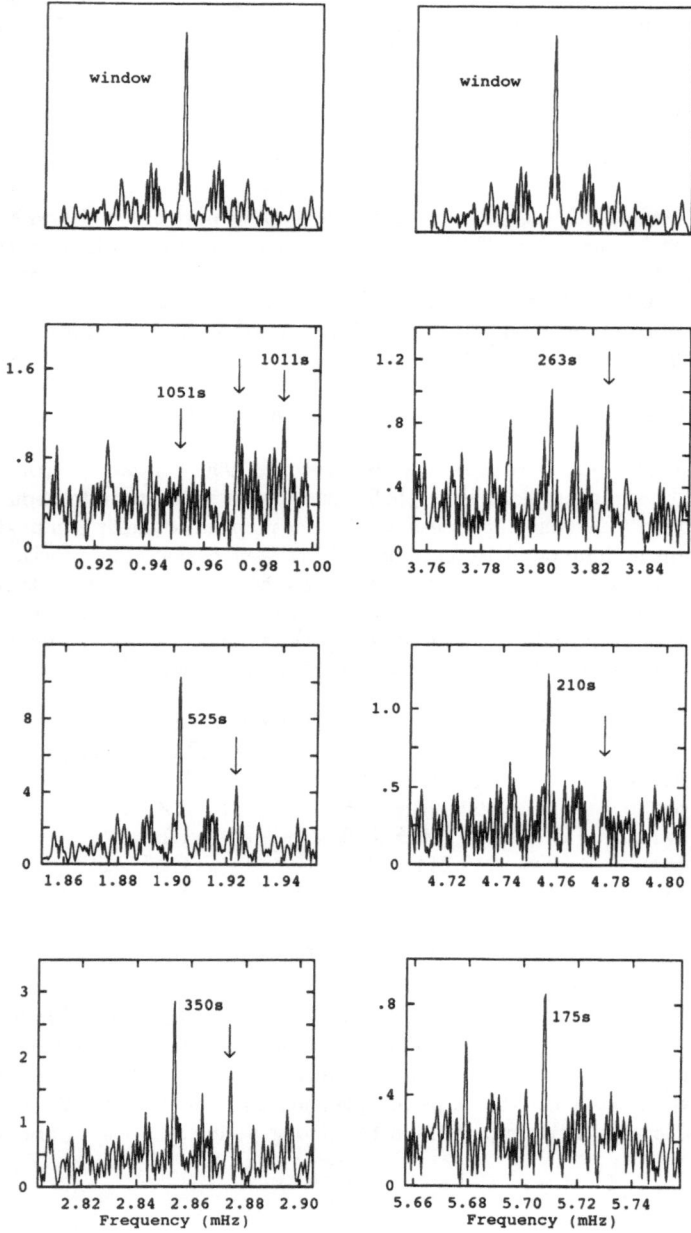

Figure 3: Parts of the amplitude spectrum near the possible fundamental period (1051 s) and its harmonics. The window function is shown twice in the two upper panels.The identified sidebands are marked with arrows without labels.

curve will change dramatically from night to night or within a few hours, because of the beat between the periods.

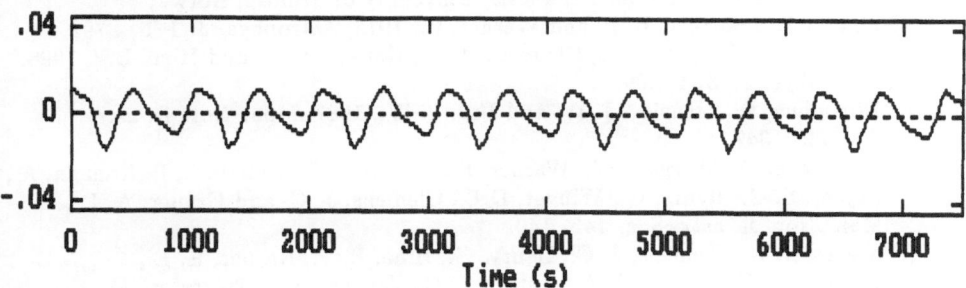

Figure 4: A light curve generated from the set of harmonics as observed in 1990. The ordinate shows the flux relative a mean value.

2.3 THE FINE-STRUCTURE OF THE HARMONICS

By comparing the fine-structure of the harmonic peaks with the window function (Fig 3) we find low amplitude sidebands at the positions of the possible fundamental period and its 4 lowest harmonics. The frequency spacing is the same for all of the periods: $+20.7$ μHz, and not increasing with the sequential number of the harmonics as we might expect. The sidebands are marked with small arrows in Figure 3. The presence of strong sidebands, and even missing fundamental periods, is observed in some of the intermediate polars (Warner 1986). However, sidebands of constant frequency difference to the harmonic peaks have not been reported before.

3. Conclusions

The WET has again produced a unique data set, consisting of 143 hrs of photometry, which has given new information about the IBWD system AM CVn. The FT of the lightcurve is dominated by 5 harmonic peaks, where the strongest has a period of 525.6 s. These peaks are present through the whole run. We show that the lightcurve can be simulated using this set of harmonic frequences, without the need of a fundamental period of 1051 s. A peak at the fundamental period is not detected. The previously strong peak at 1011 s has almost disappeared.

We have discovered constant frequency sidebands to 4 of the harmonic peaks and to the non-detected fundamental period. The sidebands have a constant frequency difference of 20.7 μHz. This is a new and unexpected feature.

The analysis of this unique data set will continue and, we hope, will lead us to an acceptable model for the system.

References

Emanuelsen, P.-I., 1990, Master thesis, University of Tromsø, Norway.

Faulkner, J., Flannery, B. P. and Warner, B., 1972, Astrophys. J. **175**, L79.

Nather, R. E., Winget, D. E, Clemens. J. C., Hansen, C. J. and Hine, B.P, 1990, Astrophys. J., **361**, 309.

O'Donoghue, D., Menzies, J. W. and Hill, P. W., 1987, Mon. Not. R. astr. Soc, **227**, 347.

O'Donoghue, D., Wargau, W., Warner, B., Kilkenny, D., Martinez, P. Kanaan, A., Kepler, S. O., Henry, G., Winget, D.E., Clemens, J. C. and Grauer, A. D., Mon. Not. R. astr. Soc, **245**, 140.

Provencal J. L., Clemens, J. C., Henry, G., Hine, B. P., Nather, R. E., Winget, D. E., Wood, M .A., Kepler, S. O, Vauclair, G., Chevreton, M., O'Donoghue, D., Warner, B., Grauer, A. and Ferrario, L., 1989, in "White Dwarfs", (G. Wegner, ed.) Lecture Notes in Pysics no 328, Springer-Verlag.

Provencal J. L., Clemens, J. C., Henry, G., Hine, B. P., Nather, R. E., Winget, D. E., Wood, M .A., Kepler, S. O, Vauclair, G., Chevreton, M., O'Donoghue, D., Warner, B., Grauer, A. and Ferrario, L., 1991, this Workshop.

Solheim, J.- E., Robinson, E. L., Nather, R. E. and Kepler, S. O., 1984, Astron. Astrophys.**135**, 1.

Solheim, J.- E. and Kjeldseth-Moe, O. 1987, Astrophys. and Space Sci. **131**, 785.

Warner, B. and Robinson, E. L., 1972, Mon. Not. R. astr. Soc. **159**, 101.

Warner, B. 1986, Mon. Not. R. astr. Soc. **219**, 347.

Winget, D. E., 1991, this Workshop.

DISCUSSION

SHIPMAN :

Do any of your explanations of this system account for the absence of a 1051 sec period?

SOLHEIM :

We are not yet sure about any explanations for the missing 1051s period in the FT. We think it means that we have 2 identical or nearly identical pulses per period -or we may have to accept a period of 525s.

BARSTOW :

Have you made any radial velocity measurements on this system in addition to the high speed photometry.

SOLHEIM :

Yes, at Lick Observatory with the Wampler scanner, and at INT with the IDS. The results show no detectable velocity variations with any of the periods known.

SHIPMAN :

When you talk about a "planet", do you mean a low-mass stellar core which is the remnant of a low mass-star which has been "turned inside out" by accretion?

WOOD :

You suggested the possibility that the mass losing object is a gas giant; however this does not seem reasonable to me because a gas giant should not form so close to a star, should not survive the star's red giant phase, and would in any event be below the peak in the mass-radius relation and so not persist in transfering mass. Finally, even if such a configuration did exist, wouldn't you expect there to be hydrogen and many other contaminents in the spectrum?

SOLHEIM :

One interpretation of the 20.7 microHz sidebands is that they show the period of the secondary beeing 13.4 hours. The density must be about $1g/cm^3$ -and its composition the final remains of a low mass star. However mass transfer is needed to keep the disk we observe in the system -so this interpretation is not likely. The outer hydrogen rich parts of such a star may have disappeared long time ago.

IUE OBSERVATIONS OF V803 Cen IN HIGH AND LOW STATES

A.M.Ulla, J.-E.Solheim.

Institute of Mathematical and Physical Sciences
University of Tromsø, 9000 Tromsø, Norway

ABSTRACT. We present here a brief analysis of all IUE spectra currently available of the *Interacting Binary White Dwarf* system V803 Cen (also called AE1). The spectra presented are low resolution, and a comparison between spectra corresponding to the high and low states of the object is made. All previous spectroscopy in the visible-range, published in the literature for this object is also reviewed.

1. INTRODUCTION

V803 Cen is a southern hemisphere, faint, blue, strongly variable object, with unususal spectral characteristics. It was discovered by *Elvius (1975)* but remained unstudied for several years until *Westin (1980)* took more visible-range spectra of it, and other authors later on also reported relevant spectroscopic and photometrical results for this object *(i.e. O'Donoghue et al. 1987; Kepler et al. 1989; O'Donoghue and Killkenny 1989; O'Donoghue et al. 1990).*

V803 Cen, together with AM CVn, PG1346+082 and GP Com constitute the so-called group of *Interacting Binary White Dwarf* (IBWD) systems, as a subclass of the Nova-Like Cataclysmic Variables (CV) *(e.g. la Dous 1989).* The main peculiarity of these four objects is the absence of H in their spectra, which makes them differ considerably from the vast majority of H-rich CVs. Their visible-range spectra are almost exclusively composed of broad and shallow HeI lines, superimposed on flat or blue continua.

The IBWD model was proposed for the first time for AM CVn by *Faulkner et al. (1972).* This model assumes that an extremely low mass He white dwarf (DBWD) overflows its Roche lobe in an ultra-short period close binary system, and transfers, through the inner Lagragian point, matter that is accreted by another DBWD (the primary object) via an accretion disc. The same model has been also proposed for the other three closely related objects GP Com *(Nather et al. 1981)*, PG1346+082 *(Wood et al. 1987)*, and V803 Cen *(O'Donoghue et al. 1987).*

It is worth noticing that, in the case of GP Com only, is the orbital period of the system known *(Nather et al. 1979)*, while for the other three objects this is still an open question. However, from photometrical analyses it is known that AM CVn, PG1346+082, and V803 Cen display periodicities of the order of ~1000 seconds, which are thought to be their orbital periods or closely related to them *(O'Donoghue et al. 1990).*

In the particular cases of PG1346+082 and V803 Cen, these objects also display erratic changes in brightness in a time-scale of hours. These changes are accompanied by a transition in magnitudes which can be as big as from V~13 to V~17. For V803 Cen particularly, these considerable changes of the magnitude V, are accompanied by rather small changes of color *(Westin 1980).*

Our IUE study of V803 Cen is aimed at making a comparison of the spectroscopic properties of this object when in low (V~17) and high (V~13) states, for the first time in the ultraviolet-range

441

G. Vauclair and E. Sion (eds.), White Dwarfs, 441–448.
© 1991 *Kluwer Academic Publishers.*

of the electromagnetic spectrum, as has been reported before by other authors for the visible-range.

We first begin in Section 2 with a brief review of all the previous visible-range spectroscopic analyses of V803 Cen that have been published, and we are aware of. We continue with a description of the IUE observations, and a presentation of some results in Section 3. We finish with a summary of conclusions in Section 4.

2. PREVIOUS SPECTROSCOPIC STUDIES IN THE VISIBLE-RANGE

The previous literature containing spectroscopic studies of V803 Cen in the visible-range of the electromagnetic spectrum, that we are aware of, are the published papers by *Elvius (1975)*, *Westin (1980)*, *O'Donoghue et al. (1987)*, *Kepler et al. (1989)*, *O'Donoghue and Kilkenny (1989)* and *O'Donoghue et al. (1990)*. We will try in this section to briefly summarise the main spectroscopic results and conclusions reported by these authors, prior to presenting some of our preliminary results obtained for the same object in the ultraviolet-range.

Elvius (1975) reported spectra with rather blue continua, without any emission lines and with only some broad, shallow absorption lines due to HeI. No H or CII lines were detected in these spectra. On one of the nights, the strongest feature was a double line at approximately λ 4100 and 4115 Å. *Elvius* also noticed that there was a considerable change in the spectrum from one night to the next, and this behaviour was shown more than once by the object, which this author compared to MV Sgr, classified as a variable R Cr Borealis star. From the data taken by *Elvius*, it seems that V803 Cen spends most of its time near maximum brightness. Preliminary measurements of radial velocity gave uncertain values around -118 km/s, that would correspond to a distance of 6 kpc for this object, which, if so, should be affected by an appreciable extinction.

We should note in passing, that we do not correct our IUE spectra from interstellar extinction, because there is no appreciable depression in the spectra at the λ ~2200 Å band.

Westin (1980) noticed spectra of V803 Cen that partly resembled those of DBWDs, with very broad and diffuse lines, almost exclusively of HeI in absorption. The line intensities were variable and many of the lines appeared to be blended. One of the absorption features, present at λ ~4130 Å, was assumed to arise from SiII. In no case however, did the spectra of V803 Cen show any evidence of emission lines. The broadened HeI absorption lines varied in strength from very strong to very weak or absent. Also, wavelength measurements were uncertain, but indicated large radial velocities that varied in time, over the range +180 km/s to -180 km/s. From the data considered by *Westin*, V803 Cen seems to spend most of its time near minimum brightness.

Eight spectra of V803 Cen at B~15.4 were recorded in 55 minutes by *O'Donoghue et al. (1987)*, covering the spectral region λ 3500-5500 Å. These authors noticed that the signal to noise ratio was too low in any one spectrum to allow any useful comments, so the spectra were added separately to give a resulting one which showed no obvious absorption or emission lines, except maybe for the HeI λ4471 Å line weakly in emission. This spectrum, similar to the low state spectrum of PG1346+082 from *Wood et al. (1987)*, revealed a blue continuum, and the black body distribution that most closely matched it, has T ~ 10000-15000 K.

Kepler et al. (1989) performed spectrophotometry of V803 Cen in high (B~13.5), intermediate (B~14) and low (B~17.2) states, from λ 3800 to 7100 Å. These authors concluded that essentially all the observed absorption lines were due to HeI, with a possibility for HeII at λ6560 Å in the high state spectrum. The spectra at B~17.2 had low signal to noise ratio and showed no evidence for lines in absorption or emission. No emission lines were detected in either of the spectra, as well as no significant departures from black body distributions, the corresponding temperatures ranging from 7315 ± 73 K at B~17.2 to 25776 ± 682 K at B~13.5. For *Kepler et al. (1989)* therefore, the observations were consistent with the IBWD model, where most of the optical light comes from the accretion disc. This disc must be optically thick even in the faint state, to explain the absence of emission lines noticed in their study.

O'Donoghue and Kilkenny (1989), took spectra of V803 Cen in March 1987 (when the object was at B~16.8) and April 1987 (at B~13). In their spectra there were moderately broad, shallow HeI absorption lines superimposed on a blue continuum, when the object was in high state. These lines became weaker as the system's luminosity decreased, until in the low state the spectrum was featureless, with the possible exception of some weak emission lines. Although the signal to noise ratio of the low state spectrum was very low, these authors pointed out that there was no evidence for a DBWD absorption spectrum, when the system was in the low state. The mean high state spectrum contained no H and, with the exception of some weak lines at λ ~4625 and ~4673 Å possibly due to OII and/or NII, was dominated by HeI lines. These HeI lines in the high state were much shallower than in DBWDs, and the gravity dependent lines such as λ 3705, 3867 and 4143, 4388 Å, were stronger in V803 Cen than in DBWDs. Stark-broadening in the atmosphere of the accretion disc, was proposed by the authors as the mechanism responsible for the line widths detected, as the FWHMs of the lines were apparently inconsistent with rotational Doppler-broadening. All strong lines in the spectra of V803 Cen had associated forbidden components, and the strong unblended lines showed a small but distinct asymmetry affecting the line core. This asymmetry was variable, but a search for rapid variability in radial velocity or line profile shape failed to detect any periodicity with a semi-amplitude limit of 16 km/s in the 0.1-2.5 mHz (400-10000 sec.) range. To account for the asymmetry of the line profiles, they noticed that as there was no evidence for any emission filling-in of the absorption lines of V803 Cen, either in the form of an S-wave, or as the system declined in luminosity, such a phenomenon could be ruled out as a valid explanation. Instead, the authors proposed that these asymmetries can be understood in terms of the formation of the absorption lines in a non-circular disc. The axis of this disc would rotate in the inertial frame with a much slower period than the orbital period of the system, in the same way as has been proposed for some Dwarf Novae in outburst. This non-circular disc, is also compatible with a tidal explanation for the ~1000 seconds periodicity detected in the system.

Finally, visible-range spectroscopy of V803 Cen was performed by *O'Donoguhe et al. (1990)*, in coordination with high-speed photometry observations as part of the first campaign of the *Whole Earth Telescope* (WET) *(Nather 1989)* project, for the same object. The IUE low state observations that we analyse in Section 3, were taken at the same time. All visible-range spectra were obtained when V803 Cen was near 17th magnitude, except one that was taken when the system was brighter than 14th magnitude. This high state spectrum showed broad, shallow HeI absorption lines. The final reduced spectra were not flux calibrated, due to spurious turnovers in the ultraviolet for the instrumental sensitivity calibration. No significant differences were found among spectra taken the same night, so they were summed together to produce mean spectra for each night, with better signal to noise ratio. Emission lines of HeI could be seen in each of these averaged spectra. The authors also suspected that the emission HeI lines were occasionally double, and noticed significant differences in the strengths of the lines from night to night, possibly correlated with the system's luminosity. The widths of the emission lines were similar to those of the HeI absorption lines in high state, supporting the same mechanism, Stark-broadening, as being responsible for them. The authors pointed out that the lines were optically thick, this being supported by the fact that the ratio I_{5051}/I_{4922} was close to unity. No significant radial velocity variations were found exceeding ~40 km/s.

3. IUE OBSERVATIONS AND RESULTS

Three IUE low resolution spectra of V803 Cen taken in 1988 when the object was in low state (V~17), together with five more low resolution spectra taken in February 1990, when it was in high state (V~13.5), constitute all the information that, concerning ultraviolet spectroscopy, currently exists for this object. The February 1990 IUE observation was scheduled as target of opportunity and it was performed when the system was found to be in its high state *(O'Donoghue, private communication)*. This was essential to make possible the ultraviolet low/high states comparison intended. Photometry done the nights before and after this IUE run gave the same magnitude

(V=13.58; *O'Donoghue, private communication*), and the object was assumed to be constant during the IUE observations. Table 1. shows the log of IUE observations.

TABLE 1. *Log of IUE observations.*

FILE NAME	DATE	EXPOSURE TIME	STATE	OBSERVERS
SWP33084	88/072	275 min.	low	Solheim/Wesemael/Wood
SWP33090	88/074	567 min.	"	"
LWP12843	88/072	90 min.	"	"
SWP38268	90/059	25 min.	high	Ulla
SWP38269	"	90 min.	"	"
SWP38270	"	98 min.	"	"
LWP17436	"	40 min.	"	"
LWP17437	"	40 min.	"	"

Since no strong spectral features for this object, are expected in the wavelength region ~2000-3200 Å where the long wavelength camera of the IUE satellite operates, the LWP spectra were given short exposure times, which were enough to account for the continuum slope in this spectral range. The first SWP high state spectrum mentioned in Table 1 (SWP38268) is underexposed (25 minutes), which makes it appear different from the other two SWP spectra acquired the same night, SWP38269 and SWP38270.

The reduction and analysis have been performed following standard procedures, using STAR-LINK software. The H Ly α line, being geocoronal, was removed from all SWP spectra.

In Figure 1, we present as an example, spectra corresponding to the low state of the object (SWP33084;LWP12843, bottom) and to the high state (SWP38269;LWP17436, top), to show the flux level transition that the system undergoes when passing from magnitude ~17 to magnitude ~13.5.

The SWP38269 and SWP38270 spectra taken hours apart during the same night in February 1990, show some small differences between them, while the two SWP33084 and SWP33090 low state spectra taken two days apart in 1988, are very different from one another. Some of these differences may partly be due to the long exposures needed, and the low signal to noise ratio in these low state spectra. For example, NV λ ~1240 Å and NIV λ ~1284 Å are the most important features clearly evident in emission in SWP33084. In SWP33090, the NV line is weaker, and the NIV line has considerably decreased in strength. Also the continuum level appears lower in the SWP33090 spectrum. In the high state spectra, the NV λ ~1240 Å absorption line is very weak, and a possible identification for NIV λ 1284 Å is very doubtful. Also in the high state spectra, the main absorption feature is placed around 1302 Å which has been identified as SiIII, and the broadest one, around 1400 Å as SiIV. Identification of HeII at λ ~1640 Å is doubtful in the low state. In the high state, we find a shallow, broad, absorption line red shifted 1200±400 km/s.

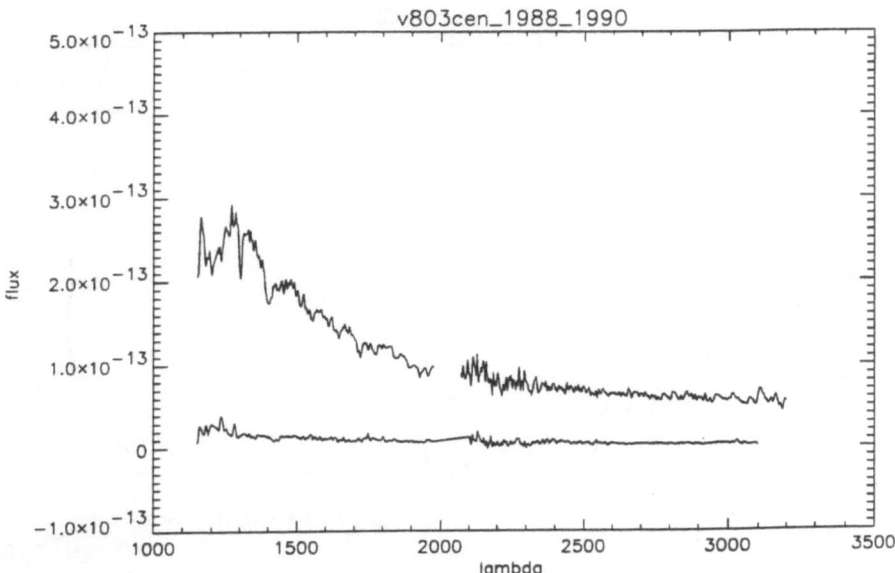

Figure 1. *Comparison between IUE high state spectra of V803 Cen (top: SWP38269,LWP17436) taken in 1990, and low state spectra of the same object (bottom: SWP33084,LWP12843) taken in 1988. "lambda" is in Å, and "flux" is in ergs/cm^2/sec/Å.*

In general, broad, asymmetric, shifted spectral features, are present in the IUE spectra of V803 Cen. Lines of NIV, NV, and CIV, in absorption in the high state, and in emission in the low state, are seen. The full lines-identification, profiles-fitting, and line-shifts measurements, have not yet been completed. However, calculations of radial velocities for the main lines such as NV, SiIII, SiIV and CIV, give values ranging from -1800 km/s to +1200 km/s. These velocities are much larger than those detected for lines in the visible-range (e.g.*Elvius, 1975* and *Westin, 1980*). Simultaneous blue and red shifted components are also present in the same —mainly high state— spectra (e.g. SWP38269).

Some of the above mentioned details can be seen in Figure 2, where a comparison between two SWP spectra of V803 Cen, the top one corresponding to high state and the bottom one to the low, is presented. In this Figure, it is possible to see how the absorption lines present in the high state spectrum turn into emission in the low state. This transition is accompanied, in some of the cases, by a relative shift in the wavelength position (notice for example CIV around λ1548 Å).

We will conclude the present discussion saying some words about the continua of the presented IUE spectra. A single black body distribution does not seem to fit all the covered ultraviolet range, especially for the spectra corrreponding to the high state, where two power laws (of the form $F_\lambda \propto \lambda^{-\alpha}$) with indices $\alpha_1 = 2.44$, and $\alpha_2 = 1.07$ are required to account for the continuum slope of the spectra in the short wavelength range (1200-2000 Å) the former, and the long wavelength range (2000-3200 Å), the latter. A single power law with index $\alpha \sim 1.7$ provides a good fit to the low state continuum.

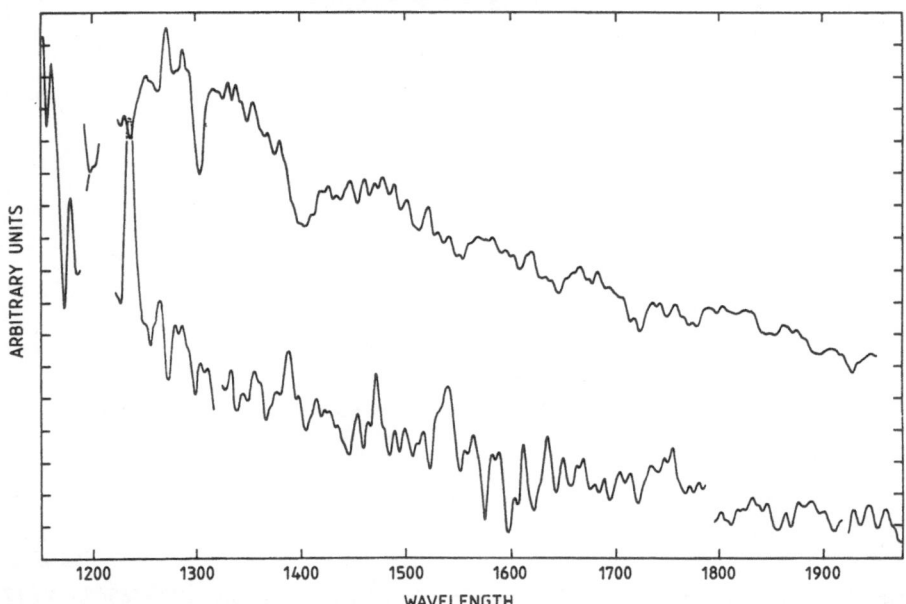

V803 Cen: swp38269, swp33090

ARBITRARY UNITS

WAVELENGTH

Figure 2. *Comparison between one IUE SWP high state spectrum of V803 Cen (top: SWP38269) taken in 1990, and one low state spectrum of the same object (bottom: SWP33090) taken in 1988. "Wavelength" is in Å.*

The $\alpha_1 = 2.44$ index obtained is not far from the 2.33 theoretically predicted value for a standard optically thick disc. We have to take, nevertheless, inclination considerations into account, before drawing definitive conclusions concerning this point. It has been proposed by several authors (e.g. *O'Donoghue and Kilkenny, 1989*), that the disc in V803 Cen is optically thick and of low inclination. This is supported by our UV absorption line-profiles.

Although we have no detailed temperature calculations yet, the previous result may indicate that a hot source contributes to the steepening of the continuum slope towards short wavelengths, when the object reaches high state; the corresponding temperature for it being at least ~50000 K. An extension of the IUE spectra here studied in the visible-range, may put constraints on the temperatures determination if several sources are present in the system.

4. CONCLUSIONS

We have presented all available spectroscopic information for V803 Cen in the ultraviolet-range of the electromagnetic spectrum, as well as a summary of the main results previously obtained by other authors in the visible-range, for the same object. We have detected in the IUE spectra we studied, the presence of metal lines such as NIV, NV, SiIII, SiIV and CIV, some of which are also characteristic of Dwarf Novae. Some of these lines clearly change from emission into absorption, when the object undergoes a ~4 magnitudes transition from low (V~17) to high (V~13.5) state.

The lines' intensities can vary dramatically from one observation to the next (two days apart). Strong shifts in wavelegth are also detected and seem to indicate that a wide range of velocities are present in the system. Strong line variability and doubling, and shifts in wavelength, have previously also been reported for the visible-range lines of this object. However, the velocities obtained from the ultraviolet lines are considerably larger than those reported for the lines in the visible-range. We do not as yet have accurate values for temperatures, but the clear presence of a hot source towards short ultraviolet wavelengths when the object reaches high state, seems to favour a possible explanation in terms of some kind of outburst in an accretion regime, in an ultra-short period binary, as specified by the IBWD model proposed for V803 Cen.

Acknowledgements

We are grateful to the staff at the IUE ESA Tracking Station at VILSPA (Spain) for help during observations, and analysis of the data. We acknowledge the use of the ULDA data base in Uppsala Observatory (Sweden). We thank Dr. Kjeldseth-Moe for useful discussions, and Drs. Larsson, Hall and Rasilla for their comments on this manuscript. We also thank Drs. Kepler, Larsson and O'Donoghue for reporting the magnitude of V803 Cen so we could target it in the high state. This research was supported by a travel grant from the Norwegian Research Council for Science and the Humanities. AMU thanks "Caja General de Ahorros de Canarias" and "Gobierno de Canarias" for the concession of a fellowship, inside the "Research Fellowships Programme for Postgraduates", 1988-1990.

References

Elvius,A., (1975). *Astr.Ap.* 44,117.

Faulkner,J., Flannery,B.P. and Warner,B., (1972). *Ap.J.Letters* 175,L79.

Kepler,S.O., Steiner,J.E. and Jablonski,F., (1989). In: "White Dwarfs", *IAU Coll. 114*, 443, G.Wegner (ed.), Springer-Verlag, New York.

la Dous,C., (1989). "Dwarf Novae and Nova-Like Stars", in *Cataclysmic Variables and Related Objects*, M.Hack (ed.), NASA/CNRS Monograph Series.

Nather,R.E., (1989). In: "White Dwarfs", *IAU Coll. 114*, 109, G.Wegner (ed.), Springer-Verlag, New York.

Nather,R.E., Robinson,E.L. and Stover,J.R., (1979). In: "White Dwarfs and Variable Degenerated Stars", *IAU Coll. 53*, 453, H.M.Van Horn and V.Weideman (eds.), Univ. of Rochester Press.

Nather,R.E., Robinson,E.L. and Stover,J.R., (1981). *Ap.J.* 244,269.

O'Donoghue,D. and Kilkenny,D., (1989). *M.N.R.A.S.* 236,319.

O'Donoghue,D., Menzies,J.W. and Hill,P.W., (1987). *M.N.R.A.S.* 227, 347.

O'Donoghue,D., Wargau,W., Warner,B., Kilkenny,D., Martinez,P., Kanaan,A., Kepler;S.O., Henry,G., Winget,D.E., Clemens,J.C. and Grauer,A.D., (1990). *M.N.R.A.S.* 245,140.

Westin,B.A.M., (1980). *Astr.Ap.* 81,74.

Wood,M.A., Winget,D.E., Nather,R.E., Hessman,F.V., Liebert,J., Kurtz,D.W., Wesemael,F., and Wegner,G., (1987). *Ap.J.* 313,757.

DISCUSSION

BARSTOW :

Have you detected CII in your spectra? and do the carbon features also turn from absorption in high state into emission in low state?

ULLA :

We have not detected CII, but yes CIV at around λ 1550 Å, in absorption in the high state and in emission in the low. If the corresponding feature we have identified as CIV in emission in low state is correct, it appears considerably blue shifted.

SION :

Which lines are blue shifted? and are they in the high state spectrum or in the low state spectrum?

SOLHEIM :

In the low state we find blue shifted emission lines of NV, NIV, and CIV in addition to red shifted absorption lines. In the high state, we see blue shifts for NV and SiIII and red shifts for the other lines.

BUES :

Your IUE spectra are similar to ours of VYScl. In its low state (1983, Nov.) V $\approx$ 18.5, strong emission lines of SiIV occur, whereas in 1986, when the high state (V $\approx$ 12. 5) is reached again, weak absorption features are seen. This object is hydrogen-rich and the optical spectrum is dominated by hydrogen lines and the Balmer jump in emission. Sometimes HeI lines are seen in absorption. The orbital period, obtained by Cowley and Hutchings (1984) is 3h98.

WHOLE EARTH TELESCOPE OBSERVATIONS OF PG1346+082

J.L. PROVENCAL, D.E. WINGET, R.E. NATHER, J.C. CLEMENS
Department of Astronomy and McDonald Observatory
The University of Texas at Austin
B.P. HINE
Nasa Ames Research Center
Moffet Field, California
G. HENRY
Tennessee State University
Nashville, Tennessee
S.O. KEPLER
Instituto de Fisica
Universidade Federal Do Rio Grande Do Sul
Brazil
G. VAUCLAIR, M. CHEVRETON
Observatoire Midi-Pyrénées et Observatoire de Meudon
France
D.O'DONOGHUE, B. WARNER
Department of Astronomy
University of Cape Town
South Africa
D.A. GRAUER
Department of Physics and Astronomy
University of Arkansas at Little Rock
Lilia FERRARIO
Australian National University
Australia

ABSTRACT. We report the latest results of continuing analysis of Whole Earth Telescope data of PG1346+082, acquired in March 1988. We present the light curve of the data set and its Fourier Transform. Examination of the dominant short-term photometric variation at 1490 seconds has revealed a band of power consisting of fifteen separate frequencies. We present the list of frequencies and discuss the status of our investigation of their nature.

1. Introduction

PG1346+082 is a member of the group of four known interacting binary white dwarf systems (Wood *et al.* 1987). IBWDs are believed to consist of two helium

G. Vauclair and E. Sion (eds.), White Dwarfs, 449–456.
© 1991 *Kluwer Academic Publishers.*

white dwarfs of extreme mass ratio transferring material via an accretion disk. The system is a photometric variable with a four magnitude variation (13.6 to 17.2 in V) on a four to five day timescale. Superimposed on this large variation is a short timescale oscillation with a period of 1490 seconds. PG1346+082 also exhibits flickering, the classic signature of mass transfer. The optical spectrum consists of broad absorption lines of neutral helium during outburst and weak emission features at all other times. No hydrogen has been detected.

2. Data Analysis

Whole Earth Telescope coverage of PG1346+082 spanned more than two weeks (see Provencal *et al.*, 1988 for Journal of Observations). The acquired light curve is presented in Figure 1. During the observing run, PG1346+082 varied from magnitude 14.3 to 17.2 in V. Figure 2 displays the magnitude of the system throughout the run as a function of time. The system is capable of changing its brightness by over a magnitude on timescales of hours.

The power spectrum of the entire WET data set is dominated by a complex band of power around 1490 seconds, with power also present near 735 seconds. In addition, a multitude of small but significant peaks are distributed throughout the FT (Figure 3).

One of our principal goals is to identify the physical origin of this power. Examination of the nature and behaviour of the dominant band of power at 1490 seconds in an important first step towards realizing this goal. For example, is the band composed of separate, coherent frequencies, or is it a manifestation of some process such as amplitude or phase modulation of a few independant variations? If the result is the latter, the variations must originate in the mass accreting white dwarf, as the disk is incapable of supporting coherent oscillations. If the former is true, the band of power may represent the binary period, or the rotation period of the mass accreting star. Perhaps neither is true, and the band of power is the result of quasiperiodic oscillations in the accretion disk. The behaviour of the power at 1490 seconds will allow us to eliminate some of these possibilities.

The large magnitude variations present in Figure 1 suggest that amplitude modulation must play a role in the origin of the band of power at 1490 seconds. To minimize this possibility, the data set was broken into three groups by magnitude: low state, middle state and high state. Throughout most of the run PG1346+082 remained in low state. Indeed, the sytem never attained its maximum brightness. For this reason, the analysis reported here has concentrated on PG1346 in low state.

The low state amplitude FT, defined as the square root of the power spectrum, of the region around 1490 seconds is displayed in Figure 4. The lower curve in Figure 4 is a mirror image of the spectral window: the pattern produced in the FT by a single sinusoid sampled exactly as the data.

The first step toward identifying the mechanism responsible for this band of power at 1490 seconds is to identify the frequencies composing it. We decoded this

451

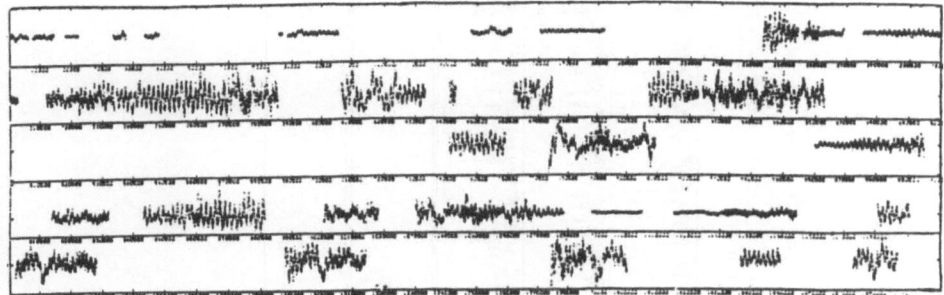

Figure 1: Light Curve of PG1346+082 obtained by the Whole Earth Telescope in March 1988.

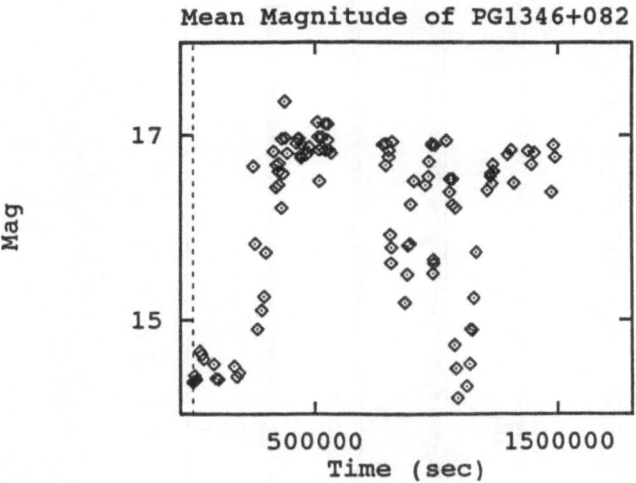

Figure 2: The magnitude of PG1346+082 during the WET campaign.

452

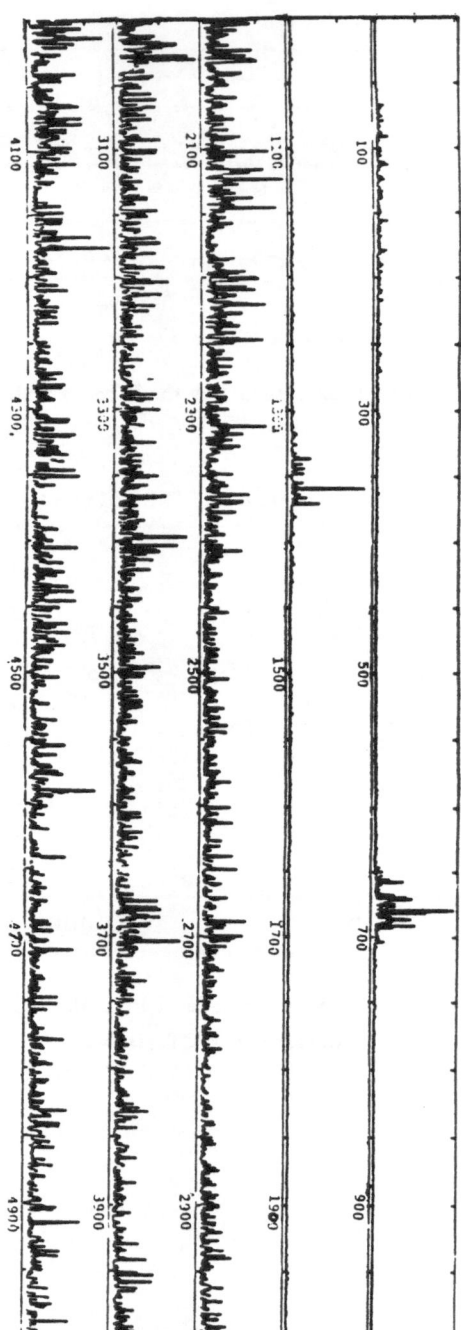

Figure 3: The Fourier power transform
of the WET data set.

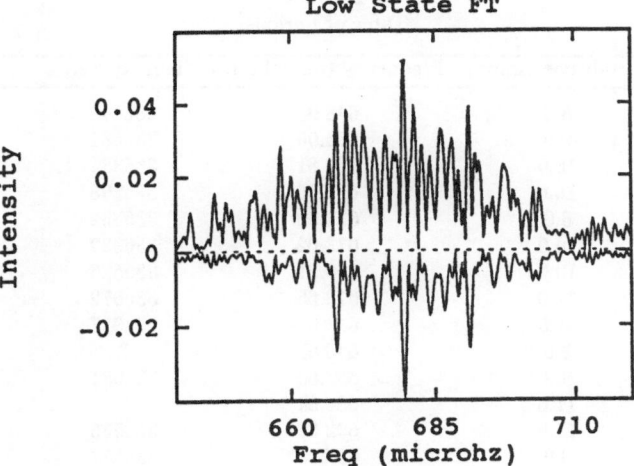

Figure 4: Amplitude FT and window of
the low state data in the region of 1490 seconds.

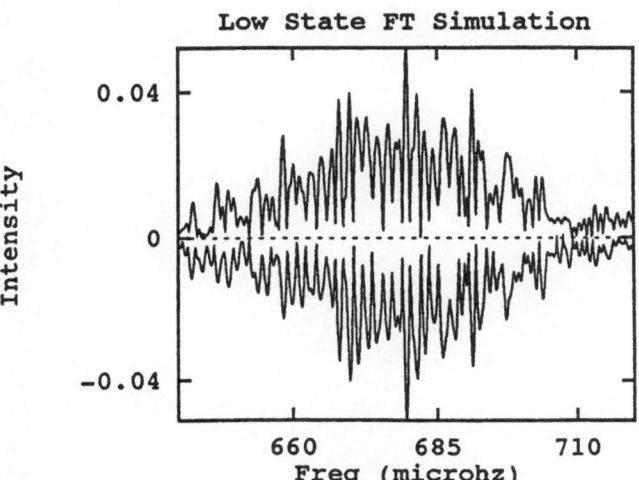

Figure 5: The low state amplitude FT
and simulation.

Table 1
Table of Periods

Period	Amplitude (mmg)	Frequency (microhertz)	Time of Zero
1540.93	8.0	648.90	336548
1529.03	9.0	654.00	336581
1492.95	15.0	669.81	335331
1489.74	16.0	671.26	335298
1488.50	6.0	671.82	335832
1485.84	16.0	673.02	336922
1478.61	10.0	676.31	335526
1471.35	37.0	679.65	335672
1457.69	11.5	686.02	334317
1456.49?	3.0	686.58	?
1454.63	10.0	687.60	335981
1452.77	11.0	688.34	?
1444.35	11.0	692.35	334536
1434.21	8.0	697.25	334617
1432.23?	3.0	698.21	?

region of power by employing two separate methods. This first involved synthesizing the original FT from single sine functions. This was an integral, iterative method in which the original transform was compared to a simulated FT composed of frequencies sampled exactly as the data. Comparison of this simulated FT with the original indicated where additional frequencies could be added to improve the simulation. The second, differential method involved a deconvolution, or spectral stripping, technique. The largest amplitude peak in the data was fit with a single sinusoid. This sinusoid was then subtracted from the data. The process was repeated for the next largest amplitude peak. Extreme care was taken to subtract only peaks that were single and to obtain each peak's phase via nonlinear least squares analysis before it was removed.

Table 1 lists the 15 periods, frequencies and amplitudes found from these two methods. Figure 5 displays the original low state amplitude spectrum. The lower curve is the mirror image of the best fit derived by these methods. We believe this decomposition into individual amplitudes and frequencies is unique within the limits imposed by noise.

Having established the frequencies composing the band of power, we next addressed the question of their nature: is each independant, or can we explain them as amplitude or phase modulation of a small number of frequencies? A blow is dealt to amplitude modulation as a source of the band of power by a careful examination of the timescales involved. The FT of a single frequency that is undergoing periodic amplitude modulation will consist of two peaks separated by the modulation frequency, centered on the original single frequency. Therefore, there are two approaches to consider. The first is to assume that the two closest frequencies in the band are produced by amplitude modulation. The modulation period required in this case is longer than the run length. The second approach is to assume that the bandwidth, with due consideration of the width of the spectral window, is an indication of the modulation period. The modulation period required in this case is approximately 0.35 seconds. Although PG1346+082 is capable of changing its magnitude on a timescale of hours, the system's timescale to complete an outburst is several days.

We were unable to find a reasonable phase modulation to explain the band of power at 1490 seconds. We have investigated how the changing phase of a sinusoid will be manifested in the Fourier transform: both continuous and sudden phase changes place additional power only in the aliases present in the spectral window. New, independant peaks are not created. Some but not most of the fifteen periods found in PG1346+082 lie on or near an alias of another peak.

These results led to an investigation of the coherency of the dominant frequency at 1471 seconds. The fit of a single sinusoid with a period of 1471.35 seconds maintains phase to within ten percent throughout the run, despite the large magnitude variations.

The amplitude of the band of power around 1490 seconds as a function of time suggests that it arises in a portion of the system that maintains a constant luminosity even during outburst. The amplitude of the band decrease in proportion to the increase in magnitude.

3. Conclusions

The intensive coverage provided by WET has led to significant progress toward understanding this complex system. The 1490 second oscillation is not a single periodicity but rather a band of power comprised of as many as fifteen separate periodicities. The dominant frequency, at 1471.35 seconds, is coherent over the run length. This is the first time a stable periodicity has been definitively detected in this system. Also, the behaviour of the magnitude of the band of power indicates that it arises on the white dwarf.

The behaviour of PG1346+082 cannot be explained by a simple binary model. Given the example of the H-rich cataclysmic variables, variations at the orbital period or the rotation period of either star could be expected. In the case of PG1346+082, we are left with an unknown mechanism for the remaining periodicites. The long term behaviour of these frequencies will be instrumental in resolving the question of their origin. It is interesting to note, as suggested by Wood et al. (1986), that the temperature of the accretor places it within the DB instability strip, where helium partial ionization will drive pulsation. The additional periodicities in PG1346+082 could be nonradial g-modes such as are observed in single helium rich white dwarfs of similar temperature.

References

Kepler, S.O., Robinson, E.L., Nather, R.E., 1982 *Ap. J.*, **254**, 676-682.

Nather, R.E., Winget, D.E., Clemens, J.C., Hansen, C.J., Hine, B.P., 1990, *Ap. J.*, **361**, 309-317.

Provencal, J.L., Clemens, J.C., Henry, G., Hine, B.P., Nather, R.E., Winget, D.E., Wood, M.A., Kepler, S.O., Vauclair, G., Chevreton, M., O'Donoghue, D., Warner, B., Grauer, A.D., Ferrario, L., *IUA Colloquiumm No 114*, 1988, p. 296.

Press, W.H., Flannery, B.P., Teukolsky, S.A., Vettering, W.T., **Numerical Recipes**, Cambridge University Press, New York, 1987.

Robinson, E.L., in *IUA Colloquium 53, White Dwarfs and Variable Degenerate Stars*, ed. H.M. Van Horn and V. Weidemann (Rochester, N.Y.: University of Rochester), p. 343, 1979.

Wood, M.A., Winget, D.E., Nather, R.E., Hessman, Frederic, V., Liebert, James, Kurtz, D.W., Wesemael, F., Wegner, G., 1987, *Ap. J.*, **313**, 757-771.

ON THE ORIGIN OF LMXRBS: THE ONEMG CASE

J. ISERN[1,2], R. CANAL[3,2], J. LABAY[3,2]

1) Centre d'Estudis Avançats de Blanes (C.S.I.C.)

2) Grup d'Astrofísica (IEC)

3) Departament d'Astronomia i Meteorologia (Universitat de Barcelona)

ABSTRACT: The ignition of Ne-O in mass accreting white dwarfs, formed in close binary systems by mass loss from stars in the range of $8 \, M_\odot \leq M \leq 12 \, M_\odot$, is preceded by electron captures on ^{24}Mg and ^{24}Na. Electron captures on ^{20}Ne (or maybe on ^{24}Mg and ^{24}Na) are the triggering mechanism of explosive ignition at densities in the vicinity of 9.5×10^9 g/cm^3. The outcome depends on the adopted propagation velocity of the thermonuclear burning front. Burning fronts propagating with hydrodynamical velocities($\geq 0.02 c_s$) lead to the total disruption of the star or, in some cases, to the formation of an "iron" white dwarf. Burning fronts propagating conductively lead to the formation of a neutron star.

1. Introduction

The standard mechanism to form a neutron star is the collapse of the iron core of a massive star followed by a type II supernova explosion. This mechanism is commonly admitted for massive x-ray binary sources but not for the low-mass ones. The reason is that in the last case more than half of the mass of the system is ejected during the explosion and the system could no survive the event. Three mechanisms have been

457

G. Vauclair and E. Sion (eds.), White Dwarfs, 457–464.

proposed to explain the presence of a neutron star in a low-mass X-ray binary system : i) capture of a non degenerate star by a previously formed neutron star. ii) Standard type II supernova explosion in a triple system. iii) Accretion induced collapse of a white dwarf without explosion.

The first mechanism assumes that a neutron star is produced by the core collapse of a single massive star and later capture of a nondegenerate companion: Clark (1975), Fabian et al (1975), Sutantyo (1975). This mechanism can certainly work in globular cluster cores where the number of stars per unit of volume is high and the relative velocity of the stars is small. However, it is not adequate in the case of galactic bulge sources due to the small stellar density and the high relative velocities. In this case, the energy to be dissipated by tidal forces would be of the order of the binding energy of the nondegenerate star.

The second scenario involves a massive close binary accompanied at large distance by a late-type dwarf (Eggleton and Verbunt, 1986). Formation of a low-mass X-ray binary would then involve three steps. First, the massive pairs evolve into a massive X-ray binary through mass transfer and a supernova explosion. Second, the remaining massive star evolves until Roche lobe is overflown and a common envelope forms. The neutron star then spirals into the center of the red giant, and a Thorne-Zytkov object forms (Thorne and Zytcov, 1975), i.e.-a red supergiant with a neutron star core. Third, long-term expansion of theenvelope should eventually lead to a common envelope phase with the late-type dwarf. The latter object would then spiral in, and the envelope would likely be lost (its mass having been previously reduced by wind emission, and its structure being distended and loosely bound). A close binary consisting of a neutron star plus a low-mass companion might finally result.

The last mechanism assumes an accreting massive white dwarf in a binary system that manages to delay the central thermonuclear runaway up to densities high enough to permit to electron captures on the incinerated material to overcome the explosion and induce the collapse

(Schatzman 1974). There are two ways to delay the explosion to high densities. One is to cool down the explosive mixture, as in the case of CO white dwarfs (see Canal, Isern, Labay (1990) for an extensive review). The other is to consider less explosive mixtures, as in case of ONeMg white dwarfs (Miyaji et al, 1980).

2. The Evolution of ONeMg Cores

Stars with masses $M \geq 8$ $M_{\odot}$ ignite carbon nonexplosively (Woosley and Weaver 1986). Quasihydrostatic carbon burning then leads to growth of a core made of an admixture of ^{16}O, ^{20}Ne and ^{24}Mg. The mass range 8 $M_{\odot} \leq M \leq 12$ $M_{\odot}$ poses however a distinct problem, since the electrons become degenerate prior to Ne-O ignition (Nomoto 1984). In the upper half of this mass (10 $M_{\odot} \leq M \leq 12$ $M_{\odot}$) this leads to Ne flashes but not to explosive disruption nor dynamical collapse, and evolution proceeds up to Si burning and growth of an electron degenerate "Fe-Ni" (nuclear statistical equilibrium) core (Woosley and Weaver 1986). In the lower half of the mass range (8 $M_{\odot} \leq M \leq 10$ $M_{\odot}$), before the point of Ne ignition is reached electron Fermi energies at the center of the star become larger than the thresholds for electron captures on ^{24}Mg (and shortly afterwards on ^{24}Na) first, and later on ^{20}Ne. These electron captures have a double effect: they lower the electron mole number Y_e (and with it Chandrasekhar's mass, which is proportional to Y_e^2), and on the other hand they heat up the plasma, eventually inducing Ne-O ignition (Miyaji et al 1980; Miyaji and Nomoto 1987). Ignition densities are $\approx 10^{10}$ g/cm^3. Thus, electron degeneracy is not removed until the material has been processed to NSE and a hydrodynamical burning front propagates outwards from the center of the star.

If these stars are members of a close binary system they can avoid temporarily this fate by removing their helium layer when they expand in the red giant phase (Nomoto 1984, Iben and Tutukov 1986) and becoming an ONeMg white dwarf. Their existence is indicated by the observation of "Ne novae" (Truran and Livio 1986). Afterwards, during its evolution to the red giant stage, the companion will start transferring matter to the white dwarf and a situation rather similar to that described in the case of the ONeMg core of an isolated star will occur.

The exact density at which explosive ignition is triggered is crucial in determining the fate of these cores. Depending on it, either electron captures on NSE material remove energy and pressure fast enough for gravitational collapse to ensue or, on the contrary, the energy released by the spreading burning completely disrupts the star. Core masses corresponding to densities $\approx 10^{10}$ g/cm^3 are close to the Chandrasekhar's mass and small differences can lead to opposite outcomes.

Exact values of Ne-O ignition densities depend on the criterion adopted for the onset of convective instability, as well as on the relevant electron-capture rates. The instabilities found by these cores before reaching the Chandrasekhar's mass are the electron captures on ^{24}Mg ($\rho_{Mg} = 4 \times 10^9$ g/cm^3), on ^{20}Ne ($\rho_{Ne} = 9.1 \times 10^9$ g/cm^3) and on ^{16}O ($\rho_O = 1.9 \times 10^{10}$ g/cm^3). During such captures, a high quantity of entropy is produced (the temperature increases due to the γ-ray production and to the distortion of the Fermi distribution) and matter tends to be unstable. Although the situation is rather complex (Mochkovitch 1983), it is possible to make two extreme approximations: Either to apply the Schwarzschild criterion or to apply the Ledoux criterion for convection.

If the Schwarzschild criterion is applied, the entropy generated by electron captures induces the formation of a convective zone that transports the energy very efficiently. As the capture proceeds, the star gradually contracts until a density of 2×10^{10} g/cm^3 is reached. At this point, oxygen ignites and the burning propagates through all the star. In this case, a collapse to a neutron star is guaranteed for any value of the burning front velocity (Miyaji et al 1980). If the Ledoux criterion is applied, the Y_e-gradient inhibits convection and a strong local heating is produced. Due to the thermal neutrino emission, the star just scapes to ignition induced by electron captures on ^{24}Mg, but cannot avoid the ignition due to electron captures on ^{20}Ne at a density of ~ 9.5x10^9 g/cm^3. Therefore, from this discussion it could be thought that the central ignition would happen somewhere between 0.95 and 2x10^{10} g/cm^3.

Electron captures, however, were computed following the gross theory of β-decay by Miyaji et al (1980). Takahara et al (1989) have examined the electron capture rates using the best electronic wave functions available and have found that the rates agree within a factor 2 or 3 except for electron captures on ^{24}Na, which are underestimated by a factor of 10 for densities higher than the threshold density. Thus, if the Ledoux criterion is applied, the picture previously described does not change very much because electron captures happen in the neighborhood of the threshold. But if the Schwarzschild criterion is applied, the central regions are provided with fresh ^{24}Mg and as the density increases, electron captures on ^{24}Na occur at densities higher than the threshold and the ignition of oxygen would probably happen at densities between 4 and 9.5x10^9 g/cm^3.

As the outcome of the ignition at 2x10^{10} g/cm^3 should always be a neutron star, it seems reasonable to examine the consequences of a central runaway at a density of 9.5x10^9 g/cm^3.

3. Models and Results

Calculations start from a ONeMg core with mass M = 1.34 M$_\odot$, central density ρ_c = 9.5x10^9 g/cm^3, central temperature (before explosive ignition) T_c = 2.3x10^8 K, and a uniform chemical composition (but in the innermost layers) X_O = 0.12, X_{Ne} = 0.72, and X_{Mg} = 0.12. In order to be consistent with previous evolution, ^{24}Ne is substituted to ^{24}Mg in the region from center to the point where electron Fermi energy equals the threshold energy for the corresponding capture. The equation of state for the ion component of the plasma is taken from Ichimaru, Iyetomi and Ogata (1988). For the electron component, an ideal Fermi gas plus electron-positron pairs was adopted. The electron-capture rates on NSE material are calculated from the expressions of Epstein and Arnett (1975), recalibrated by comparison with the more recent rates of Fuller, Fowler, and Newman (1982). Hydrodynamic burning propagation is simulated through a parameterized burning front velocity: $v_b = Fc_s[1-\exp(-r/R_0)]$, where c_s is the local sound velocity, r the distance to the center, and F and R_0 are adjustable parameters (Woosley 1990). In the case of a carbon deflagration, the suggested values are $F = 0.5$ and $R_0 = 2 \times 10^7$ cm. In Table 1 are displayed the models explored here. In all of them the front velocity rises more slowly than in the aforementioned carbon case. This has been adopted in order to somehow reflect the smaller specific energy released in completely burning the ONeMg mixture as compared to the CO case. In fact, the velocity in the center should be equal to the conductive velocity up to the point where the aforementioned expression gives equal or higher velocity, and therefore our calculation favors collapse even more.

TABLE 1: Model characteristics

Model	F	$R_0(10^7 cm)$	$M_{inc}(M_\odot)$	$M_{Ni}(M_\odot)$	$E_{kin}(10^{51}erg)$	$M_{rem}(M_\odot)$
A	0.3	2.0	1.03	0.42	0.59	
B	0.3	5.0	0.85	0.32	0.41	
C	0.3	10.0			0.14	1.24
D	0.15	10.0				1.34

Models A and B are clearly disrupted in a supernova like explosion. They might correspond to some type of SNI explosion when taking place in a ONeMg white dwarf or in a single star having previously lost its H-rich envelope (otherwise, in the last case, a SNII would result). Model D (an extremely slow deflagration) recontracts after the initial bounce to eventually collapse to a neutron star. In the four last columns of Table 1 are given the incinerated masses, the ^{56}Ni masses, the kinetic energies, and remnant masses, respectively. The difference in nickel masses between models A and B not only arises from the difference in the respective masses of incinerated material but also from larger neutronization of the central layers in model B, the last producing more nonradioactive ^{56}Fe.

Model C shows a frontier behavior: it expands and ejects $\simeq 0.10\ M_\odot$ to recontract afterwards to a white dwarf of smaller mass and a composition made of Fe-peak nuclei. This case would give a dim outburst but it would produce a "Fe" white dwarf. The existence of these objects might be indicated by "Fe novae" such as Nova Mus 1983 (Freitas Pacheco and Codina 1983), or V1370 Aql (Snejders et al 1984). It should be noted, in this respect, that explosive ejection of $\simeq 0.1\ M_\odot$ of material would not disrupt the close binary system (Taam and Fryxell 1984).

4. Conclusions

These calculations show that the collapse to a neutron star is not likely in the case of ONeMg cores igniting O-Ne at densities smaller than 10^{10} g/cm^3 unless the burning front propagates with velocities smaller than $0.02C_s$. These low velocities seem excluded in the ONeMg case: they can only be associated with burning propagation within a central solid core and such a core cannot exist even in a ONeMg white dwarf, since at the above densities previous electron captures on ^{24}Mg and ^{24}Na would already have melted it.

ONeMg cores remain, nonetheless, extremely interesting objects. The physics of their explosive ignition may well be a key to such phenomena as SNIa and SNIb outbursts, or perhaps to a new type of nova.

Acknowledgements: This work has been supported in part by CICYT grants PB87-0304, PB87-0147 and PB87-0150.

5. References.

Canal R., Isern J., Labay J. 1990, Ann. Rev. Astron. Astrophys. <u>28</u>, 183.

Clark G.W. 1975, Astrophys. J. (Letters) <u>199</u>, L143.

Eggleton P.P., Verbunt F. 1986, MNRAS <u>44</u>, 227.

Epstein R.E., Arnett W.D. 1975, Astrophys. J. <u>201</u>, 202.

Fabian A.C., Pringle J.E., Rees M.J. 1975, MNRAS <u>172</u>, 15P.

Freitas Pacheco J.A., Codina S.J. 198,3 MNRAS <u>214</u>, 481.

Iben I., Tutukov A.V. 1986, Astrophys. J. <u>311</u>, 753.

Ichimaru S., Iyetomi H., and Ogata S. 1988, Astrophys. J. (Letters), <u>334</u>, L17.

Miyaji S., Nomoto K. 1987, Astrophys. J. <u>318</u>, 307.

Miyaji S., Nomoto K., Yokoi K., Sugimoto D. 1980, Publ. Astron. Soc. Japan <u>32</u>, 303.

Mochkovitch R. 1984, in "Problems of Collapse and Numerical Relativity", ed. D.Banzel and M.Signore (Reidel, Dordrecht), p.125.

Nomoto K. 1984, Astrophys. J. <u>277</u>, 791.

Schatzman E. 1974, Presented at Int. Sch. Cosmol. Grav. Erice, Italy.

Snejders M.A.J. et al 1984, MNRAS <u>211</u>, 7B.

Sutantyo W. 1975, Astron. Astrophys. <u>44</u>, 227.

Taam R.E. , Fryxell B.A. 1984, Astrophys. J. <u>279</u>, 166.

Takahara M. et al. 1989, Nucl. Phys. <u>A504</u>, 167.

Thorne K.S., Zytkov A.N. 1975. Astrophys.J (Letters) <u>174</u>, L143.

Truran J., Livio M. 1986, Astrophys. J. (Letters) <u>199</u>, L19.

Woosley S.E., Weaver T.A. 1986, Ann Rev. Astron. Astrophys. <u>24</u>, 205.

INDEX

Abundances 235, 238, 249, 253, 278, 317, 322, 324, 327, 334, 346, 372, 417, 420

Accretion 229, 230, 275, 277, 295, 317, 333, 372, 417, 419

Asteroseismology 129, 159, 167, 193, 205, 219

Atmospheric parameters 53, 56, 75, 219, 361

Attractor 175

Binaries 279, 379, 395, 409, 432, 457

Birth rates 67

Cataclysmic variables 417, 441

Chaos 134, 167, 178, 185

Classification 111, 380

Convection 78, 159, 162, 213, 230, 257, 278, 305, 460

Cooling 5, 47, 67, 146, 153, 200, 396, 399, 420, 421

Core composition 148, 153, 459

DB variables 257, 259, 305, 314

Diffusion 235, 236, 251, 333, 372, 417

Einstein 17, 18, 29, 32, 235, 373

Evolution 143, 146, 424, 459

Evolutionary tracks 1, 153

Exosat 17, 18, 29, 32, 104, 235, 373

Fourier (transform, spectrum) 135, 145, 185, 187, 435

Galaxy (age) 89, 95, 396

Gravity 346, 410

Gravitational redshift 84, 238, 343, 381, 388

Hubble Space Telescope 369, 381

International Ultraviolet Explorer (IUE) 32, 153, 160, 231, 235, 237, 249, 259, 353, 369, 381, 418, 441

Light curves 175, 185, 432, 435

Line broadening 318, 353

Luminosity function 89, 122

Magnetic white dwarfs 126, 285

Masses 75, 122, 388, 396, 410, 419

Mixing 76, 230, 305, 346, 365, 372

Mode trapping 137, 147, 196, 199, 208, 212, 365

Model atmospheres 17, 21, 32, 44, 162, 163, 258, 267, 285, 287, 320, 347

Neutron stars 457

NLTE model atmospheres 222, 243

Nonradial pulsations 129, 130, 144, 146, 205, 211, 219

Novae 417, 419, 420, 426

PG1159 stars 17, 18, 39, 104, 219

Phase diagram 5, 7

Planetary nebulae (Central stars) 1, 17, 29, 39

Planetary nebulae (Central stars) : individuals

  – A7 39, 44
  – EGB6 39, 44
  – IW1 40
  – Jn1 40, 44
  – K1-16 17, 18, 220
  – K2-2 41
  – NGC246 19
  – NGC7293 19, 29, 39
  – PW1 44
  – S188 44
  – VV47 39, 40
  – WDHS1 41, 44

Rosat 99, 371
Rotational spitting 136

Space density 67
Stratification 129, 137, 154, 194, 245, 267, 347
Strömgren sphere 299
Subdwarf B stars 53, 55, 409
Subdwarf O stars 39, 53, 409
Supernovae 409, 410, 457
Surveys 109, 121, 379, 382

Wavelet 185
Wind 236, 245
White dwarfs : individuals
  – AM CVn 137, 431
  – BPM17088 261
  – BPM27606 287
  – BPM70524 261
  – CoD-38 10980 374, 375
  – G107-70 372
  – G117-B15A 138, 143, 153
  – G191-B2B 106
  – G191-16 167, 185, 187
  – G207-9 365
  – G226-29 365
  – G29-38 365
  – G35-26 275
  – G74-7 324
  – G77-50 322
  – GD40 375
  – GD154 139
  – GD165 138
  – GD190 261
  – GD323 347
  – GD358 138, 261
  – GD394 375
  – H1504+65 18, 104, 220, 375
  – HL TAU-76 167
  – HS1234+4811 125
  – HS1254+3430 126
  – HZ 43 106
  – KPD0005+5106 18, 34, 375
  – L745-46A 327, 375
  – LP790-29 285
  – PG1144+005 18, 220
  – PG1159-035 18, 135, 205, 220, 223
  – PG1346+082 433
  – PG1351+489 175, 178
  – PG1424+535 220, 223
  – PG1520+525 223
  – PG1707+427 139, 223
  – Procyon B 372
  – 40 Eri B 343
  – Ross 640 327
  – Sirius B 106
  – V471 Tauri 105, 375
  – V803 Cen 433, 441

Thermal pulses 1

Whole Earth Telescope 129, 134, 143, 206, 219, 432, 443

ZZ Ceti stars 126, 144, 153, 159, 162, 167, 185, 193, 211, 305, 361